Springer Series on
Atoms+Plasmas

17

Editor: G. Ecker

Springer

Berlin
Heidelberg
New York
Barcelona
Budapest
Hong Kong
London
Milan
Paris
Tokyo

Springer Series on

Atoms+Plasmas

Editors: G. Ecker   P. Lambropoulos   I. I. Sobel'man   H. Walther
Managing Editor: H. K. V. Lotsch

V. N. Tsytovich

Lectures on Non-linear Plasma Kinetics

Translation edited by D. ter Haar

With 36 Figures

 Springer

Professor Vadim N. Tsytovich

General Physics Institute, Vavilova str. 38, 117942 Moscow, Russia

Translator: Professor D. ter Haar

P. O. Box 10, Petworth, West Sussex, GU28 0RY, United Kingdom

Series Editors:

Professor Dr. Günter Ecker

Ruhr-Universität Bochum, Fakultät für Physik und Astronomie,
Lehrstuhl Theoretische Physik I, Universitätsstrasse 150,
D-44801 Bochum, Germany

Professor Peter Lambropoulos, Ph. D.

Max-Planck-Institut für Quantenoptik,
D-85748 Garching, Germany, and
Foundation for Research and Technology – Hellas (FO.R.T.H.),
Institute of Electronic Structure & Laser (IESL),
University of Crete, PO Box 1527, Heraklion, Crete 71110, Greece

Professor Igor I. Sobel'man

Lebedev Physical Institute, Russian Academy of Sciences,
Leninsky Prospekt 53, 117333 Moscow, Russia

Professor Dr. Herbert Walther

Sektion Physik der Universität München, Am Coulombwall 1,
D-85748 Garching/München, Germany

Managing Editor: Dr.-Ing. Helmut K.V. Lotsch

Springer-Verlag, Tiergartenstrasse 17, D-69121 Heidelberg, Germany

Die Deutsche Bibliothek - CIP-Einheitsaufnahme

Cytovič, Vadim N.:
Lectures on nonlinear plasma kinetics / V. N. Tsytovich.
Transl. ed. by D. ter Haar. - Berlin ; Heidelberg ; New York :
Springer, 1995
(Springer series on atoms + plasmas ; 17)
ISBN-13:978-3-642-78904-5

NE: GT   CIP data applied for

ISBN-13:978-3-642-78904-5   e-ISBN-13:978-3-642-78902-1
DOI: 10.1007/978-3-642-78902-1

Typesetting: Camera ready copy from the translator using a Springer T_EX macro package
SPIN 10064268   54/3144 - 5 4 3 2 1 0 - Printed on acid-free paper

Preface

Non-linear plasma physics is a very fast developing field which has many applications, the number of which is increasing daily. Other branches of physics are interested in this field, not only because many of the developments of non-linear physics first were met with in plasma physics, but also because the development of methods for dealing with non-linear problems is extremely fast and it is also often possible to use plasma problems to illustrate fundamental non-linear phenomena.

During the last ten years I have given lectures on non-linear plasma kinetics to students of the Moscow Institute for Physics and Technology and I have found it necessary to change my presentation each year. As I want to present new developments as well as old established results, when new discoveries appear, it has been necessary for me not only to fit those in at the appropriate places, but also to rearrange the old material to take into account the effect of the new results; sometimes I have had to change the main presentation completely. It looks to me that this procedure was finally convergent, and I hope that the material as given here is presented in the simplest possible way, without loss of generality.

I feel that I have also been able to present the material in such a way that throughout the book there are just two leading ideas. The first one is that I am showing the very close relations between the different types of non-linear interactions and that they are all based on a simple picture of the physical nature of the interactions. Many aspects of this line of reasoning are never properly emphasised and some of them are still not widely used by researchers in this field. The second leading idea is even more general: I show that collective effects produce drastic changes in all known wave-particle interaction processes, such as particle collisions, wave scattering, and the emission of waves. I try to emphasise that in a plasma the corresponding cross-sections depend strongly on the distributions of other particles and waves and that this is a collective effect which can change the cross-sections, sometimes by several orders of magnitude. Even research workers in the field of non-linear plasma physics are not well acquainted with these two ideas. It is clear from research papers that these two points are not well understood. Moreover, students starting to learn the techniques of non-linear plasma physics from the original literature are buried by details about particular kinds of interactions and lose sight of the general ideas. I have tried to avoid this difficulty by concentrating on the fundamental ideas for a single type of collective motion – which is also probably the simplest one: electrostatic collective motions.

When I tried to write out my lectures I had to follow a logical presentation and this led me automatically in some cases to problems which had not yet been solved, or even stated properly. In this way my own research profited from the writing of this book. I have had to interrupt my work on the book, pursue my research, and publish various research papers showing, I think, the importance of these new processes for many problems in non-linear plasma dynamics. After that I incorporated the new results in my lectures to post-graduate students. Since this material fitted very nicely in my presentation, I could not exclude it from the book, when I tried to present the new material in the simplest possible way. Moreover, I felt that these problems arose from my desire to follow a logical presentation when preparing the book. The fact that the book contains a discussion of these new problems does not mean that it is a mixture of a monograph and a textbook. On the other hand, the material presented here should also be of interest to active reserach workers in the field.

This book was edited and improved by Professor Dik ter Haar, to whom I am extremely grateful for his efforts which made the publication of this book possible. I am also grateful to him for many helpful suggestions and discussions of the problems to which this book is devoted.

V.N. Tsytovich

Table of Contents

1 Non-linear Plasma Kinetics in Modern Physics

1.1 Introduction

Modern physics is based on non-linear concepts. Relatively new terms, such as wave-wave interactions, coherent structures, collective phenomena, self-organisation, solitons, chaos, are used extensively in many branches of physics, and even outside phyics – for instance, in biophysics and chemistry. The meaning of these terms as well as their description are similar in the different branches of research. It is also true that often the same phenomenon was discovered independently in quite different areas of physics.

In all these investigations plasma physics plays an exceptional rôle for several – physical, historical, technological, and even psychological – reasons. The result has been that non-linear research in plasma physics has been ahead of such research in other branches of non-linear physics. Since plasma physics was leading the research in non-linear physics, it became an important source for results and methods in this field to be used in other fields of physics for solving old problems. As an example we may mention research in a field, which is a typical non-linear one and which is at the same time of the utmost practical importance: the occurrence and fast development of turbulence in non-equilibrium systems. We shall state below in detail what we mean by the term turbulence. For the moment it suffices to note that it involves a strong, random motion excited by a regular, strongly non-linear source. Turbulent phenomena were first encountered in hydrodynamic motions, for example, the random eddies excited in gases or liquids by fast moving objects. Although hydrodynamic turbulence is a problem related to many practical questions, such as the construction of water pipes or the flow round airplane wings, the physical understanding and description of turbulence has been at a very simple level. Applications were made empirically and notwithstanding great efforts no real progress in fundamental research in this field was made for a very long time. This continued until the concept of turbulence was generalised and developed for plasma physics. The huge development related to problems in plasma physics made it possible to understand and solve the basic problems of hydrodynamic turbulence. This was done only recently by using methods developed by research workers in the field of plasma physics.

Let us now discuss why and how non-linear plasma physics became one of the major problems of modern physics and why its results are at the present time applied so widely. First of all there are physical reasons. Most plasmas are collections of free positive and negative charges which are electrically neutral on average. It is not always realised how huge the electric fields created by these charges would be, if there were no overall electrical neutrality. Consider a spherical volume of electrons with a 1 cm radius and ask what the density would have to be to create a potential energy at its surface equal to the electron rest mass energy $m_e c^2$ (where m_e is the electron mass and c the speed of light), that is, close to 1 MeV. Since $e^2/m_e c^2 \sim 10^{-13}$ cm, if e is the elementary charge – which is characteristic for the charge of the particles in a typical plasma – the density should be of the order of 10^{13} cm^{-3}. The strength of the electric field in this example would be of the order of 10^6 V/cm. This means that, if there were some perturbation in a plasma with a density of 10^{13} cm^{-3} in which 1 cm^3 of electrons were displaced over a distance of 1 cm, an electric field of the order of 10^6 V/cm would be created. This is a very strong field and would lead to strongly non-linear effects. This means that a plasma is a strongly non-linear medium; its non-linearities can easily be excited and may be accompanied by very strong electric fields.

Another physical reason why a plasma can be used to test general non-linear ideas is that most plasmas are very simple systems consisting of free particles which interact with one another through the very simple Coulomb law, the potential of which is inversely proportional to the distance between the particles as long as they move with non-relativistic velocities. One of the most important properties discovered for such systems of charged particles is that if there are many charged particles present this will change not only the interaction between the particles themselves, but also all other non-linear interactions. These changes are usually called *collective effects*. They are known not only for plasmas, but also in other physical systems. One of the best known of such collective effects is the appearance of effective masses for the particles in semiconductors and metals. The renormalisation problems in elementary particle physics are of the same nature. However, a plasma gives us the opportunity to study collective properties in a very simple system. This can help us to understand the physical reasons why the cross-sections of very important processes change if there are other particles present.

Yet another reason for the fast development of non-linear plasma physics was the need to find answers to questions arising from rapidly developing applications. The best known of those are the problems arising in controlled fusion research and in space physics research. It is interesting to note that plasma physics before the start of extensive controlled fusion and space research did not attract many researchers, since it was primarily connected with the physics of gas discharges. This field was mainly empirical and, strictly speaking, could hardly be called a scientific field of research. Only recently, by using developments in non-linear plasma physics, has it become possible to understand some aspects of the behaviour of such a complicated system as the partially ionised plasma of a gas discharge and the appearance of various dissipative self-organising structures in gas discharges. Of course, there were some indications of collective processes in a plasma, such as the anomalously fast dissipation of a beam of fast electrons, as long ago as 1929 in the pioneering research in this field by Langmuir; however, these indications were dismissed as rather exotic phenomena.

In the early research in the field of controlled fusion the plasma was considered to be a very simple state of matter and nobody expected that it was necessary to have any detailed knowledge of its fundamental physical properties in order to solve the practical problems. However, it soon became clear that a plasma can behave unexpectedly and that there appear collective non-linear motions some of which create collective fields. Moreover, it turned out that these collective effects, in fact, are the main factors determining the behaviour of this state of matter.

Research on the properties of the plasmas found in space and, in general, in astrophysics also was for a long time in a primitive state. In this field only very simple concepts were used which were based upon the motion of charged particles in an external magnetic field and on magnetohydrodynamics. Here again, it was found in many space projects that this simple-minded description did not tally with the observations. Stated concisely one can say that before the early sixties research workers in these fields did not realise the general importance of non-linear processes in plasma physics; they did not consider the plasma as a system which most clearly showed the general nature of non-linear properties. Research workers in other fields, such as biology, chemistry, hydrodynamics, or acoustics, also did not realise the importance of non-linear effects. However, the sudden enormous increase in activity in controlled fusion and space research raised many questions and required a better understanding of the observed behaviour of plasmas. The successful development in plasma physics in the early sixties was due to a large extent to the fact that emphasis was put on research on the fundamental non-linear properties of plasmas. One can therefore consider the sixties as the starting period for modern plasma physics. It took five or six years to construct the basic foundations of non-linear plasma physics and this has remained an extremely active research field. The rate of development in this field continues to be large and there is a continuous flood of new results. This affects not only those fields of research which we have already mentioned, but also many other fields and we shall try to discuss this in the later sections of the present chapter. We may just mention at this point that several of the spacecraft programmes contain as a necessary element measurements of the plasma particle distributions and of the fields produced by collective motions in the plasma. As only a negligible part, such as the dust particles and the planets, of the matter in our Universe is not ionised so that the Universe consists mainly of a plasma with its concomitant collective behaviour, the better understanding of the nature of plasmas is extremely important for our understanding of the Universe.

A third important reason why non-linear plasma kinetics has become very important is a historical one. Important research of a general nature was carried out by some outstanding physicists. It was fortunate that several such people were working on these problems at the same time and that they therefore had to compete with one another. This is perhaps not surprising, since it often happens that when a new field of physics becomes important many of the best physicists start working in it.

If we want to say in a few words what the essential features of the new developments are, we can state that interactions between particles and collective non-linear motions are the most important interactions rather than those of individual particles with one another. The collective motions themselves are governed by their own, complicated non-linear interactions so that it is necessary to take a dynamic view of the collective motions. This approach is called *non-linear kinetics* and it is the main subject of the present book. We must note that these problems also include the questions of the occurrence of randomness in the system, of the collective nature of dissipation, as well as of coherent structures and self-organisation processes. An important aspect is that one must understand such a state of matter as a turbulent state and that one must also understand the relation between regular motions and structures, on the one hand, and random motions (chaos), on the other hand.

1.2 Controlled Fusion Research and Non-linear Plasma Physics

It is not possible in a few words to characterise the many investigations in the field of controlled fusion. We shall restrict ourselves to describing those aspects which are of a more general physical nature. The thermonuclear output depends on the confinement time and on the temperature and plasma density reached. The time during which the energy is confined is determined by the rate at which plasma particles "jump" from one magnetic field line to another, that is, the rate at which they diffuse across the confining magnetic field. Numerous investigations have shown that the observed confinement times cannot be explained by diffusion purely due to particle collisions, but that they are related to the action of some collective non-linear structures on the particle motion. It appears that as a result the particle "jumps" are several orders of magnitude faster than if they were due solely to particle collisions. At one time research workers in this field were even thinking that this effect might prevent the construction of a reasonable size reactor. Fortunately, this was too pessimistic a point of view and at this moment in time the biggest installations have already achieved the necessary conditions for break-even, that is, they have, by using a helium-tritium mixture, achieved temperatures and plasma densities sufficient for the start of the burning of a controlled thermonuclear reaction. The size of these installations in which these results have been obtained is, however, much larger than if the binary collisions were determining the energy confinement times. The observed confinement times for electrons are larger than if due to binary collisions by two orders of magnitude, and for protons the discrepancy is a factor four, or maybe even fifteen. This effect is called *anomalous diffusion*; it is due to the interaction between the particles and random, magnetic and electrostatic, fields created by non-linear motions in the plasma. On the other hand, coherent magnetic structures, which appear as the result of the excitation and damping of collective modes, are also important in the determination of the confinement time. This problem has not yet been fully solved and to get a complete solution one needs a better understanding of the relations and balances between coherent structures and collective motions.

Plasma heating is, like confinement, related to many collective phenomena; these lead to the observed worsening of the confinement time during the heating process.

1.3 Laser Plasmas; Laser–Matter Interactions and Collective Plasma Processes

Laser physics is a highly non-linear field, since non-linear processes figure naturally in the behaviour of strong coherent laser radiation. Non-linear laser physics, also called non-linear optics, is largely concerned with the interactions between coherent modes and the excitation of coherent harmonics. The non-linearities in laser physics are used for many purposes, such as frequency multiplication or the creation of harmonics of the laser radiation. However, strong laser radiation often also excites fluctuations over a broad frequency range. Those collective modes which are excited by the laser radiation can transfer their energy to fast electrons and this, in turn, can have important consequences for the problem of laser controlled fusion.

Laser controlled fusion, or *inertial fusion*, is the term used to describe the programme to obtain fusion through the irradiation of micron-size small solid targets with strong, pulsed laser radiation. The evaporation and subsequent ionisation of the surface material of the target by the laser radiation creates a *plasma corona* around the target. This corona is heated by the laser radiation. However, the thermonuclear output is proportional to the square of the density so that the reactions must occur before the target material expands. This means that the material must be compressed before it is heated. This compression is achieved by the same laser pulse that produces the corona. At the present moment in time compressions achieved are by a factor 10, or perhaps a bit more, in the linear dimensions, which means a factor 10^3, or better, in density. After the compression, however, collective fields start to play a significant rôle. They transfer energy to fast electrons which then can penetrate into the centre of the target, heat it, stop the compression, and force the target to explode. This was the reason for the failure of the large programmes which had considerable financial backing. As a result most research workers in laser fusion left this field and started to look at other schemes in which the collective non-linear plasma processes do not play such an important rôle.

Another important non-linear process decreasing the efficiency of the heating of the plasma corona by laser radiation is the scattering of this radiation by the collective modes in the plasma which have been excited by the same laser pulse. The scattering turns out to be mainly back-scattering and the reflection coefficient may turn out to be as high as 90%. This means that due to collective effects the laser radiation may be reflected before it reaches the target.

Due to these two effects – the collective acceleration of fast electrons and the collective reflection of the laser radiation – the output from the very large laser-fusion installation, "Shiva", was several orders of magnitude smaller than had been expected when collective effects were not taken into account. This example show that even for purely practical reasons knowledge of the fundamental non-linear properties of a plasma is important. It is not true that such knowledge was not used when the experiments were planned, but the experimentalists did not believe the theoretical predictions to be a hundred per cent reliable and hoped that actual experiments would be more complicated than any theoretical model and would confound the theoretical pessimism. The general "ideological" point of view that collective phenomena always occur in a plasma had been neglected.

It was only after these disappointing experimental results had been obtained that large computer experiments – computer simulations – were performed. These also showed the big rôle played in this problem by collective effects. Taking into account the principle that a plasma is as a rule a highly non-linear and turbulent state of matter one expects that these collective phenomena should play an important part in the planning of experiments. One can also expect that in the case when the energy input into the plasma is sufficiently large the plasma itself will find a channel to produce strong collective effects.

Similar effects will appear whenever strong laser radiation interacts with matter, whatever its form, since the creation of a plasma corona and the excitation of strong collective fields are general properties of such an interaction. We shall in what follows describe concrete mechanisms for the excitation of collective plasma fields by laser radiation.

One of the aims of the presentation in this book is to give a clear picture of the above-mentioned "ideology" and to give many examples of how its use can help to avoid mistakes of the kind described earlier in this section. We shall in the next chapter give physical arguments which are a basis for this general "ideology". We have already given two examples from important fields of physics, but there are many more which can be given. Together they show the leading rôle played by plasma physics in the development of present-day physics in general.

1.4 Powerful Relativistic Electron Beams

Research on powerful relativistic electron beams is relatively new. It started in the early seventies when a technique for producing cold electron emission in high-voltage – MV or higher – devices was developed. At the present time the common aim of experimentalists is to produce a compact source of "relativistic" electrons with energies in the range of 100 to 500 keV and currents in the range of 200 to 400 kA, although the attained values for the largest devices are tenths of an MeV and tenths of an MA. The pulse length of the beam ranges from tenths of nanoseconds to several milliseconds. To some extent such beams compete with laser beams, although many of their properties are different.

The most common subjects for research on relativistic electron beams are their interaction with matter – plasmas and other targets – and the generation of strong radiation. Relativistic electron beams are often more efficient than laser beams for producing collective perturbations when interacting with a plasma. The energy transferred to the plasma is determined by the non-linear interactions between various collective plasma modes and by the plasma turbulence created by the beam. The emission due to these collective modes can be used to produce powerful radiation sources in the short- wavelength radiofrequency range. The efficiency of the transformation of the beam energy into the radiated energy is determined by the non-linear processes in the plasma.

An interesting new effect for powerful relativistic beams is the appearance of large self-fields of a collective nature. Sometimes most of the energy transferred by the beam is the self-field energy rather than the energy of the beam particles. In this sense the beam would be a bunch of collective fields rather than a bunch of particles. However, the fields and the particles are coupled in a self-consistent state so that it is impossible to separate fields from particles. In the rest frame of the beam the energy of the collective fields is larger than the particle energy; in this frame the particle energy is connected with their oscillations at right angles to the beam velocity in the collective fields produced by themselves. This occurs because the particles excite the fields coherently – the strength of the field is proportional to the square of the particle number rather than to its first power. Not only the collective fields excited when the beam interacts with the plasma but also the changes in the collective self-fields of the beam are important.

There is another consequence of the fact that the beam particle velocities are close to the speed of light. As a result the collective magnetic fields may be of the same order of magnitude as the collective electric fields. Since the magnetic forces lead to self-contraction – the *pinch effect* – and the electric forces to self-repulsion it becomes possible for charged electron beams to appear when both magnetic and electric forces are present and of the same order of magnitude. One can easily see from the estimate given in the introduction of this chapter that the collective fields of such beams may be very huge.

1.5 Powerful Accelerators and Collective Acceleration Methods

There are two trends in the development of conventional accelerators: on the one hand, an increase in the energy of the accelerated particles and, on the other hand, an increase in the number of particles accelerated per pulse, or an increase in the accelerated particle flux. If the number of accelerated particles is increased collective effects may become important. The first such effect is the creation of collective electrostatic fields by bunches of accelerated particles. A second effect is the appearance of instabilities which produce collective fields, as in a plasma. Accelerator physics tends to use a terminology which differs from that of plasma physics. However, the physics of collective phenomena in powerful accelerators is analogous to that in plasma physics and sometimes a simple change in terminology will reduce a problem in accelerator physics to the corresponding one in plasma physics. This is not surprising since both fields of research deal with very simple systems consisting of a collection of charged particles. Knowledge of non-linear plasma physics and of the above-mentioned "ideology" is therefore important also in accelerator physics.

If the number of accelerated particles increases one must use plasma acceleration since the collective fields become so strong that there may occur an automatic partial charge compensation. The idea thus arises to use a plasma for particle acceleration. Such an accelerator is called a collective accelerator since it uses collective plasma fields to accelerate the particles. The main problem for collective accelerators is the presence of non-linear interactions between collective plasma fields as they may lead to their randomisation. Random fields also accelerate particles – the creation of fast particles in laser-fusion experiments is a good example for showing that such a process can be efficient – but if one wants to accelerate a few particles to a very high energy regular fields are much more efficient than random fields with the same energy. The stochastisation of collective fields is a general non-linear property of them. Great efforts to regularise collective fields which are excited by various sources, such as strong particle or radiation beams, have not given any encouraging results. Recent studies have therefore tended to use collective fields during a short period of time, before they become randomised.

The randomisation process depends both on the strength of the field and on the bandwidth of the collective fields which have been excited. Hence, acceleration is possible if the source exciting the collective fields has a narrow bandwidth. Coherent laser radiation has been proposed for the excitation of the collective fields and such accelerators are called *laser accelerators*. The experimental set-up consists of two coherent laser pulses propagating in opposite directions with frequencies which differ by the frequency of the collective plasma mode which is to be excited. This kind of acceleration is due to the non-linear interactions of the two laser beams with the collective plasma mode. Laser accelerators have been studied in many laboratories and both the excitation of strong collective fields as well as the acceleration of electrons have been obtained. There exists a different collective acceleration scheme which uses for the excitation of a strong collective mode in the plasma short, pulsed, powerful relativistic particle beams. In that case there will exist at a certain distance behind the front of the beam a regular collective field – the randomisation processes are convective in this case and develop in space rather than in time. This kind of accelerator is called a *wake-field accelerator* since the collective field which is excited appears like a wake behind the bunched beam. Wake-field accelerators have also been studied in many laboratories.

We see that the study of collective accelerators meets with one of the fundamental problems of non-linear physics – the problem of the randomisation by non-linear interactions. In how far this problem is solved determines how long a time interval can be used for the particle acceleration and hence the maximum particle energy which can be achieved. In this way experiments on collective accelerators are related not only to the physics of powerful lasers and powerful relativistic beams but also to the fundamentals of non-linear interactions.

1.6 Dynamic Chaos, Self-organisation, and Turbulence

The discovery that in a completely dynamic system a random state can appear, that is, that dynamic chaos exists, is a new revolution in natural philosophy, comparable in its significance to the other revolutions at the beginning of the present century. It was found that non-linear dynamic systems, even those with a small number of degrees of freedom, could become stochastic. The predictions of dynamic chaos are important not only in physics, but also in economics and sociology.

Physicists are accustomed to describe systems with a large number of degrees of freedom using statistical methods, although they had applied them mainly to systems in or close to statistical equilibrium. It is clear that dynamic chaos would be important for non-linear plasmas, since often in a plasma a few collective or external coherent modes may be excited leading to a system far from equilibrium. Laser accelerators are just one example of such systems.

In a plasma far from equilibrium it can often happen that many kinds of collective non-linear motions are excited. The problem is how to describe them either dynamically or statistically. The answer to this problem given by dynamic chaos is that in many cases there will be a mixture of dynamic and random motions. This means that phase space is not completely filled with random motions, but that there remain islands of regular motions. Due to non-linear interactions the energy in the case of a system far from equilibrium will flow in phase space from one part to another. When most of phase space is occupied by random motions in the case where such fluxes are present we call the state *turbulent*. We can visualise the simplest turbulent state as one in which random motions of different sizes are superimposed upon one another.

The larger the energy fluxes in the turbulent state, the larger the rate of dissipation since the fluxes will finally reach the regions where there is dissipation. This means that in order to have stationary turbulence one needs a constant source of excitation. The larger the fluxes, the larger also the rate at which the energy can flow into regions in phase space where the motion is still regular. As a result the amplitude of the collective motions in the regularity domain will increase and the non-linear interactions between the collective motions inside the regular region will become important. As a rule these will form some non-linear structures. If there is dissipation in the regular region, regular structures can exist, or grow, only provided the energy flux into the regular region exceeds some threshold value. The structures formed in the regular region are usually called dissipative structures in this case. They have something in common with living biological creatures, and the energy flux into the regular region is analogous to food. The process of the formation of regular dissipative structures in almost random systems is called a *self-organisation process*. The formation of regular dissipative structures is accompanied by an increase in the dissipation in the system as a whole and, roughly speaking, "the increase in entropy" due to the enhanced dissipation, enlarges the possibility of a decrease in entropy in a small phase space volume needed to create regular dissipative structures. This again is similar to what happens in the case of living creatures.

We see thus that an increase in the energy input in such open systems not only increases randomness and dissipation, but also enhances the opportunity for creating regular dissipative structures. This scenario is at the present time the main idea behind current research in this field and we may expect that many new details will be found in future investigations. However, even at this moment a great deal has been achieved which makes it possible to sketch the main features of self-organisation and its close connection with the problems of turbulent states.

Many of the investigations in this field are carried out by mathematicians or mathematically inclined physicists; on the other hand, for applications to plasma physics one needs research workers with a good physical intuition. We hope that such research will be done in the near future since many plasma problems are waiting for such studies. The problem of turbulence in plasma theory is a topic studied intensely and so far many important results have been obtained theoretically and verified experimentally. We shall try to discuss these problems in what follows. Turbulence for which self-organised dissipative structures cannot appear is called *weak turbulence*, whereas turbulence in which coherent dissipative structures are formed is called *strong turbulence*. Particle and laser beams, provided they are sufficiently powerful, usually excite strong turbulence. On the other hand, in the case of weak beams it is possible that weak turbulence is excited.

An important question for the dissipative structures which are excited in a plasma is that of the physical mechanisms for dissipation. We shall discuss this problem in detail in what follows. The dissipative structures are known as *cavitons* in the case of high-frequency collective plasma motions. These have been observed and studied experimentally. In many cases of plasma turbulence experiments the potential differences due to the collective fields have been measured as functions of the position in the plasma. This signal can often suddenly change from being almost random to regular; this means that one comes from a random region into a region with regular structures. This picture is very common not only in measurements of plasma turbulence, but also in studies of other kinds of turbulence – for instance, in investigations of fluid hydrodynamics. This so-called *intermittency* phenomenon is obviously an indication of the presence of regular structures. Altogether, the laser-plasma and beam-plasma interactions are strongly connected with problems of plasma turbulence, self-organisation, and the creation of dissipative structures.

Apparently, turbulence in a tokamak is strong turbulence. The structures created in this case are mainly magnetic structures. Observations show the presence of *magnetic islands* in a "sea" of random magnetic fields superimposed upon the external regular magnetic field. It seems that the regular structures and the random magnetic fields are in some kind of equilibrium, similar to those described above in general terms. The presence of such self-organising structures as magnetic islands is the basis for a global plasma self-organisation process in tokamaks. The anomalous transport properties observed in tokamaks are due to the global self-organisation process. This problem is at the present time discussed at great length, but we are still waiting for the development of a detailed description of the dynamics of turbulence and self-organisation in tokamaks. This seems an excellent topic for good physicists to apply their efforts to.

1.7 Vortex Turbulence and Self-organisation

Circular motions are components of many of the collective motions. This is like the motion in a *vortex*. A plasma in a strong magnetic field can create vortex motions in planes perpendicular to the direction of the field. Such vortices form in tokamaks. The ions and the electrons do not follow each other exactly in these motions; that is, they form vortex currents which create magnetic field structures which ultimately leads to the formation of the magnetic islands we discussed earlier. They determine the plasma confinement. We see thus that vortex motions and vortex structures are very common in tokamaks.

Vortices are a general element of the motion in an incompressible liquid or gas. We can easily see this; it is a consequence of the fact that in an incompressible liquid any liquid element moving from one place to another must be replaced by another liquid element. Because of continuity such a motion can occur only along closed curves, that is, it should lead to vortices. It can be shown that any movement in an incompressible liquid can be represented as a movement of a collection of vortices.

A plasma in a strong magnetic field moves as an incompressible liquid in a plane perpendicular to it. A plasma moves also as an incompressible liquid in the absence of a magnetic field in the limit of very large scales and very long time intervals. The development of the vortex turbulence under those conditions is the same as in liquids.

Vortex turbulence can be two or three-dimensional. In the latter case the direction of the vortices can be arbitrary, but in the former case all the vortices are directed in the same direction (the direction of the vortex is defined as the orientation of its axis of rotation). Two-dimensional vortices appear in a thin layer of an incompressible gas – or liquid – and their axes of rotation are at right angles to the plane of the layer. All planets, including the Earth, have a thin layer of atmosphere; the cyclones and anticyclones in the atmosphere are the two-dimensional vortices. Self-organisation processes in the vortex turbulence in the atmosphere can create either large coherent vortex structures which determine the weather, or small powerful structures like tornados. Similar vortex structures appear in oceans. This means that atmospheric and oceanic physics can have many features in common with tokamak physics. Sometimes one can even use the same description and one can go from describing the one to describing the other simply by changing the nomenclature. This is another example showing the strong connections between different branches of research when one is considering non-linear behaviour.

There is another link – with the physics of galaxy formation. The stars and gas in galaxies can form flat disks which may behave as a two-dimensional incompressible gas. The observed spiral structure of a galaxy can thus be considered as self-organised coherent vortex structures.

The general properties of turbulence and self-organisation in incompressible hydrodynamics are related to three-dimensional vortices. In developed turbulence dissipative structures can be formed only by small three-dimensional vortices which may form either a vortex line structure or a curved vortex structure – called a vortex worm. This is possible because the axis of rotation of an individual vortex is not fixed in the three-dimensional case. These dissipative structures can play an important rôle in the determination of the rate of turbulent dissipation. This is an important and broad field of research in hydrodynamics at the present time.

1.8 Self-organisation in a Gas Discharge Plasma

A low-temperature plasma can usually be confined by walls; in that case surface phenomena due to plasma-wall interactions are very important. There are many physical and chemical processes which occur close to the surface of the wall and this makes the system much more complicated than a magnetically confined plasma. This was one of the reasons why it took a long time for gas-discharge physics to reach a high level. However, the presence of many processes which can occur simultaneously and the big rôle played by collisions because of the low temperature open many channels for the development of turbulence and self-organisation processes. The presence of various structures in gas discharges such as *striations* have been well known for a long time. Recently they have been explained as being self-organised dissipative structures. Striations are localised regions with higher temperatures and lower densities surrounded by regions with lower temperatures and higher densities. Recent research has shown that if the plasma in a gas discharge is subjected to strong radio-frequency radiation there appear many different structures in it. The formation of these structures is related to the ionisation by strong radiation.

There are many other self-organising processes in the physics of a gas-discharge plasma. Amongst them we may mention the so-called *Kwarzchawa structures* which are elongated along the currents in a gas discharge and which are related to magnetic-field structures. Another kind of structures are the double layers appearing close to the surface of the discharge. When we are considering surface effects we must also mention chemical reactions in the plasma and various structures related to plasma chemistry. One obvious effect is the possible appearance of large currents produced by the fast ions, which are the product of chemical reactions, leaving the surface. The magnetic structures produced by these ions may become self-organised and form magnetic structures which can localise and confine the gas discharge close to the surface.

1.9 Non-linear Processes in Dusty Plasmas

If a plasma contains a certain amount of dust particles its surface area will be greatly increased. Usually such dust particles are solid particles of about a micron in size which exist together with the ionised gas – the plasma – in the same volume. In the case when the dust particles have a fractal structure – which seems to be extremely probable – their surface areas will be very large. However, even if this were not the case, the presence of a small amount of dust particles can increase surface effects. In the case where the material of the dust particle is chemically active this can lead to many collective effects: magnetic structures can be self-organised and a ball discharge may occur – this has by some research workers been proposed as a mechanism for the appearance of ball lightning.

The dust particles in a plasma are usually highly non-linear objects since plasma currents interacting with a dust particle will charge it sometimes with a charge of 10^4 or more electron charges. Chemical processes or the effect of external radiation on the dust surface can also charge the particles. For example, charging can occur due to the photo-effect or secondary electron emission. The presence of such strongly non-linear structures as dust particles surrounded by a cloud of plasma particles in a strongly non-linear stage has many important consequences, such as an increase in the absorption of waves and anomalously strong scattering of waves. The scattering cross-section may increase by a factor of the order of the square of the charge of the dust particle in terms of the electron charge. This kind of scattering of a radio signal by dust particles has apparently been observed in the lower ionosphere at heights of the order of 80 km. The presence of dust is probably due to pollution. The detection of dust at those heights is important as a possible means of detecting vortex motion in the upper atmosphere. In turn, this is important for understanding solar-terrestrial relations and for discovering how activity on the Sun may affect atmospheric motions. Dusty-plasma physics has many applications in laboratory physics and technology, in space physics, and in astrophysics. Among laboratory applications we may mention the problem of electronics and plasma film etching, when the product of the etching may form charged dust particles, levitating in a double layer above the surface of the sample; after the etching process is finished they are a source for natural contamination. Among space-physics applications we may mention the formation of planetary systems, the physics of planetary rings, and first and foremost cometary physics, since the comets are large sources for dust fluxes. Among astrophysical applications we may mention that there are many regions in space which contain a lot of dust, especially in regions where star formation takes place, in cosmic-maser regions, in supernova-explosion regions, and in the centre of our Galaxy.

1.10 Non-linearities of the Ionospheric Plasma

It is well known that the ionosphere is a weakly ionised layer (or layers), which is (are) essential for radio communications over long distances, surrounding the Earth at a height of the order of hundreds of kilometers. Non-linear effects in the ionosphere were discovered quite some time ago, but it is only recently that significant attention has been paid to studies of various non-linear structures in the ionosphere and to measurements of collective field strengths. On the one hand, there exists a huge current structure in the shape of a ring current; on the other hand, many small spatial structures have now also been observed in the ionosphere. The latter are produced by non-stationary heating and ionisation and by the penetration of fast particles from the magnetosphere. This penetration of fast particles can give rise to a gas discharge with a beautiful and complicated structure, the so-called *aurora.*

Many research groups have in recent years developed ionospheric experiments. In those experiments powerful radio emitters are used to send pulsed radio signals to create collective fields in the ionosphere, to heat it, and to accelerate particles in it. By analysing the reflected signal one can detect the various processes occurring in the ionosphere. When the ionosphere is close to the threshold for the production of a gas discharge one can artificially create an aurora. The phenomena occurring in the regions of the ionosphere where the incident signal is reflected are similar to those appearing when laser radiation interacts with a plasma. As in the case of laser-plasma interactions most of the incident energy is transferred to fast particles. Other experiments carried out on the ionosphere are aimed at a local heating of the plasma and the formation of an ionised region to create an artificial mirror for reflecting short-wavelength radiation.

1.11 Space Physics Research

Measurements of the properties of space plasmas by satellites are at the present time oriented to measuring the strengths of the collective fields and the energy distributions of the plasma particles. The region called the magnetosphere is a region bounded by the surface on which the pressure of the solar wind is balanced by the magnetic pressure of the Earth's magnetic field. In the direction of the Sun this boundary is produced by a shock wave, the so-called *bow shock.*. The position of the bow shock is not rigorously fixed since the solar-wind pressure varies, but it is usually at nine to ten Earth radii. In the direction away from the Sun there exists a magneto-tail which extends over distances much larger than that of the bow shock. Measurements of the collective plasma field in the magnetosphere and in the solar wind (outside the magnetosphere) have been carried out in numerous flights and they have shown the existence of many collective plasma motions over a broad frequency range.

The magnetosphere is filled with fast particles; part of them penetrate from the solar wind into the magnetosphere, but many are accelerated in the magnetosphere by collective plasma fields. Some of them penetrate to lower heights – in polar regions – and produce the aurora. This acceleration is similar to that occurring in the earlier mentioned laser-plasma interactions – although different collective motions may be responsible, especially those connected with the magnetospheric magnetic fields. Many parts of the magnetosphere – especially the bow-shock region – are strongly turbulent.

Many problems of magnetospheric physics are intimately connected with the problem of plasma turbulence. One of the satellite programmes was even called "Turbulence". The particle distributions in the solar wind and in the magnetosphere are non-equilibrium distributions; this is another argument for assuming that the action of the plasma fields may occur in regions different from those where the measurements are performed. Because of the turbulence excited by the solar wind at the boundary of the magnetosphere, the magnetosphere (of most of the planets with a magnetic field) becomes a source in space for strong radiation in the long-wavelength range – the kilometer range. Kilometric radiation is emitted from regions close to the polar region of the magnetosphere; it is due to the turbulence excited by accelerated particles falling onto the lower atmosphere and producing auroras. Kilometric radiation from other planets is a spectacular effect which cannot be seen directly from the Earth's surface as it is reflected outwards by the ionosphere.

Recently several space experiments have been performed in which strong plasma turbulence was artificially excited. In one of those experiments barium clouds were injected into the magnetosphere; they were quickly ionised by solar radiation and formed expanding clouds of a denser plasma, forcing the magnetospheric magnetic fields out from the region they occupy and thus creating many turbulent plasma fields. These fields can then be measured and analysed by satellites which follow behind the barium cloud emitting ones. Collective plasma fields are observed both on the boundary of the expanding cavern structure and inside it. It is found that the excitation of collective plasma fields is accompanied by fast-particle acceleration. This is consistent with the idea of a turbulent plasma and also can be explained qualitatively.

Another kind of experiment which is carried out at the present time is connected with exciting plasma turbulence by a strong radio-transmitter operating on the satellite. In this way recent space experiments use the whole arsenal of the ideas of modern plasma physics both in their planning and when analysing the data obtained.

1.12 Plasma Astrophysics

Plasma astrophysics is involved in many fields, among them the physics of stars, especially of the Sun, the physics of star creation from the interstellar medium, the physics of galactic nuclei, the so-called *quasars*, and the physics of magnetic stars, especially that of the magnetospheres of compact neutron stars, the so-called *pulsars*. Of course, all theoretical work is based on astronomical observations, but these lead in many cases to generally accepted hypotheses about the physical processes taking place in those astronomical objects. Clearly all these objects consist of a plasma and the general "ideology" of the non-linear behaviour of plasmas plays here an even more important rôle than in the physics of laboratory plasmas. A few of the phenomena can be modelled in laboratories.

The physics of the Sun's atmosphere is a large area for research. Even in the "quiet" Sun there are many interesting plasma processes, but most research effort is devoted to such explosive phenomena as the *solar flares*. The collective plasma fields which occur in those explosions produce both fast accelerated particles as well as the so-called *sporadic emission* at various wavelengths. The explosions are followed by the appearance of various very turbulent magnetic structures above the Sun which are produced by convection. The complicated structures of the magnetic tubes and magnetic spots are apparently due to some self-organisation processes in which dissipation takes place through the emission of various electromagnetic and plasma waves. The explosions are accompanied by shock waves propagating outwards in the solar atmosphere and by the ejection of fast particles and plasma clouds into the solar wind. The fast, almost relativistic particles leaving the Sun produce a sporadic radio-emision which can be detected and is interpreted as an emission process due to the strong plasma turbulence which is excited by fast particles in the upper solar atmosphere. The plasma clouds and the fast particles reach the Earth's atmosphere and produce significant perturbations of its magnetic structure which are known as *magnetic storms*. The additional ionisation produced in the Earth's atmospheres by solar explosions can appear as low down as 10 km above the Earth's surface. It is important to investigate solar explosions if one is concerned about the safety of cosmonauts.

Many recent investigations have shown that the quiet Sun is also very turbulent and the formation of the solar wind can be explained only if anomalous turbulent processes are taken into account. The formation and heating of the solar corona is due to the anomalous dissipation of random collective magnetic fields.

As to other stars, most attention has been paid to pulsar magnetospheres which – according to modern ideas – are filled with a turbulent electron-positron plasma. Many details of the pulsar emission can be explained using modern theoretical ideas about the non-linear interactions between the collective modes in the pulsar magnetosphere.

Powerful turbulent processes leading to a very strong emission also occur in such astronomical systems as galactic nuclei or the regions of star formation – where sometimes cosmic masers are observed.

1.13 Cosmic-Ray Physics. Gravitational Fields

Cosmic rays are the highly relativistic particles which impinge on the Earth with a constant flux. The highest energies in cosmic rays reach 10^{20} eV. The cosmic-ray intensity as function of their energy decreases rather slowly – as a power law – which shows that they are in a highly non-equilibrium state. The main problem about the origin of cosmic rays is to understand how they can be accelerated to such high energies and why their spectral energy distribution is so highly non-equilibrium. The main "ideology" of non-linear plasma physics can again be applied in this case and the answer should be that they were accelerated in a strongly non-equilibrium turbulent plasma in regions of space where the most violent explosions occur. These explosions are the stellar explosions – the so-called *supernova explosions* – and the violent explosions in galactic nuclei. This is now a generally accepted scenario.

Another important problem is the observed isotropy of the cosmic rays. A large amount of research has also been devoted to this problem and at the present time it is known that an anisotropic cosmic-ray distribution creates collective fields in the interstellar plasma which isotropise the cosmic-ray distribution. Cosmic rays play an important rôle in space as their energy density is comparable to the energy density of the cold plasma component and to the magnetic field energy density in space.

We showed in the Introduction that small charge separations in a plasma can create large collective fields. This is the reason why a plasma usually remains quasi-neutral. However, a plasma containing relativistic particles can be charged, if the particles produce currents. The attractive forces between parallel currents of relativistic particles can almost cancel the repulsive forces between them due to the charges of the particles. A small amount of charges of the opposite sign can therefore make the system a stationary one. Charged plasmas have been studied in laboratories where relativistic beams are injected into magnetic traps. In this case a balance is reached through the centrifugal forces due to the rotation of the relativistic particles in a plane at right angles to the direction of the magnetic field of the trap.

The properties of a charged plasma are quite similar to those of a system of rotating particles which attract one another, which corresponds to gravitational interactions. This analogy leads to a new direction of researcch connected with the collective properties of gravitational interactions; many results from plasma physics could be transferred to this field of research. Important new results could be obtained and they could be applied especially to the problems of planetary rings.

Modern non-linear plasma physics is also used in many other branches of physics. We may mention solid-state physics, even apart from solid-state plasma physics, surface physics, even apart from plasma etching, and plasma chemistry. We have already mentioned that similar descriptions are also used in biology, chemistry, and even economics.

1.14 Outline of the Remainder of this Book

The short review given in the present chapter should have given an impression of the importance of the field to be discussed in the main part of the book. What we have discussed so far should have left the feeling that many problems in different fields of research show a great similarity; for example, the problem of particle acceleration by plasma turbulence alone has many applications in different fields of research. The aim of the present lectures is to produce a general feeling for how all above-mentioned problems can be stated and for what is the physical background and basis for the general "ideology" of non-linear collective phenomena. A great number of books and reviews have been devoted to many of the actual problems which we touched upon above and we give at the end of this chapter a list of such books and reviews. However, in what follows we discuss some general concepts which can be used for all of them. We have tried to avoid hiding these general ideas beneath the complicated details of the specific problems. We hope that these lectures can serve as a guide to the general physics which should never be buried under the complications of the field of interest and that they can then be applied to specific problems. In the short review we gave in this chapter we have as a rule tried to avoid oversimplifications, but all the same we attempted to describe in a few words the ideas of numerous investigations, however difficult this is, in fact. Those readers who finish this course of lectures are recommended to return to the present introductory chapter. Many aspects of the problems should then have become much clearer. We recommend that only after this the reader should turn to the description of particular problems in the reviews and books referred to at the end of this chapter.

In the remainder of the book, we start in Chap. 2 reminding ourselves of the general properties of collective plasma motions, emphasising those points which will be important in what follows. Already in this first stage we try to formulate a general "ideology" which, in fact, is the result and the summary of many years of study by a great many research workers. In Chap. 3 we introduce the probability description for interactions between strong plasma waves. This is applied throughout the book to many other non-linear processes. We continue a discussion of this description in Chap. 4 with the aim of showing physical reasons why the cross-sections for particle collisions in a plasma differ from those known for the collisions of isolated particles. We find a very simple way to describe the results; these results are usually obtained by rather complicated procedures and several books exist to describe results which in Chap. 4 are obtained in a few lines. Chapter 5 is devoted to a general and rather simple description of non-linear interactions; this is the basis of the descriptions of all non-linear processes in what follows. The description given enables us to show clearly that there exist profound relations between different non-linear processes in a plasma.

In Chap. 6 we return to a probability description for scattering processes and we show very simply why scattering cross-sections in a plasma differ by many orders of magnitude from those in vacuo. The concept of plasma turbulence is introduced in Chap. 7; we determine turbulent spectra and discuss applications to laboratory and space plasmas. Chapter 8 is devoted to non-linear decay processes and we show clearly how they are related to other processes. Our derivations are somewhat different from those often used and this enables us to establish connections which are not so clear in other approaches. In Chap. 9 we consider strong non-linear interactions which appear when a powerful energy source is applied to a plasma. We describe the non-dissipative non-linear structures which play an important rôle in many applications. Chapter 10 deals with dissipative structures in plasmas, with self-organisation, and with strong turbulence.

The importance of a new dissipation mechanism, connected with inhomogeneities in coherent non-linear structures, is stressed in Chap. 11. We show the necessary conditions for the development of a non-linear maser effect and indicate the kind of quantities which can be adiabatically conserved in non-linear interactions. Chapter 12 discusses important problems connected with non-linear interactions between particle fluctuations and waves; this enables us to find an important collective effect in bremsstrahlung which is discussed in Chap. 13. In that chapter we return to the probability description and again show simply why the probability for bremsstrahlung in a plasma is so different from that found in most textbooks where bremsstrahlung is discussed without taking into account collective effects. The last chapter, Chap. 14, is devoted to a brief description of those physical processes which can be important in more sophisticated considerations. Emphasis is put on the physical aspects of applying the general concepts to particular problems. Throughout the book we give examples from laboratory or space plasma physics of practical applications of the general concepts.

There remains one important point to be made. The material of the present chapter not only serves as an introduction, but it is also used in the chapters that follow. We have given in this chapter a physical explanation of many problems and also gave some definitions. Although this was done in a somewhat superficial way, they will be used in what follows. Some later estimates will follow from the discussion of the first chapter. Hence, we do not recommend skipping the present chapter when one starts reading this book. The material presented here is important for all later discussions. We had in mind that this way of presenting the material is helpful for a better understanding of the subject, by returning to problems several times – each time at a higher level.

1.15 Recommended Literature

The following list is not a list of references, but we list some monographs and reviews which we recommend; in them one can find references to the original papers.

A.I. Akhiezer, I.A. Akhiezer, R.V. Polovin, A.G. Sitenko, and K.N. Stepanov, Collective Oscillations in a Plasma, Pergamon Press, Oxford, 1967.

A.I. Akhiezer, I.A. Akhiezer, R.V. Polovin, A.G. Sitenko, and K.N. Stepanov, Plasma Electrodynamics, Pergamon Press, Oxford, 1975.

A.F. Aleksandrov, L.S. Bogdankevich, and A.A. Rukhadze, Principles of Plasma Electrodynamics, Springer Verlag, Heidelberg, 1972.

H. Alfvén and K.-G. Fälthammer, Cosmic Electrodynamics, Clarendon Press, Oxford, 1963.

L.A. Artsimovich, Controlled Thermonuclear Reactions, Pergamon Press, Oxford, 1979.

L.A. Artsimovich and R.Z. Sagdeev, Plasma Physics for Physicists, Benjamin, New York, 1985.

R. Balescu, Statistical Mechanics of Charged Particles, North-Holland, Amsterdam, 1965.

D. Bekefi, Radiation Processes in Plasmas, Wiley, New York, 1966.

R.C. Davidson, Methods in Nonlinear Plasma Theory, Academic Press, New York, 1972.

R.C. Davidson, Theory of Nonneutral Plasma, Benjamin, Reading, Mass., 1974.

G. Ecker, Theory of Fully Ionized Plasma, Academic Press, New York, 1972.

V.L. Ginzburg, Propagation of Electromagnetic Waves in a Plasma, Pergamon Press, Oxford, 1970.

V.L. Ginzburg and A.A. Rukhadze, Waves in Magnetoactive Plasmas, Handb. Physik **49/2** (1972).

V.L. Ginzburg and V.N. Tsytovich, Transition Radiation and Transition Scattering, Adam Hilger, Bristol, 1990.

D. ter Haar and V.N. Tsytovich, Modulational Instability in Astrophysics, Phys. Rept. **73**, 175 (1978).

S. Ichimaru, General Principles of Plasma Physics, Benjamin, Reading, Mass., 1973.

B.B. Kadomtsev, Plasma Turbulence, Academic Press, New York, 1965.

B.B. Kadomtsev, Tokamak Plasma Physics, Adam Hilger, Bristol, 1994.

S.A. Kaplan, S.B. Pikel'ner, and V.N. Tsytovich, Plasma Physics of Solar Atmosphere, Phys. Repts. **15**, 1 (1974).

S.A. Kaplan and V.N. Tsytovich, Plasma Astrophysics, Pergamon Press, 1973.

V.I. Karpman, Nonlinear Waves in Dispersive Media, Pergamon Press, Oxford, 1975.

Yu.L. Klimontovich, The Statistical Theory of Non-Equilibrium Processes in a Plasma, Pergamon Press, 1969.

N.A. Krall and A.W. Trivelpiece, Principles of Plasma Physics, McGraw-Hill, New York, 1973.

D.B. Melrose, Plasma Astrophysics, Gordon and Breach, New York, 1980.

A.B. Mikhailovskii, Theory of Plasma Instabilities, Consultants Bureau, New York, 1974.

M.V. Neslin, Dynamics of Beams in Plasmas, Springer Verlag, Heidelberg, 1991.

N. Rostoker and M. Reiser (Eds), Collective Methods of Acceleration, Harwood Academic Publ., Chur, 1979.

D.J. Rous and M. Clark, Plasma and Controlled Fusion, Wiley, New York, 1960.

L.I. Rudakov and V.N. Tsytovich, Strong Langmuir Turbulence, Phys. Repts. **40**, 1 (1978).

R.Z. Sagdeev and A.A. Galeev, Nonlinear Plasma Theory, Benjamin, New York, 1969.

A.G. Sitenko, Fluctuations and Nonlinear Interactions of Waves in a Plasma.

L. Spitzer, Physics of Fully Ionized Gases, Interscience, New York, 1962.

T.H. Stix, Theory of Plasma Waves, McGraw-Hill, New York, 1962.

S.G. Thornhill and D. ter Haar, Langmuir Turbulence and Modulational Instability, Phys. Repts **43**, 43 (1978)

V.N. Tsytovich, Nonlinear Effects in Plasma, Plenum Press, New York, 1970.

V.N. Tsytovich, An Introduction to the Theory of Plasma Turbulence, Pergamon Press, Oxford, 1972.

V.N. Tsytovich, Theory of Turbulent Plasma, Consultants Bureau, New York, 1977.

J. Weiland and H. Wilhelmsson, Coherent Non-Linear Interaction of Waves in Plasmas, Pergamon Press, Oxford, 1977.

D.G. Wentzel, Plasma Instabilities in Astrophysics, Gordon and Breach, New York, 1969.

2 Collective Plasma Oscillations

2.1 Elementary Description
of Collective Plasma Oscillations

To start with we shall remind our readers of the elementary description of collective plasma oscillations. We must emphasise that plasma oscillations are excited by the collective motion of many particles, which through simultaneous displacements create a field, rather than by a single particle. Let us consider a quasi-neutral, homogeneous plasma in which at any point the charge of the electrons is cancelled by the charge of the ions. Consider a layer of thickness d and displace the electrons in this layer over a distance $\delta x = d$ (see Fig.2.1). It is important that all the electrons in the layer δx are displaced over the same distance, as only in this case will the field due to these displacements be created collectively.

As a result of the displacement of the electrons the ion charge is no longer cancelled, that is, the ion charge is "bare" and a positive charge appears, although the ions were not shifted. The electric field E which appears due to the charge separation will attract the electron layer back towards its original position. After having shifted the electron layer we now let the layer move under the action of the field which has appeared, and assume that the heavy ions do not move. At the moment the electrons reach their initial positions they will have a maximum velocity and although at that moment they exactly cancel the ion charge, the electron layer will continue to move until it reaches the same distance from the position of electric neutrality as when it started to move, but on the opposite side. The electron layer then starts to move back, and so on. In this way the so-called *Langmuir oscillations* are produced. It is not difficult to find the frequency of these oscillations. The field strength produced by a plane sheet is $E = 4\pi\sigma$, where σ is the surface charge density, $\sigma = -en_0\delta x$, with n_0 the unperturbed electron density and e the electron charge. We can write the equation of motion for the layer in the form

$$m_e \frac{d^2}{dt^2} \delta x = -eE = -4\pi e^2 n_0 \delta x = -m_e \omega_{pe}^2 \delta x, \qquad (2.1)$$

where

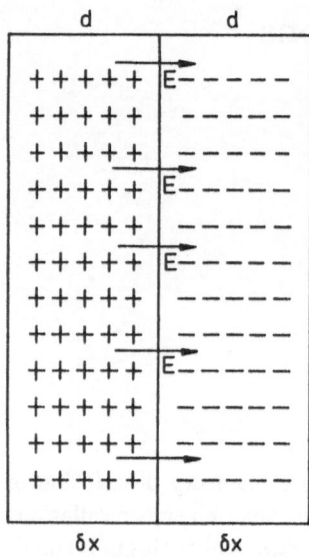

Fig. 2.1. Langmuir oscillation produced by the displacement of an electron layer over a distance equal to its thickness

$$\omega_{\text{pe}}^2 = \frac{4\pi n_0 e^2}{m_{\text{e}}}. \tag{2.2}$$

Here m_{e} is the electron mass and ω_{pe} the *electron plasma frequency* or *electron Langmuir frequency*. Let us briefly discuss the domain of the validity of such a picture and the mechanisms which can destroy the synchronism of the electron motions and thus destroy the collective nature of the fields which are created. First of all, it is clear that such perturbations can travel as waves. However, in Eq.(2.1) the frequency is the same for all electrons and the electric field is uniform. We can consider another case when the layer is rather thick while the electron displacement δx is much smaller than the thickness d of the layer. A uniform field will be created within the whole thickness d since the displacement of the electrons will create two thin layers of width δx at the boundaries of a thick neutral layer (see Fig.2.2).

Within the layer of thickness d the electrons will oscillate synchronously although their amplitude at each point of the thick layer could be small. It seems clear that outside a layer of thickness d such perturbations will excite perturbations with a wavelength of the order of d. As we can imagine such layers with arbitrary large thicknesses – as long as the plasma is homogeneous for such thicknesses – the wavelength of plasma waves can be arbitrarily long. Therefore one cannot see any reason why large-size collective motion of electrons of a large size should be destroyed as long as the plasma is homogeneous. On the other hand, for short wavelengths the thermal motion of the particles – which we have neglected so far – may be important. Let us assume that the electron velocity distribution is thermal. Let $v_{T\text{e}}$ be the root

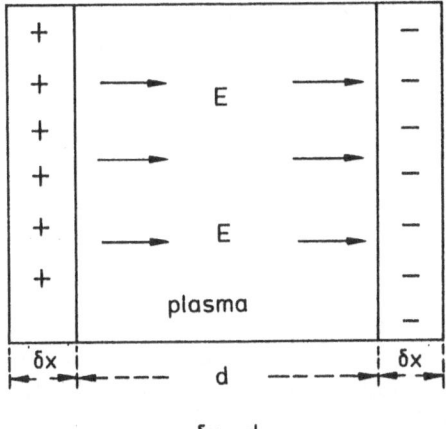

$\delta x \ll d$

Fig. 2.2. Langmuir oscillation produced by the displacement of an electron layer over a distance much less than its thickness

mean square of the component of the electron velocity in an arbitrary direction, and let T_e be the electron temperature in energy units, that is, we put Boltzmann's constant equal to unity. In that case v_{Te} is the *electron thermal velocity* which satisfies the equation

$$v_{Te} = \sqrt{\frac{T_e}{m_e}}. \tag{2.3}$$

The thermal distribution for the velocity component, v_x, along the arbitrarily chosen direction – which we choose as the x-direction – is proportional to $\exp(-v_x^2/v_{Te}^2)$. During a quarter of the plasma period the electrons will in the plasma oscillations move over a distance d whereas due to the thermal motion they will move during the same time interval over a distance

$$d_e \approx \frac{v_{Te}}{\omega_{pe}}. \tag{2.4}$$

This means that thermal motion will destroy the coherent motion in the layer if $d \ll d_e$. The quantity d_e, defined by Eq.(2.4) with the equal rather than the approximate equal sign, is called the *electron Debye radius*. We have thus found that plasma waves must have wavelengths, λ, which are longer than the Debye radius:

$$d_e \ll \frac{\lambda}{2\pi} \leqslant \infty. \tag{2.5}$$

The phase velocities of the waves,

$$v_{ph} = \frac{\omega}{k} \approx \frac{\omega_{pe}}{k} = \omega_{pe}\frac{\lambda}{2\pi}, \tag{2.6}$$

where $k = 2\pi/\lambda$ is the wavenumber, should thus lie in the range

$$v_{Te} \ll v_{ph} \leqslant \infty, \tag{2.7}$$

and the wavenumbers correspondingly in the range

$$0 \leqslant k \ll \frac{1}{d_e}. \tag{2.8}$$

It turns out that the last inequality is not as strong as it is written and waves can exist even for $k \cong 1/3d_e$.

The coherent motion of the electrons can be disturbed also by collisions between the electrons and the ions. Let the average frequency of these collisions be ν_{ei}. We can then formulate another necessary condition for the existence of coherent electron motions:

$$\omega_{pe} \gg \nu_{ei}. \tag{2.9}$$

Strictly speaking, an exactly linear coherent motion does not exist. It always has a finite amplitude so that only approximately can one consider the oscillations to be linear. Let us denote the amplitude of the field strength by $\mathbf{E}$ and the amplitude of the potential in the wave by ϕ ($\mathbf{E} = -\nabla\phi$). Weak non-linearity corresponds to the case where the maximum displacement of the electrons in the wave field is much smaller than the wavelength:

$$\delta x_{max} \approx \frac{e\delta E_{max}}{m_e \omega_{pe}^2} \ll \frac{\lambda}{2\pi} \approx \frac{1}{k}; \qquad \delta E_{max} \approx k\phi_{max}, \tag{2.10}$$

or

$$e\delta\phi_{max} \ll \frac{m_e \omega_{pe}^2}{k^2} = m_e v_{ph}^2. \tag{2.11}$$

We shall see below that sometimes a stronger inequality should be satisfied:

$$e\delta\phi_{max} \ll m_e v_{Te}^2 = T_e. \tag{2.12}$$

This one can be understood, if the non-linear interactions cover waves with all wavelengths including the smallest possible one.

2.2 Some Definitions. Debye Screening

Plasma can be defined as a state of matter. A plasma contains charges of opposite sign, but the quasi-neutrality property is not the main property of a plasma: we have mentioned earlier that charged plasmas can also exist. One of the main properties of a plasma is that it consists of almost free, moving charges, that is, that the average kinetic energy of the particles is much larger than the average energy of the particle interactions.

Let us now consider a fully ionised, quasi-neutral plasma consisting of non-relativistic electrons of charge e and ions of charge $-Ze$; the electron density n_e will then be related to the ion density n_i through the relation

$$n_e = Zn_i. \tag{2.13}$$

The average interaction of the electrons is determined by their average distance apart, $n_e^{-1/3}$, and the necessary condition that the electron gas behaves as a plasma is

$$e^2 n_e^{1/3} \ll T_e, \tag{2.14}$$

while the analogous condition for the ion gas is

$$Z^2 e^2 n_i^{1/3} \ll T_i, \tag{2.15}$$

where T_e and T_i are, respectively, the electron and ion temperatures. We introduce the *average ion thermal velocity* v_{Ti} by the equation (m_i is the ion mass)

$$v_{Ti} = \sqrt{\frac{T_i}{m_i}}, \tag{2.16}$$

the *ion plasma frequency* or *ion Langmuir frequency* ω_{pi} by the relation

$$\omega_{pi}^2 = \frac{4\pi n_i Z^2 e^2}{m_i}, \tag{2.17}$$

and the *ion Debye radius* d_i by the relation

$$d_i = v_{Ti}/\omega_{pi}. \tag{2.18}$$

Ions become important for low-frequency oscillations in the case when, because of their thermal motion, the electron motion is no longer synchronised – when the wavelength is much shorter than the electron Debye radius – but the thermal motion of the ions is not yet able to destroy the collective ion motion, that is, when we have

$$d_i \ll \lambda \ll d_e, \qquad d_i \ll d_e, \tag{2.19}$$

or

$$T_i \ll ZT_e. \tag{2.20}$$

In this case ion oscillations with frequencies close to ω_{pi} can exist with wavelengths in a restricted range, determined by Eq.(2.19).

The quantities d_e and d_i determine the distance over which the field of an arbitrary, not too large (see below) charge which is placed in the plasma will be screened. Indeed, let ϱ_q be the charge density corresponding to a total (extra) charge q which produces only small perturbations $\delta\varrho_e$ and $\delta\varrho_i$ of the electron and ion charge densities, respectively. The equation for the electrostatic potential ϕ is then

$$\nabla^2\phi = -4\pi\varrho_q - 4\pi\delta\varrho_e - 4\pi\delta\varrho_i, \tag{2.21}$$

where we have

$$\delta\varrho_e = en_e\left[e^{-e\phi/T_e} - 1\right], \qquad \delta\varrho_i = -Zen_i\left[e^{Ze\phi/T_i} - 1\right]$$

in the case of a static distribution of electrons and ions at thermal equilibrium in the potential ϕ. In the case of a sufficiently small charge q so that

$$|e\phi| \ll T_e, \qquad |Ze\phi| \ll T_i, \tag{2.22}$$

we can linearise the expressions for $\delta\varrho_e$ and $\delta\varrho_i$,

$$\delta\varrho_e \approx -\frac{e^2n_e\phi}{T_e}, \qquad \delta\varrho_i \approx -\frac{Z^2e^2n_i\phi}{T_i}, \tag{2.23}$$

so that Eq.(2.21) becomes

$$\nabla^2\phi - \frac{\phi}{d^2} = -4\pi\varrho_q, \tag{2.24}$$

where d is the *Debye screening radius* – without the qualification "electron" or "ion" –

$$\frac{1}{d^2} = \frac{1}{d_e^2} + \frac{1}{d_i^2}. \tag{2.25}$$

For a point charge when we have $\varrho_q = q\delta(\mathbf{r})$ we find

$$\phi = \frac{q}{r}e^{-r/d}. \tag{2.26}$$

We must emphasise that Eq.(2.26) was obtained for a charge at rest and is strictly speaking valid only for that case. In the case of a moving charge the electron and ion Boltzmann distributions may not have enough time to become established. This is the case when the velocity of the charge is higher than the thermal velocity. Equation (2.26) is thus valid provided the velocity of the charge is lower than both the ion and the electron thermal velocities. If the velocity of the charge is higher than the ion, but lower than the electron thermal velocity,

$$v_{Ti} \ll v \ll v_{Te}, \tag{2.27}$$

the screening is determined by the electron Debye radius. The left-hand side of inequality (2.27) needs one correction: instead of v_{Ti} one should write $v_{Ti}d_e/d_i$. Indeed, the Boltzmann distribution does not have enough time to become established the case when $v_{Ti} \ll v$, but the ions can dynamically screen the charge for $v \ll v_{Ti}d_e/d_i$. Dynamical screening will be unimportant when

$$v \gg \frac{v_{Ti}d_e}{d_i} = \sqrt{\frac{ZT_e}{m_i}}. \tag{2.28}$$

One can find this criterion from the following estimates. A wave with wavenumber k will act on an ion with a frequency $kv = v/d_e$, due to the Doppler shift of the frequency. The ion will therefore not be able to produce polarisation, if this frequency is much larger than ω_{pi}, that is, if

$$v \gg \omega_{pi}d_e = v_s, \tag{2.29}$$

where

$$v_s \equiv \frac{v_{Te}\omega_{pi}}{\omega_{pe}} = \frac{v_{Ti}d_e}{d_i} = \sqrt{\frac{ZT_e}{m_i}}. \tag{2.30}$$

The quantity v_s is called the *ion sound velocity*. The reason for this terminology is the following. Consider not an external charge but a fluctuation of the charge density in the plasma; this will be screened in the same way as an external charge, with one exception: it will not be screened, but propagate if the screening by the electrons is exactly cancelled by the dynamical screening by the ions. Such a perturbation is thus able to propagate in the plasma with the ion-sound velocity v_s. This means that a wave with a phase velocity equal to the ion-sound velocity,

$$v_{ph} = v_s; \qquad v_{ph} = \frac{\omega}{k}, \tag{2.31}$$

can propagate in the plasma. Such a wave is called an *ion-sound wave*. Denoting the frequency of the ion-sound wave by ω_k^s we can write

$$\omega_k^s = kv_s. \tag{2.32}$$

Both ion-sound waves and ion oscillations with a frequency ω_{pi} can exist only if inequality (2.20) is satisfied. For a hydrogen plasma we have $Z = 1$ and this inequality becomes

$$T_e \gg T_i. \tag{2.33}$$

Returning to the problem of the screening of the charge field we conclude that this screening is determined by the electrons when

$$v_s \ll v \ll v_{Te}. \tag{2.34}$$

It should be emphasised that the screening charge always has the opposite sign of that of the charge which is placed in the plasma, independent of what that sign is. This means that in the case of a positive external charge the screening charge will be negative; this negative charge is produced by an excess of electrons and a depletion of ions. On the other hand, a positive charge placed in the plasma will produce a positive screening charge, produced by an excess of ions and a depletion of electrons. Any external charge, whatever its sign, will thus be screened both by ions and by electrons (see Fig.2.3)

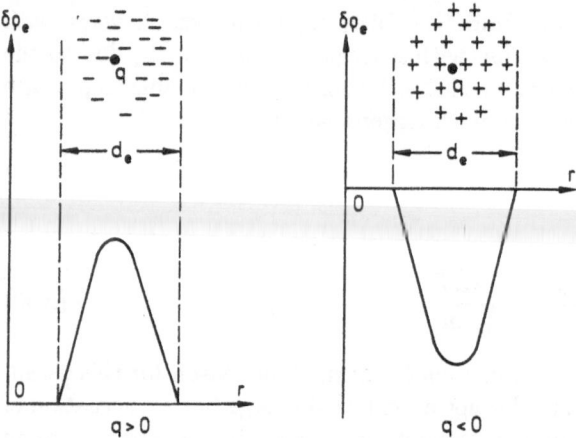

Fig. 2.3. Screening of a positive charge and of a negative charge by electrons and ions

If a force is applied to the inserted charge, this force will also act on the screening charges. As we shall see later on, this is very important for the scattering of waves in a plasma and for the bremsstrahlung of a plasma. The statement that an external charge of either sign could be screened both by electrons and by ions is correct only in the linear approximation when the perturbations of the electrons and of the ions are small. It is clear that if a positive external charge q is very large it will be necessary to remove all electrons from its vicinity to compensate for its charge. To estimate the maximum value q_{max} of the charge for which we can still apply linear screening, we take into account the important rôle played by the value of the potential at a distance from the charge of the order of the Debye radius. In that way we get

$$\left| \frac{q_{max}}{e} \right| < \frac{T_e d_e}{e^2}, \quad \frac{T_i d_i}{Z e^2}. \tag{2.35}$$

So far we have considered the charge q to be an external charge. The question arises of what happens if we consider the charge q to be one of the charges in the plasma. It seems obvious to expect that such a charge will be screened in the same way as an external charge. The important problem arises then: How can it be that each charge in the plasma, on the one hand, is screened by the other charges while, on the other hand, each charge also takes part in the screening of the other charges? We shall give an answer to this question in what follows.

2.3 Number of Particles in the Debye Sphere as a Fundamental Plasma Parameter

We now introduce a dimensionless parameter N_d equal to the total number of particles in the Debye sphere; this parameter plays a very important rôle in the whole of plasma physics. The *Debye sphere* is a sphere with the Debye radius as its radius. We first introduce the *number of electrons in the (electron) Debye sphere*, N_{de},

$$N_{de} = \frac{4\pi n_e d_e^3}{3}, \tag{2.36}$$

and the *number of ions in the (ion) Debye sphere*, N_{di},

$$N_{di} = \frac{4\pi n_i d_i^3}{3}, \tag{2.37}$$

If the difference between N_{de} and N_{di} is insignificant we shall use the notation N_d and simply call it the number of particles in a Debye sphere. We shall see that there are several reasons why this so-called *plasma parameter* is so important.

First reason for the importance of the plasma parameter

Our definition of a plasma as a state of matter which we gave at the beginning of § 2.2 leads necessarily to the inequality

$$N_d \gg 1. \tag{2.38}$$

Indeed, Eqs.(2.14) and (2.15) can, omitting factors of the order of unity, be written in the form

$$N_{de}^{2/3} \gg 1, \qquad N_{di}^{2/3} \gg 1. \tag{2.39}$$

An important point, however, is that N_d is not just a large value, but it is an extremely large value. We shall give some examples illustrating this statement.

Let us start with an example where the value of N_d is not very large. This is the example of a *laser plasma* in a laser-fusion programme experiment. The plasma density decreases because the irradiated target expands. Since the starting density can be higher than the solid-state density the plasma frequency is rather high. For an electron density $n_e \approx 10^{20}$ cm^{-3} we have $\omega_{pe} \approx 5 \times 10^{14}$ s^{-1}; if we take the electron temperature to be $T_e \approx 3$ keV, we find $v_{Te} \approx 10^9$ cm s^{-1} and $d_e = v_{Te}/\omega_{pe} \approx 2 \times 10^{-6}$ cm, and thus $N_{de} \approx 10^3$. For other possible parameters one can get $N_{de} \approx 10^2$.

Let us give another example: a tokamak with parameter values which put it close to ignition – conditions achieved recently in the biggest tokamaks. We take $n_e \approx 2 \times 10^{13}$ cm^{-3}, $T_e \approx 10$ keV, whence $\omega_{pe} \approx 3 \times 10^{11}$ s^{-1}, $v_{Te} \approx 2 \times 10^9$ cm s^{-1}, and $d_e \approx 10^{-2}$ cm, and thus finally $N_{de} \approx 3 \times 10^7$. For a different set of parameters we find $N_{de} \approx 3 \times 10^6$.

In the *solar atmosphere* at a height of the order of 5×10^{10} cm above the visible surface of the Sun where the most powerful explosions – the solar flares – occur, we have $n_e \approx 10^8$ cm^{-3}, $T_e \approx 1$ keV, $\omega_{pe} \approx 5 \times 10^8$ s^{-1}, $v_{Te} \approx 5 \times 10^8$ cm s^{-1}, $d_e \approx 1$ cm, and $N_{de} \approx 10^8$. Somewhat lower in the solar atmosphere we have $N_{de} \approx 10^7$ and somewhat higher $N_{de} \approx 3 \times 10^8$.

In the *interplanetary plasma* of the solar wind close to the Earth we have $n_e \approx 9$ cm^{-3}, $\omega_{pe} \approx 8 \times 10^5$ s^{-1}, $v_{Te} \approx 5 \times 10^7$ cm s^{-1}, $d_e \approx 10^2$ cm, and $N_{de} \approx 3 \times 10^6$.

For the hot regions of the *interstellar plasma* we have $n_e \approx 10^{-2}$ cm^{-3}, $\omega_{pe} \approx 5 \times 10^4$ s^{-1}, $v_{Te} \approx 3 \times 10^8$ cm s^{-1}, $d_e \approx 10^4$ cm, and $N_{de} \approx 10^{10}$. Other possible estimates for the various parameters can give even higher values up to $N_{de} \approx 10^{12}$, and for the relativistic electron-positron plasma in the *pulsar magnetospheres* N_{de} can be even much higher.

The results of all these estimates can be expressed in the strong inequality

$$N_d \ggg 1. \tag{2.40}$$

The triple inequality sign means that

$$N_d = 10^k, \quad k \gg 1.$$

The physical meaning of these estimates is that in a real plasma the condition that the state is a plasma state, in the sense we have defined such a state, is fulfilled with a large margin and that usually the kinetic energy of the particles is larger by many orders of magnitude than the potential energy of the particle interactions. With a high degree of accuracy the plasma particles can be considered to be free particles.

Second reason for the importance of the plasma parameter

The inequality $N_d \gg 1$ implies that the time needed for a plasma to reach thermal equilibrium is much longer than a characteristic plasma period. The electron-ion collision frequency ν_{ei} (see Eq.(2.9)) determines the time for the thermalisation of the electrons; it is of the order of $n_e v_{Te} \sigma$, where σ is the cross-section for electron-ion collisions. We can put $\sigma = \pi r^2$ where r would be a length of the order of the distance at which the electron kinetic and potential energies are equal:

$$\frac{Ze^2}{r} \approx T_{\mathrm{e}}.$$

Hence we find approximately

$$\frac{\nu_{\mathrm{ei}}}{\omega_{\mathrm{pe}}} \approx \frac{Z^2}{N_{\mathrm{de}}} \lll 1. \tag{2.41}$$

In fact, the time for the equalisation of the ion and electron temperatures is even much longer than the one we get from the estimate (2.41). The estimate (2.41) shows also that a plasma is to a good accuracy *collisionless*. Related to this is the fact that the collisional electric conductivity of a plasma is extremely high. In fact, if we equate the friction force, $\nu_{\mathrm{ei}} m_{\mathrm{e}} u$, of drifting electrons, with a drift velocity u, to the force eE exerted by an external electric field, we find for the current density j the equation

$$j = e n_{\mathrm{e}} u = \frac{e^2 n_{\mathrm{e}} E}{m_{\mathrm{e}} \nu_{\mathrm{ei}}} = \sigma E, \tag{2.42}$$

where here σ is the electrical conductivity of the plasma. From Eq.(2.42) we find for σ

$$\sigma = \frac{\omega_{\mathrm{pe}}^2}{4\pi \nu_{\mathrm{ei}}} \approx \frac{\omega_{\mathrm{pe}} N_{\mathrm{de}}}{4\pi}. \tag{2.43}$$

Since the electrical conductivity has N_{de} in the numerator, it is a very large quantity. For instance, the electrical conductivity of a gas-discharge plasma of density 10^{10} cm^{-3} is of the same order of magnitude as that of copper. In many ways a plasma can often be considered to be a *superconductor* in the sense that we have $\sigma \rightarrow \infty$.

Third reason for the importance of the plasma parameter

The energy density of thermal plasma waves turns out to be much smaller than the thermal particle energy density, and their ratio is again determined by $1/N_{\mathrm{d}}$. Let W be the energy density of the plasma waves and W_T its value for a plasma in thermal equilibrium. To find W_T we use the Rayleigh-Jeans law which states that for thermal equilibrium at a temperature T each degree of freedom has an energy T. In the case of Langmuir oscillations we thus have

$$W_T = \int \frac{T\, d^3\mathbf{k}}{(2\pi)^3} \approx \frac{T k_{\mathrm{max}}^3}{6\pi^2} \approx \frac{T}{6\pi^2 d^3}. \tag{2.44}$$

This equation gives us an overestimate of W_T since k_{max} is usually smaller than d and rather about $0.3d$. Leaving the estimate (2.44) as a maximum estimate we have

$$\frac{W_T}{nT} = \frac{1}{6\pi^2 N_d} \ll 1. \tag{2.45}$$

Equation (2.45) means that in thermal equilibrium the particles have a much larger energy density than the waves; the particles can thus be a large energy reservoir for waves in the case when the plasma is in a non-equilibrium state – for example, if the particles have a non-thermal distribution. It is difficult to imagine that in the case when the particles are in a non-equilibrium state while the waves are at their thermal level there would not be a channel of interactions between them which would transfer energy from the particles to the waves, seeing the huge difference in energy densities of particles and waves. Most laboratory plasmas are created by sending an energy pulse into a gas and it is very difficult to create the plasma in an initial state which is an equilibrium state with the accuracy given by the small ratio (2.45). Even in those experiments when a plasma is created in the maximally "quiet" way by emission from a heated cathode, the level of the initial fluctuations is superthermal, even though it is much lower than for other mechanisms to generate a plasma. In all existing experiments we thus have usually

$$W \gg W_T. \tag{2.46}$$

Experimentalists usually characterise the non-equilibrium nature of the plasma state by the mean fluctuations in the plasma density. Because of the equation of continuity a change in density in an electron wave is characterised by $\delta n_e/n_e \approx \delta v_e/v_{ph}$, and from the equations of motion for the electrons we have

$$\delta v_e \approx \frac{eE}{m_e\omega_{pe}}, \qquad \frac{(\delta n_e)^2}{n_e^2} \approx \frac{E^2}{4\pi n_e m_e v_{ph}^2}. \tag{2.47}$$

For Langmuir waves half of the energy is in the electrostatic field and half in in the particle motion; we have thus

$$W \approx \frac{E^2}{4\pi},$$

and since the minimum phase velocities are equal to v_{Te}, it follows that

$$\frac{(\delta n_e)^2}{n_e^2} \approx \frac{W}{n_e T_e}. \tag{2.48}$$

This means that in thermal equilibrium we have

$$\frac{(\delta n_e)^2}{n_e^2} \leq \frac{1}{6\pi^2 N_{de}}. \tag{2.49}$$

We have seen that often N_{de} is a large quantity. For N_{de} in the range from 10^5 to 10^6 Eq.(2.49) gives a relative level of fluctuations from 10^{-6} to 10^{-7}. Such a low level of fluctuations is very difficult to reach in real experiments.

On the other hand, since we have $N_d \ggg 1$ the relaxation of the initial fluctuations to thermal equilibrium is very slow. Therefore the initial plasma state will correspond to $W \gg W_T$. The oscillations are, however, usually weakly non-linear, which is described by Eq.(2.10). As we have $\delta\phi \approx E/k$, the condition for weak non-linearity can be written in the form

$$\frac{e^2 \delta\phi^2}{m_e^2 v_{ph}^4} \approx \frac{E^2}{4\pi n_e m_e v_{ph}^2} \ll 1, \tag{2.50}$$

or, since the minimum possible value of v_{ph} is v_{Te},

$$\frac{W}{n_e T_e} \ll 1. \tag{2.51}$$

Combining this with Eq.(2.46) we see that there exists a very broad range of parameters for which, on the one hand, the oscillations are weakly non-linear and, on the other hand, superthermal:

$$\frac{1}{6\pi^2 N_{de}} \ll \frac{W}{n_e T_e} \ll 1. \tag{2.52}$$

Fourth reason for the importance of the plasma parameter

Even small deviations of the particle distribution from a thermal equilibrium distribution can often open a channel for the transfer of energy from the particles to the plasma oscillations. Let ε be a measure of the relative deviation of the particle distribution from equilibrium and let $\varepsilon\omega_{pe}$ be an estimate for the growth rate of the conversion of particle to wave energy. Since the process which prevents this conversion is due to the damping of the oscillations the rate of which is given by $\nu_{ei} \approx \omega_{pe}/N_{de}$, we see that conversion can happen provided

$$\varepsilon \gg \frac{1}{N_{de}}, \tag{2.53}$$

which is a very weak inequality; this means that even a very small deviation of the particle distribution from equilibrium can start the conversion of the particle energy into plasma oscillations.

Concluding this section we shall give an estimate for the maximum value of a charge q which can be screened by plasma particles. We shall see that it is also determined by N_d. According to Eq.(2.35) this maximum charge can be expressed as

$$\left| \frac{q_{max}}{e} \right| \approx 4\pi N_{de}, 4\pi N_{di}. \tag{2.54}$$

Such large charges can, indeed, exist on dust particles in a plasma. Since the electrons have a larger thermal velocity than the ions, electron rather than ion thermal currents will charge a dust particle until its charge will stop the electron current onto the particle. A dust particle – or any other solid particle in a plasma – will have a so-called *floating potential*, $e\phi \approx T_e$, and thus the dust particle will carry a charge $q = q_{max}$. In this case the distributions of the electrons – or ions – around the dust particle will be non-linear and Eq.(2.54) can be used only for order of magnitude estimates. Such highly charged dust particles are, indeed, observed in laboratory and space plasmas. Sometimes the charge on the dust particle will be sufficient to break it up. This process, in fact, regulates the size distribution of dust particles in space.

We finally must mention that N_d is proportional to $T^{3/2}$ and inversely proportional to $n^{1/2}$, that is, the higher the temperature, the larger the value of N_d and the lower the density, the larger the value of N_d.

2.4 Kinetic Theory
of Linear Collective Plasma Oscillations

We shall now consider a quantitative kinetic theory of collective plasma oscillations, using the simplest example of electrostatic waves in the absence of a magnetic field and of plasma inhomogeneities. In what follows our main purpose is to use this mode to illustrate general plasma properties for the simplest possible example, but one which is one of the most important ones for applications. As long as one considers non-relativistic particles and neglects magnetic-field generation by the plasma, one can restrict the discussion to electrostatic fields alone. We shall use a kinetic description of the particles and shall neglect their binary collisions.

Particles of the kind α, with $\alpha =$ e, i, respectively, for electrons and ions, will be characterised by a distribution function f_p^α which gives the probability for a particle of kind α to have a momentum within the range $\mathbf{p}$ to $\mathbf{p} + d\mathbf{p}$, normalised to unit space volume and unit phase volume, $d^3\mathbf{p}/(2\pi)^3$. The density n_α of particles of kind α, that is the number of particles of kind α per cm^3 is then given by the equation

$$n_\alpha = \int \frac{f_p^\alpha \, d^3\mathbf{p}}{(2\pi)^3}. \tag{2.55}$$

We shall omit the indices α whenever there is no danger of confusion. The particle distributions can fluctuate. There are two possible kinds of fluctuations. The first are the normal fluctuations for a large system of particles. In the case of thermal equilibrium they are the thermal fluctuations. However, we consider all possible particle distributions including non-equilibrium ones. They all have the usual statistical fluctuations which are not necessarily thermal but which can still be determined for free particles provided the average distribution is known (see below). These fluctuations exist in the absence of collective fields.

Another kind of fluctuations appears when the collective fields are random; those fluctuations are proportional to the strength of the collective fields. We shall use the same $\langle \cdots \rangle$ sign for the averages of fluctuations of either the first or the second kind, or of both kinds together. We do not think that this will lead to any misunderstandings.

We denote the average distribution function by $\Phi_{\mathbf{p}}$:

$$\Phi_{\mathbf{p}} = \langle f_{\mathbf{p}} \rangle, \tag{2.56}$$

and the fluctuating part by $\delta f_{\mathbf{p}}$:

$$\delta f_{\mathbf{p}} = f_{\mathbf{p}} - \Phi_{\mathbf{p}}. \tag{2.57}$$

The equation for $f_{\mathbf{p}}$ is an equation of continuity in phase space and follows directly from the Liouville theorem:

$$\frac{\partial}{\partial t} f_{\mathbf{p}} + \left(\mathbf{v} \cdot \frac{\partial}{\partial \mathbf{r}} \right) f_{\mathbf{p}} + e \left(\mathbf{E} \cdot \frac{\partial}{\partial \mathbf{p}} \right) f_{\mathbf{p}} = 0. \tag{2.58}$$

We have neglected here the Lorentz force, assuming that the fields are purely electrostatic. The meaning of Eq.(2.58) can also be seen by noting that its solution can be described by characteristics which according to Eq.(2.58) satisfy the equations:

$$\frac{d\mathbf{r}}{dt} = \mathbf{v}, \qquad \frac{d\mathbf{p}}{dt} = e\mathbf{E}, \tag{2.59}$$

which are nothing but the equations of motion for individual particles in the electric field $\mathbf{E}$. In other words, Eq.(2.58) describes the change in the particle distribution corresponding to each particle moving along a trajectory determined by the existing electric field.

Let us now consider an initial stationary and uniform distribution, $f_{\mathbf{p}}^{(0)}$, which, if there are no electric fields, obviously satisfies Eq.(2.58). Let us now find the first-order correction to this distribution function which is linear in the field strength $\mathbf{E}$. We shall denote it by $f_{\mathbf{p}}^{(1)}$. We then have

$$\frac{\partial f_{\mathbf{p}}^{(1)}}{\partial t} + \left(\mathbf{v} \cdot \frac{\partial}{\partial \mathbf{r}} \right) f_{\mathbf{p}}^{(1)} = - \left(e\mathbf{E} \cdot \frac{\partial}{\partial \mathbf{p}} \right) f_{\mathbf{p}}^{(0)}. \tag{2.60}$$

Before solving this equation, let us consider a case where the electric field is random, that is, where we have $\langle \mathbf{E} \rangle = 0$. Averaging Eq.(2.58) we find

$$\frac{\partial}{\partial t} \Phi_{\mathbf{p}} + \left(\mathbf{v} \cdot \frac{\partial}{\partial \mathbf{r}} \right) \Phi_{\mathbf{p}} = -e \left\langle \left(\mathbf{E} \cdot \frac{\partial}{\partial \mathbf{p}} \right) \delta f_{\mathbf{p}} \right\rangle, \tag{2.61}$$

and subtracting this averaged equation from the original one we find

$$\frac{\partial}{\partial t}\delta f_{\mathbf{p}} + \left(\mathbf{v}\cdot\frac{\partial}{\partial \mathbf{r}}\right)\delta f_{\mathbf{p}} = -e\left(\mathbf{E}\cdot\frac{\partial}{\partial \mathbf{p}}\right)\Phi_{\mathbf{p}}$$

$$-e\left[\left(\mathbf{E}\cdot\frac{\partial}{\partial \mathbf{p}}\right)\delta f_{\mathbf{p}} - \left\langle\left(\mathbf{E}\cdot\frac{\partial}{\partial \mathbf{p}}\right)\delta f_{\mathbf{p}}\right\rangle\right]. \tag{2.62}$$

We shall later on use Eqs.(2.61) and (2.62) to describe non-linear aspects of fluctuations for various problems. If we use an approximation which is linear in the fluctuations we can neglect the terms within the square brackets in Eq.(2.62) as they contain products of a fluctuating field and a fluctuating distribution function; we then have

$$\frac{\partial}{\partial t}\delta f_{\mathbf{p}} + \left(\mathbf{v}\cdot\frac{\partial}{\partial \mathbf{r}}\right)\delta f_{\mathbf{p}} = -e\left(\mathbf{E}\cdot\frac{\partial}{\partial \mathbf{p}}\right)\Phi_{\mathbf{p}}. \tag{2.63}$$

This equation is the same as Eq.(2.60), apart from the notation, for the case when the average distribution function is homogeneous and stationary. We shall be interested in that case and we shall solve Eq.(2.63) bearing in mind that $\Phi_{\mathbf{p}}$ can play the rôle of the initial distribution function and $\delta f_{\mathbf{p}}$ that of a first-order correction to it. In the general case the average distribution function may vary in time and space. The results obtained from Eq.(2.63) assuming that the average distribution function is homogeneous and stationary are approximately valid also when this distribution function varies slowly in time and space as compared to the variation of $\delta f_{\mathbf{p}}$ in time and space.

We shall often in what follows use a Fourier representation; in other words, we shall expand the various quantities in plane "waves" – we shall in what follows use the term wave when there is a definite wavenumber dependence of the frequency, whereas in the Fourier expansion such a dependence does not exist as the wavenumber and the frequency are independent variables. For the electric field we use the expansion:

$$\mathbf{E}(\mathbf{r},t) = \int \mathbf{E}_{\mathbf{k},\omega}\,\mathrm{e}^{\mathrm{i}(\mathbf{k}\cdot\mathbf{r})-\mathrm{i}\omega t}\,d^4k, \qquad d^4k \equiv d^3k\,d\omega. \tag{2.64}$$

Similar expansions will be used for other quantities, including $\delta f_{\mathbf{p}}$. Sometimes we shall use instead of the two subscripts ω and $\mathbf{k}$ a single one, $k = \{\mathbf{k}, \omega\}$, using a *four-vector notation*. We remind ourselves that we use an italic letter, $k = |\mathbf{k}|$ to denote the *absolute magnitude of the vector* $\mathbf{k}$.

The Fourier transform of Eq.(2.63) now becomes

$$\mathrm{i}\big(\omega - (\mathbf{k}\cdot\mathbf{v})\big)\,\delta f_{\mathbf{p},k} = e\left(\mathbf{E}_k\cdot\frac{\partial\Phi_{\mathbf{p}}}{\partial \mathbf{p}}\right), \tag{2.65}$$

where, for instance, we have written simply $\mathbf{E}_k$ for $\mathbf{E}_{\mathbf{k},\omega}$. We shall solve Eq.(2.65) by simply dividing by $\big(\omega - (\mathbf{k}\cdot\mathbf{v})\big)$, but in the general case we should add to the solution the solution of the homogeneous equation,

$$\delta\big(\omega - (\mathbf{k}\cdot\mathbf{v})\big), \tag{2.66}$$

which corresponds to Eq.(2.65) with a zero right-hand side. We shall see in what follows that this is, indeed, necessary when considering the natural statistical fluctuations which are present even when there are no electrical fields present.

Dividing Eq.(2.65) by $(\omega - (\mathbf{k} \cdot \mathbf{v}))$ we obtain just the term produced by the field:

$$\delta f_{\mathbf{p},\mathbf{k}} = \frac{e\left(\mathbf{E}_{\mathbf{k}} \cdot \frac{\partial \Phi_{\mathbf{p}}}{\partial \mathbf{p}}\right)}{i(\omega - (\mathbf{k} \cdot \mathbf{v}))}. \tag{2.67}$$

We shall show that even for this expression we can properly deal with the problem of the pole $\omega = (\mathbf{k} \cdot \mathbf{v})$ only by using the causality principle which shows that one should add a well defined multiple of the δ-function (2.66) if one wants to calculate the contributions from the pole of (2.67) as a principal value. However, the pole does not play an important rôle in our discussion in the present section and we shall continue this discussion when we consider the Landau damping problem later on.

From Eq.(2.67) we see the change in the particle distribution function due to an electric field perturbation. However, the field itself is determined by the perturbations in the particle distributions from the *Poisson equation*:

$$\text{div}\mathbf{E} = 4\pi\varrho = 4\pi \sum_{\alpha} \int e_{\alpha} \delta f_{\mathbf{p}}^{\alpha} \frac{d^3\mathbf{p}}{(2\pi)^3}, \tag{2.68}$$

where ϱ is the total *charge density*, $\sum e_{\alpha} n_{\alpha}$.

The Fourier transform of Eq.(2.68) is

$$i(\mathbf{k} \cdot \mathbf{E}_{\mathbf{k}}) = 4\pi \sum_{\alpha} \int e_{\alpha} \delta f_{\mathbf{p},\mathbf{k}}^{\alpha} \frac{d^3\mathbf{p}}{(2\pi)^3}. \tag{2.69}$$

In our present discussion we restrict ourselves to electrostatic fields which can be expressed as the gradient of some potential; this means that their Fourier transforms are proportional to $\mathbf{k}$. In other words these fields are longitudinal – their polarisation vector is directed along the direction in which the field propagates. It is therefore useful to introduce a unit polarisation vector, $\mathbf{k}/|\mathbf{k}| = \mathbf{k}/k$, and write

$$\mathbf{E}_{\mathbf{k}} = E_{\mathbf{k}} \frac{\mathbf{k}}{k}. \tag{2.70}$$

Substituting Eq.(2.65) into Eq.(2.69) and using Eq.(2.70) we obtain the equation

$$\varepsilon_{\mathbf{k}} E_{\mathbf{k}} \equiv D_{\mathbf{k}} = 0, \tag{2.71}$$

where $\varepsilon_{\mathbf{k}} \equiv \varepsilon_{\mathbf{k},\omega}$ is the *plasma dielectric permittivity* (for longitudinal perturbations):

$$\varepsilon_{\mathbf{k}} = 1 + \sum_{\alpha} \frac{4\pi e_{\alpha}^2}{k^2} \int \frac{1}{\omega - (\mathbf{k} \cdot \mathbf{v})} \left(\mathbf{k} \cdot \frac{\partial \Phi_{\mathbf{p}}^{\alpha}}{\partial \mathbf{p}} \right) \frac{d^3\mathbf{p}}{(2\pi)^3}. \qquad (2.72)$$

The dispersion equation for longitudinal collective oscillations corresponds to putting the dielectric permittivity equal to zero:

$$\varepsilon_{\mathbf{k}} \equiv \varepsilon_{\mathbf{k},\omega} = 0. \qquad (2.73)$$

The solutions, $\omega_{\mathbf{k}}^{\sigma}$, of Eq.(2.73) give the frequency as as definite function of the wavenumber. In principle there can be different branches of solutions; these are indicated by the superscript σ. We have already seen that there exist *Langmuir oscillations* and *ion-sound oscillations*; they each form one branch: we use $\sigma = \ell$ for Langmuir waves and $\sigma = $ s for ion-sound waves.

2.5 Kinetic Theory of Langmuir and Ion–Sound Plasma Oscillations

It is desirable, before giving a description of the various plasma waves, to find an approximate expression for the dielectric permittivity. Although the distribution function $\Phi_{\mathbf{p}}$ at this moment is arbitrary, we shall assume that we can introduce some average velocity which for a thermal distribution would be the same as the thermal velocity. We denote the square root of the average of the square of the particle velocity of an arbitrary distribution function by v_T. If $\omega \gg kv_T$ we can expand the denominator which occurs in the expression for the plasma dielectric permittivity:

$$\frac{1}{\omega - (\mathbf{k} \cdot \mathbf{v})} \simeq \frac{1}{\omega} + \frac{(\mathbf{k} \cdot \mathbf{v})}{\omega^2} + \frac{(\mathbf{k} \cdot \mathbf{v})^2}{\omega^3} + \frac{(\mathbf{k} \cdot \mathbf{v})^3}{\omega^4} + \dots . \qquad (2.74)$$

We shall see in a moment that it is necessary to take so many terms in the expansion; this is because the first term gives a zero result for any distribution function – the dielectric permittivity contains in that case an integral of the derivative of the distribution function over all momenta – while the third term gives zero for any distribution function which is an even function of the momenta. In the first approximation we take into account only the first two terms in the expansion. We then get for the case when there is just one kind of particle

$$\varepsilon_{\mathbf{k}} = 1 + \frac{4\pi e^2}{k^2\omega^2} \int (\mathbf{k} \cdot \mathbf{v}) \left(\mathbf{k} \cdot \frac{\partial \Phi_{\mathbf{p}}}{\partial \mathbf{p}} \right) \frac{d^3\mathbf{p}}{(2\pi)^3} \cong 1 - \frac{4\pi ne^2}{m\omega^2}. \qquad (2.75)$$

Let us consider the Langmuir oscillations in this approximation. Since they are high-frequency oscillations we can neglect the contributions from the ions to the dielectric permittivity so that we have

$$\varepsilon_{\mathbf{k}} = \varepsilon_{\mathbf{k},\omega} \simeq \varepsilon_{\mathbf{k}}^e \simeq 1 - \frac{\omega_{pe}^2}{\omega^2} = 0. \tag{2.76}$$

The solution $\omega_{\mathbf{k}}^\ell$ of this equation will in this approximation be independent of $\mathbf{k}$:

$$\omega_{\mathbf{k}}^\ell = \omega_{pe}. \tag{2.77}$$

This is the same result which was obtained earlier in the elementary description of the Langmuir oscillations. We have, as stated earlier, denoted the Langmuir oscillations by a superscript ℓ. The k-dependence of $\omega_{\mathbf{k}}^\ell$ corresponds now to a straight line parallel to the k-axis (see Fig.2.4).

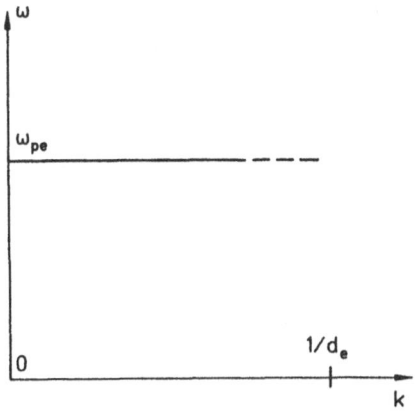

Fig. 2.4. The dispersion curve for the Langmuir oscillations in a cold plasma

In this approximation the group velocity $d\omega_{\mathbf{k}}/d\mathbf{k}$ is equal to zero. To find the group velocity one must take into account the next two terms in the expansion (2.74) – bearing in mind that the third term gives zero for a distribution which is even in the momenta. We find for electrons – the expression will be valid for any type of particles provided $\omega \gg kv_T$, if we omit the label e characterising the electrons –

$$\varepsilon_{\mathbf{k},\omega} \simeq 1 - \frac{\omega_{pe}^2}{\omega^2} - \frac{3k^2(v_T^e)^2\omega_{pe}^2}{\omega^4} = 0, \qquad \omega \gg kv_T^e, \tag{2.78}$$

where v_T^e is an effective average velocity which for an arbitrary distribution is defined by the relation

$$\left(v_T^e\right)^2 = \frac{\int v^2 \, \varPhi_{\mathbf{p}}^e \, d^3\mathbf{p}/(2\pi)^3}{3\int \varPhi_{\mathbf{p}}^e \, d^3\mathbf{p}/(2\pi)^3}. \tag{2.79}$$

For a thermal distribution we have $\left(v_T^e\right)^2 = v_{Te}^2 = T_e/m_e$. We can easily solve Eq.(2.78). Since the last term in Eq.(2.78) is small, we can substitute

$\omega^2 = \omega_{pe}^2$ in it to a first approximation. The simplest way to proceed is, in fact, to make this substitution in only one factor ω^2. We then get

$$\omega_k^2 \simeq \omega_{pe}^2 + 3k^2 v_{Te}^2,$$

or

$$\omega_k^\ell \simeq \omega_{pe} + \frac{3k^2 v_{Te}^2}{2\omega_{pe}}. \tag{2.80}$$

In this approximation the group velocity is not equal to zero, but we have

$$\mathbf{v}_{gr} = \frac{d\omega_k^\ell}{dk} = \frac{3k v_{Te}^2}{\omega_{pe}}, \tag{2.81}$$

and

$$v_{gr} = \frac{3v_{Te}^2}{v_{ph}}. \tag{2.82}$$

Since the phase velocity of Langmuir waves lies in the range $v_{Te} < v_{ph} < \infty$, we have for the group velocity the range

$$0 < v_{gr} < 3v_{Te}. \tag{2.83}$$

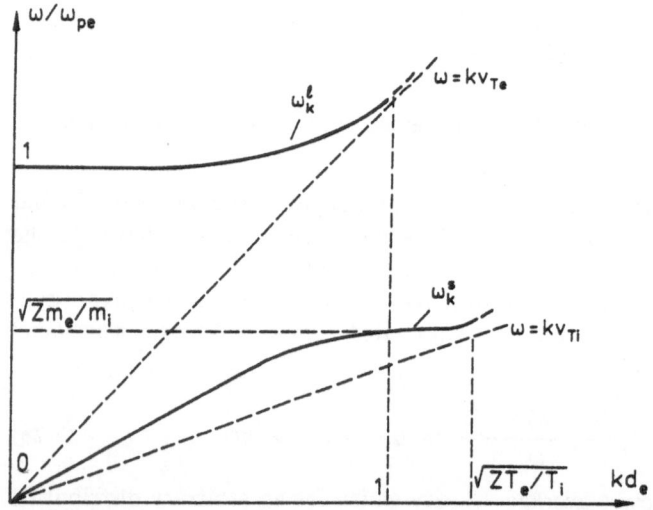

Fig. 2.5. The dispersion curves for the Langmuir and the ion-sound oscillations in a warm plasma

We show in Fig.2.5 the k-dependence of ω_k^ℓ, as given by Eq.(2.80). We also show in this figure the ion-sound wave branch. To find its dispersion equation we need to know an approximate expression for the dielectric permittivity for any type of particle in the opposite limit when $\omega \ll kv_T$. We shall again consider an arbitrary particle distribution and for $\omega \ll kv_T$ we write down the following approximate expression for the dielectric permittivity

$$\varepsilon_{\mathbf{k},\omega} = 1 + \frac{4\pi e^2}{k^2} \int \frac{1}{(\mathbf{k}\cdot\mathbf{v})} \left(\mathbf{k}\cdot\frac{\partial\Phi_{\mathbf{p}}}{\partial\mathbf{p}}\right) \frac{d^3\mathbf{p}}{(2\pi)^3}. \tag{2.84}$$

For an isotropic distribution we have $(\mathbf{k}\cdot\partial\Phi_{\mathbf{p}}/\partial\mathbf{p}) = (\mathbf{k}\cdot\mathbf{v})(\partial\Phi_{\mathbf{p}}/\partial v)/mv$ and we define for such an isotropic distribution a new average velocity, V_T, by the equation

$$\frac{1}{V_T^2} = \left[\int \frac{1}{v}\frac{\partial\Phi_{\mathbf{p}}}{\partial v}\frac{d^3\mathbf{p}}{(2\pi)^3}\right] \bigg/ \left[\int \Phi_{\mathbf{p}}\frac{d^3\mathbf{p}}{(2\pi)^3}\right]. \tag{2.85}$$

For a thermal distribution we have $V_T^2 = v_T^2 = T/m$. Using the definition (2.85) we can write the approximate expression for the permittivity in the form

$$\varepsilon_{\mathbf{k},\omega} \simeq 1 + \frac{\omega_p^2}{k^2 v_T^2}, \qquad \omega_p^2 = \frac{4\pi n e^2}{m}, \qquad \omega \ll kV_T. \tag{2.86}$$

In the case of Langmuir oscillations we need consider only the contribution from the electrons and we have

$$\varepsilon_{\mathbf{k},\omega} = 1 + \frac{\omega_{pe}^2}{k^2 V_{Te}^2}. \tag{2.87}$$

The right-hand side of this equation is always positive which means that Langmuir oscillations cannot exist if $\omega \ll kV_{Te}$. The dielectric permittivity (2.87) also describes the Debye screening by electrons. For non-equilibrium distributions the electron Debye radius is thus defined by the expression

$$d_e = V_{Te}/\omega_{pe}. \tag{2.88}$$

We shall now consider the *ion-sound oscillations*. Let us consider the following range of phase velocities for a non-equilibrium distribution:

$$kv_{Ti} \ll \omega \ll kv_{Te}. \tag{2.89}$$

To get the contribution to the dielectric permittivity from the electrons we can use Eq.(2.87) and for the contribution from the ions we can use the opposite limit which corresponds to Eq.(2.78) with a change of the index e to i:

$$\varepsilon_{\mathbf{k},\omega} \simeq 1 + \frac{\omega_{pe}^2}{k^2 V_{Te}^2} - \frac{\omega_{pi}^2}{\omega^2} - \frac{3k^2(v_T^i)^2}{\omega^2} = 0. \tag{2.90}$$

The term which takes into account the thermal motion of the ions is significant only when $\omega \simeq \omega_{pi}$, $k \gg 1/d_e$ when the contribution from the electrons becomes negligible. In that case we obtain

$$\omega_k^s \simeq \omega_{pi} + \frac{3k^2(v_T^i)^2}{2\omega_{pi}}. \tag{2.91}$$

This part of the ion-sound branch is shown in Fig.2.5 for $kd_e > 1$. If $k \ll 1/d_i$ and ω is of the order of ω_{pi} or much smaller, we can neglect the contribution from the thermal motion of the ions and we get

$$\omega_k^s \simeq \frac{\omega_{pi}}{\sqrt{1 + \omega_{pe}^2/k^2 V_{Te}^2}}. \tag{2.92}$$

When we have $k \ll 1/d_e = \omega_{pe}/V_{Te}$ the ion-sound branch is approximately a sound wave:

$$\omega_k^s \simeq k v_s. \tag{2.93}$$

In the $k \gg 1/d_e$ case the frequency of the ion-sound branch is close to the ion plasma frequency:

$$\omega_k^s \simeq \omega_{pi}. \tag{2.94}$$

The ion-sound velocity in a non-equilibrium plasma is determined by the average velocity V_{Te} which may differ from the thermal velocity:

$$v_s = V_{Te} \sqrt{\frac{m_e}{m_i}}. \tag{2.95}$$

We should make one important remark in connection with the solutions of the equation $\varepsilon_{k,\omega} = 0$. We understand by ω_k here and in what follows only the positive solution of this equation. This means that the general solution will be

$$\omega = \pm \omega_k. \tag{2.96}$$

The reason for defining ω_k to be positive is that later on we shall consider the plasma waves to be *quasi-particles* with an energy and a momentum: the energy of such a quasi-particle will be positive and equal to $\hbar\omega_k$ and its momentum equal to $\hbar k$. We may also mention the analogy between the Langmuir spectrum and a particle spectrum,

$$\varepsilon_p = mc^2 + \frac{p^2}{2m}. \tag{2.97}$$

It follows from Eq.(2.80) that the plasma frequency plays the rôle of the rest-mass energy and the thermal term that of the kinetic energy of a Langmuir *plasmon*.

To illustrate the rôle of non-equilibrium distributions we shall consider a case corresponding to a situation often found in experiments, when one observes a tail of accelerated particles. Although both the main part of the distribution and the fast particle tail may be non-thermal, we shall consider here a rough model in which the distribution function can be approximated by a bi-thermal distribution. We thus consider the case when the particle distribution is the sum of a thermal distribution of temperature T_e and density n_e and a second thermal distribution of "hot" particles with a temperature T_h and a density n_h. We then have

$$m_e v_{Te}^2 = \frac{n_e T_e + n_h T_h}{n_e + n_h}, \qquad m_e V_{Te}^2 = \frac{(n_e + n_h) T_e T_h}{n_e T_h + n_h T_e}. \qquad (2.98)$$

In the case when $n_e \gg n_h$ and $T_e \ll T_h$, while the pressures are approximately equal, $n_e T_e \simeq n_h T_h$, the first average velocity, v_{Te}, is larger than the second one, V_{Te}, but only by a factor $\sqrt{2}$; on the other hand, in the case when $n_e T_e \ll n_h T_h$, the first one is determined by T_h – and we have $m_e v_{Te}^2 = n_h T_h / n_e$ –, while the second one is determined by T_e – and we have $m_e V_{Te}^2 = T_e$. This example shows that the difference between the two velocities should be taken into account, even if the density of the "hot" component is relatively low as compared to that of the main part of the distribution.

Another possibility is that rather than a hot component, a cold one (n_h, T_h) is superimposed upon the main component (n_e, T_e) so that we have $n_e \gg n_h$ and $T_e \gg T_h$. In that case we have $m_e v_{Te}^2 = T_e$, but if $n_h T_e \gg n_e T_h$, the Debye radius will be determined by another quantity, namely, $m_e V_{Te}^2 = n_e T_h / n_h$, that is, it will be much smaller than the estimate obtained by using the temperature T_e.

Another important feature is that the presence of a small number of hot electrons may change the dispersion of the Langmuir waves. Let us again assume that $T_h \gg T_e$ and $n_h \ll n_e$ and let us denote the average thermal velocities of the two distributions by $v_{Te} = \sqrt{T_e/m_e}$ and $v_{Th} = \sqrt{T_h/m_e}$. We shall now consider the following range of phase velocities:

$$k v_{Te} \ll \omega_{pe} \ll k v_{Th}. \qquad (2.99)$$

We can then use for the hot component the Debye screening approximation so that we are led to the following dispersion relation:

$$\varepsilon_{\mathbf{k},\omega} = 1 - \frac{\omega_{pe}^2}{\omega^2} - \frac{3k^2 (v_T^e)^2}{\omega^2} + \frac{n_h \omega_{pe}^2}{n_e k^2 v_{Th}^2} = 0. \qquad (2.100)$$

If we have $k \ll \omega_{pe}(v_{Te} v_{Th})^{-1/2}(n_h/n_e)^{-1/4}$, which corresponds to values of k which are not necessarily very small since the density ratio occurs only to the one-fourth power, the dispersion term due to the the fast particles – the hot component – will dominate, and we find:

$$\omega_{\mathbf{k},\omega}^\ell = \omega_{pe} - \frac{n_h \omega_{pe}^3}{2 n_e k^2 v_{Th}^2} = 0. \qquad (2.101)$$

Such a spectrum of Langmuir oscillations which is inversely proportional to k^2 is called a *Langmuir oscillation spectrum with an inverse dispersion*. Such

a spectrum can often be created in a plasma with a fast particle component –
we mentioned earlier that the fast-particle-acceleration phenomenon is very
common both in laboratory and in space plasmas.

At the end of this section we want to say a few words about non-
electrostatic, that is, electromagnetic, waves in an isotropic plasma. Their
phase velocity is of the order of or even larger than the velocity of light and
for non-relativistic particles one can neglect thermal effects. The refractive
index will then be equal to

$$\sqrt{\varepsilon(\omega)} \simeq \sqrt{1 - \frac{\omega_{pe}^2}{\omega^2}}.$$

Denoting the branch of the electromagnetic waves by $\omega_{\mathbf{k}}^t$ we can write

$$\frac{\omega_{\mathbf{k}}^t}{k} = \frac{c}{\sqrt{\varepsilon(\omega)}}, \tag{2.102}$$

and we find

$$\left(\omega_{\mathbf{k}}^t\right)^2 = \omega_{pe}^2 + k^2 c^2. \tag{2.103}$$

This last equation shows that the frequency of the electromagnetic waves is
always larger than the plasma frequency, the phase velocity larger than the
velocity of light, and the group velocity less than the velocity of light. We
show in Fig.2.6 both the branch $\omega_{\mathbf{k}}^t$ of the electromagnetic waves and the
branch $\omega_{\mathbf{k}}^\ell$ of the Langmuir waves.

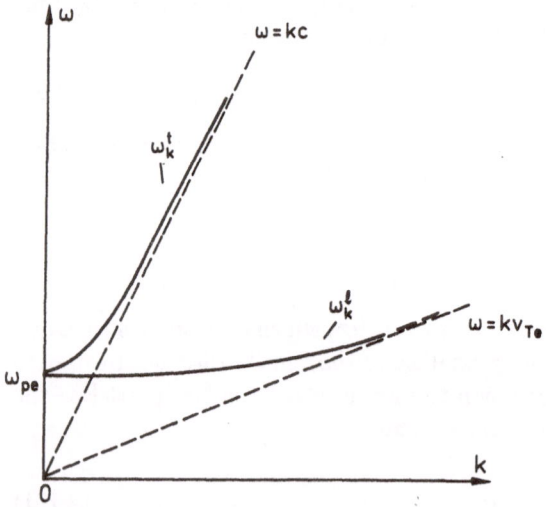

Fig. 2.6. The dispersion curves for Langmuir and for transverse waves

2.6 Landau Damping

Let us start with a sketch of the history of the problem. First of all, it is clear nowadays that *Landau damping* is a fundamental effect in plasma physics which has been verified in many laboratory experiments and has been used for the interpretation of a large amount of data from experiments or space observations. The problem, when it was first discussed by Landau, was not as clear as it is now. As we mentioned earlier, the problem is that when we solve the linear equation for the perturbation of the particle distribution function we can always add an arbitrary solution of the homogeneous equation; in particular, we can add a solution which is such that it leads to the dielectric permittivity having an imaginary part, which means damping. However, there remains the uncertainty that one can add any arbitrary solution of the homogeneous equation. Landau considered an initial value problem and had used Laplace transforms which meant that there were definite rules for finding the path in the complex ω-plane which is going round the $\omega = (\mathbf{k} \cdot \mathbf{v})$ pole. He then found that there was damping present. The expression we wrote down for the dielectric permittivity of the plasma contained a singularity for $\omega = (\mathbf{k} \cdot \mathbf{v})$. This singularity does not introduce any complications in the derivation of the dielectric permittivity, as long as we define how one should calculate integrals and how one should go around the pole. One possibility is to consider the integral to be a principal value integral and then there is no damping.

We shall use the causality principle to find out what kind of solution of the homogeneous equation, proportional to $\delta(\omega - (\mathbf{k} \cdot \mathbf{v}))$, one should add to find the correct imaginary part of the dielectric permittivity. Let us introduce phenomenologically some damping of the particle distribution which allows the particles to "forget" the past state, and then we shall find the limit when this damping tends to zero – we can imagine this damping to be an interaction of the plasma particles with a large system which is not a thermostat. We can introduce on the right-hand side of the equation for the particle distribution function instead of zero a term $-\epsilon f_{\mathbf{p}}$. Later on we shall take the limit as $\epsilon \to 0$. Instead of the denominator $\omega - (\mathbf{k} \cdot \mathbf{v})$ in the expression for the dielectric permittivity we then get a denominator $\omega - (\mathbf{k} \cdot \mathbf{v}) + i\epsilon$. In order that the causality principle be valid it is necessary that ϵ is positive. We shall indicate this by asking for the limit $\epsilon \to +0$. We can write

$$\frac{1}{\omega - (\mathbf{k} \cdot \mathbf{v}) + i\epsilon} = \frac{\omega - (\mathbf{k} \cdot \mathbf{v})}{(\omega - (\mathbf{k} \cdot \mathbf{v}))^2 + \epsilon^2} - i\frac{\epsilon}{(\omega - (\mathbf{k} \cdot \mathbf{v}))^2 + \epsilon^2}, \quad (2.104)$$

and in the limit as $\epsilon \to +0$ we get

$$\frac{1}{\omega - (\mathbf{k} \cdot \mathbf{v}) + i\epsilon} \xrightarrow[\epsilon \to +0]{} \mathcal{P}\frac{1}{\omega - (\mathbf{k} \cdot \mathbf{v})}$$
$$- i\pi\delta(\omega - (\mathbf{k} \cdot \mathbf{v})) \equiv \frac{1}{\omega - (\mathbf{k} \cdot \mathbf{v}) + i0}, \quad (2.105)$$

where $\mathcal{P}$ indicates that the principal value of any integral involving this expression must be taken. We shall in what follows use the notation with i0 in the denominator to indicate the limit as $\epsilon \to 0$. With the help of Eq.(2.105) we can find the imaginary part of the dielectric permittivity:

$$\text{Im}\{\varepsilon_{\mathbf{k},\omega}\} = -\frac{4\pi^2}{k^2} \sum_\alpha e_\alpha^2 \int \delta(\omega - (\mathbf{k}\cdot\mathbf{v})) \left(\mathbf{k}\cdot\frac{\partial\Phi_{\mathbf{p}}}{\partial\mathbf{p}}\right) \frac{d^3\mathbf{p}}{(2\pi)^3}. \quad (2.106)$$

Earlier in this section when solving the dispersion relation we neglected the imaginary part of the dielectric permittivity, that is, we solved the equation

$$\text{Re}\{\varepsilon_{\mathbf{k},\omega}\} = 0. \quad (2.107)$$

This makes sense when the damping is small. We can improve the solution we found by taking weak damping into account. We now write $\omega = \omega_{\mathbf{k}} + i\gamma_{\mathbf{k}}$ with $\gamma_{\mathbf{k}} \ll \omega_{\mathbf{k}}$, where $\omega_{\mathbf{k}}$ satisfies the equation

$$\text{Re}\{\varepsilon_{\mathbf{k},\pm\omega_{\mathbf{k}}}\} = 0. \quad (2.108)$$

In the complete equation $\varepsilon_{\mathbf{k},\omega} = 0$ we now take into account both $\gamma_{\mathbf{k}}$ and $\text{Im}\{\varepsilon_{\mathbf{k},\omega}\}$:

$$\begin{aligned}
\varepsilon_{\mathbf{k},\omega} &= \varepsilon_{\mathbf{k},\omega_{\mathbf{k}}+i\gamma_{\mathbf{k}}} = \text{Re}\,\varepsilon_{\mathbf{k},\omega_{\mathbf{k}}+i\gamma_{\mathbf{k}}} + i\,\text{Im}\,\varepsilon_{\mathbf{k},\omega_{\mathbf{k}}+i\gamma_{\mathbf{k}}} \\
&\simeq \text{Re}\,\varepsilon_{\mathbf{k},\omega_{\mathbf{k}}} + i\,\text{Im}\,\varepsilon_{\mathbf{k},\omega_{\mathbf{k}}} + i\gamma_{\mathbf{k}}\left.\frac{\partial}{\partial\omega}\text{Re}\{\varepsilon_{\mathbf{k},\omega}\}\right|_{\omega=\omega_{\mathbf{k}}}.
\end{aligned} \quad (2.109)$$

From this equation we find the following expression for the damping:

$$\gamma_k = -\left[\text{Im}\,\varepsilon_{\mathbf{k},\omega_{\mathbf{k}}}\right] \Big/ \left[\left.\frac{\partial}{\partial\omega}\text{Re}\{\varepsilon_{\mathbf{k},\omega}\}\right|_{\omega=\omega_{\mathbf{k}}}\right]. \quad (2.110)$$

We should mention that in the imaginary part of ε we have to put $\omega = \omega_{\mathbf{k}}$ which means that it contains the delta-function $\delta(\omega_{\mathbf{k}} - (\mathbf{k}\cdot\mathbf{v}))$. As long as the frequency is different from $\omega_{\mathbf{k}}$ the imaginary part does not correspond to a conservation law for an elementary process of a Vavilov-Cherenkov interaction between particles and waves, but it does so when $\omega = \omega_{\mathbf{k}}$. The *Vavilov-Cherenkov effect* is also called the *Cherenkov effect*, especially outside Russia. The correct name is Vavilov-Cherenkov effect since S.I.Vavilov played an important rôle in its discovery. Normally this name is used in the Soviet literature, but as the other name is shorter it is also used sometimes; we shall use both names. We note that we have thus for the Cherenkov emission of plasma waves the relation

$$\omega_{\mathbf{k}} = (\mathbf{k}\cdot\mathbf{v}). \quad (2.111)$$

One can, in fact, find this relation from a simple consideration of energy and momentum conservation in the process in which a particle emits a wave. Let $\varepsilon_{\mathbf{p}}$ and $\mathbf{p}$ be the initial energy and momentum of the particle and let

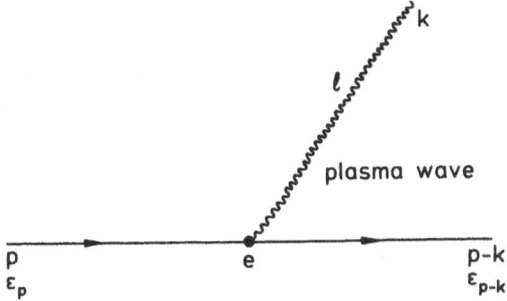

Fig. 2.7. Resonance interaction between a charged plasma particle and a plasma wave

$\hbar\omega_k$ and $\hbar k$ be the energy and momentum of the emitted wave; the final momentum and energy of the particle will then be $\mathbf{p} - \hbar\mathbf{k}$ and $\varepsilon_{\mathbf{p}-\hbar k}$ (see Fig.2.7)

As the momentum $\hbar\mathbf{k}$ of the wave is small as compared to the particle momentum we can expand $\varepsilon_{\mathbf{p}-\hbar k}$ in terms of $\hbar\mathbf{k}$ and we then find from the energy conservation law, $\varepsilon_{\mathbf{p}} = \varepsilon_{\mathbf{p}-\hbar k} + \hbar\omega_k$, the relation

$$\left(\mathbf{k}\cdot\frac{\partial\varepsilon_\mu}{\partial\mathbf{p}}\right) \equiv (\mathbf{k}\cdot\mathbf{v}) = \omega_{\mathbf{k}}. \tag{2.112}$$

This is just Eq.(2.111). We can therefore conclude that Landau damping and Cherenkov emission should be closely connected – and we shall see later on that Landau damping describes stimulated Cherenkov emission and absorption processes.

From a classical point of view Eq.(2.111) expresses the condition for the particle and the wave to be in resonance. If the particle velocity is in the same direction as the wave velocity, it states that the phase velocity of the wave should be equal to the particle velocity. However, if the particle is moving at some angle to the wave velocity, the component of the wave velocity along the particle velocity appears in the resonance condition. In both cases the wave field is static in the frame in which the particle is at rest. The resonance we are discussing here is called *wave-particle resonance*.

From the point of view of a wave-particle resonance we can consider Landau damping as follows. If the velocity of the particle is somewhat different from the resonant velocity, the particle can be trapped by the wave or not. The particle has therefore a certain degree of "freedom" and the motion of particle close to resonance can be said to be in some sense unstable – small changes in the initial conditions may lead to quite different final results. This means that the motion may become random and methods used in the theory of dynamical chaos prove that this, indeed, will happen. The uncertainty can be solved here by appealing to the causality principle. This principle introduces irreversibility; the physical reason for this is the indeterminate nature of the particle motion for velocities close to the resonance velocity. This problem is thus related to the general problem of the relations between regular and random motions. Let us mention in this connection that the use of retarded potentials in electrodynamics also involves arguments of causality and irreversibility. In this sense, the use of the causality principle when calculating Landau damping has the same basis as the use of retarded potentials in electrodynamics.

We shall now write down an explicit expression for the Landau damping of Langmuir waves and of ion-sound waves for the simplest case of thermal particle distributions.

Langmuir waves are Landau damped by electrons. After substuting a thermal electron distribution into (2.106) one can easily, by using the δ-function integrate over the component of the particle velocity parallel to the wavevector $\mathbf{k}$. The integration over the other velocity components is trivial and involves the normalisation of the distribution function. Using the fact that $\partial \mathrm{Re}\{\varepsilon\}/\partial \omega \simeq 2/\omega_{\mathrm{pe}}$ we finally find

$$\gamma_{\mathbf{k}} = -\sqrt{\frac{\pi}{8}}\, \omega_{\mathrm{pe}}\, \frac{\omega_{\mathrm{pe}}^3}{k^3 v_{Te}^3}\, \exp\left[-\frac{\omega_{\mathrm{pe}}^2}{2k^2 v_{Te}^2} - \frac{3}{2}\right]. \tag{2.113}$$

The damping becomes large when $k \approx 1/d_{\mathrm{e}}$; this is in accordance with the absence of waves with $k > 1/d_e$. When the phase velocity becomes smaller, more particles can be in resonance with waves and the Landau damping becomes stronger. When the phase velocity is large as compared to the thermal velocity the Landau damping is exponentially small.

Landau damping of ion-sound waves by ions is important only when $k \geqslant 1/d_i$ and one can find an expression for it from Eq.(2.113), changing the indices e to i. At first sight Landau damping of ion-sound by electrons could be strong since the phase velocity of the waves is much smaller than the electron thermal velocity. However, this is not the case. The resonance condition for electrons interacting with an ion-sound wave can be satisfied only for the relatively small fraction of the electrons which move almost perpendicular to the wave, since the phase velocity of the wave is much smaller than the particle velocity. This statement is correct whatever the direction of the wave propagation, although for different propagation directions different particles will be in resonance with the wave. For an isotropic wave distribution all particles can therefore take part in the damping. However, for a given wave only particles in a narrow cone perpendicular to the propagation direction of the wave can be in resonance. The opening angle of this cone is determined by the ratio of the sound velocity to the electron thermal velocity, that is, by a factor which is approximately equal to $\sqrt{Zm_e/m_i}$. From this one concludes that the Landau damping rate should be smaller that the ion-sound frequency approximately by a factor $\sqrt{Zm_e/m_i}$. Indeed, an exact calculation of the appropriate integral gives the result:

$$\gamma_{\mathbf{k}}^{\mathrm{s}} = -\sqrt{\frac{\pi}{8}}\,\omega_{\mathbf{k}}^{\mathrm{s}}\,\sqrt{\frac{Zm_{\mathrm{e}}}{m_{\mathrm{i}}}}\,\frac{1}{\left(1 + k^2 d_{\mathrm{e}}^2\right)^{3/2}}. \tag{2.114}$$

Concluding this section we wish to state once again that all experiments which have so far been performed completely confirm the theoretical expressions for the Landau damping of the plasma modes which we have described here.

2.7 Beam-Plasma Instability

The sign of $\gamma_{\mathbf{k}}$ is always negative for a thermal particle distribution. It is also negative for any isotropic distribution which does not need to be thermal. Indeed, for such a distribution we have

$$\left(\mathbf{k}\cdot\frac{\partial\Phi_p}{\partial\mathbf{p}}\right) = \frac{(\mathbf{k}\cdot\mathbf{v})}{v}\frac{\partial\Phi_p}{\partial p} = \frac{\omega_{\mathbf{k}}}{v}\frac{\partial\Phi_p}{\partial p}, \tag{2.115}$$

and integrating over the angles in Eq.(2.106) gives

$$\gamma_{\mathbf{k}} = \frac{\omega_{\mathbf{k}}}{k^3\left(\partial\varepsilon_{\mathbf{k},\omega}/\partial\omega\right)_{\omega=\omega_{\mathbf{k}}}}\sum_{\alpha}e_{\alpha}^2\int_{m_{\alpha}\omega_{\mathbf{k}}/k}\frac{1}{v^2}\frac{\partial\Phi_p^{\alpha}}{\partial p}\,p^2\,dp. \tag{2.116}$$

Putting $p = m_{\alpha}v$ we find

$$\gamma_{\mathbf{k}} = -\frac{\omega_{\mathbf{k}}}{k^3\left(\partial\varepsilon_{\mathbf{k},\omega}/\partial\omega\right)_{\omega=\omega_{\mathbf{k}}}}\sum_{\alpha}e_{\alpha}^2 m_{\alpha}^2\Phi_{m_{\alpha}\omega_{\mathbf{k}}/k}^{\alpha}, \tag{2.117}$$

which is always negative, provided $\partial\varepsilon/\partial\omega > 0$. For an instability to occur anisotropy is needed.

One of the most interesting anisotropic distributions is a weak beam of electrons in a plasma. Let us consider the case where all the particles in the beam are moving in the same direction, say, the x-direction, and let the average velocity, v_{b}, of the electrons in the beam be significantly higher than the electron thermal velocity, $v_{T\mathrm{e}}$, the beam density, n_{b}, be much less than the average plasma density, n_{e}, and the spread of the beam electron velocity, δv_{b}, be much less than the average beam velocity: $\delta v_{\mathrm{b}} \ll v_{\mathrm{b}}$. This type of distribution is shown in Fig.2.8.

To find the instability growth rate for waves propagating along the direction of the beam we integrate the expression for $\gamma_{\mathbf{k}}$, with $k = k_x$, over the components of the momentum of the beam particles perpendicular to the direction in which the beam propagates ($p_x = p = m_{\mathrm{e}}v$) and introduce a beam-particle distribution function Φ_v^{b}, normalised with respect to the beam velocity v:

$$\int\Phi_{\mathbf{p}}^{\mathrm{b}}\frac{d^3\mathbf{p}}{(2\pi)^3} = \int\Phi_v^{\mathrm{b}}\,dv = n_{\mathrm{b}}. \tag{2.118}$$

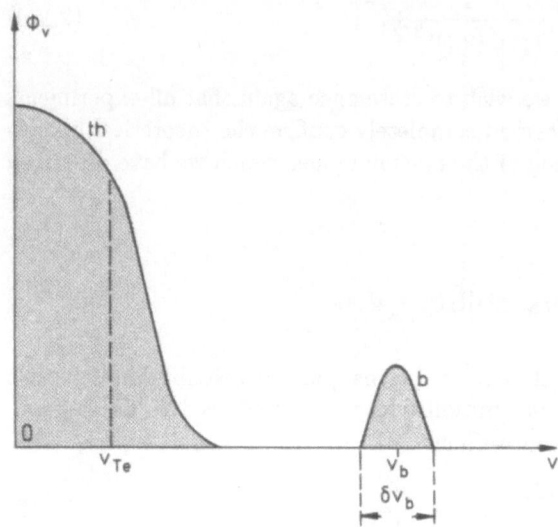

Fig. 2.8. Electron distribution function with thermal particles (th) and an electron beam (b)

Since for Langmuir waves $\partial\varepsilon/\partial\omega$ is equal to $2/\omega_{pe}$ we find from Eqs.(2.110) and (2.106):

$$\gamma_k^b = \frac{2\pi^2 e^2 \omega_{pe}}{k m_e} \int \delta(\omega_{pe} - kv) \frac{\partial\Phi_v^b}{\partial v} \, dv$$

$$= \frac{2\pi^2 e^2 \omega_{pe}}{k^2 m_e} \frac{\partial\Phi_v^b}{\partial v}\bigg|_{v=\omega_{pe}/k} . \tag{2.119}$$

An instability occurs in the velocity range where the derivative of the beam-particle distribution is positive. Note that the result given here is valid for positive velocities; for negative velocities, where the derivative of the thermal distribution is positive, the thermal particles are not interacting with waves propagating in the opposite direction and the sign of γ_k is the opposite of that given by Eq.(2.119).

One can estimate the order of magnitude of the growth rate from Eq.(2.119) by putting

$$\frac{\partial\Phi_v^b}{\partial v} \cong \frac{\Phi_v^b}{\delta v_b} \cong \frac{n_b}{(\delta v_b)^2}, \qquad \Phi_v^b \delta v_b \cong n_b. \tag{2.120}$$

The last estimate follows from the normalisation (2.118). Omitting a numerical factor $\frac{1}{2}\pi$, since the estimates cannot pretend an accuracy as high as that, we find

$$\gamma_k^b \simeq \omega_{pe} \frac{n_b}{n_e} \frac{v_b^2}{(\delta v_b)^2}. \tag{2.121}$$

Excitation of waves will occur only, if this growth rate is higher than the damping due to binary electron-ion collisions, $\gamma_b > \nu_{ei} \approx \omega_{pe}/N_{de}$. The threshold for this instability is thus very low:

$$\frac{n_b}{n_e} > \frac{(\delta v_b)^2}{v_b^2} \frac{1}{N_{de}}. \tag{2.122}$$

In the case when δv_b is of the order of v_b – the maximum possible spread – we still have

$$\frac{n_b}{n_e} > \frac{1}{N_{de}}. \tag{2.123}$$

For the parameter ε, which we introduced earlier in § 2.3, which characterises a small deviation of the particle distribution from its equilibrium value,

$$\varepsilon = \frac{n_b}{n_e}, \tag{2.124}$$

we obtain the relation given earlier.

This example clearly shows that very small deviations of a distribution from that in thermal equilibrium will lead to a rapid excitation of plasma oscillations. For such an excitation it is sufficient that we have $\varepsilon > 1/N_{de}$.

Problems

1. Find the angular dependence of the growth rate of the instability of a one-dimensional beam with $v_b \gg v_{Te}$, $n_b \ll n_e$, and $\delta v_b \ll v_b$.
2. Find the growth rate of the ion-sound instability for a plasma in which there is a current with electrons and ions both having a thermal distribution with a relative drift velocity u for which $v_{Ti} \ll u \ll v_{Te}$.
3. Show that any isotropic relativistic particle distribution is stable against the excitation of plasma waves.
4. Find the growth rate for the excitation of Langmuir and of ion-sound waves for the case of slightly anisotropic particle distributions $\Phi_p = \Phi_{|p|}^{(0)} + (\mathbf{n} \cdot \mathbf{p})\Phi_{|p|}^{(1)}$, where $\mathbf{n}$ is a unit vector in the direction of the anisotropy.
5. Repeat the exercise of Prob. 4 for the case of highly relativistic particles – cosmic rays $(p \gg mc)$ – for which the distribution is a power law: $\Phi_{|p|}^{(0)} \propto |\mathbf{p}|^{-\gamma}$, $\Phi_{|p|}^{(1)} \propto |\mathbf{p}|^{-\mu}$.

3 Kinetics of Random Collective Excitations. Quasi-linear Interactions

3.1 Introduction. Distribution Function for Plasma Oscillations

We start with a general discussion which raises important questions which, at first sight, seem to have obvious answers but which on further consideration turn out to be far from simple to answer. Let us accept that a typical plasma state is a state where many collective oscillations are excited at a level which is sufficiently high for their interaction to be non-linear. In that case it is reasonable to use for these collective oscillations a kinetic description which is similar to the one used for particles. Indeed, for an ensemble of a great number of oscillations it is reasonable, as it is for an ensemble of a large number of particles, not to try to follow the development of each oscillation – its phase and its amplitude – separately and in detail. We are used to the necessity of a statistical description when we are dealing with particles; the usual argument for this is that both from a mathematical and from a physical point of view it is impossible to follow each particle separately. In fact, if all particles interact strongly with one another, chaos will ensue and any initial conditions will soon be forgotten: it is physically impossible to follow the trajectory of any individual particle. Even if we forget the quantal dualism of waves and particles, oscillations are in this sense not very different from particles: as in the case of an ensemble of a large number of particles, an ensemble of a great number of oscillations should also show a statistical behaviour, if the interactions between them lead to randomisation processes. Recent investigations of dynamical chaos have confirmed that even in a system with a small number of degrees of freedom, non-linear interactions between these degrees of freedom will lead to random behaviour. We are here interested in systems with a very large number of degrees of freedom. For this case it is extremely probable that random behaviour will occur if the oscillations interact with one another, that is, we shall find chaotic behaviour of the oscillations. This problem still needs a special analysis which as yet has not been completed; we mentioned in Chap.1 that regular and chaotic motions have a tendency to coexist, to support one another, and to form some self-organised structures. We shall start from the hypothesis that randomness is established in a system of a great many, interacting oscillations and we shall attempt to describe the oscillations statistically. We then are faced with the usual question of what part of the phase space of the oscillations is occupied by chaotic oscillations and what part by regular oscillations. The answer will depend on the strength of the fields. For the moment we shall consider that part in which random behaviour is established. Later on, in subsequent chapters, we shall consider the interaction between random and regular oscillations.

First of all, we must know how to define a distribution function for random oscillations, how to measure it, and how to write down a kinetic equation describing the evolution of such a distribution function. A field is said to be a random field, if its average value vanishes:

$$\langle \mathbf{E} \rangle = 0. \tag{3.1}$$

Quadratic combinations of fields will, of course, not necessarily be zero. In the general case, a field will have both a regular and a random component:

$$\mathbf{E} = \langle \mathbf{E} \rangle + \delta \mathbf{E}. \tag{3.2}$$

A trivial consequence of (3.2) is that $\langle \delta \mathbf{E} \rangle = 0$. For the moment we shall consider the simplest case when there is no regular component, and we then have $\delta \mathbf{E} = \mathbf{E}$. We shall now show how to describe random fields statistically, introducing a distribution function which depends on their wavevector. We can introduce such a distribution using the analogy with the particle distribution function $\Phi_{\mathbf{p}}$. Let us remind ourselves that the average number of particles, n, per unit volume was defined by the equation

$$n = \int \Phi_{\mathbf{p}} \frac{d^3 \mathbf{p}}{(2\pi)^3}, \tag{3.3}$$

while the average energy, E, per unit volume and the average momentum, $\mathbf{P}$, of the particles per unit volume are defined by the relations

$$E = \int \varepsilon_{\mathbf{p}} \Phi_{\mathbf{p}} \frac{d^3 \mathbf{p}}{(2\pi)^3}; \qquad \mathbf{P} = \int \mathbf{p} \Phi_{\mathbf{p}} \frac{d^3 \mathbf{p}}{(2\pi)^3}. \tag{3.4}$$

For waves it is natural to define their distribution function $N_{\mathbf{k}}$ through their average energy, W, per unit volume. By analogy with the first of Eqs.(3.4) we shall write

$$W = \int \omega_{\mathbf{k}} N_{\mathbf{k}} \frac{d^3 \mathbf{k}}{(2\pi)^3}. \tag{3.5}$$

From a quantum point of view the energy of a single oscillation is equal to $\hbar \omega_{\mathbf{k}}$. The energy defined by (3.5) differs from the quantum expression by a factor $\hbar$. However, in what follows we shall consider only classical effects so that it is not reasonable to introduce the quantum constant. In fact, W will be a purely classical energy. On the other hand, we have already seen that using a quantum description, and subsequently taking the classical limit, clarifies the physical meaning of the conservation laws, and we shall encounter several other examples of this kind. The simplest way of joining the two approaches is to use units in which Dirac's constant $\hbar$ is put equal to unity.

We could equally have defined the distribution function of the oscillations through the average momentum, $\mathit{\Pi}$, of the waves per unit volume by analogy with the average momentum of the particles per unit volume:

$$\mathit{\Pi} = \int \mathbf{k}\, N_{\mathbf{k}}\, \frac{d^3\mathbf{k}}{(2\pi)^3}. \tag{3.6}$$

From a quantum point of view this expression is trivial, but one should take into account that classically the momentum of waves can be calculated independent of their energy and that both the momentum and the energy must take into account the particle momentum moving with the wave. Moreover, for the electrostatic waves, which we are considering here, the field momentum is zero, and for Langmuir waves, for instance, half of their energy is electrostatic energy and the other half is the energy of the particle motion in the waves. The question now is whether the distribution function defined using the wave energy and that defined using the wave momentum are the same. They are, indeed, the same and we shall prove this several times in what follows.

Finally, we can introduce the average number, N, of waves per unit volume, again using the analogy with particles:

$$N = \int N_{\mathbf{k}}\, \frac{d^3\mathbf{k}}{(2\pi)^3}. \tag{3.7}$$

We can immediately write down one equation for the distribution function, namely, the equation describing Landau damping. Since the time dependence of the amplitude of a wavefield which is damped (or amplified) can be described by an exponent $e^{\gamma t}$, and since both the energy density and $N_{\mathbf{k}}$ are proportional to the square of the amplitude, their time dependence will be described by the exponential $e^{2\gamma t}$. Therefore, the equation describing the Landau damping of the waves can be written as follows:

$$\frac{dN_{\mathbf{k}}}{dt} = 2\gamma_{\mathbf{k}}^{\mathrm{L}} N_{\mathbf{k}}. \tag{3.8}$$

In what follows we shall use for the Landau damping rate either the above notation or simply $\gamma_{\mathbf{k}}$, whenever this cannot lead to any misunderstandings.

3.2 Correlation Functions for Random Fields

We shall now consider the problem of how to measure the distribution functions of waves. Experimentally, it is, for instance, possible to find the local time dependence of the potential, using data from probe measurements. Of course, there is a general problem for all such measurements, namely, how to find the actual value of the potential in a plasma without the perturbations introduced by the probe; this, however, is a separate problem and we shall assume that it has been solved. There are also other experimental methods for determining the local time dependence of the potential, for instance, using laser or radio scattering. For our purposes it is sufficient that we may assume that this time-dependent local potential has been measured correctly and that it is known. If the measured potential is a random one, the time dependence will be a complicated function of the time, something like the curve shown in Fig.3.1, which may have an average value ϕ_0 as well as fluctuations around this average value.

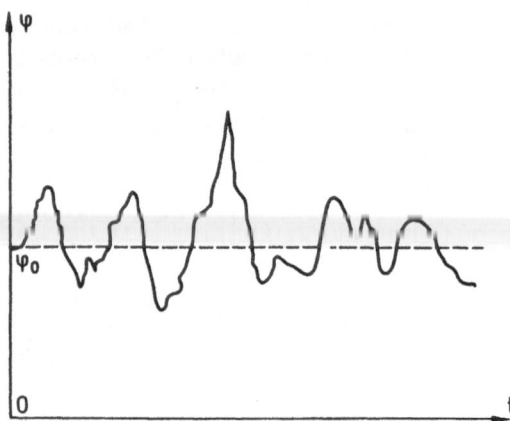

Fig. 3.1. A particular realisation of a random potential in the plasma as function of time

Not every complicated function of the kind shown in Fig.3.1 is necessarily a random one. One set of data from measurements one can call a sample of such a set of measurements. If we repeat the measurements with the same starting conditions we get another sample – another realisation. One can average over these realisations. The question is to decide whether the measured function is random or not, if possible from just one set of measurements. Modern theoretical concepts of chaos give a relatively simple prescription for solving this problem of randomness – and even how to find the degree of randomness – of an observed signal, but we shall just mention that this problem can in principle be solved, without going into it in any more detail. In the case of a random potential we can determine its characteristics by calculating the so-called *auto-correlation function* $K(t, t')$ of the potential:

$$K(t, t') = \langle \phi(t)\phi(t') \rangle = \langle \phi(t)\phi(t + \tau) \rangle, \qquad t' = t + \tau. \tag{3.9}$$

In the case of a stationary system the result of the measurements will be independent of the particular moment from which we start to count the time, that is, the auto-correlation function should depend solely on the time difference $t' - t = \tau$.

One can also measure the potential at different positions in space as well as at different times and calculate more complicated correlation functions – these are sometimes called *cross-correlation functions* but we shall call them simply correlation functions:

$$
\begin{aligned}
K(\mathbf{r}, t; \mathbf{r}', t') &= \langle \phi(\mathbf{r}, t)\phi(\mathbf{r}', t') \rangle \\
&= \langle \phi(\mathbf{r}, t)\phi(\mathbf{r} + \rho, t + \tau) \rangle, \qquad \mathbf{r}' = \mathbf{r} + \rho.
\end{aligned} \tag{3.10}
$$

For a system which is homogeneous on average the correlation function will depend only on $\mathbf{r}' - \mathbf{r} = \rho$, rather than on both $\mathbf{r}$ and $\mathbf{r}'$ separately.

One of the best set of such measurements was carried out by using a movable probe in a plasma discharge under stable conditions and measurements of the potential were made in more than 10^6 points; a powerful computer was needed to calculate the correlation function from these data. In our discussions we shall assume that refined measurements of this kind have already been carried out in our system. The question then is how to use the data to calculate the distribution function of the waves, $N_\mathbf{k}$.

We shall now consider the case of a potential which is on average stationary and homogeneous so that we may assume that the correlation function depends only on ρ and τ. The conclusions we reach will be valid also when there are slow changes in time and space, that is, the changes in time are small during a period of the waves and the changes in space are small over a distance equal to the wavelength of the waves. If there are such slow changes $N_\mathbf{k}$ will still depend on the time t and the coordinate $\mathbf{r}$.

Let us consider the Fourier components $\phi_{\mathbf{k},\omega}$ of the potential ϕ:

$$
\phi(\mathbf{r}, t) = \int \phi_{\mathbf{k},\omega} \, e^{i[(\mathbf{k}\cdot\mathbf{r}) - \omega t]} \, d^3k \, d\omega. \tag{3.11}
$$

In terms of the Fourier components we can write the correlation function (3.10) in the form

$$
\begin{aligned}
&\langle \phi(\mathbf{r}, t)\phi(\mathbf{r}', t') \rangle \\
&= \int \langle \phi_{\mathbf{k},\omega}\phi_{\mathbf{k}',\omega'} \rangle \, e^{i[(\mathbf{k}\cdot\mathbf{r}) + (\mathbf{k}'\cdot\mathbf{r}') - \omega t - \omega' t']} \, d^3k \, d^3k' \, d\omega \, d\omega'.
\end{aligned} \tag{3.12}
$$

For the case we are considering at the moment the right-hand side should depend only on $\mathbf{r} - \mathbf{r}'$ and $t - t'$; this is possible only, if we have $\mathbf{k}' = -\mathbf{k}$ and $\omega' = -\omega$, which means that the average value $\langle \phi_{\mathbf{k},\omega}\phi_{\mathbf{k}',\omega'} \rangle$ should be proportional to $\delta(\mathbf{k} + \mathbf{k}')\delta(\omega + \omega')$:

$$
\langle \phi_{\mathbf{k},\omega}\phi_{\mathbf{k}',\omega'} \rangle = |\phi|^2_{\mathbf{k},\omega} \, \delta(\mathbf{k} + \mathbf{k}')\delta(\omega + \omega'). \tag{3.13}
$$

This equation can be used as a definition of $|\phi|^2_{\mathbf{k},\omega}$; it must be positive, since on the left-hand side of Eq.(3.13) there appears, if we take the δ-function into account, a factor $\phi_{\mathbf{k}',\omega'} = \phi_{-\mathbf{k},-\omega} = \phi^*_{\mathbf{k},\omega}$.

We can also introduce correlation functions of the field components since for electrostatic fields they follow directly from those of the potentials:

$$\mathbf{E} = -\nabla\phi, \qquad \mathbf{E}_{\mathbf{k},\omega} = -i\mathbf{k}\phi_{\mathbf{k},\omega}; \tag{3.14}$$

$$\langle E_{i;\mathbf{k},\omega} E_{j;\mathbf{k}',\omega'} \rangle = k_i k_j |\phi|^2_{\mathbf{k},\omega} \delta(\mathbf{k}+\mathbf{k}')\delta(\omega+\omega'). \tag{3.15}$$

Often the correlation function of fields is defined as follows:

$$\langle E_{i;\mathbf{k},\omega} E_{j;\mathbf{k}',\omega'} \rangle = \frac{k_i k_j}{k^2} |E|^2_{\mathbf{k},\omega} \delta(\mathbf{k}+\mathbf{k}')\delta(\omega+\omega'). \tag{3.16}$$

In that case we clearly have

$$|E|^2_{\mathbf{k},\omega} = k^2 |\phi|^2_{\mathbf{k},\omega}.$$

In fact, the correlation functions $|E|^2_{\mathbf{k},\omega}$ and $|\phi|^2_{\mathbf{k},\omega}$ are just the Fourier components of the corresponding correlation function of the fields and of the potential, treated as functions of ρ and τ.

Let us now assume that the correlation function $|E|^2_{\mathbf{k},\omega}$ is known from measurements. We have alrady mentioned that not all random fields are fields of random waves. In the case when the random fields are, indeed, the fields of random waves there should appear a sharp peak in the measured spectrum of the correlation functions close to the frequency $\pm\omega_{\mathbf{k}}$ of the collective oscillations. The width $\delta\omega_{\mathbf{k}}$ of these peaks will in actual cases be finite and determined by slow processes such as the damping of the waves, their non-linear interactions, and the slow dependence of $|E|^2_{\mathbf{k},\omega}$ on time and space in the case when the correlation function itself is slightly inhomogeneous and slowly dependent on the time. We shall discuss many of these effects later on. We give in Fig.3.2 a qualitative sketch of $|E|^2_{\mathbf{k},\omega}$ as function of ω for a given $\mathbf{k}$.

It is important that we can only speak about oscillations in the case when $\delta\omega_{\mathbf{k}} \ll \omega_{\mathbf{k}}$. There are cases when there are peaks in the correlation functions when there is no turbulence while they disappear in the presence of turbulence. In this case turbulence smoothens out the oscillations, that is, in the presence of turbulence it is no longer possible to speak about these oscillations. We shall assume that the oscillations exist, that is, that $\delta\omega_{\mathbf{k}} \ll \omega_{\mathbf{k}}$. To a first approximation we can treat these peaks as being very narrow and describe them by a δ-function. To express this explicitly we must take into account that the field strength $\mathbf{E}(\mathbf{r},t)$ is a real, rather than a complex, quantity so that we have for the Fourier components the relation

$$\mathbf{E}_{-\mathbf{k},-\omega} = \mathbf{E}^*_{\mathbf{k},\omega}, \tag{3.18}$$

and thus

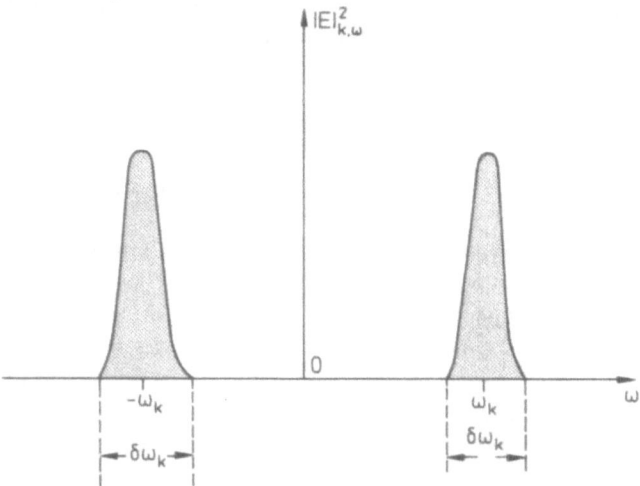

Fig. 3.2. Frequency dependence of the correlation function of random fields in a plasma

$$|E|^2_{-\mathbf{k},-\omega} = |E|^2_{\mathbf{k},\omega}. \tag{3.19}$$

If we now denote the coefficient of $\delta(\omega - \omega_{\mathbf{k}})$, which approximately describes the right-hand peak in Fig.3.2, by $|E|^2_{\mathbf{k}}$, the coefficient of $\delta(\omega + \omega_{\mathbf{k}})$, which describes the left-hand peak in Fig.3.2), will be determined by Eq.(3.19) and be equal to $|E|^2_{-\mathbf{k}}$. We find thus

$$|E|^2_{\mathbf{k},\omega} = |E|^2_{\mathbf{k}}\,\delta(\omega - \omega_{\mathbf{k}}) + |E|^2_{-\mathbf{k}}\,\delta(\omega + \omega_{\mathbf{k}}). \tag{3.20}$$

From the experimental data we can thus find in this way the quantity $|E|^2_{\mathbf{k}}$. The question now arises how to use it to calculate the distribution function of the waves $N_{\mathbf{k}}$. To do this we shall use Eq.(3.5) for the energy of random oscillations.

3.3 Energy Density of Random Oscillations

We can write down a general expression for the energy density of electrostatic fields, using its time derivative $(1/4\pi)\big(\mathbf{E} \cdot (\partial/\partial t)\mathbf{D}\big)$:

$$W = \int_{-\infty}^{t} \frac{1}{4\pi} \left\langle \left(\mathbf{E} \cdot \frac{\partial}{\partial t}\mathbf{D}\right) \right\rangle dt. \tag{3.21}$$

Here $\mathbf{D}$ is the electric induction: $\mathbf{D}_{\mathbf{k}} = \varepsilon_{\mathbf{k}}\mathbf{E}_{\mathbf{k}}$; $\mathbf{k} = \{\mathbf{k}, \omega\}$. We have written down here not an expression for the energy density itself, but for its average value, averaged over a statistical ensemble, since the statistical average corresponds to the average energy density W of random oscillations which we introduced earlier. Substituting into (3.21) the Fourier expansion for the field $\mathbf{E}$ and the induction $\mathbf{D}$ we find:

$$W = -\mathrm{i} \int_{-\infty}^{t} dt \int \frac{d^4k\, d^4k'}{8\pi} \left\langle (\mathbf{E_k} \cdot \omega' \varepsilon_{\mathbf{k'}} \mathbf{E_{k'}}) \right.$$
$$\left. + (\mathbf{E_{k'}} \cdot \omega \varepsilon_{\mathbf{k}} \mathbf{E_k}) \right\rangle e^{\mathrm{i}[(\{\mathbf{k}+\mathbf{k'}\}\cdot\mathbf{r})-(\omega+\omega')t]}, \quad d^4k = d^3k\, d\omega. \quad (3.22)$$

We have written down an expression which is symmetric in k and k' where we have again used k for the four-dimensional vector and d^4k for a volume element in four-dimensional k-space. We carry out the integration over the time assuming that there are no fields as $t \to -\infty$,

$$\int_{-\infty}^{t} dt\, e^{-\mathrm{i}(\omega+\omega')t} \longrightarrow \frac{1}{-\mathrm{i}(\omega+\omega')}\, e^{-\mathrm{i}(\omega+\omega')t}.$$

Energy has, strictly speaking, no definite meaning in a dissipative system. We shall restrict ourselves here to the case when the imaginary part of $\varepsilon_{\mathbf{k}}$ is negligibly small. We have then $\mathbf{k} + \mathbf{k'} = 0$ for the case where the fields are on average stationary and homogeneous, and as a result we get an indeterminate expression, since both the numerator and the denominator vanish because for real values of $\varepsilon_{\mathbf{k}}$ we have $\varepsilon_{-\mathbf{k}} = \varepsilon_{\mathbf{k}}$. This indeterminacy can be removed by taking the limit:

$$\frac{\omega' \varepsilon_{\mathbf{k'}} + \omega \varepsilon_{\mathbf{k}}}{\omega' + \omega} \longrightarrow \frac{\partial}{\partial \omega}(\omega \varepsilon_{\mathbf{k}}), \quad \mathbf{k'} \to -\mathbf{k}, \quad (3.23)$$

and we then get

$$W = \int d^4k\, \frac{|E|^2_{\mathbf{k},\omega}}{8\pi} \frac{\partial(\omega\varepsilon_{\mathbf{k}})}{\partial \omega}. \quad (3.24)$$

Substituting (3.20) into (3.24) and using the fact that we have $\varepsilon_{\mathbf{k},\omega} = 0$ for the oscillations and that $N_{\mathbf{k}}$ and W are related through Eq.(3.5) we obtain

$$W = \int \frac{d^3k}{4\pi} \omega_{\mathbf{k}} |E|^2_{\mathbf{k}} \left.\frac{\partial \varepsilon_{\mathbf{k}}}{\partial \omega}\right|_{\omega=\omega_{\mathbf{k}}} = \int \frac{d^3k}{(2\pi)^3} \omega_{\mathbf{k}} N_{\mathbf{k}}, \quad (3.25)$$

from which we finally get the following relation between the correlation function of the fields and the distribution function of the waves:

$$N_{\mathbf{k}} = 2\pi^2 |E|^2_{\mathbf{k}} \left.\frac{\partial \varepsilon_{\mathbf{k},\omega}}{\partial \omega}\right|_{\omega=\omega_{\mathbf{k}}}. \quad (3.26)$$

We have thus found not only a definite expression for the distribution function of the waves but also a procedure for measuring it. It is important that this distribution function was introduced for the case of random waves.

3.4 Stimulated Vavilov-Cherenkov Emission and Landau Damping

We now introduce the *probability, $w_{\mathbf{p}}(\mathbf{k})$, for stimulated Vavilov-Cherenkov emission per unit time* and per unit phase volume $d^3\mathbf{k}/(2\pi)^3$ of a wave with its momentum $\mathbf{k}$ in the range from $\mathbf{k}$ to $\mathbf{k}+d\mathbf{k}$ and with energy $\omega_{\mathbf{k}}$ (we use again units such that $\hbar = 1$) by a particle of momentum $\mathbf{p}$ and energy $\varepsilon_{\mathbf{p}}$. We shall keep open the possibility that there are several kinds of particles, indicated by a superscript α, and several kinds of waves, indicated by a superscript σ; this means that, in general, we may be dealing with probabilities $w_{\mathbf{p}}^{\alpha,\sigma}(\mathbf{k})$. After the emission the particle will have a momentum $\mathbf{p}-\mathbf{k}$ and an energy $\varepsilon_{\mathbf{p}-\mathbf{k}}$; this means that the emission process corresponds to a transition of a particle from an energy level $\varepsilon_{\mathbf{p}}$ to an energy level $\varepsilon_{\mathbf{p}-\mathbf{k}}$ (see Fig.3.3).

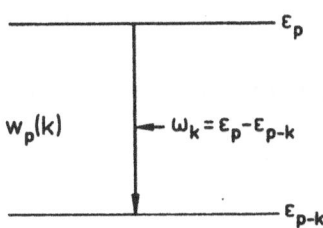

Fig. 3.3. Energy level diagram corresponding to a process of resonant wave emission

We can now use the *Einstein relation* between the probabilities for stimulated and spontaneous emission processes which states that for the stimulated process one must introduce an extra factor $N_{\mathbf{k}}$. Using this relation we find for the probability for an stimulated transition from the level $\mathbf{p}$ to the level $\mathbf{p}-\mathbf{k}$ the expression $N_{\mathbf{k}}w_{\mathbf{p}}(\mathbf{k})$ and the same for the probability for an stimulated transition from the level $\mathbf{p}-\mathbf{k}$ to the level $\mathbf{p}$. We can then calculate the rate $(dN_{\mathbf{k}}/dt)_+$ at which the number of waves increases due to the stimulated and spontaneous emission processes; this rate is proportional to the number of particles $\Phi_{\mathbf{p}}$ in the level $\mathbf{p}$ from which the emission occurs. To obtain the total transition rate we must sum over all possible particle momenta which can lead to the emission of a wave with momentum $\mathbf{k}$:

$$\left(\frac{dN_{\mathbf{k}}}{dt}\right)_+ = \int w_{\mathbf{p}}(\mathbf{k})(N_{\mathbf{k}}+1)\,\Phi_{\mathbf{p}}\,\frac{d^3\mathbf{p}}{(2\pi)^3}. \tag{3.27}$$

The rate $(dN_{\mathbf{k}}/dt)_-$ at which the number of waves decreases is determined by the number of particles in the level $\mathbf{p}-\mathbf{k}$ from which absorption of a wave with momentum $\mathbf{k}$ can take place:

$$\left(\frac{dN_{\mathbf{k}}}{dt}\right)_- = -\int w_{\mathbf{p}}(\mathbf{k})N_{\mathbf{k}}\,\Phi_{\mathbf{p}-\mathbf{k}}\,\frac{d^3\mathbf{p}}{(2\pi)^3}. \tag{3.28}$$

The balance between emission and absorption gives us an equation of motion or kinetic equation for the number of waves:

$$\frac{dN_{\mathbf{k}}}{dt} = \left(\frac{dN_{\mathbf{k}}}{dt}\right)_{+} + \left(\frac{dN_{\mathbf{k}}}{dt}\right)_{-} = q_{\mathbf{k}} + 2\gamma_{\mathbf{k}}N_{\mathbf{k}}, \tag{3.29}$$

where $q_{\mathbf{k}}$ describes the spontaneous Vavilov-Cherenkov emission process; it is an expression which is independent of the number of waves $N_{\mathbf{k}}$:

$$q_{\mathbf{k}} = \int w_{\mathbf{p}}(\mathbf{k})\,\Phi_{\mathbf{p}}\,\frac{d^3p}{(2\pi)^3}. \tag{3.30}$$

The second term on the right-hand side of Eq.(3.29), which is proportional to the number of quanta $N_{\mathbf{k}}$, describes the Landau damping (compare Eq.(3.8)). Taking into account the fact that $k \ll p$, we find

$$\gamma_{\mathbf{k}} = \frac{1}{2}\int w_{\mathbf{p}}(\mathbf{k})\left(\mathbf{k}\cdot\frac{\partial\Phi_{\mathbf{p}}}{\partial\mathbf{p}}\right)\frac{d^3p}{(2\pi)^3}. \tag{3.31}$$

Comparing this expresion with Eqs.(2.110) and (2.113) for Landau damping we find the probability for Vavivilov-Cherenkov emission of longitudinal plasma waves (superscript σ) by particles of the kind α:

$$w_{\mathbf{p}}^{\alpha,\sigma}(\mathbf{k}) = \frac{8\pi^2}{k^2}\frac{e_\alpha^2}{\left|\partial\varepsilon_{\mathbf{k}}/\partial\omega\right|_{\omega=\omega_{\mathbf{k}}^\sigma}}\,\delta\left(\omega_{\mathbf{k}}^\sigma - (\mathbf{k}\cdot\mathbf{v})\right). \tag{3.32}$$

The same expression can be obtained from the power Q^{sp} of the spontaneous emission:

$$Q^{\mathrm{sp}} = \int \omega_{\mathbf{k}}\left(\frac{dN_{\mathbf{k}}}{dt}\right)^{\mathrm{sp}}\frac{d^3k}{(2\pi)^3} = \int q_{\mathbf{k}}\omega_{\mathbf{k}}\frac{d^3k}{(2\pi)^3}$$

$$= \int \omega_{\mathbf{k}}w_{\mathbf{p}}(\mathbf{k})\frac{d^3k}{(2\pi)^3}\,\Phi_{\mathbf{p}}\frac{d^3p}{(2\pi)^3}. \tag{3.33}$$

This expression contains the emission power of a single particle:

$$Q_{\mathbf{p}} = \int \omega_{\mathbf{k}}w_{\mathbf{p}}(\mathbf{k})\frac{d^3k}{(2\pi)^3}, \tag{3.34}$$

averaged over the particle distribution:

$$Q = \int Q_{\mathbf{p}}\Phi_{\mathbf{p}}\frac{d^3p}{(2\pi)^3}. \tag{3.35}$$

Equation (3.34) has a simple meaning: the power emitted by a single particle is equal to the probability for the emission multiplied by the energy of the emitted wave and integrated over the phase volume of the emitted waves.

The results of the present section show that Landau damping can be interpreted simply as stimulated Cherenkov emission of longitudinal waves. However, the way we derived this result shows that our present discussion contains not only an interpretation of results which had been obtained already earlier, but also a generalisation of them. Indeed, the kinetic equation for the distribution function of the waves implies that both the wave distribution function N_k and the particle distribution function Φ_p vary slowly in time and space over times of the order of the period and distances of the order of the wavelength of the waves. However, in calculating the Landau damping we had assumed that Φ_p was the initial particle distribution function. We had earlier mentioned that this function may vary slowly, but we have now seen that this slow variation is not arbitrary, but follows directly from the stimulated emission and absorption processes.

We may thus state that, to begin with, we postulated a kinetic equation for the waves starting from a general idea about the existence of transition probabilities and then found those probabilities from a correspondence principle. In fact, we can use the general equation to formulate an initial value problem when the particle distribution is changed slightly and its change is determined by the initial distribution function. This was the problem solved by Landau. Following this we were able to use an expression obtained by Landau to find the emission probability which can then be substituted into more general equations which describe the changes in the wave and particle distribution functions due to the same emission pocess. This *generalised concept of Landau damping* is called *quasi-linear Landau damping*. The meaning of this term is that in this approach the particle distribution may depend on the wave intensity and therefore the dependence of the damping on the wave intensity will no longer be linear. In the quasi-linear approach the particle distribution will change because of the emission and absorption processes since the population of the energy levels changes when stimulated processes are taking place. All these changes are governed by the same probability.

3.5 Quasi-linear Equation for the Particle Distribution

If we want to write down a kinetic equation for the evolution of the particle distribution function, taking into account the stimulated and spontaneous emission of waves in the Vavilov-Cherenkov effect, we must look for the balance equation for particles of a given momentum **p**. We consider thereto emission processes involving waves with a given momentum **k** and then integrate over all possible **k**. Waves with a momentum **k** can be emitted both by a particle with momentum **p** and by a particle with momentum **p** + **k** and can be absorbed by a particle with momentum **p** − **k** and by a particle with momentum **p** (see Fig.3.4).

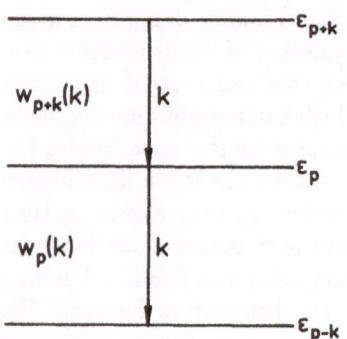

Fig. 3.4. Change in particle energy in resonant wave emission processes

The probability is determined by the particle momentum in the upper energy level. The transitions between the states with momenta $\mathbf{p} + \mathbf{k}$ and $\mathbf{p}$ therefore have the probability $w_{\mathbf{p}+\mathbf{k}}(\mathbf{k})$, while those between the state with momenta $\mathbf{p}$ and $\mathbf{p} - \mathbf{k}$ have the probability $w_{\mathbf{p}}(\mathbf{k})$. The sum of the stimulated and spontaneous emission has a factor $N_{\mathbf{k}} + 1$ and the stimulated absorption has a factor $N_{\mathbf{k}}$. The rate of decrease of the distribution function corresponding to particles in the level with momentum $\mathbf{p}$ will thus be

$$\left(\frac{d\Phi_{\mathbf{p}}}{dt}\right)_{-} = -\int \left[w_{\mathbf{p}}(\mathbf{k})(N_{\mathbf{k}}+1)\Phi_{\mathbf{p}} + w_{\mathbf{p}+\mathbf{k}}(\mathbf{k})N_{\mathbf{k}}\Phi_{\mathbf{p}}\right] \frac{d^3\mathbf{k}}{(2\pi)^3}. \quad (3.36)$$

and the rate of increase of the same distribution function will be:

$$\left(\frac{d\Phi_{\mathbf{p}}}{dt}\right)_{+} = +\int \left[w_{\mathbf{p}}(\mathbf{k})N_{\mathbf{k}}\Phi_{\mathbf{p}-\mathbf{k}}\right.$$

$$\left. + w_{\mathbf{p}+\mathbf{k}}(\mathbf{k})(N_{\mathbf{k}}+1)\Phi_{\mathbf{p}+\mathbf{k}}\right] \frac{d^3\mathbf{k}}{(2\pi)^3}. \quad (3.37)$$

When writing down the balance equation we shall use an expansion in the ratio of the wave to the particle momentum; we then find a diffusion equation (remember that we sum over repeated indices):

$$\frac{d\Phi_{\mathbf{p}}}{dt} = \left(\frac{d\Phi_{\mathbf{p}}}{dt}\right)_{+} + \left(\frac{d\Phi_{\mathbf{p}}}{dt}\right)_{-} = \frac{\partial}{\partial p_i} D_{ij} \frac{\partial}{\partial p_j} \Phi_{\mathbf{p}} + \frac{\partial}{\partial p_i} F_i \Phi_{\mathbf{p}}. \quad (3.38)$$

The first term on the right-hand side of Eq.(3.38) corresponds to the stimulated and the second term to the spontaneous processes, and

$$D_{ij} = \int k_i k_j w_{\mathbf{p}}(\mathbf{k}) N_{\mathbf{k}} \frac{d^3\mathbf{k}}{(2\pi)^3}, \quad (3.39)$$

$$F_i = \int k_i w_{\mathbf{p}}(\mathbf{k}) \frac{d^3\mathbf{k}}{(2\pi)^3}. \quad (3.40)$$

It is clear that the stimulated processes will dominate if the energy density of the waves is sufficiently large, to be precise, if it is significantly higher than the level of the thermal fluctuations. This is, in fact, quite a weak condition:

$$\frac{W}{nT} \gg \frac{1}{N_{\mathrm{d}}}. \tag{3.41}$$

If (3.41) holds one can neglect the spontaneous processes both in the equation for the distribution function of the waves and in the equation for the distribution function of the particles. Such a system of equations is called a set of *quasi-linear equations*. The total time derivatives in those equations can be written in the form:

$$\frac{d\Phi_{\mathbf{p}}}{dt} = \frac{\partial\Phi_{\mathbf{p}}}{\partial t} + \left(\mathbf{v} \cdot \frac{\partial\Phi_{\mathbf{p}}}{\partial\mathbf{r}}\right); \qquad \frac{dN_{\mathbf{k}}}{dt} = \frac{\partial N_{\mathbf{k}}}{\partial t} + \left(\mathbf{v}_{\mathrm{gr}} \cdot \frac{\partial N_{\mathbf{k}}}{\partial\mathbf{r}}\right), \tag{3.42}$$

where $\mathbf{v}_{\mathrm{gr}} = d\omega_{\mathbf{k}}/d\mathbf{k}$ is the group velocity of the oscillations. We can thus write the set of quasi-linear equations in the form:

$$\frac{\partial\Phi_{\mathbf{p}}}{\partial t} + \left(\mathbf{v} \cdot \frac{\partial\Phi_{\mathbf{p}}}{\partial\mathbf{r}}\right) = \int \left(\mathbf{k} \cdot \frac{\partial}{\partial\mathbf{p}}\right) w_{\mathbf{p}}(\mathbf{k}) N_{\mathbf{k}} \left(\mathbf{k} \cdot \frac{\partial\Phi_{\mathbf{p}}}{\partial\mathbf{p}}\right) \frac{d^3\mathbf{k}}{(2\pi)^3}; \tag{3.43}$$

$$\frac{\partial N_{\mathbf{k}}}{\partial t} + \left(\mathbf{v}_{\mathrm{gr}} \cdot \frac{\partial N_{\mathbf{k}}}{\partial\mathbf{r}}\right) = \int w_{\mathbf{p}}(\mathbf{k}) N_{\mathbf{k}} \left(\mathbf{k} \cdot \frac{\partial\Phi_{\mathbf{p}}}{\partial\mathbf{p}}\right) \frac{d^3\mathbf{p}}{(2\pi)^3}. \tag{3.44}$$

We said that the spontaneous terms can be neglected if the level of the wave fluctuations is significantly higher than the thermal level. This does not mean, however, that on the thermal level all fluctuations are connected with spontaneous Cherenkov emission. There are also other fluctuations of the same order of magnitude, such as, for example, those due to the fields of individual particles when they approach one another closely. This means that whereas spontaneous processes describe effects at the level of thermal fluctuations, they do not describe all the thermal fluctuations. Hence, it is not possible to take into account only the spontaneous processes and neglect other fluctuations of the same order of magnitude. On the other hand, if we take into account only the stimulated processes we get a closed, self-consistent set of equations. These equations satisfy the conservation laws of the total energy of the waves and the particles and of the total momentum of the waves and the particles. These quantities were defined earlier by Eqs.(3.4) to (3.6). We have

$$\frac{d}{dt}(E + W) = 0; \qquad \frac{d}{dt}(\mathbf{P} + \boldsymbol{\Pi}) = 0. \tag{3.45}$$

The validity of the first conservation law can be proved by multiplying Eq.(3.43) by $\varepsilon_{\mathbf{p}}$ and integrating the resulting equation over $\mathbf{p}$, multiplying Eq.(3.44) by $\omega_{\mathbf{k}}$ and integrating the resulting equation over $\mathbf{k}$, and finally adding the two expressions obtained in that way. The second conservation law is obtained similarly, multiplying Eq.(3.43) by **p**, integrating over **p**, multiplying Eq.(3.44) by **k**, integrating over **k**, and finally adding the two results. This result clearly proves the earlier statement that the wave momentum is given by Eq.(3.6), or the statement that defining the wave distribution, starting from the expressions for the wave energy and the wave momentum, gives the same result. The conservation laws (3.45) can be used to prove that in an unstable case energy is, indeed, transferred from particles to waves; this can be used to find the amount of energy transferred to the waves for the case when the initial and the final particle distributions are known.

Sometimes it is useful to write the quasi-linear equations in terms of the correlation functions of the random fields. This can be done by substituting the probability (3.32) into Eqs.(3.43) and (3.44) and using (3.26) to express the distribution function of the waves in term of the correlation function of the waves. The quasi-linear set of equations then will have the form

$$
\frac{d\Phi_{\mathbf{p}}}{dt} = 2\pi e^2 \int |E|_{\mathbf{k}}^2 \frac{d^3 k}{k^2} \left(\mathbf{k} \cdot \frac{\partial}{\partial \mathbf{p}} \right) \delta(\omega_{\mathbf{k}} - (\mathbf{k} \cdot \mathbf{v})) \left(\mathbf{k} \cdot \frac{\partial}{\partial \mathbf{p}} \right) \Phi_{\mathbf{p}}, \quad (3.46)
$$

$$
\frac{d|E|_{\mathbf{k}}^2}{dt} = \frac{8\pi^2 e^2 |E|_{\mathbf{k}}^2}{k^2 (\partial \varepsilon_{\mathbf{k}}/\partial \omega)_{\omega = \omega_{\mathbf{k}}}} \int \delta(\omega_{\mathbf{k}} - (\mathbf{k} \cdot \mathbf{v})) \left(\mathbf{k} \cdot \frac{\partial \Phi_{\mathbf{p}}}{\partial \mathbf{p}} \right) \frac{d^3 p}{(2\pi)^3}. \quad (3.47)
$$

In this form the equations do not show any trace of a quantum description; we are obviously dealing with completely classical effects. However, we still think that using a partial quantum description has some advantages, as was illustrated above. For example, it is not so easy to find an expression for the wave momentum, when there is no field momentum – as happens in the case of electrostatic waves – but it is trivial from a quantum point of view.

3.6 Quasi-linear Relaxation of Beams in a Plasma

In a quantum description one can with advantage use the analogy between amplification in a two-level system and the instability of a beam in a plasma. An inverse population in a two-level system leads to an instability which is the basis of quantum amplifiers and generators. It is well known that irradiation of a two-level system can lead to the equalisation of the populations and to a saturation of the amplification. Similar processes can take place in the case of a beam plasma instability. The relation $\partial \Phi_p / \partial p > 0$ which is necessary for a beam plasma instability corresponds to a population inversion, since in this case the number of particles with higher energies is larger than the number of particles with lower energies. Just as in a two-level system the populations tend to become equalised, when the beam plasma instability develops there is a diffusion of the beam distribution function in momentum space in such a way that the derivative $\partial \Phi_p / \partial p$ decreases. This is described by the quasi-linear equations. In order to find a qualitative picture of the various stages of beam relaxation in a plasma we first must answer the question why an isotropic distribution is stable even though there can be parts of the distribution function with a positive derivative, that is, with an inverse population. The reason is that, indeed, these parts of the distribution contribute to the excitation of plasma oscillations, but for the same wavevector there are also particles which correspond to a negative derivative – since the distribution function describes a finite number of particles, for high velocities it must have a negative derivative. The part of the distribution function with a positive derivative gives amplification, but the part with a negative derivative leads to damping. For isotropic distributions damping always dominates, and this is the reason why an isotropic distribution is always stable. For an anisotropic beam the situation is quite different since the waves with a given wavevector interact only with particles with a well defined momentum. Therefore, a wave with a given wavevector is either damped or amplified. The waves for which the derivative of the initial distribution is positive are the first to be excited. Particles will start to diffuse towards lower momenta, diminishing the derivative of the distribution function. Particles with momenta lower than those of the initial beam distribution start to excite waves with smaller phase velocities which means that there is a continuing spread of the beam distribution towards lower velocities. One might say that there is some kind of wave in velocity space which propagates towards lower velocities and which spreads out the beam distribution. This process will terminate when this wave reaches the thermal distribution since at that moment a rapidly increasing Landau damping will prevent the waves from growing. We show in Fig.3.5 the subsequent stages of the evolution of the beam distribution function for times $t_3 > t_2 > t_1$, where t_1 corresponds to the initial distribution.

This evolution of the beam distribution function will finally lead to the formation of a plateau, between the beam velocity v_b and the thermal velocity v_{Te}, where the distribution function will be constant. From the way we normalise the distribution function we have

$$\int \Phi_v \, dv \; = \; n_b, \tag{3.48}$$

where the integration is over the velocities $v \gg v_{Te}$ (we are assuming here that $v_b \ll v_{Te}$ and $\delta v_b \ll v_b$).

Let $\Phi_v(\infty)$ be the distribution function which is finally established after the quasi-linear relaxation process is finished; this will be a constant between v_{Te} and v_b – bear in mind that we have assumed that $v_b \gg v_{Te}$. From the law of conservation of the total number of resonant (beam) particles we then find, using Eq.(3.48),

$$\Phi_v(\infty) \; = \; \frac{n_b}{v_b}. \tag{3.49}$$

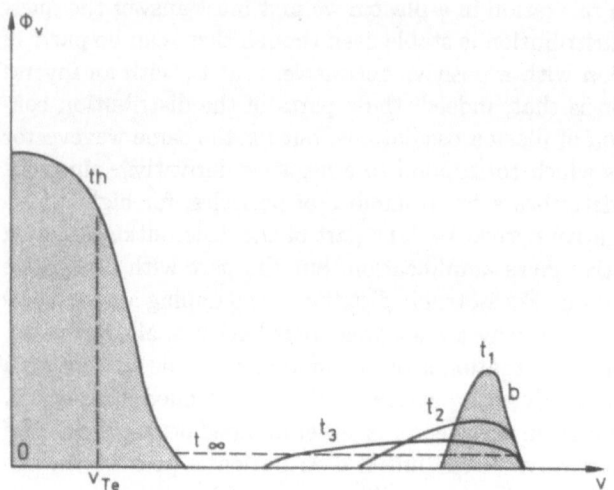

Fig. 3.5. Evolution of a beam particle distribution during quasi-linear relaxation

Is it possible to answer the question of what part of the initial beam energy has been converted into plasma wave energy during the quasi-linear relaxation process? The initial energy density was

$$E(0) \cong \tfrac{1}{2} n_b m_e v_b^2, \tag{3.50}$$

where we have taken into account the fact that $\delta v_b \ll v_b$. The final beam energy density when the plateau has been established will be

$$E(\infty) = \int \tfrac{1}{2} m_e v^2 \Phi_v(\infty) \, dv \simeq \tfrac{1}{6} n_b m_e v_b^2. \tag{3.51}$$

Using the energy conservation law we find for the energy transferred to the plasma oscillations during the quasi-linear relaxation the expression

$$W(\infty) = E(0) - E(\infty) = \tfrac{1}{3} n_b m_e v_b^2 = \tfrac{2}{3} E(0). \tag{3.52}$$

The energy converted into plasma oscillations is thus comparable to the initial beam energy.

The one-dimensional model used to derive this conclusion does not take into account waves excited at some angle to the direction of the beam velocity. These waves have a lower growth rate. Their effect on the particle distribution, which is felt mainly after the one-dimensional relaxation has been finished, will proceed to change the beam distribution function producing a more isotropic distribution until the degree of anisotropy will be small and below the threshold at which excitation takes place. After that the excited waves will be only absorbed by the beam. The inverse of the characteristic time τ_{ql} for the whole relaxation process will not greatly differ from the rate at which wave excitation takes place in the final stage of the relaxation process when the spread of the beam distribution will be of the order of the initial value of the beam velocity:

$$\frac{1}{\tau_{\text{ql}}} \approx \omega_{\text{pe}} \frac{n_{\text{b}}}{n_{\text{e}}}. \tag{3.53}$$

For many applications this is a very short time.

We can also answer the question of what will be the final distribution of the waves over wavenumbers as a result of the one-dimensional development of the quasi-linear relaxation. For this purpose we introduce the energy distribution of the oscillations normalised with respect to their phase velocity, as follows:

$$\int W_{v_{\text{ph}}} \, dv_{\text{ph}} = W \simeq \int \frac{dk}{2\pi} |E|_k^2 = \int \omega_{\text{pe}} \, dv_{\text{ph}} \frac{|E|_k^2}{2\pi v_{\text{ph}}^2}, \tag{3.54}$$

where we have used the fact that $v_{\text{ph}} = \omega_{\text{pe}}/k$, $k = k_x$, and

$$W_{v_{\text{ph}}} = \frac{k^2 |E|_k^2}{2\pi \omega_{\text{pe}}}, \qquad |E|_k^2 = \frac{2\pi v_{\text{ph}}^2}{\omega_{\text{pe}}} W_{v_{\text{ph}}}. \tag{3.55}$$

It is useful to introduce such a distribution since in the one-dimensional case the resonance condition has the form $v_{\text{ph}} = v$ so that in all expressions the quantity W_v enters. Using the particle and wave distributions which we have introduced we can write the quasi-linear system of equations (3.46) and (3.47) in the form

$$\frac{\partial}{\partial t} \Phi_v = \frac{\pi \omega_{\text{pe}}}{n_{\text{e}} m_{\text{e}}} \frac{\partial}{\partial v} v W_v \frac{\partial}{\partial v} \Phi_v, \tag{3.56}$$

$$\frac{\partial}{\partial t} W_v = \frac{\pi \omega_{\text{pe}}}{n_{\text{e}}} v^2 W_v \frac{\partial}{\partial v} \Phi_v. \tag{3.57}$$

From these equations it follows that

$$\frac{\partial}{\partial t} \Phi_v = \frac{1}{m_{\text{e}}} \frac{\partial}{\partial v} \frac{1}{v} \frac{\partial}{\partial t} W_v. \tag{3.58}$$

This equation can be integrated with respect to time and we find

$$\Phi_v(\infty) - \Phi_v(0) = \frac{1}{m_{\text{e}}} \frac{\partial}{\partial v} \frac{1}{v} \left(W_v(\infty) - W_v(0) \right). \tag{3.59}$$

Neglecting the energy $W_v(0)$ of the initial oscillations as compared to their final energy $W_v(\infty)$ and considering the range of velocities well above v_{Te}, but below v_{b}, we find

$$m_{\text{e}} \Phi_v(\infty) v = m_{\text{e}} n_{\text{b}} \frac{v}{v_{\text{b}}} = \frac{1}{v} W_v(\infty), \tag{3.60}$$

where we have used Eq.(3.49), or

$$W_v(\infty) = m_e n_b \frac{v^2}{v_b}. \qquad (3.61)$$

From this equation we find again the result which we obtained earlier from the energy conservation law:

$$W(\infty) = \int W_v(\infty)\, dv = \tfrac{1}{3} m_e n_b v_b^2. \qquad (3.62)$$

Using Eq.(3.55) we can write the spectrum (3.61) as a function of the wavenumber:

$$W_k(\infty) = \frac{|E|_k^2(\infty)}{2\pi} = W_{v_{ph}}(\infty) \frac{v_{ph}^2}{\omega_{pe}} = \frac{m_e n_b \omega_{pe}^3}{v_b} \frac{1}{k^4}. \qquad (3.63)$$

This spectrum is valid for phase velocities which are neither close to the initial beam velocity nor close to the electron thermal velocity:

$$v_{Te} \ll \frac{\omega_{pe}}{k} = v_{ph}, \qquad \delta v_b \ll \frac{\omega_{pe}}{k} - v_b = v_{ph} - v_b. \qquad (3.64)$$

This spectrum which is created after the quasi linear relaxation and which is proportional to $1/k^4$ is a power-law spectrum and its main energy is concentrated in the range of phase velocities close to the initial beam velocity. We shall show later on that there is a non-linear process which can compete with the quasi-linear relaxation and which slows down the beam relaxation process in the plasma and changes the spectrum of the waves excited by the beam. The criterion for whether the non-linear or the quasi-linear process will dominate depends on the beam velocity – and under certain conditions on the beam density.

3.7 Quasi-linear Description and Fluctuating Fields

We shall now consider the fluctuations stimulated by strong random plasma fields and we shall neglect the natural statistical fluctuations which exist even when there are no strong plasma fields present. We shall be interested in the changes in the particle distribution function Φ_p which are linear in the intensity, that is, the square of the amplitude, of these fields. The equation for these changes is a generalisation of the quasi-linear equation for the average distribution function. Let us return to the general description of fluctuations and write down Eq.(2.61) for the average distribution function Φ_p and Eq.(2.67) for the fluctuating part δf_p of the distribution function which is linear in the strength of the fluctuating field $\mathbf{E}$ – we assume that there is no regular field present:

$$\frac{\partial}{\partial t}\varPhi_{\mathbf{p}} + \left(\mathbf{v}\cdot\frac{\partial}{\partial \mathbf{r}}\right)\varPhi_{\mathbf{p}} = -e\left\langle\left(\mathbf{E}\cdot\frac{\partial}{\partial \mathbf{p}}\right)\delta f_{\mathbf{p}}\right\rangle, \qquad (3.65)$$

$$\delta f_{\mathbf{p},\mathbf{k}} = \frac{e\left(\mathbf{E}_{\mathbf{k}}\cdot\frac{\partial \varPhi_{\mathbf{p}}}{\partial \mathbf{p}}\right)}{\mathrm{i}\big(\omega - (\mathbf{k}\cdot\mathbf{v}) + \mathrm{i}0\big)}. \qquad (3.66)$$

We have here taken the causality principle into account – as we did when discussing Landau damping – and considered only that part of the fluctuations which are stimulated by the fields and neglected those which are independent of the field – we shall consider those in detail in the next chapter. We are assuming that the field strength is sufficiently strong so that we are able to neglect the effects connected with the natural fluctuations in the particle distribution which occur even when there are no fields present. This is similar to the neglect of spontaneous processes when compared to the stimulated ones. We consider the average distribution function in Eq.(3.66) to be slowly varying in time and space as long as these variations are very small over both a characteristic period and a characteristic wavelength of the field fluctuations. Substituting (3.66) into (3.65) we find:

$$\frac{d\varPhi_{\mathbf{p}}}{dt} = -\frac{e^2}{\mathrm{i}}\frac{\partial}{\partial p_i}\int\frac{d^4k\,d^4k'}{\omega - (\mathbf{k}\cdot\mathbf{v}) + \mathrm{i}0}$$
$$\times\langle E_{i,\mathbf{k}}E_{j,\mathbf{k}'}\rangle\frac{\partial\varPhi_{\mathbf{p}}}{\partial p_j}\mathrm{e}^{\mathrm{i}[(\{\mathbf{k}+\mathbf{k}'\}\cdot\mathbf{r})-(\omega+\omega')t]}. \qquad (3.67)$$

We shall assume that to a first approximation the correlation functions of the fields satisfy the conditions for a stationary and homogeneous system. They will then contain factors $\delta(\omega + \omega')$ and $\delta(\mathbf{k} + \mathbf{k}')$ – see Eq.(3.16). We thus get

$$\frac{d\varPhi_{\mathbf{p}}}{dt} = -\frac{e^2}{\mathrm{i}}\frac{\partial}{\partial p_i}\int\frac{d^3k\,d\omega}{\omega - (\mathbf{k}\cdot\mathbf{v}) + \mathrm{i}0}\frac{k_ik_j}{k^2}|E|^2_{\mathbf{k},\omega}\frac{\partial\varPhi_{\mathbf{p}}}{\partial p_j}, \qquad (3.68)$$

where we have used the fact that

$$|E|^2_{-\mathbf{k},-\omega} = |E|^2_{\mathbf{k},\omega}. \qquad (3.69)$$

By carrying out a change of variables $\mathbf{k} \to -\mathbf{k}$, $\omega \to -\omega$ we can check that the real part of the resonant denominator does not make a contribution to expression (3.69) and we therefore find

$$\frac{d\varPhi_{\mathbf{p}}}{dt} = \pi e^2\frac{\partial}{\partial p_i}\int\frac{k_ik_j}{k^2}|E|^2_{\mathbf{k},\omega}\,\delta\big(\omega - (\mathbf{k}\cdot\mathbf{v})\big)\frac{\partial\varPhi_{\mathbf{p}}}{\partial p_j}\,d^3k\,d\omega. \qquad (3.70)$$

This equation is more general than the quasi-linear equation as it contains an arbitrary correlation function rather than the one which describes the plasma oscillations – which has a peak at the frequency $\omega_{\mathbf{k}}$ of the oscillations. This equation can therefore be used in the case when strong non-linear interactions smooth out these peaks of the linear waves. That case would correspond to strong turbulence. If we substitute into (3.70) the expression (3.20) valid for plasma oscillations with a peak at ω_k, we get the quasi-linear equation which we derived earlier in a different way:

$$\frac{d\Phi_p}{dt} = 2\pi e^2 \frac{\partial}{\partial p_i} \int \frac{k_i k_j}{k^2} |E|^2_k \, \delta(\omega_k - (\mathbf{k} \cdot \mathbf{v})) \frac{\partial \Phi_p}{\partial p_j} \, d^3\mathbf{k}. \qquad (3.71)$$

Problems

1. Find the power of the spontaneous Vavilov-Cherenkov emission of Langmuir and ion-sound waves for a particle moving in a plasma with a velocity v which is much higher than v_{Te}.
2. Langmuir waves have a relatively narrow spectrum ($\delta k \ll k$) with phase velocities corresponding to the tail of a thermal distribution. Consider the problem of the saturation of Landau damping.
3. A weak beam of particles is injected into a half-space occupied by a plasma. Find the distance over which the one-dimensional quasi-linear relaxation of the beam occurs and find the energy density of the plasma oscillations which are excited in the process. Assume that $n_b \ll n_e$, $v_b \gg v_{Te}$, and $\delta v_b \ll v_b$.

4 Fluctuations and Particle Collisions

4.1 Introduction.
Averaged Distributions and Fluctuations

An important problem in plasma physics is the description of particle collisions and another problem is the question of what is the nature of the statistical fluctuations in a plasma when there are no external fields or strong collective oscillations. At first sight these two problems are unrelated but, in fact, they are related to one another. In discussing fluctuations when there are no fields present we shall consider the case of an arbitrary particle distribution which could be any non-equilibrium distribution. If the distribution is a thermal distribution function the fluctuations will be the thermal fluctuations. On the other hand, in the case of a non-equilibrium distribution the fluctuations are not thermal and we can call them the natural fluctuations of a large system of free particles. We shall define this term more precisely in what follows. For the moment we just mention that these fluctuations are closely connected with particle collisions in the plasma. We shall describe these relations in the present chapter and we shall show that collisions between particles in a plasma are quite different from collisions between particles in the absence of other plasma particles.

Landau was the first to consider particle collisions in a plasma and their effect on the distribution function of the plasma particles. He assumed that one may use for those collisions the known cross-sections for charged particle collisions when there is no plasma. This was a natural assumption at that time (1937). Landau derived an additional term $I_{\mathbf{p}}^{col}$ in the kinetic equation for the particle distribution function; it describes the change in the distribution function due to collisions and is called the *Landau collision integral*. It usually stands on the right-hand side of the kinetic equation:

$$\frac{\partial f_{\mathbf{p}}}{\partial t} + \left(\mathbf{v} \cdot \frac{\partial f_{\mathbf{p}}}{\partial \mathbf{r}}\right) + \left(e\mathbf{E} \cdot \frac{\partial f_{\mathbf{p}}}{\partial \mathbf{p}}\right) = I_{\mathbf{p}}^{col}. \tag{4.1}$$

Equation (4.1) differs from the kinetic equation (2.58) which we considered earlier by the presence of the collision integral on the right-hand side. One might think therefore that Eq.(2.58) is an approximation and Eq.(4.1) the more general one. This, however, is not correct. To see this we once more write down Eq.(2.58):

$$\frac{\partial f_{\mathbf{p}}}{\partial t} + \left(\mathbf{v} \cdot \frac{\partial f_{\mathbf{p}}}{\partial \mathbf{r}} \right) + \left(e\mathbf{E} \cdot \frac{\partial f_{\mathbf{p}}}{\partial \mathbf{p}} \right) = 0. \tag{4.2}$$

It is not very difficult to see that Eq.(4.2) is quite general and takes into account the particle collisions. Indeed, we say that charged particles collide when one of them approaches another one closely and the electric field of one of the colliding particles changes the trajectory of the other particle. In Eq.(4.2) all fields are taken into account, including those which lead to particle collisions. The solution of the problem was found by Bogolyubov who proposed a chain of related systems of correlation functions and showed how one can "derive" Eq.(4.1) from Eq.(4.2). It is impossible that Eqs.(4.1) and (4.2) are correct simultaneously. It appears that the meaning of the distribution functions in these two equations is different. The misunderstanding lies in that we are not using the correct notation. Let $f_{\mathbf{p}}$ be the exact particle distribution function which takes into account the natural fluctuations of the distribution, while $\Phi_{\mathbf{p}}$ is the distribution function, averaged over those fluctuations. The correct equation (4.1) should contain the latter distribution function, that is, it should be written in the form

$$\frac{\partial \Phi_{\mathbf{p}}}{\partial t} + \left(\mathbf{v} \cdot \frac{\partial \Phi_{\mathbf{p}}}{\partial \mathbf{r}} \right) + \left(e\mathbf{E} \cdot \frac{\partial \Phi_{\mathbf{p}}}{\partial \mathbf{p}} \right) = I_{\mathbf{p}}^{\text{col}}. \tag{4.3}$$

The method developed by Bogolyubov makes it possible to obtain Eq.(4.3) from Eq.(4.2). It is natural that after averaging the fluctuating fields – which appear when particles approach each other closely – are smoothened out, but the average effect of these fluctuations is described by the appearance of a collision integral on the right-hand side of Eq.(4.3). The fields in Eqs.(4.2) and (4.3) are different: in Eq.(4.2) $\mathbf{E}$ is the exact field, including the fluctuating fields, while in Eq.(4.3) $\mathbf{E}$ is the averaged field.

There exists now a large literature devoted to calculating the collision integrals, using averaging over the fluctuations. In fact, for a large system there is no other consistent method for solving the problem. We shall here give a very simple description of this procedure, bearing in mind that the physical interpretation of the final result will be used whenever we consider non-linear and emission processes. Moreover, it turns out that the same physical effects appear in non-linear interactions as in collisions.

The main question in the case of collisions is whether it is possible to use for the particle collisions in a plasma the well known cross-sections for binary Coulomb particle collisions when there are no other particles present. A doubt arises because the fluctuations in a plasma are produced by all the particles. Before proceeding further we shall give a brief summary of the results to be obtained in the present section. The main result is that there are essential differences between the cross-sections for collisions in a plasma and the well known ones for free particles when there are no other particles present; this result is not connected with the interactions between the particles but with collective effects produced by the other particles. We shall see later that a similar effect exists in the case of non-linear interactions. At first sight this result may seem to be surprising since the interactions and collisions are weak because the parameter N_d is large ($N_d \gg 1$). On the other hand, this same parameter, which shows that the collisions are rare, also shows that there are many particles inside the Debye sphere inside which the field is not screened so that some of them could affect the collision processes of other particles through the field they produce. The screening of the fields of the colliding particles by other particles in the Debye sphere is a collective effect produced by these particles and it obviously may change the collision cross-sections. However, the screening effect is produced on average so that fluctuations may be important. Landau pointed out that the collision integral calculated using the Coulomb collision cross-sections diverges so that from physical arguments it is necessary to introduce a cut-off for the integration. The collision integral which takes the collective effects during collisions into account is called the *Landau-Balescu collision integral*; it carries Landau's name because he was the first to consider collisions in a plasma and to take into account some of the collective effects – without, however, proving their existence – and it carries Balescu's name as he was the first to calculate the effect of particle collisions taking all collective effects into account and showing that there are other effects besides those leading to the Debye screening of the fields of the colliding particles. We should add that it would not be correct to say that Balescu's only merit was to prove what Landau had already intuitively taken into account – even if we concentrate on the problem of Debye screening.

We are dealing here with a physical, rather than just a mathematical problem. In fact, it looks to be relatively unimportant from the point of view of actual calculations since the cut-off parameter enters the result under a logarithm sign so that its actual value does not greatly affect the calculations. However, there are, first of all, examples of other cross-sections, important for non-linear interactions, where these cross-sections change by several orders of magnitude due to collective effects. Secondly, the whole problem is of a general physical importance since the effect of changing cross-sections because of collective effects permeates the whole of modern plasma physics, especially non-linear plasma physics. To their surprise research workers found similar, but even more important effects in non-linear interactions. When the effect was first found for collisions it was thought to be specifically a phenomenon connected with collisions and no great attention was paid to the fact that the collision cross-sections had been drastically changed due to the presence of other particles – the total effect, when integrated over all impact parameters, did not depend strongly on collective effects, since it led only to a factor under a logarithm sign, even though for some impact parameters the cross-section depended strongly on the collective effects. In fact, Landau considered it just to be a reason for applying a cut-off procedure. However, after similar effects were discovered in such phenomena as non-linear interactions and bremsstrahlung, it was realised that these effects are fundamental in nature. We shall attempt to emphasise throughout all our discussions this idea of a fundamental importance of the effect – even at the present time this is not always fully realised.

One can only understand this fundamental effect in the framework of a theory of the fluctuations. First of all, when one describes a large system of particles it is necessary to use some averaging procedures which implies that we renounce the idea of giving a description of the details, since this is impossible for a large system. Using an average description implies that there may be deviations from it: these are the fluctuations. Without fluctuations the system would be homogeneous and stationary and it would not produce any fields or emit or scatter waves.

If we use an average description for the particles we can get any effective interactions between them only through the fluctuations. Mathematically the problem reduces to deriving Eq.(4.3) from a fluctuation theory in a form such that the collision integral on the right-hand side contains only the average distribution function Φ_p. It turns out that the physical meaning of Φ_p is somewhat different from what one might think: it does not describe single particles, but a collection of particles – put somewhat differently: it describes the particles together with the effect of their interactions, that is, it describes *"dressed" particles* or *quasi-particles*. We shall see in what follows that such quasi-particles are particles surrounded by a *screening cloud*, sometimes called a screening shell. In some approximation such a cloud is a Debye screening cloud, but the rate of screening may depend on the particle velocity, that is, it may be dynamic screening. We shall prove these statements in what follows; at the moment we made them in order to have a general picture in our minds before turning to a detailed mathematical description.

We must mention here an important point: the averaging cannot change the total number of particles so that the number of screened ("dressed") particles must be equal to the total number of particles. The question then arises: where do the particles which screen the fields of other particles come from? The answer is that the screening is produced by the same particles during their fluctuations. The real, bare particles take part in both motions, the averaged one and the fluctuating one. During their fluctuations the particles are able to screen the field of other particles which thus become quasi-particles. We have thus reached a stage where we need to change our picture of a plasma as a collection of free particles. In fact, it is better to consider it as a collection of classical neutral atoms, since the field of each paricle is screened. The "dressed" particles behave in many aspects like neutral atoms. In this context it is unimportant that the shell which screens the central charge is produced statistically by other particles which means that it moves closer to or further away from the central particle all the time. The behaviour of this classical atom is similar to the normal neutral atoms where the screening shell – the electron shells – are constantly present around the central charge. On the other hand, there certainly are some differences, namely the screening is a dynamic screening and depends on the particle velocity. What we said a moment ago about the classical atom referred to a central charge at rest. When its velocity increases the screening shell disappears and for large velocities the quasi-particle will behave as a free "undressed" or "bare" particle – the shell is stripped off when the velocity increases. The concept of screened charges, provided their velocity is sufficiently low, is quite a fundamental one. Even recently it was not simple to realise that this is a general concept. We shall attempt to use it consistently and to show how it also drastically changes non-linear interactions. From the point of view of a mathematical description the result is exact, as it is derived by averaging over the fluctuations.

4.2 Fluctuations of Independent Particles

Let there be an average particle distribution function $\Phi_{\mathbf{p}}$ which may be any non-equilibrium distribution function and let us consider the statistical fluctuations in this system of particles. These fluctuations will not be the thermal fluctuations since the distribution function is assumed not to be the thermal one. To begin with we shall consider the particles to be free and completely independent statistical elements of a large system of particle which together constitute a non-equilibrium state of the plasma. This implies that we assume that the particles do not produce any fields, since otherwise they would become dependent upon one another through these fields. For the moment we therefore consider the particles to be uncharged. We shall see later that the correlations due to their being charged are, indeed, weak.

For N statistically independent elements – N may be the total number of particles in some volume in the plasma – the mean square of the fluctuations $\langle (\delta N)^2 \rangle$ is equal to $\langle N \rangle$, that is, we have

$$\langle (\delta N)^2 \rangle \; = \; \langle N \rangle, \tag{4.4}$$

where $\delta N = N - \langle N \rangle$ and the pointed brackets indicate averages. This relation is used extensively in statistical physics. We shall use it here to find the square of the fluctuations in the distribution function, $\delta f_{\mathbf{p}}^{\alpha}$, where α denotes the kind of particles we are dealing with (electrons, ions, ...). As the N statistically independent elements in Eq.(4.4) we shall consider particles of either type α or of type β which have their momenta in the range $\mathbf{p}$ to $\mathbf{p} + d\mathbf{p}$. This quantity is just $f_{\mathbf{p}}^{\alpha}$ and its average value is equal to $\Phi_{\mathbf{p}}^{\alpha}$. From Eq.(4.4) it follows that the average of the square of $\delta f_{\mathbf{p}}^{\alpha}$ is proportional to $\Phi_{\mathbf{p}}^{\alpha}$. Let us write this equation in a more general form, considering the average of the product of $f_{\mathbf{p}}^{\alpha}$ and $f_{\mathbf{p}'}^{\beta}$, and taking into account that this average will vanish if either the kinds of particles or the momenta are different:

$$\langle \delta f_{\mathbf{p}}^{\alpha(0)} \delta f_{\mathbf{p}'}^{\beta(0)} \rangle \propto \Phi_{\mathbf{p}}^{\alpha} \delta_{\alpha,\beta} \delta(\mathbf{p} - \mathbf{p}'). \tag{4.5}$$

We have introduced here the superscript (0) to indicate that the particles are completely independent and do not create fields. To find the multiplying coefficient on the right-hand side of Eq.(4.5) one must be careful, since Eq.(4.4) was written for the case where we considered the total number of particles in a given volume, whereas $\delta f_{\mathbf{p}}$ and $\Phi_{\mathbf{p}}$ are normalised with a weight function $d^3\mathbf{p}/(2\pi)^3$.

If the particles are uncharged $\delta f_{\mathbf{p}}^{\alpha}$ should satisfy the equation

$$\frac{\partial \delta f_{\mathbf{p}}^{\alpha(0)}}{\partial t} + \left(\mathbf{v} \cdot \frac{\partial \delta f_{\mathbf{p}}^{\alpha(0)}}{\partial \mathbf{r}}\right) = 0. \tag{4.6}$$

It is easy to write down the solution of this equation, if we introduce Fourier components,

$$\delta f_{\mathbf{p},\mathbf{k},\omega}^{\alpha(0)} \equiv \delta f_{\mathbf{p},\mathbf{k}}^{\alpha(0)}, \qquad k = \{\mathbf{k},\omega\},$$

and we have

$$-i\left(\omega - (\mathbf{k} \cdot \mathbf{v})\right) \delta f_{\mu,k}^{\alpha(0)} = 0. \tag{4.7}$$

From this equation it follows that

$$\delta f_{\mathbf{p},k}^{\alpha(0)} \propto \delta\left(\omega - (\mathbf{k} \cdot \mathbf{v})\right). \tag{4.8}$$

In order to be able actually to carry out the averaging over fluctuations we want to use the average of the product of the Fourier components $\delta f_{\mathbf{p},k}^{\alpha(0)}$ and $\delta f_{\mathbf{p}',k'}^{\beta(0)}$ of the distribution functions rather than Eq.(4.5). We must bear in mind that the correlation functions of any two random quantities which on average are stationary and homogeneous will be proportional to $\delta(\omega + \omega')\delta(\mathbf{k}+\mathbf{k}')$. We have already used this when discussing correlation functions of random fields. At the moment it is sufficient for us that the average value $\langle \delta f_{\mathbf{p},k}^{\alpha(0)} \delta f_{\mathbf{p}',k'}^{\beta(0)} \rangle$ must be proportional to $\delta(\mathbf{k}+\mathbf{k}')$. Together with the relation (4.8) this will lead to proportionality to $\delta(\omega + \omega')$. Indeed, we have

$$\delta\left(\omega - (\mathbf{k} \cdot \mathbf{v})\right) \delta\left(\omega' - (\mathbf{k}' \cdot \mathbf{v})\right) \delta(\mathbf{k} + \mathbf{k}')$$
$$= \delta\left(\omega - (\mathbf{k} \cdot \mathbf{v})\right) \delta\left(\omega' + (\mathbf{k} \cdot \mathbf{v})\right) \delta(\mathbf{k} + \mathbf{k}')$$
$$= \delta\left(\omega - (\mathbf{k} \cdot \mathbf{v})\right) \delta(\omega + \omega') \delta(\mathbf{k} + \mathbf{k}').$$

For the average we are interested in we have thus found the following factors:

$$\langle \delta f_{\mathbf{p},k}^{\alpha(0)} \delta f_{\mathbf{p}',k'}^{\beta(0)} \rangle \propto \Phi_{\mathbf{p}}^{\alpha} \delta_{\alpha,\beta} \delta(\mathbf{p} - \mathbf{p}') \delta\left(\omega - (\mathbf{k} \cdot \mathbf{v})\right) \delta(\mathbf{k} + \mathbf{k}') \delta(\omega + \omega'). \tag{4.9}$$

We shall now prove that the normalisation condition leads to a numerical factor on the right-hand side of Eq.(4.9) which is equal to 1. There is a difference between Eq.(4.9) which contains Fourier components relating to an infinite volume and Eq.(4.4) which is valid for a finite volume, such as, for instance, a cube of edge length L. To find the numerical factor of Eq.(4.9) we must use a Fourier expansion in a finite volume of linear size L and then consider the limit as $L \to \infty$. This is easy enough and we restrict ourselves here to some comments which follow immediately from the way we have written Eq.(4.9). The average value of the fluctuations in the number of particles in a volume L^3 can be found by multiplying the expression, corresponding to Eq.(4.9) but in a finite volume, by $(L^6/(2\pi)^6)e^{i\{([\mathbf{k}+\mathbf{k}']\cdot\mathbf{r})-(\omega+\omega')t\}}$ and integrating over $\mathbf{p}$, $\mathbf{p}'$, $\mathbf{k}$, and $\mathbf{k}'$. On the right-hand side the integral $\int \Phi_{\mathbf{p}} (d^3\mathbf{p}/(2\pi)^3) L^3$ will give the total number of particles in the volume L^3; the integral over $\mathbf{p}'$ is calculated by using $\delta(\mathbf{p}-\mathbf{p}')$; the presence of the factor $\delta(\mathbf{k}+\mathbf{k}')\delta(\omega+\omega')$ ensures that the right-hand side is independent of $\mathbf{r}$ and t; the integration over ω is carried out using the factor $\delta(\omega - (\mathbf{k}\cdot\mathbf{v}))$; and there is left only an integral (sum) over $\mathbf{k}$ with a factor $L^3/(2\pi)^3$. In fact, these sums (over the three components of $\mathbf{k}$) are over $2\pi n_i/L$, where the n_i are integers and the i correspond to the three coordinate axes; in the Fourier series representation the factor $L^3/(2\pi)^3$ thus disappears and we are left with the sum

$$\sum_{\mathbf{n}} 1,$$

which is equal to 1 since there is only a contribution from $\mathbf{n} = 0$ because of the independence of the summand on the coordinates; the factor we have been calculating is therefore equal to 1. We have thus proved that for independent particles averaging over the fluctuations can be carried out using the relation

$$\langle \delta f_{\mathbf{p},\mathbf{k}}^{\alpha(0)} \delta f_{\mathbf{p}',\mathbf{k}'}^{\beta(0)} \rangle = \Phi_{\mathbf{p}}^{\alpha}\delta_{\alpha,\beta}\delta(\mathbf{p} - \mathbf{p}')\delta(\omega - (\mathbf{k}\cdot\mathbf{v}))\delta(\mathbf{k}+\mathbf{k}')\delta(\omega+\omega'). \quad (4.10)$$

We emphasise that in Eq.(4.10) the distribution function $\Phi_{\mathbf{p}}$ is arbitrary and that the fluctuations which we are considering are non-thermal; however, in the limit of a thermal distribution Eq.(4.10) describes the thermal fluctuations. We shall sometimes use the term *natural statistical fluctuations* or just statistical fluctuations to describe the fluctuations we have just considered in a system of independent particles which is not in equilibrium.

4.3 Fluctuations of Charged Particles

Charged paricles create fields. Non-relativistic particles create electrostatic fields which can be found from the Poisson equation

$$\text{div}\,\mathbf{D} = 4\pi\varrho; \qquad i(\mathbf{k}\cdot\varepsilon_{\mathbf{k}}\mathbf{E}_{\mathbf{k}}) = 4\pi\varrho_{\mathbf{k}}; \qquad \mathbf{k} = \{\mathbf{k},\omega\}. \qquad (4.11)$$

Let $\mathbf{E}_{\mathbf{k}}^{(0)}$ denote the field produced by particle fluctuations and $\varrho_{\mathbf{k}}^{(0)}$ the charge density of these fluctuations:

$$\varrho_{\mathbf{k}}^{(0)} = \sum_\beta e_\beta \int \delta f_{\mathbf{p}',\mathbf{k}}^{\beta(0)} \frac{d^3\mathbf{p}'}{(2\pi)^3}, \tag{4.12}$$

$$\mathbf{E}_{\mathbf{k}}^{(0)} = \frac{4\pi\mathbf{k}\varrho_{\mathbf{k}}^{(0)}}{ik^2\varepsilon_{\mathbf{k}}} = \frac{4\pi\mathbf{k}}{ik^2\varepsilon_{\mathbf{k}}} \sum_\beta e_\beta \int \delta f_{\mathbf{p}',\mathbf{k}}^{\beta(0)} \frac{d^3\mathbf{p}'}{(2\pi)^3}. \tag{4.13}$$

We must now take into account that these fields will change the particle fluctuations. The equation which should be used in the case of charged particles is not Eq.(4.6), but

$$\frac{\partial \delta f_{\mathbf{p}}^\alpha}{\partial t} + \left(\mathbf{v} \cdot \frac{\partial \delta f_{\mathbf{p}}^\alpha}{\partial \mathbf{r}}\right) = -e_\alpha \left(\mathbf{E} \cdot \frac{\partial \Phi_{\mathbf{p}}^\alpha}{\partial \mathbf{p}}\right). \tag{4.14}$$

We have neglected here terms which are non-linear in the fluctuations since in the case of natural statistical fluctuations of the particle distribution their relative order of magnitude will be $1/N_{\mathrm{de}}$. If we needed them, we could, in principle, consider them using perturbation theory; we refer to Chap.13 where we are considering bremsstrahlung in a plasma. However, at the moment we need only the first approximation and we shall use Eq.(4.14). In terms of the Fourier components we have

$$i\left(\omega - (\mathbf{k} \cdot \mathbf{v})\right) \delta f_{\mathbf{p},\mathbf{k}}^\alpha = e_\alpha \left(\mathbf{E}_{\mathbf{k}}^{(0)} \cdot \frac{\partial \Phi_{\mathbf{p}}^\alpha}{\partial \mathbf{p}}\right). \tag{4.15}$$

We have substituted here the field $\mathbf{E}_{\mathbf{k}}^{(0)}$ as we are interested only in effects which are linear in $\delta f_{\mathbf{p},\mathbf{k}}^{(0)}$. Equation (4.15) contains a term linear in $\delta f_{\mathbf{p},\mathbf{k}}^{(0)}$ which was already taken into account when we calculated $\mathbf{E}_{\mathbf{k}}^{(0)}$ since the dielectric permittivity contains the linear plasma response. We therefore need only take into account an additional term in the fluctuating part of the distribution function.

The solution of Eq.(4.15) will contain the solution of the homogeneous equation, which corresponds to particles without a charge; this solution is equal to $\delta f_{\mathbf{p},\mathbf{k}}^{\alpha(0)}$. We should add to this the solution of the equation with the right-hand side proportional to the field strength. Using the causality principle we then find

$$\delta f_{\mathbf{p},\mathbf{k}}^\alpha = \delta f_{\mathbf{p},\mathbf{k}}^{\alpha(0)} + \frac{e_\alpha}{i(\omega - (\mathbf{k} \cdot \mathbf{v}) + i0)} \left(\mathbf{E}_{\mathbf{k}}^{(0)} \cdot \frac{\partial \Phi_{\mathbf{p}}^\alpha}{\partial \mathbf{p}}\right). \tag{4.16}$$

After substituting expression (4.13) for the field into Eq.(4.16) we find an expression for the fluctuation part of the particle distribution function which is linear in $\delta f_{\mathbf{p},\mathbf{k}}^{\alpha(0)}$:

$$\delta f_{\mathbf{p},\mathbf{k}}^\alpha = \delta f_{\mathbf{p},\mathbf{k}}^{\alpha(0)} - \frac{e_\alpha \left(\mathbf{k} \cdot \frac{\partial \Phi_{\mathbf{p}}^\alpha}{\partial \mathbf{p}}\right)}{k^2\varepsilon_{\mathbf{k}}(\omega - (\mathbf{k} \cdot \mathbf{v}) + i0)} \sum_\beta e_\beta \int \delta f_{\mathbf{p}',\mathbf{k}}^{\beta(0)} \frac{d^3\mathbf{p}'}{(2\pi)^3}. \tag{4.17}$$

This expression shows that the fluctuations in the distribution function of particles of the kind α depend on those of particles of the kind β, but for the same values of $\mathbf{k}$ and ω. This mutual effect on each other's fluctuations does not occur if either of the charges e_α or e_β is zero.

Let us recall that when we derived the quasi-linear equation we used a relation similar to Eq.(4.16), namely the equation

$$\delta f^\alpha_{\mathbf{p},\mathbf{k}} = \frac{e_\alpha}{\mathrm{i}(\omega - (\mathbf{k}\cdot\mathbf{v}) + \mathrm{i}0)}\left(\mathbf{E}^\sigma_\mathbf{k}\cdot\frac{\partial\Phi^\alpha_\mathbf{p}}{\partial\mathbf{p}}\right), \qquad (4.18)$$

where $\mathbf{E}^\sigma$ is the field of strong plasma oscillations of the kind σ. In the general case we must take into account both types of fluctuations:

$$\delta f^\alpha_{\mathbf{p},\mathbf{k}} = \delta f^{\alpha(0)}_{\mathbf{p},\mathbf{k}} + \frac{e_\alpha}{\mathrm{i}(\omega - (\mathbf{k}\cdot\mathbf{v}) + \mathrm{i}0)}\left(\left[\mathbf{E}^\sigma_\mathbf{k} + \mathbf{E}^{(0)}_\mathbf{k}\right]\cdot\frac{\partial\Phi^\alpha_\mathbf{p}}{\partial\mathbf{p}}\right). \qquad (4.19)$$

In the case where one considers both types of fluctuations one should bear in mind that the fluctuations of the strong collective oscillations exist only close to the frequency $\omega^\sigma_\mathbf{k}$ within a width $\delta\omega^\sigma_\mathbf{k}$, whereas the natural statistical fluctuations exist also outside this range and in regions where the oscillations do not exist. Inside the lines of the collective oscillations the natural statistical fluctuations of the particle distribution function describe the spontaneous emission processes and they are thus usually small as compared to the other fluctuations if the power of the oscillations is high. The natural statistical fluctuations lead to a collision integral while the strong oscillations lead to the quasi-linear integral. Those integrals can be added, only approximately, if we can in the integration over the range where the collective oscillations occur neglect the spontaneous emission processes while in the other regions, where the collective oscillations cannot exist, we can take into account only the natural statistical fluctuations.

4.4 Landau-Balescu Collision Integral

We shall start with the equation for the average distribution function, retaining only the field $\mathbf{E}^{(0)}$:

$$\frac{\partial\Phi^\alpha_\mathbf{p}}{\partial t} + \left(\mathbf{v}\cdot\frac{\partial\Phi^\alpha_\mathbf{p}}{\partial\mathbf{r}}\right) = -e_\alpha\left\langle\left(\mathbf{E}^{(0)}\cdot\frac{\partial\delta f^\alpha_\mathbf{p}}{\partial\mathbf{p}}\right)\right\rangle = I^{\mathrm{col},\alpha}_\mathbf{p}$$
$$= I^{\mathrm{col},\alpha(1)}_\mathbf{p} + I^{\mathrm{col},\alpha(2)}_\mathbf{p}. \qquad (4.20)$$

We shall use for $\delta f^\alpha_\mathbf{p}$ the expression which is linear in $\delta f^{\alpha(0)}_\mathbf{p}$, that is, expression (4.16). The collision integral $I^{\mathrm{col},\alpha}_\mathbf{p}$ has been split into two parts, the first part containing the first term on the right-hand side of Eq.(4.16) and the second one containing the second term. We have therefore

$$I_{\mathbf{p}}^{\text{col},\alpha(1)} = -e_\alpha \left\langle\!\!\left\langle \left(\mathbf{E}^{(0)} \cdot \frac{\partial \delta f_{\mathbf{p}}^{\alpha(0)}}{\partial \mathbf{p}} \right) \right\rangle\!\!\right\rangle. \tag{4.21}$$

Before writing down the second part of the collision integral we change to the Fourier transforms. We then find for the two parts of the collision integral:

$$I_{\mathbf{p}}^{\text{col},\alpha(1)} = -e_\alpha \frac{\partial}{\partial p_i} \int \left\langle E_{i,\mathbf{k}'}^{(0)} \delta f_{\mathbf{p},\mathbf{k}}^{\alpha(0)} \right\rangle$$
$$\times\, e^{i\{([\mathbf{k}+\mathbf{k}']\cdot\mathbf{r})-(\omega+\omega')t\}} d^4k\, d^4k', \tag{4.22}$$

$$I_{\mathbf{p}}^{\text{col},\alpha(2)} = -e_\alpha \frac{\partial}{\partial p_i} \int \left\langle E_{i,\mathbf{k}'}^{(0)} \frac{e_\alpha}{i(\omega - (\mathbf{k}\cdot\mathbf{v}) + i0)} E_{j,\mathbf{k}}^{(0)} \frac{\partial \Phi_{\mathbf{p}}^\alpha}{\partial p_j} \right\rangle$$
$$\times\, e^{i\{([\mathbf{k}+\mathbf{k}']\cdot\mathbf{r})-(\omega+\omega')t\}} d^4k\, d^4k', \tag{4.23}$$

Since the correlation functions contain factors $\delta(\mathbf{k} + \mathbf{k}')$ and $\delta(\omega + \omega')$, integration over $\mathbf{k}$ leads to putting the exponential functions equal to unity. We can now substitute expression (4.13) for the field in Eq.(4.22) and we find

$$I_{\mathbf{p}}^{\text{col},\alpha(1)} = e_\alpha \frac{\partial}{\partial p_i} \sum_\beta e_\beta \int \frac{4\pi k_i}{ik^2\varepsilon_{-\mathbf{k}}} \left\langle \delta f_{\mathbf{p}',-\mathbf{k}}^{\beta(0)} \delta f_{\mathbf{p},\mathbf{k}}^{\alpha(0)} \right\rangle d^4k\, \frac{d^3p'}{(2\pi)^3}. \tag{4.24}$$

We now use Eq.(4.10) for averaging over the fluctuations and the result is

$$I_{\mathbf{p}}^{\text{col},\alpha(1)} = 4\pi e_\alpha^2 \frac{\partial}{\partial p_i} \int \frac{k_i}{ik^2\varepsilon_{-\mathbf{k}}} \delta\!\left(\omega - (\mathbf{k}\cdot\mathbf{v})\right) \Phi_{\mathbf{p}}^\alpha \frac{d^3k\, d\omega}{(2\pi)^3}. \tag{4.25}$$

The integration in Eq.(4.25) is over all possible frequencies and wavevectors. However, it is interesting first to consider the range over which collective oscillations can occur, that is, the range for which $\varepsilon_{\mathbf{k}} = 0$; this range corresponds to one where the denominator in (4.25) tends to zero. Since $\varepsilon_{-\mathbf{k}} = \varepsilon_{\mathbf{k}}^*$ the only contribution to (4.25) from that range comes from the imaginary part of $1/\varepsilon_{-\mathbf{k}}$, as can be verified easily by making a change of variables $\mathbf{k} \to -\mathbf{k}$. If to satisfy the causality principle we take into account a small dissipative imaginary part of $\varepsilon_{\mathbf{k}}$ we find a way to calculate the principal part of the integral for $\varepsilon_{\mathbf{k}} \simeq 0$:

$$\text{Im}\,\frac{1}{\varepsilon_{-\mathbf{k}}} \simeq \pi \frac{\omega}{|\omega|} \delta(\text{Re}\{\varepsilon_{\mathbf{k}}\}) = \pi \frac{\delta(\omega - \omega_{\mathbf{k}}) - \delta(\omega + \omega_{\mathbf{k}})}{(\partial \text{Re}\{\varepsilon_{\mathbf{k}}\}/\partial\omega)_{\omega=\omega_{\mathbf{k}}}}. \tag{4.26}$$

Substituting this expression into Eq.(4.25) and in the second term changing the integration variable from $\mathbf{k}$ to $-\mathbf{k}$ we get

$$I_{\mathbf{p}}^{\text{col},\alpha(1)} = \left(\frac{\partial}{\partial \mathbf{p}} \cdot [\mathbf{F}_{\mathbf{p}} \Phi_{\mathbf{p}}^\alpha] \right);$$

$$\mathbf{F}_{\mathbf{p}} = 8\pi^2 e_\alpha^2 \int \frac{\mathbf{k}\, d^3k}{k^2(2\pi)^3} \frac{\delta\!\left(\omega_{\mathbf{k}} - (\mathbf{k}\cdot\mathbf{v})\right)}{(\partial \varepsilon_{\mathbf{k}}/\partial\omega)_{\omega=\omega_{\mathbf{k}}}}. \tag{4.27}$$

Since the imaginary part of ε_k is small we have substituted ε_k for $\mathrm{Re}\,\varepsilon_k$ in Eq.(4.27). One checks easily that Eq.(4.27) is the same as the equation we obtained earlier for the spontaneous Vavilov-Cherenkov emission (see Eqs.(3.32) and (3.40)). Equation (4.22) thus contains the term corresponding to the spontaneous emission of waves. However, selecting from the whole integration domain only that part which corresponds to oscillations is, of course, incorrect; other parts of the integration domain over the frequency and the wavevector may make contributions which are of the same order of magnitude or even larger and, indeed, this appears to be the case. We therefore write down a general expression for $\mathrm{Im}\,1/\varepsilon_{-k}$:

$$\mathrm{Im}\,\frac{1}{\varepsilon_{-k}} = -\frac{\mathrm{Im}\{\varepsilon_{-k}\}}{|\varepsilon_k^2|} = \frac{\mathrm{Im}\{\varepsilon_k\}}{|\varepsilon_k^2|}, \tag{4.28}$$

together with the expression for the imaginary part of the dielectric permittivity which we have used earlier,

$$\mathrm{Im}\{\varepsilon_k\} = -4\pi^2 \sum_\beta \frac{e_\beta^2}{k^2} \int \delta\big(\omega - (\mathbf{k}\cdot\mathbf{v}')\big) \left(\mathbf{k}\cdot\frac{\partial \Phi_{\mathbf{p}'}^\beta}{\partial \mathbf{p}'}\right) \frac{d^3\mathbf{p}'}{(2\pi)^3}. \tag{4.29}$$

Substituting these expressions into Eq.(4.25) we get a general expression for $I_{\mathbf{p}}^{\mathrm{col},\alpha(1)}$:

$$I_{\mathbf{p}}^{\mathrm{col},\alpha(1)} = -2\sum_\beta e_\alpha^2 e_\beta^2 \frac{\partial}{\partial p_i} \int \frac{k_i k_j\, d^3\mathbf{k}}{k^4 |\varepsilon_{\mathbf{k},(\mathbf{k}\cdot\mathbf{v})}|^2}$$

$$\times\ \delta\big((\mathbf{k}\cdot\mathbf{v}) - (\mathbf{k}\cdot\mathbf{v}')\big) \frac{d^3\mathbf{p}'}{(2\pi)^3}\, \Phi_{\mathbf{p}}^\alpha\, \frac{\partial \Phi_{\mathbf{p}'}^\beta}{\partial p_j}. \tag{4.30}$$

We have here already integrated over the frequency ω so that we could everywhere substitute $(\mathbf{k}\cdot\mathbf{v})$ for ω.

In order to calculate the second part, (4.23), of the collision integral we must know the correlation function of the fields of the particle fluctuations. Substituting expression (4.13) for the fields we get

$$\left\langle E_{i,\mathbf{k}'}^{(0)} E_{j,\mathbf{k}}^{(0)} \right\rangle = \frac{16\pi^2 k_i k_j}{k^4 |\varepsilon_k|^2} \sum_{\beta,\beta'} e_\beta e_{\beta'} \int \left\langle \delta f_{\mathbf{p}',\mathbf{k}'}^{\beta(0)} f_{\mathbf{p}'',\mathbf{k}}^{\beta'(0)} \right\rangle \frac{d^3\mathbf{p}'\, d^3\mathbf{p}''}{(2\pi)^6}. \tag{4.31}$$

We have used here the fact that we may put $\mathbf{k}' = -\mathbf{k}$ in the factor in front of the integral, since there will ultimately be a factor $\delta(\mathbf{k} + \mathbf{k}')$. We now substitute expression (4.10) for the correlation function on the right-hand side of Eq.(4.31) and we find

$$\left\langle E_{i,\mathbf{k}'}^{(0)} E_{j,\mathbf{k}}^{(0)} \right\rangle = \frac{16\pi^2 k_i k_j}{k^4 |\varepsilon_k|^2 (2\pi)^6} \sum_\beta e_\beta^2$$

$$\times \int \delta\big(\omega - (\mathbf{k}\cdot\mathbf{v}')\big)\, \Phi_{\mathbf{p}'}^\beta\, \delta(\mathbf{k} + \mathbf{k}')\, d^3\mathbf{p}'. \tag{4.32}$$

We must now substitute Eq.(4.32) into Eq.(4.23) and take into account that by carrying out a change of variables $\mathbf{k} \to -\mathbf{k}$, $\mathbf{k}' \to -\mathbf{k}'$ we can see that only the imaginary part of $1/(\omega - (\mathbf{k}\cdot\mathbf{v}) + i0)$ contributes. As this imaginary part is equal to $-\pi\delta(\omega - (\mathbf{k}\cdot\mathbf{v}))$ we finally get

$$I_{\mathbf{p}}^{col,\alpha(2)} = 2\sum_{\beta} e_{\alpha}^2 e_{\beta}^2 \frac{\partial}{\partial p_i} \int \frac{k_i k_j\, d^3\mathbf{k}}{k^4 |\varepsilon_{\mathbf{k},(\mathbf{k}\cdot\mathbf{v})}|^2}$$
$$\times\, \delta((\mathbf{k}\cdot\mathbf{v}) - (\mathbf{k}\cdot\mathbf{v}')) \frac{d^3\mathbf{p}'}{(2\pi)^3} \Phi_{\mathbf{p}'}^{\beta} \frac{\partial \Phi_{\mathbf{p}}^{\alpha}}{\partial p_j}. \tag{4.33}$$

As in Eq.(4.30) we have here taken into account that due to the δ-function the frequency ω can everywhere be replaced by $(\mathbf{k}\cdot\mathbf{v})$. We shall discuss the physical meaning of Eq.(4.33) in what follows.

The complete equation for the averaged distribution function with the Landau-Balescu collision integral on the right-hand side of the equation is thus

$$\frac{\partial \Phi_{\mathbf{p}}^{\alpha}}{\partial t} + \left(\mathbf{v}\cdot\frac{\partial \Phi_{\mathbf{p}}^{\alpha}}{\partial \mathbf{r}}\right) = 2\sum_{\beta} e_{\alpha}^2 e_{\beta}^2 \frac{\partial}{\partial p_i} \int \frac{k_i k_j\, d^3\mathbf{k}}{k^4 |\varepsilon_{\mathbf{k},(\mathbf{k}\cdot\mathbf{v})}|}$$
$$\times\, \delta((\mathbf{k}\cdot\mathbf{v}) - (\mathbf{k}\cdot\mathbf{v}')) \frac{d^3\mathbf{p}'}{(2\pi)^3} \left[\Phi_{\mathbf{p}'}^{\beta} \frac{\partial \Phi_{\mathbf{p}}^{\alpha}}{\partial p_j} - \Phi_{\mathbf{p}}^{\alpha} \frac{\partial \Phi_{\mathbf{p}'}^{\beta}}{\partial p_j'}\right]. \tag{4.34}$$

In our discussion of Eq.(4.34) we shall be concerned with its physical consequences. If there exists an averaged electric field $\mathbf{E}$, we must add a term $e_{\alpha}(\mathbf{E}\cdot(\partial/\partial\mathbf{p})\Phi_{\mathbf{p}}^{\alpha}$ to the left-hand side of Eq.(4.34).

4.5 Structure of the Collision Integral. Examples

Let us consider some of the conclusions which can be drawn from the expression we have found for the collision integral.

1. *The kinetic equation for the average particle distribution function.* The kinetic equation for the average particle distribution function $\Phi_{\mathbf{p}}^{\alpha}$ with a collision integral on the right-hand side of the equation is a complicated integral equation for $\Phi_{\mathbf{p}}^{\alpha}$, since the collision term does not only contain the product of the distributions functions of the colliding particles – which is as should have been expected – but it also contains the distribution function in the denominator through the dielectric permittivity:

$$\varepsilon_{\mathbf{k},(\mathbf{k}\cdot\mathbf{v})} = 1 + \sum_{\alpha'} \frac{4\pi e_{\alpha'}^2}{k^2}$$
$$\times \int \frac{1}{(\mathbf{k}\cdot\mathbf{v}) - (\mathbf{k}\cdot\mathbf{v}') + i0} \left(\mathbf{k}\cdot\frac{\partial \Phi_{\mathbf{p}'}^{\alpha'}}{\partial \mathbf{p}'}\right) \frac{d^3\mathbf{p}'}{(2\pi)^3}. \tag{4.35}$$

One can find the collision integral which Landau derived from the general expression by taking the limit as $\varepsilon_{\mathbf{k},(\mathbf{k}\cdot\mathbf{v})} \simeq 1$. Whether it is possible to use the resultant expression as an approximation for any particular distribution needs a special study since $\varepsilon_{\mathbf{k}}$ depends strongly on the distribution function and that may be far from an equilibrium one. In the case when the velocities of interest are of the order of the average velocities of the particles of the same kind ($v' \approx v \approx v_T$) we can get an approximate expression from Eq.(4.35): $\varepsilon_{\mathbf{k},(\mathbf{k}\cdot\mathbf{v})} \approx 1 + 1/k^2d^2$. Hence we can only put the dielectric permittivity approximately equal to unity for $k \gg 1/d$. Estimates for other kinds of collisions – for instance, collisions between fast particles – will be different. In other words, it is not always possible to put $\varepsilon_{\mathbf{k},(\mathbf{k}\cdot\mathbf{v})}$ equal to 1 and when it is possible it can only be done approximately. In that case the collision integral takes the form found by Landau and we get:

$$\frac{d\Phi_{\mathbf{p}}^{\alpha}}{dt} = 2\sum_{\beta} e_{\alpha}^2\,e_{\beta}^2\,\frac{\partial}{\partial p_i} \int \frac{k_i k_j\,d^3\mathbf{k}\,d^3\mathbf{p}'}{k^4(2\pi)^3}$$

$$\times\,\delta\big((\mathbf{k}\cdot\mathbf{v}) - (\mathbf{k}\cdot\mathbf{v}')\big) \left[\Phi_{\mathbf{p}'}^{\beta}\frac{\partial\Phi_{\mathbf{p}}^{\alpha}}{\partial p_j} - \Phi_{\mathbf{p}}^{\alpha}\frac{\partial\Phi_{\mathbf{p}'}^{\beta}}{\partial p_j'}\right]. \qquad (4.36)$$

2. *The Landau collision integral.* We shall now consider the Landau collision integral and write it in a slightly simpler form, introducing the tensor

$$I_{i,j} = 2 \int \frac{k_i k_j\,d^3\mathbf{k}}{k^4}\,\delta\big((\mathbf{k}\cdot\mathbf{u})\big), \qquad (4.37)$$

where $\mathbf{u}$ is the relative particle velocity,

$$\mathbf{u} = \mathbf{v} - \mathbf{v}'. \qquad (4.38)$$

The only vector determining the tensor $I_{i,j}$ is $\mathbf{u}$ and the tensor can therefore be constructed by using just the unit tensor $\delta_{i,j}$ and the tensor $u_i u_j/u^2$. Since we must have $I_{i,j}u_i = I_{i,j}u_j = 0$, the only allowed combination of these two tensors can be $\delta_{i,j} - u_i u_j/u^2$, so that we find

$$I_{i,j} = I(u)\left(\delta_{i,j} - \frac{u_i u_j}{u^2}\right). \qquad (4.39)$$

From Eq.(4.37) we find for $I(u)$ the expression

$$I(u) = \tfrac{1}{2}I_{i,i}$$

$$= \int \frac{d^3\mathbf{k}}{k^2}\,\delta\big((\mathbf{k}\cdot\mathbf{u})\big) = 2\pi \int_{k_{\min}}^{k_{\max}} \frac{dk}{ku} = \frac{2\pi}{u}\,\ln\frac{k_{\max}}{k_{\min}}. \qquad (4.40)$$

We have introduced the cut-offs $k_{\max}$ and $k_{\min}$ since formally the integral is divergent. This was also done by Landau. The lower cut-off $k_{\min}$ can easily be found – or rather estimated – by bearing in mind that we have used the approximation $\varepsilon_{\mathbf{k},(\mathbf{k}\cdot\mathbf{v})} \approx 1$ in deriving the Landau collision integral. We therefore put

$$k_{\min} \simeq \frac{1}{d_e}. \tag{4.41}$$

The value of $k_{\max}$ is also determined by bearing in mind the approximations made. We remember that we neglected the non-linear effects in the fields $E^{(0)}$. The criterion that these effects are small will be $W^{(0)}/nmv^2 \ll 1$; here $W^{(0)}$ is the average energy density in the field $E^{(0)}$ and we must remember that for thermal distributions we must replace the average particle energy density nmv^2 by nT. The value of $W^{(0)}$ can be found from Eq.(4.32):

$$\frac{W^{(0)}}{nmv^2} = \int \frac{1}{4nmv^2} \left\langle E^{(0)}_{\mathbf{k}'} E^{(0)}_{\mathbf{k}} \right\rangle d^4k\, d^4k'$$

$$\simeq \frac{4\pi}{mv^2} \sum_\beta e_\beta^2 \frac{n_\beta}{n} \int \frac{d^3\mathbf{k}}{k^2(2\pi)^3} \simeq \frac{e^2}{mv^2} k_{\max} \simeq 1. \tag{4.42}$$

Hence we find for $k_{\max}$ the relation

$$k_{\max} \simeq \frac{mv^2}{e^2}. \tag{4.43}$$

The physical meaning of this relation is that for $k = k_{\max}$ the kinetic energy $\frac{1}{2}mv^2$ of the colliding particles is equal to their potential energy $e^2/r_{\min} = e^2 k_{\max}$. Another restriction on the maximum possible value of k arises from quantum effects which we have neglected. Quantum effects would become important if k would become of the order of the reciprocal of the de Broglie wavelength $\hbar/mv$; this means that we have another upper limit $k'_{\max}$:

$$k'_{\max} \simeq \frac{mv}{\hbar}. \tag{4.44}$$

In the collision integral we must substitute the smaller of $k_{\max}$ and $k'_{\max}$. We find that $k'_{\max}$ is smaller than $k_{\max}$ for high velocities such that $e^2/\hbar v < 1$.

The logarithm occurring in Eq.(4.40) is usually called the *Coulomb logarithm* and denoted by Λ. For $e^2/\hbar v \ll 1$ we have

$$\Lambda = \ln \frac{k_{\max}}{k_{\min}} \simeq \ln \frac{T}{2\hbar\omega_{\text{pe}}}, \tag{4.45}$$

and for $e^2/\hbar v \gg 1$ we have

$$\Lambda = \ln \frac{k_{\max}}{k_{\min}} \simeq \ln 3N_d. \tag{4.46}$$

For applications it is important that the Coulomb logarithm is usually a large number:

$$\Lambda \gg 1. \tag{4.47}$$

Using Eq.(4.40) we can now write the collision integral tensor (4.39) in the form

$$I_{i,j} = \frac{2\pi\Lambda}{u}\left(\delta_{i,j} - \frac{u_i u_j}{u^2}\right). \tag{4.48}$$

3. *Fast electron energy losses due to electron-ion collisions.* As an example of the use of the Landau collision integral we shall calculate the energy losses due to electron-ion collisions of fast electrons with velocities much higher than the thermal velocity. For fast electrons we have $u \simeq v$ and the term in the Landau collision integral (4.36) containing the derivative of the ion distribution function is smaller than the one containing the derivative of the electron distribution function by a factor m_e/m_i. Therefore we can write

$$\frac{d\Phi_{\mathbf{p}}^e}{dt} = 2\pi\Lambda e^2 e_i^2 n_i \frac{\partial}{\partial p_i}\left[\frac{1}{v}\left(\delta_{i,j} - \frac{v_i v_j}{v^2}\right)\right]\frac{\partial\Phi_{\mathbf{p}}^e}{\partial p_j}. \tag{4.49}$$

We can immediately find the average change in the electron momentum from Eq.(4.49):

$$\frac{d\mathbf{P}_e}{dt} = \frac{d}{dt}\int \mathbf{p}\Phi_{\mathbf{p}}^e \frac{d^3\mathbf{p}}{(2\pi)^3}$$

$$= -2\pi\Lambda e^2 e_i^2 n_i \int \frac{d^3\mathbf{p}}{(2\pi)^3}\left[\frac{1}{v}\frac{\partial\Phi_{\mathbf{p}}^e}{\partial\mathbf{p}} - \frac{\mathbf{v}}{v^3}\left(\mathbf{v}\cdot\frac{\partial\Phi_{\mathbf{p}}^e}{\partial\mathbf{p}}\right)\right]. \tag{4.50}$$

Integrating by parts we find

$$\frac{d}{dt}\int \mathbf{p}\Phi_{\mathbf{p}}^e \frac{d^3\mathbf{p}}{(2\pi)^3} = -4\pi\Lambda e^2 e_i^2 n_i \int \frac{\mathbf{v}\,d^3\mathbf{p}}{(2\pi)^3 v^3 m_e}\Phi_{\mathbf{p}}^e. \tag{4.51}$$

We denote the average value of any quantity G averaged over the distribution of particles of kind α by $\overline{G}_\alpha$:

$$\overline{G}_\alpha = \frac{1}{n_\alpha}\int G\Phi_{\mathbf{p}}^\alpha \frac{d^3\mathbf{p}}{(2\pi)^3}. \tag{4.52}$$

We can now write Eq.(4.51) in the form

$$\frac{d\overline{\mathbf{P}}_e}{dt} = -\frac{\omega_{pe}^4\Lambda}{4\pi n_e}\overline{\left[\frac{\mathbf{P}}{v^3}\right]}_e, \tag{4.53}$$

where for the sake of simplicity we have taken $e_i = -e$ and $n_i = n_e$.

From Eq.(4.53) it follows that if the particle velocities are of the order of v_{Te}, the characteristic time $\tau_{ei} = 1/\nu_{ei}$ for the dissipation of electron momenta will be of the order of

$$\nu_{ei} = \frac{\omega_{pe}}{3N_{de}}\Lambda. \tag{4.54}$$

We see that the collision frequency ν_{ei} is determined by the large Coulomb logarithm Λ.

4. *Fast electron energy losses due to spontaneous emission of Langmuir oscillations.* We shall now consider the momentum losses of fast electrons due to the spontaneous emission of Langmuir oscillations and we shall show that they are usually much smaller than those due to electron-ion collisions.

We use Eq.(4.27) to calculate the change in the average electron momentum due to the emission of Langmuir waves:

$$\frac{d}{dt} \int \mathbf{p} \Phi_{\mathbf{p}}^e \frac{d^3\mathbf{p}}{(2\pi)^3} = -4\pi e^2 \omega_{pe}^2 \int \frac{\mathbf{k} \, d^3\mathbf{k}}{(2\pi)^3 k^2} \delta\big(\omega_{pe} - (\mathbf{k} \cdot \mathbf{v})\big) \frac{\Phi_{\mathbf{p}}^e \, d^3\mathbf{p}}{(2\pi)^3}. \quad (4.55)$$

After integrating over the angles in the integral over $\mathbf{k}$ we find

$$\frac{d}{dt} \int \mathbf{p} \Phi_{\mathbf{p}}^e \frac{d^3\mathbf{p}}{(2\pi)^3} = -e^2 \omega_{pe}^2 \int \Phi_{\mathbf{p}}^e \frac{\mathbf{p} \, d^3\mathbf{p}}{(2\pi)^3 v^3} \int_{\omega_{pe}/v}^{k_{max}} \frac{dk}{k}. \quad (4.56)$$

The expression we have just obtained is very similar to the one we found for momentum losses due to electron-ion collisions (see Eq.(4.51)). In the expression for the losses due to the emission of Langmuir waves another logarithm enters, namely

$$\ln \frac{k_{max} v}{\omega_{pe}}. \quad (4.57)$$

This logarithm is not very large as for Langmuir waves k cannot be larger than ω_{pe}/v_{Te}. Putting $k_{max} \simeq \omega_{pe}/v_{Te}$ we get

$$\frac{d\overline{\mathbf{p}}_e}{dt} = -\frac{\omega_{pe}^4}{4\pi n_e} \overline{\left[\frac{\mathbf{p}}{v^3} \ln \frac{v}{v_{Te}} \right]_e}, \quad (4.58)$$

This means that the losses due to wave emission are usually much smaller than the losses due to collisions. This confirms what we said earlier that spontaneous emission should be treated together with other effects of the fluctuations responsible for collisions.

5. *The general structure of the collision integral.* We shall now consider the general structure of the collision integral. Even formally this structure is significantly different from that of the quasi-linear collision integal. Indeed, the collision integral (4.34) can be written as a sum of collision integrals describing collisions of particles of type α with particles of type β:

$$I_{\mathbf{p}}^{col,\alpha} = \sum_{\beta} I_{\mathbf{p}}^{col,\alpha,\beta}, \quad (4.59)$$

$$I_{\mathbf{p}}^{col,\alpha,\beta} = \frac{\partial}{\partial p_i} D_{i,j}^{\alpha,\beta} \frac{\partial}{\partial p_j} \Phi_{\mathbf{p}}^{\alpha} + \frac{\partial}{\partial p_i} F_i^{\alpha,\beta} \Phi_{\mathbf{p}}^{\alpha}, \quad (4.60)$$

where

$$D_{i,j}^{\alpha,\beta} = 2e_\alpha^2 e_\beta^2 \int \frac{k_i k_j \, d^3\mathbf{k}}{k^4 |\varepsilon_{\mathbf{k},(\mathbf{k}\cdot\mathbf{v})}|^2} \, \delta\big((\mathbf{k}\cdot\mathbf{v}) - (\mathbf{k}\cdot\mathbf{v}')\big) \, \Phi_{\mathbf{p}'}^\beta \frac{d^3\mathbf{p}'}{(2\pi)^3}, \qquad (4.61)$$

$$F_i^{\alpha,\beta} = -2e_\alpha^2 e_\beta^2 \int \frac{k_i \, d^3\mathbf{k}}{k^4 |\varepsilon_{\mathbf{k},(\mathbf{k}\cdot\mathbf{v})}|^2}$$

$$\times \, \delta\big((\mathbf{k}\cdot\mathbf{v}) - (\mathbf{k}\cdot\mathbf{v}')\big) \left(\mathbf{k} \cdot \frac{\partial \Phi_{\mathbf{p}'}^\beta}{\partial \mathbf{p}'}\right) \frac{d^3\mathbf{p}'}{(2\pi)^3}. \qquad (4.62)$$

In the quasi-linear integral there also appears a diffusion term, but now we have an additional term with a friction force $\mathbf{F}$. This kind of term is negligible in the quasi-linear equation in the case of strong oscillations. In the case of the collision integral, however, the two terms are of the same order of magnitude. This leads to the conclusion that for the case of the collision integral there may be a stationary solution for which $I_{\mathbf{p}}^{\mathrm{col},\alpha,\beta} = 0$, whereas there is no such solution – except the trivial one where there are no oscillations – in the case of the quasi-linear integral. Indeed, for a stationary solution the diffusion term in the collision integral may be balanced by the friction force:

$$D_{i,j}^{\alpha,\beta} \frac{\partial \Phi_{\mathbf{p}}^\alpha}{\partial p_j} + F_i^{\alpha,\beta}\Phi_{\mathbf{p}}^\alpha = \mathrm{const.} \qquad (4.63)$$

This equation can be satisfied by a distribution function which decreases when the momenta become large whereas the quasi-linear equation for an isotropic distribution has only solutions which are constant when the momenta become large, which is a solution which has no meaning. For the quasi-linear equation only a one-dimensional model can lead to a plateau, but such a plateau is unstable against three-dimensional relaxation. It turns out that a stationary solution can only be reached in the quasi-linear description, if there is a constant source of oscillations or if the energy of the oscillations is equal to zero. On the other hand, the collision integral equation has stationary solutions, possibly several ones. In fact, the solutions of Eq.(4.63) with a non-vanishing constant right-hand side correspond to the case when there is a non-vanishing particle flux in momentum space. This means that somewhere in momentum space there must be a particle source and somehere else a particle sink. If there are no particle sources and sinks, the right-hand side of Eq.(4.63) is equal to zero.

In the case when particles of kind β have a thermal distribution with temperature T we find from Eqs.(4.61) and (4.62) that

$$F_i^{\alpha,\beta} = \frac{v_j}{T} D_{i,j}^{\alpha,\beta}, \qquad (4.64)$$

and if at the same time the right-hand side of Eq.(4.63) is equal to zero, that equation will lead to a thermal distribution of particles of kind α with temperature T. We see thus that the collision integral equation can be satisfied by all particles having thermal distributions with the same temperature T for all of them. The same equation also describes the relaxation of non-thermal distributions to thermal equilibrium.

4.6 Conservation Laws.
Kinetic and Hydrodynamic Descriptions

The collision integral equation satisfies a number of conservation laws. One would expect that energy, momentum, and particle number conservation would be satisfied globally, that is, after integration over both momentum and configuration spaces. However, the kinetic equation with the collision integral on the right-hand side satisfies the conservation laws locally, that is, after integration over momentum space only. We shall write

$$\left[\frac{\partial \Phi_{\mathbf{p}}^{\alpha}}{\partial t}\right]^{col} \equiv I_{\mathbf{p}}^{col,\alpha} = \sum_{\beta} I_{\mathbf{p}}^{col,\alpha,\beta},$$

and introduce the notation

$$M^{\alpha} = \{n_{\alpha}, \mathbf{P}_{\alpha}, E_{\alpha}\}, \qquad m_{\mathbf{p}} = \{1, \mathbf{p}, \varepsilon_{\mathbf{p}}\},$$

and

$$\left[\frac{\partial M^{\alpha}}{\partial t}\right]^{col} = \int \sum_{\beta} m_{\mathbf{p}} I_{\mathbf{p}}^{col,\alpha,\beta} \frac{d^3\mathbf{p}}{(2\pi)^3}. \tag{4.65}$$

We see that M^{α} is either the density n_{α} of particles of the kind α, in which case the collision integral itself is integrated over the momenta in Eq.(4.65), or the local momentum $\mathbf{P}_{\alpha}$ of particles of the kind α, in which case the collision integral is first multiplied by $\mathbf{p}$ before being integrated over all momenta, or it is the total local energy E_{α} of particles of the kind α, and then the collision integral is multiplied by ε_{α} before being integrated over all momenta. We can now find the following conservation laws:

1. *Conservation of particle numbers.* We find that the number of particles of each kind is conserved:

$$\left[\frac{\partial n_{\alpha}}{\partial t}\right]^{col} = 0. \tag{4.66}$$

This conservation law follows from the fact that the collision integral can be written as the derivative with respect of the momentum of an expression which, in fact, describes the flux of the particle distribution in momentum space.

If we now consider a kinetic equation with on the right-hand side the collision integral and on the left-hand side the expression

$$\frac{\partial \Phi_{\mathbf{p}}^{\alpha}}{\partial t} + \left(\mathbf{v} \cdot \frac{\partial \Phi_{\mathbf{p}}^{\alpha}}{\partial \mathbf{r}} \right) + \left(e\mathbf{E}_{\alpha} \cdot \frac{\partial \Phi_{\mathbf{p}}^{\alpha}}{\partial \mathbf{p}} \right),$$

we get after integration over the momenta and using Eq.(4.66) a continuity equation for particles of the kind α:

$$\frac{\partial n_{\alpha}}{\partial t} + \frac{\partial}{\partial \mathbf{r}} \left(n_{\alpha} \overline{\mathbf{v}}_{\alpha} \right) = 0, \tag{4.67}$$

where $\overline{\mathbf{v}}_{\alpha}$ is the average of the velocity over the particle distribution, as defined by Eq.(4.52). The second term on the left-hand side of the continuity equation (4.67) describes the convection of particles from one part of space to another. After integrating either Eq.(4.67) or the original kinetic equation over the whole of space we find the conservation of the total number of particles of the kind α:

$$\left[\frac{\partial N_{\alpha}}{\partial t} \right]^{\text{col}} = 0, \qquad N_{\alpha} = \int n_{\alpha} \, d^3\mathbf{r}. \tag{4.68}$$

2. *Conservation of momentum.* We can now find the conservation of the total local momentum of the whole system:

$$\sum_{\alpha} \left[\frac{\partial \mathbf{P}_{\alpha}}{\partial t} \right]^{\text{col}} = 0. \tag{4.69}$$

We can prove this conservation law by multiplying the collision integral by $\mathbf{p}$, integrating over $\mathbf{p}$, making the change of variables $\alpha \to \beta$, $\beta \to \alpha$, $\mathbf{p} \to \mathbf{p}'$, and $\mathbf{p}' \to \mathbf{p}$ and integrating again. It is important that the momentum is conserved only after summing over all particle types, because momentum can be transferred from one kind of particles to another. One cannot use the kinetic equation together with the momentum conservation law (4.69) and integration over the momenta to describe the convection of momentum of one kind of particles since the momentum of one kind of particles alone is not conserved. If we performed such an integration over momenta the left-hand side would take the form of a hydrodynamic equation, but the right-hand side has a simple form only in the limit of rapid collisions, that is, in the limit of the hydrodynamic description (see below).

3. *Conservation of energy.* The total energy of the whole system is also locally conserved in collisions:

$$\sum_{\alpha} \left[\frac{\partial E_{\alpha}}{\partial t} \right]^{\text{col}} = 0. \tag{4.70}$$

This conservation law can be proved in a way, similar to the proof of the momentum conservation law. In this case one multiplies by ε_p rather than p and one also makes use of the fact that, due to the appearance of the relevant δ-function, we have $(\mathbf{k} \cdot \mathbf{v}) = (\mathbf{k} \cdot \mathbf{v}')$. As in the case of the conservation of momentum conservation of energy only occurs after summation over all the different kinds of particles, since in collisions energy can be transferred from one kind of particle to another.

After integrating over momenta we find convective energy terms on the left-hand side of the equation with heat transfer terms on the right-hand side of the equation. The latter can be given an explicit form only in the hydrodynamic limit of rapid collisions.

We can introduce further simplifications in the limiting cases of rapid or of slow collisions. The *hydrodynamic description* considers the limit of low frequencies – long time intervals when the characteristic time of the process is much longer than the inverse collision frequency – and large scales of motion – characteristic dimensions in space much larger than the mean free path due to collisions. In this limit collisions take place so fast that one may assume that at any point in space and at any time local thermal distributions are established which make the collision integral vanish locally:

$$I_p^{col,\alpha}(\Phi_p^{\alpha(0)}, \Phi_p^{\beta(0)}) = 0, \tag{4.71}$$

where $\Phi_p^{\alpha(0)}$ and $\Phi_p^{\beta(0)}$ are the local thermal distribution functions. In this limit the collision integral is the dominating term; this enables us to neglect to a first approximation the convective terms on the left-hand side of the equations. Equation (4.71) is satisfied by thermal distributions with parameters – the density and the temperature – which vary slowly in time and space. These variations can be described by the convection terms, substituting on the left-hand side of the appropriate equations the thermal distribution functions $\Phi_p^{\alpha(0)}$ and $\Phi_p^{\beta(0)}$ as a first approximation. In the collision integral the deviations from the thermal distributions must be taken into account and this leads to complicated integral equations. The approximate solutions of these equations are then substituted into the conservation equations for each kind of particles and in that way we obtain the dissipative terms in the hydrodynamic equations which describe the thermal conductivity, friction, viscosity, and so on. This is the general scheme for obtaining a hydrodynamic description of a plasma.

The *kinetic description* corresponds to the opposite limit when the characteristic frequencies of the various processes are much higher than the collision frequency and the characteristic sizes much smaller than the mean free path between collisions. In the zeroth approximation the particle distributions satisfy the equation:

$$\frac{\partial \Phi_p^{\alpha(0)}}{\partial t} + \left(\mathbf{v} \cdot \frac{\partial \Phi_p^{\alpha(0)}}{\partial \mathbf{r}} \right) + \left(e_\alpha \mathbf{E} \cdot \frac{\partial \Phi_p^{\alpha(0)}}{\partial \mathbf{p}} \right) = 0. \tag{4.72}$$

In the next approximation we take into account the collision integral and the zeroth approximation distribution function is substituted into the collision integral to obtain the corrections to the distribution function due to collisions. A good example of the use of the kinetic approach is the problem of the damping of plasma oscillations due to collisions. We shall consider the general scheme for solving this problem. We write down the kinetic equation for the electrons in a wave field $\mathbf{E}$, taking into account the collisions:

$$\frac{\partial \Phi_{\mathbf{p}}^{e}}{\partial t} + \left(\mathbf{v} \cdot \frac{\partial \Phi_{\mathbf{p}}^{e}}{\partial \mathbf{r}} \right) + \left(e\mathbf{E} \cdot \frac{\partial \Phi_{\mathbf{p}}^{e}}{\partial \mathbf{p}} \right) = I_{\mathbf{p}}^{\text{col},e,i}. \tag{4.73}$$

The collisions make only a small contribution to the plasma dielectric permittivity because we have $N_{\mathrm{d}} \gg 1$. If there are no waves the stationary distribution $\Phi_{\mathbf{p}}^{e,\text{st}}$ is homogeneous and satisfies the equation

$$I_{\mathbf{p}}^{\text{col}}(\Phi_{\mathbf{p}}^{e,\text{st}}) = 0,$$

that is, it is a thermal distribution. The perturbation $\delta \Phi_{\mathbf{p}}^{e}$ of this distribution due to the wave field satisfies the equation

$$\frac{\partial \delta \Phi_{\mathbf{p}}^{e}}{\partial t} + \left(\mathbf{v} \cdot \frac{\partial \delta \Phi_{\mathbf{p}}^{e}}{\partial \mathbf{r}} \right) + \left(e\mathbf{E} \cdot \frac{\partial \Phi_{\mathbf{p}}^{e,\text{st}}}{\partial \mathbf{p}} \right) = I_{\mathbf{p}}^{\text{col},e,i}(\delta \Phi_{\mathbf{p}}^{e}). \tag{4.74}$$

If in the first approximation we neglect the collisions we have

$$\frac{\partial \delta \Phi_{\mathbf{p}}^{e(0)}}{\partial t} + \left(\mathbf{v} \cdot \frac{\partial \delta \Phi_{\mathbf{p}}^{e(0)}}{\partial \mathbf{r}} \right) + \left(e\mathbf{E} \cdot \frac{\partial \Phi_{\mathbf{p}}^{e,\text{st}}}{\partial \mathbf{p}} \right) = 0, \tag{4.75}$$

and for the next approximation we have

$$\frac{\partial \delta \Phi_{\mathbf{p}}^{e(1)}}{\partial t} + \left(\mathbf{v} \cdot \frac{\partial \delta \Phi_{\mathbf{p}}^{e(1)}}{\partial \mathbf{r}} \right) = I_{\mathbf{p}}^{\text{col}}(\delta \Phi_{\mathbf{p}}^{e(0)}). \tag{4.76}$$

After finding $\delta \Phi_{\mathbf{p}}^{e(1)}$ we can calculate the additional contribution to the dielectric permittivity and find the damping of the plasma oscilations due to collisions. For Langmuir oscillations we find

$$\delta \varepsilon_{\mathbf{k},\omega} = i \frac{\omega_{\text{pe}}^{2}}{\omega^{3}} \nu_{\text{ei,eff}}, \tag{4.77}$$

where $\nu_{\text{ei,eff}}$ is the effective electron-ion collision frequency,

$$\nu_{\text{ei,eff}} = \frac{4}{3} \sqrt{\frac{2\pi}{m_{e}}} \frac{e^{4} Z^{2} \Lambda}{T_{e}^{3/2}} n_{i} = \frac{1}{9} \sqrt{\frac{\pi}{2}} \frac{\omega_{\text{pe}}}{N_{\text{de}}} \frac{Z^{2} n_{i} \Lambda}{n_{e}}. \tag{4.78}$$

The damping rate of the Langmuir oscillations is

$$\gamma_{\mathbf{k}}^{\ell,\text{col}} = -\tfrac{1}{2} \nu_{\text{ei,eff}}. \tag{4.79}$$

For ion-sound waves the main contribution comes from ion-ion collisions:

$$\delta\varepsilon_{k,\omega} = i \frac{8\omega_{pi}^2 k^2 v_{Ti}^2}{5\omega^5} \nu_{ii,eff}, \qquad (4.80)$$

where

$$\nu_{ii,eff} = \frac{4}{3}\sqrt{\frac{\pi}{m_i}} \frac{e^4 Z^4 \Lambda}{T_i^{3/2}} n_i = \frac{\omega_{pi}}{9\sqrt{\pi}} \frac{\Lambda}{N_{di}}, \qquad (4.81)$$

and the damping rate is determined by the equation

$$\gamma_k^{s,col} = -\frac{4}{5}\frac{k^2 v_{Ti}^2}{(\omega_k^s)^2} \nu_{ii,eff}. \qquad (4.82)$$

We leave the details of calculating these damping rates as an exercise. We merely mention that for small wavenumbers, $k \ll d_e^{-1}$, the damping rate is independent of k:

$$\gamma_k^{s,col} = -\frac{4T_i}{5T_e} \nu_{ii,eff}, \qquad (4.83)$$

whereas the frequency ω_k^s decreases with decreasing k. Nevertheless, the damping will never be strong since for small k, that is for k satisfying the relation

$$k v_s \approx \nu_{ii,eff}, \qquad (4.84)$$

the kinetic description can no longer be applied. We must bear in mind that in the kinetic regime ion-sound waves exist only when $T_e \gg T_i$ whereas in a collision-dominated regime sound waves can exist even when $T_e = T_i$. Sound waves and ion-sound waves merge into one another for wavenumbers satisfying Eq.(4.84).

The kinetic approach which can be used for a collisionless plasma is applicable in a very broad range of frequencies and wavenumbers (see Fig.4.1).

4.7 Probabilities for Particle Collisions in a Plasma

Before we start considering the physics of particle collisions in a plasma we must ask ourselves what is the reason that collisions of particles when there are other particles present differ from those for isolated particles. For this purpose it is useful to introduce the probability for particle collisions in a plasma. We should note that according to the general laws of quantum mechanics the probability per unit time for some process to occur is proportional to the square of the matrix element corresponding to this process – denoted here by M – and proportional to a δ-function describing the energy conservation law for the process (we assume that the reader is familiar with the

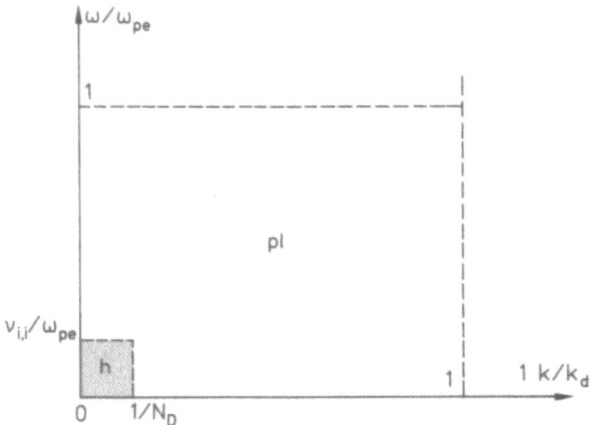

Fig. 4.1. Regions where the collisionless plasma approach (p) and the hydrodynamic approach (h) are applicable

relevant parts of quantum mechanics; we need them here only to draw analogies, since the whole description will be purely classical). Symbolically we can write the probability w for collisions in the form:

$$w \propto |M|^2 \delta(\textstyle\sum \varepsilon). \tag{4.85}$$

Although we are considering a classical collision process it is clear that we should be able to use the general quantum statement and take the classical limit and that it should be possible to introduce the probability for collisions using that limiting process. We shall return later on to the problem of how to express the collision integral in terms of this probability. For the moment we start with the statement that the collision integral should depend linearly on the probability. We then pay attention to the fact that the collision integral contains a δ-function and ask whether it expresses the conservation of energy in the collision process. Let us look at this δ-function:

$$\delta\big((\mathbf{k} \cdot \mathbf{v}) - (\mathbf{k} \cdot \mathbf{v}')\big). \tag{4.86}$$

The quantities $\mathbf{v}$ and $\mathbf{v}'$ are here the velocities of the colliding particles. Let us assume that in the collision a momentum $\mathbf{k}$ and an energy ω is transferred from one particle to another. To be more precise, let us assume that the particle with velocity $\mathbf{v}$ loses an amount of momentum $\mathbf{k}$ and an amount of energy ω, while the particle with velocity $\mathbf{v}'$ gains the momentum $\mathbf{k}$ and the energy ω. The initial energy of the first particle, with velocity $\mathbf{v}$, is thus $\varepsilon_{\mathbf{p}}$ and its final energy is $\varepsilon_{\mathbf{p}-\mathbf{k}}$. The conservation of energy and momentum for this particle can be expressed through the equation

$$\varepsilon_{\mathbf{p}} = \varepsilon_{\mathbf{p}-\mathbf{k}} + \omega. \tag{4.87}$$

In the case when the momentum transferred is small as compared to the initial particle momentum ($|\mathbf{k}| \ll |\mathbf{p}|$) this equation gives

$$\omega = (\mathbf{k} \cdot \mathbf{v}), \qquad \mathbf{v} = \frac{d\varepsilon_{\mathbf{p}}}{d\mathbf{p}}. \tag{4.88}$$

This is not a Vavilov-Cherenkov condition, as in a Vavilov-Cherenkov process a real wave is emitted with the frequency depending on the wavevector, $\omega = \omega_{\mathbf{k}}$, whereas in Eq.(4.88) we do not have such a dependence of the frequency on the wavevector. Equation (4.88) merely describes the fact that the self-field of the particle which is static in the rest frame has in the laboratory frame a frequency ω which is equal to $(\mathbf{k} \cdot \mathbf{v})$, due to the Doppler effect. Similar relations can be written for the particle with velocity $\mathbf{v}'$ which gains a momentum $\mathbf{k}$ and an energy ω:

$$\varepsilon_{\mathbf{p}'} = \varepsilon_{\mathbf{p}'+\mathbf{k}} - \omega, \tag{4.89}$$

and an expansion in terms of the small transferred momentum gives us

$$\omega = (\mathbf{k} \cdot \mathbf{v}'), \tag{4.90}$$

It is important that the ω and $\mathbf{k}$ in Eqs.(4.88) and (4.90) are the same, which means that

$$(\mathbf{k} \cdot \mathbf{v}) = (\mathbf{k} \cdot \mathbf{v}'). \tag{4.91}$$

The δ-function (4.86) thus describes, indeed, the conservation of energy in an elementary particle collision process in a plasma. We show schematically in Fig.4.2 a particle collision process involving the transfer of an amount of momentum $\mathbf{k}$ from one particle to another. We do not need to give the amount of transferred energy as it is determined by the transferred momentum: see the relation which we gave a moment ago.

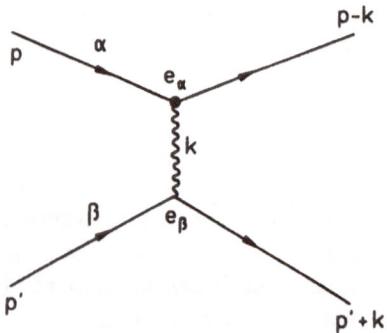

Fig. 4.2. Collision between two particles in a plasma

The transfer of energy and momentum from one particle to another takes place through the intermediary of a field which is not a real wave; however, the frequency of this field is determined by the wavevector and the particle velocity and, as in a wave, the frequency is known once we know the wavevector. This "wave" is usually called a *virtual wave*; it cannot exist if there is no particle. One can connect the field of the virtual wave with the self-field of a charge moving with a velocity $\mathbf{v}$. In the case of a unit charge we denote this field by $\mathbf{E}^{\mathbf{v}}$. Since the field we are considering is electrostatic we have $\mathbf{E}^{\mathbf{v}} = -\nabla G$ where G is the potential due to a unit charge which is also called the *Green function of the virtual field*. The charge density of a unit charge is $\varrho = \delta(\mathbf{r} - \mathbf{v}t)$ and its Fourier component is

$$\varrho_{\mathbf{k}} = \frac{\delta(\omega - (\mathbf{k} \cdot \mathbf{v}))}{(2\pi)^3}.$$

The Poisson equation gives

$$k^2 \varepsilon_{\mathbf{k},\omega} G_{\mathbf{k},\omega} = \frac{\delta(\omega - (\mathbf{k} \cdot \mathbf{v}))}{(2\pi)^3}. \tag{4.92}$$

The frequency of the field is here completely determined by the particle velocity $\mathbf{v}$ through the relation $\omega = (\mathbf{k} \cdot \mathbf{v})$. It is useful to integrate $G_{\mathbf{k},\omega}$ over ω, bearing in mind the relation between ω and $\mathbf{k}$:

$$G_{\mathbf{k}} = \int G_{\mathbf{k},\omega} \, d\omega = \frac{1}{2\pi^2 k^2 \varepsilon_{\mathbf{k},(\mathbf{k}\cdot\mathbf{v})}}. \tag{4.93}$$

One realises easily that the probability for a collision should contain this Green function $G_{\mathbf{k}}$; indeed, we have

$$|G_{\mathbf{k}}|^2 = \frac{1}{4\pi^4 k^4 |\varepsilon_{\mathbf{k},(\mathbf{k}\cdot\mathbf{v})}|^2}, \tag{4.94}$$

and the collision integral contains this expression as a factor. This gives a more translucent explanation of the reason why the dielectric permittivity in the collision frequency corresponds to the frequency $(\mathbf{k} \cdot \mathbf{v})$. It describes the dynamic screening of the fields of the colliding particles. We gave a brief discussion of the dynamic screening effect in Chap.2. However, we can see directly from Eq.(4.94) that for velocities well below the thermal velocities, $\omega = (\mathbf{k} \cdot \mathbf{v}) \ll k v_{T\alpha}$, the screening is almost static whereas for high velocities, $v \gg \omega_{\mathrm{pe}}/k$ or $k \gg \omega_{\mathrm{pe}}/v$, the screening becomes negligibly small. Considering Eq.(4.85) we can conclude that the matrix element for a collision should be proportional to the Green function G of the virtual field:

$$M \propto e_\alpha e_\beta G_{\mathbf{k}}. \tag{4.95}$$

The whole of this discussion only gives some strong indications, but no direct proof. However, we shall now find directly an equation for the probability for a collision and express it in terms of the Green function.

Let us denote the probability, which we normalise with the weight function $d^3k/(2\pi)^3$, for a collision per unit time of a particle of kind α with an initial momentum $\mathbf{p}$ with a particle of kind β with an initial momentum $\mathbf{p}'$ in which a momentum $\mathbf{k}$ within the range $\mathbf{k}, \mathbf{k} + d\mathbf{k}$ is transferred from the particle of kind α to the particle of kind β by

$$w^{\alpha,\beta}_{\mathbf{p},\mathbf{p}'}(\mathbf{k}). \tag{4.96}$$

After the collision the particles have the momenta $\mathbf{p} - \mathbf{k}$ and $\mathbf{p}' + \mathbf{k}$, respectively. The inverse process corresponds to a transition from a state with momenta $\mathbf{p} - \mathbf{k}$ and $\mathbf{p}' + \mathbf{k}$ with the transfer of a momentum $\mathbf{k}$ from the particle of kind β to the particle of kind α, that is, in the inverse process the momentum transferred from the particle of kind α to the particle of kind β is equal to $-\mathbf{k}$. The probabilities for the direct and the inverse processes are the same:

$$w^{\alpha,\beta}_{\mathbf{p}-\mathbf{k},\mathbf{p}'+\mathbf{k}}(-\mathbf{k}) = w^{\alpha,\beta}_{\mathbf{p},\mathbf{p}'}(\mathbf{k}). \tag{4.97}$$

We now expand all expressions in terms of the transferred momentum, assuming that it is small as compared to the momenta of the particles; we write

$$w^{\alpha,\beta}_{\mathbf{p},\mathbf{p}'}(\mathbf{k}) \simeq w^{\alpha,\beta(0)}_{\mathbf{p},\mathbf{p}'}(\mathbf{k}) + w^{\alpha,\beta(1)}_{\mathbf{p},\mathbf{p}'}(\mathbf{k}), \tag{4.98}$$

where the superscripts (0) and (1) denote, respectively, the zeroth and the first approximation in the transferred momentum. To zeroth approximation we have from Eq.(4.97)

$$w^{\alpha,\beta(0)}_{\mathbf{p},\mathbf{p}'}(\mathbf{k}) = w^{\alpha,\beta(0)}_{\mathbf{p},\mathbf{p}'}(-\mathbf{k}). \tag{4.99}$$

Substituting Eq.(4.98) into Eq.(4.97) we shall consider only the first-order corrections to Eq.(4.99). In the terms with $w^{\alpha,\beta(1)}_{\mathbf{p},\mathbf{p}'}(\mathbf{k})$ we can neglect the differences between $\mathbf{p} - \mathbf{k}$ and $\mathbf{p}$ and between $\mathbf{p}' + \mathbf{k}$ and $\mathbf{p}'$; in $w^{\alpha,\beta(0)}_{\mathbf{p}-\mathbf{k},\mathbf{p}'+\mathbf{k}}(\mathbf{k})$ we take into account just the first terms in the expansion in $\mathbf{k}$. The result is then

$$w^{\alpha,\beta(1)}_{\mathbf{p},\mathbf{p}'}(\mathbf{k}) - w^{\alpha,\beta(1)}_{\mathbf{p},\mathbf{p}'}(-\mathbf{k})$$
$$= -\left(\mathbf{k} \cdot \frac{\partial}{\partial \mathbf{p}}\right) w^{\alpha,\beta(0)}_{\mathbf{p},\mathbf{p}'}(\mathbf{k}) + \left(\mathbf{k} \cdot \frac{\partial}{\partial \mathbf{p}'}\right) w^{\alpha,\beta(0)}_{\mathbf{p},\mathbf{p}'}(\mathbf{k}). \tag{4.100}$$

This equation determines only that part of $w^{(1)}$ which is odd in $\mathbf{k}$; however, only that part makes a contribution to the result. For this odd part we have from Eq.(4.100)

$$w^{\alpha,\beta(1)}_{\mathbf{p},\mathbf{p}'}(\mathbf{k}) = -\tfrac{1}{2}\left(\mathbf{k} \cdot \frac{\partial}{\partial \mathbf{p}}\right) w^{\alpha,\beta(0)}_{\mathbf{p},\mathbf{p}'}(\mathbf{k}) + \tfrac{1}{2}\left(\mathbf{k} \cdot \frac{\partial}{\partial \mathbf{p}'}\right) w^{\alpha,\beta(0)}_{\mathbf{p},\mathbf{p}'}(\mathbf{k}). \tag{4.101}$$

We can now construct the balance equation for particles of the kind α. The decrease in the number of particles of kind α with momentum $\mathbf{p}$ is determined by the probability for collisions in which the momentum $\mathbf{p}$ changes multiplied by the numbers of particles of kind α and of kind β in the initial state, that is, $\Phi_{\mathbf{p}}^{\alpha}\Phi_{\mathbf{p}'}^{\beta}$, integrated over all possible values of the momentum $\mathbf{p}'$ of the particle of kind β and over all possible values of the transferred momentum $\mathbf{k}$:

$$\left[\frac{d\Phi_{\mathbf{p}}^{\alpha}}{dt}\right]_{-} = -\sum_{\beta}\int w_{\mathbf{p},\mathbf{p}'}^{\alpha,\beta}(\mathbf{k})\,\Phi_{\mathbf{p}}^{\alpha}\Phi_{\mathbf{p}'}^{\beta}\,\frac{d^3\mathbf{p}'\,d^3\mathbf{k}}{(2\pi)^6}. \tag{4.102}$$

The inverse process leads to an increase in the number of particles of kind α with momentum $\mathbf{p}$:

$$\left[\frac{d\Phi_{\mathbf{p}}^{\alpha}}{dt}\right]_{+} = \sum_{\beta} w_{\mathbf{p}-\mathbf{k},\mathbf{p}'+\mathbf{k}}^{\alpha,\beta}(-\mathbf{k})\,\Phi_{\mathbf{p}-\mathbf{k}}^{\alpha}\Phi_{\mathbf{p}'+\mathbf{k}}^{\beta}\,\frac{d^3\mathbf{p}'\,d^3\mathbf{k}}{(2\pi)^6}. \tag{4.103}$$

Introducing a change in the integration variables, $\mathbf{p}' \to \mathbf{p}' - \mathbf{k}$ and after that $\mathbf{k} \to -\mathbf{k}$ we get from Eq.(4.103):

$$\left[\frac{d\Phi_{\mathbf{p}}^{\alpha}}{dt}\right]_{+} = \sum_{\beta} w_{\mathbf{p}+\mathbf{k},\mathbf{p}'}^{\alpha,\beta}(\mathbf{k})\,\Phi_{\mathbf{p}+\mathbf{k}}^{\alpha}\Phi_{\mathbf{p}'}^{\beta}\,\frac{d^3\mathbf{p}'\,d^3\mathbf{k}}{(2\pi)^6}. \tag{4.104}$$

We use Eqs.(4.102) and (4.104) to obtain the balance equation which, after expanding in terms of $\mathbf{k}$, has the form

$$\frac{d\Phi_{\mathbf{p}}^{\alpha}}{dt} = \left[\frac{d\Phi_{\mathbf{p}}^{\alpha}}{dt}\right]_{+} + \left[\frac{d\Phi_{\mathbf{p}}^{\alpha}}{dt}\right]_{-}$$

$$= \frac{\partial}{\partial p_i}\frac{\partial}{\partial p_j}D_{ij}^{\alpha,\beta}\Phi_{\mathbf{p}}^{\alpha} + \frac{\partial}{\partial p_i}F_i'^{\alpha,\beta}\Phi_{\mathbf{p}}^{\alpha}, \tag{4.105}$$

$$D_{i,j}^{\alpha,\beta} = \tfrac{1}{2}\int k_i k_j\, w_{\mathbf{p},\mathbf{p}'}^{\alpha,\beta}(\mathbf{k})\,\Phi_{\mathbf{p}'}^{\beta}\,\frac{d^3\mathbf{p}'\,d^3\mathbf{k}}{(2\pi)^6}. \tag{4.106}$$

$$F_i'^{\alpha,\beta} = \int k_i\, w_{\mathbf{p},\mathbf{p}'}^{\alpha,\beta}(\mathbf{k})\,\Phi_{\mathbf{p}'}^{\beta}\,\frac{d^3\mathbf{p}'\,d^3\mathbf{k}}{(2\pi)^6}. \tag{4.107}$$

The equation which we have obtained can be written in the same form which we have used for the collision integral

$$\left[\frac{d\Phi_{\mathbf{p}}^{\alpha}}{dt}\right]^{col} = \frac{\partial}{\partial p_i}D_{ij}^{\alpha,\beta}\frac{\partial}{\partial p_j}\Phi_{\mathbf{p}}^{\alpha} + \frac{\partial}{\partial p_i}F_i^{\alpha,\beta}\Phi_{\mathbf{p}}^{\alpha}, \tag{4.108}$$

where

$$F_i^{\alpha,\beta} = F_i'^{\alpha,\beta} + \frac{\partial}{\partial p_j}D_{i,j}^{\alpha,\beta}. \tag{4.109}$$

In the expression for D_{ij} we can put $w = w^{(0)}$ within the approximation to which we are working, but in the expression for F_i we must also take into account the contribution from the part of $w^{(1)}$ which is odd in $\mathbf{k}$. Using Eqs.(4.101), (4.106), and (4.107) we then get

$$
\begin{aligned}
F_i^{\alpha,\beta} &= \int k_i\, w_{\mathbf{p},\mathbf{p}'}^{\alpha,\beta(1)}(\mathbf{k})\, \Phi_{\mathbf{p}'}^{\beta}\, \frac{d^3 p'\, d^3 k}{(2\pi)^6} \\
&\quad + \tfrac{1}{2} \int k_i \left(\mathbf{k}\cdot\frac{\partial}{\partial \mathbf{p}}\right) w_{\mathbf{p},\mathbf{p}'}^{\alpha,\beta(0)}(\mathbf{k})\, \Phi_{\mathbf{p}'}^{\beta}\, \frac{d^3 p'\, d^3 k}{(2\pi)^6} \\
&= \tfrac{1}{2} \int k_i\, \Phi_{\mathbf{p}'}^{\beta} \left(\mathbf{k}\cdot\frac{\partial}{\partial \mathbf{p}'} w_{\mathbf{p},\mathbf{p}'}^{\alpha,\beta(0)}\right) \frac{d^3 p'\, d^3 k}{(2\pi)^6} \\
&= -\tfrac{1}{2} \int k_i\, w_{\mathbf{p},\mathbf{p}'}^{\alpha,\beta(0)}(\mathbf{k}) \left(\mathbf{k}\cdot\frac{\partial \Phi_{\mathbf{p}'}^{\beta}}{\partial \mathbf{p}'}\right) \frac{d^3 p'\, d^3 k}{(2\pi)^6}.
\end{aligned}
\tag{4.110}
$$

In deriving Eq.(4.110) we cancelled terms containing derivatives with respect to $\mathbf{p}$ and we integrated by parts. We finally get the following expression which is the same as the one used earlier for the collision integral:

$$
\begin{aligned}
\left[\frac{d\Phi_{\mathbf{p}}^{\alpha}}{dt}\right]^{col} &= \tfrac{1}{2} \frac{\partial}{\partial p_i} \int k_i k_j\, w_{\mathbf{p},\mathbf{p}'}^{\alpha,\beta(0)}(\mathbf{k}) \\
&\quad \times \left[\Phi_{\mathbf{p}'}^{\beta} \frac{\partial \Phi_{\mathbf{p}}^{\alpha}}{\partial p_j} - \Phi_{\mathbf{p}}^{\alpha} \frac{\partial \Phi_{\mathbf{p}'}^{\beta}}{\partial \partial_j}\right] \frac{d^2 p'\, d^3 k}{(2\pi)^6}.
\end{aligned}
\tag{4.111}
$$

Comparing this with Eq.(4.34) we find the following expression for the probability for a collision:

$$
w_{\mathbf{p},\mathbf{p}'}^{\alpha,\beta(0)}(\mathbf{k}) = \frac{4 e_\alpha^2 e_\beta^2 (2\pi)^3}{k^4 |\varepsilon_{\mathbf{k},(\mathbf{k}\cdot\mathbf{v})}|^2}\, \delta\big((\mathbf{k}\cdot\mathbf{v}) - (\mathbf{k}\cdot\mathbf{v}')\big).
\tag{4.112}
$$

This result can not be obtained by methods which do not use fluctuations. The probability for collisions of particles of type α with particles of type β depends, according to Eq.(4.112), on the distribution function of all other particles through the dielectric permittivity which occurs in that formula. This is a *collective effect* in collisions. The appearance of the dielectric permittivity in the expression for the collision probability corresponds to the dynamic screening of the fields of the colliding particles. However, the number of screened particles is equal to the total number of particles. The screened particles are described by $\Phi_{\mathbf{p}}^{\alpha}$ and the exact distribution, of which $\Phi_{\mathbf{p}}^{\alpha}$ is the average, is described by $f_{\mathbf{p}}^{\alpha}$ so that we have

$$
\int \Phi_{\mathbf{p}}^{\alpha}\, \frac{d^3 p}{(2\pi)^3} = \int \langle f_{\mathbf{p}}^{\alpha}\rangle\, \frac{d^3 p}{(2\pi)^3}.
\tag{4.113}
$$

Let us remind ourselves that the particles are screened by charges of both sign. For instance, the electrons will be screened by a depletion of other electrons in the neighbourhood of the one we are considering. Because of the averaging no new electrons can appear. Due to the fluctuations each electron "has time" during the fluctuations to screen the other electrons. There are many electrons in a Debye sphere and only one electron needs to be screened. It appears that this picture of screened particles describes not only collisions but also other processes in a plasma, including the important non-linear processes. In principle, it might be the case that other processes would need a different physical scenario to describe the collective effects. One would then have to speak about "effective screening" which might be different for various processes. However, this is not the case. The picture of screened or "dressed" particles, which we found for the collision processes, can be transferred to other processes and will lead to qualitatively correct results which are the same as those obtained for them by considering fluctuations. For other processes the changes in the cross-sections for the processes considered are more dramatic than for collisions and cross-sections may be changed by many orders of magnitude by collective effects. Although the effect is not so dramatic for the collision process, the fact that the probability (cross-section) for a collision between two particles depends in Eq.(4.112) on the distribution functions of all other particles is very impressive. For low velocities the plasma particles behave as a gas of classical neutral atoms.

We must mention here that any other concepts which fail to take into account the "dressing" of the particles will be in conflict with the general physical ideas. This conclusion follows if we bear in mind that we obtained the effect of dynamic screening for the collision probability in Eq.(4.112) for any non-equilibrium particle distributions $\Phi_{\mathbf{p}}^{\alpha}$. Such a distribution might be a thermal distribution plus one particle which we might call an "external" charge. We can consider the problem of the interaction with an external charge separately and we would then obviously obtain the picture of the screening of this charge by the plasma particles. However, the charge we considered might have been any charge in the plasma and could have been called a probe charge. Since every charge is indistinguishable from the other charges in the plasma the dressing of the charges should be a general phenomenon. This is what we proved by considering fluctuations.

To conclude we shall give an expression for the collision probability (4.112) in terms of the Green function (4.94) of the virtual field:

$$w_{\mathbf{p},\mathbf{p'}}^{\alpha,\beta(0)}(\mathbf{k}) \;=\; 2\pi e_{\alpha}^2 e_{\beta}^2 (2\pi)^6 \, |G_{\mathbf{k}}|^2 \, \delta\big((\mathbf{k}\cdot\mathbf{v}) - (\mathbf{k}\cdot\mathbf{v'})\big), \tag{4.114}$$

or

$$w_{\mathbf{p},\mathbf{p'}}^{\alpha,\beta(0)}(\mathbf{k}) \;=\; 2\pi |M_{\mathbf{k}}|^2 \, \delta\big((\mathbf{k}\cdot\mathbf{v}) - (\mathbf{k}\cdot\mathbf{v'})\big), \tag{4.115}$$

$$M_{\mathbf{k}} \;=\; e_{\alpha} e_{\beta} (2\pi)^3 \, G_{\mathbf{k}}. \tag{4.116}$$

This expression is directly related to the one obtained in quantum theory; to be more precise, in the limit of a "bare" particle the matrix element (4.116) becomes the one known as the classical limit of the quantum expression. The domain of validity of the expression which we have obtained is connected with the process involving virtual longitudinal waves. The electromagnetic components of the fields produced by the particles are of order v/c where c is the velocity of light. The expression for the probability which we have obtained is thus valid provided

$$\frac{v^2}{c^2} \ll 1, \qquad \frac{v'^2}{c^2} \ll 1. \tag{4.117}$$

Appendix: Belyaev-Budker Collision Integral

In the case when the particle velocities are comparable with or even very close to the velocity of light ($\varepsilon_{\mathbf{p}} \gg mc^2$) we must take into account the electromagnetic fields generated by the particles. For an isotropic plasma one can split all fields, including the virtual fields, into longitudinal and transverse fields. The virtual longitudinal field is determined by the longitudinal dielectric permittivity $\varepsilon_{\mathbf{k},\omega}^{\ell}$ whereas the virtual transverse fields are determined by the transverse dielectric permittivity $\varepsilon_{\mathbf{k},\omega}^{t}$ where

$$\varepsilon_{ij,\mathbf{k},\omega} = \frac{k_i k_j}{k^2} \varepsilon_{\mathbf{k},\omega}^{\ell} + \left(\delta_{ij} - \frac{k_i k_j}{k^2} \right) \varepsilon_{\mathbf{k},\omega}^{t}. \tag{4.118}$$

As far as the virtual longitudinal fields are concerned all the results which we gave earlier remain valid for particles of arbitrary velocities if one makes the following substitution:

$$\varepsilon_{\mathbf{k},\omega} \rightarrow \varepsilon_{\mathbf{k},\omega}^{\ell}. \tag{4.119}$$

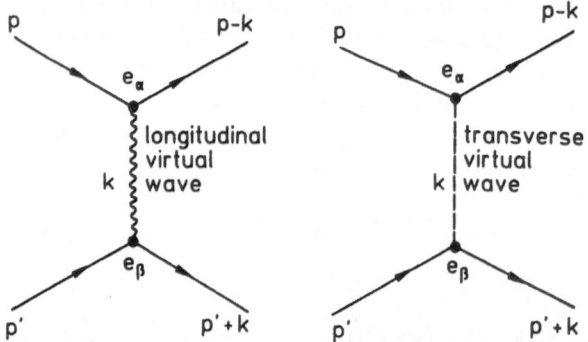

Fig. 4.3. Collision between two particles in a plasma involving either a virtual longitudinal or a virtual transverse wave

This follows from the fact that in the earlier discussion we nowhere stipulated that the particle velocities should be non-relativistic. We only made the assumption that the fields should be longitudinal. For higher velocities, close to c, the process involving a virtual transverse wave will be of the same order of magnitude as the process involving a virtual longitudinal wave. One should add the two processes with the process involving a virtual longitudinal wave being the same as before. The additional process involving a virtual transverse wave is shown in Fig.4.3. We shall give here only the final result for the processes shown in that figure:

$$w_{\mathbf{p},\mathbf{p}'}^{\alpha,\beta(\ell)}(\mathbf{k}) = \frac{4e_\alpha^2 e_\beta^2 (2\pi)^3}{k^4 |\varepsilon_{\mathbf{k},(\mathbf{k}\cdot\mathbf{v})}^\ell|^2} \delta\big((\mathbf{k}\cdot\mathbf{v}) - (\mathbf{k}\cdot\mathbf{v}')\big). \tag{4.120}$$

$$w_{\mathbf{p},\mathbf{p}'}^{\alpha,\beta(t)}(\mathbf{k}) = \frac{2e_\alpha^2 e_\beta^2 |[\mathbf{k}\wedge\mathbf{v}]|^2 |[\mathbf{k}\wedge\mathbf{v}']|^2 (2\pi)^3}{k^4 c^4 |k^2 - (\mathbf{k}\cdot\mathbf{v})^2 \varepsilon_{\mathbf{k},(\mathbf{k}\cdot\mathbf{v})}^t / c^2|^2} \delta\big((\mathbf{k}\cdot\mathbf{v}) - (\mathbf{k}\cdot\mathbf{v}')\big). \tag{4.121}$$

Equation (4.120) is the same as Eq.(4.112) apart from slight differences in notation. The transverse particle field which determines Eq.(4.121) is created by a current at right angles to $\mathbf{k}$. The charge of the particle is e_α and its current $e_\alpha \mathbf{v}$, so that the transverse component of the current will be $e_\alpha[\mathbf{v} - \mathbf{k}(\mathbf{k}\cdot\mathbf{v})/k^2]$. The vector potential $\mathbf{A}$ of the transverse field is given by the equation

$$\left(k^2 - \frac{\omega^2 \varepsilon_{\mathbf{k}}^t}{c^2}\right)\mathbf{A} = -\frac{4\pi e_\alpha}{c}\delta\big(\omega - (\mathbf{k}\cdot\mathbf{v})\big)\left(\mathbf{v} - \mathbf{k}\frac{(\mathbf{k}\cdot\mathbf{v})}{k^2}\right). \tag{4.122}$$

One can see from these expressions that one can change from the results for virtual longitudinal waves to those for virtual transverse waves by using the substitutions

$$e_\alpha \;\to\; \frac{e_\alpha}{c}\left(\mathbf{v} - \mathbf{k}\frac{(\mathbf{k}\cdot\mathbf{v})}{k^2}\right), \quad k^2 \varepsilon_{\mathbf{k},(\mathbf{k}\cdot\mathbf{v})}^\ell \;\to\; k^2 - (\mathbf{k}\cdot\mathbf{v})^2\frac{\varepsilon_{\mathbf{k},(\mathbf{k}\cdot\mathbf{v})}^t}{c^2} \tag{4.123}$$

and the relation

$$\left(\mathbf{v} - \mathbf{k}\frac{(\mathbf{k}\cdot\mathbf{v})}{k^2}\right)^2 = \frac{|[\mathbf{k}\wedge\mathbf{v}]|^2}{k^2}. \tag{4.124}$$

Using these relations we can find all the factors in Eq.(4.121) except an extra factor $\frac{1}{2}$. This factor arises as follows: the transverse field has two components and each of them corresponds to only half of the corresponding Green function. The total probability for collisions involving a virtual transverse field is the sum of two probabilities, each of which describing a collision involving one of the possible two polarisations of the virtual transverse field. As each of these polarisations corresponds to half of the Green function a factor $\frac{1}{4}$ arises for each of the two cases: hence the factor $\frac{1}{2}$.

Problems

1. Find the characteristic time for temperature equalisation for the case when the electron temperature T_e differs little from the ion temperature T_i: $\delta T \equiv T_e - T_i \ll T_e, T_i$.
2. Calculate the damping of Langmuir waves due to electron-ion collisions.
3. Calculate the damping of ion-sound waves due to ion-ion collisions.
4. Use the general expression for the friction force for a probe particle in a plasma to find the probabilities for particle collisions involving virtual longitudinal and transverse waves.
5. Use fluctuation theory to calculate the probabilities for particle collisions involving virtual transverse waves.

5 General Theory of Non-linear Interactions

5.1 Classification of Interactions

Most non-linearities considered in non-linear plasma physics are weak non-linearities. The criterion for a weak non-linearity is that the energy density of the collective oscillations should be much smaller than the particle energy density – in the case of a thermal distribution, the thermal particle energy density:

$$\frac{W}{nmv^2} \ll 1, \qquad \frac{W}{nT} \ll 1. \tag{5.1}$$

In fact, weak non-linearities occur when one can expand in the field strengths which means that sometimes one must substitute in the criterion (5.1) the phase velocity for the particle velocity. For Langmuir waves the parameter (5.1) can be written as the ratio of the square of the amplitude $v_\sim$ of the electron oscillations in the field of the wave and the square of the particle – or phase – velocity. If $(\mathbf{k} \cdot \mathbf{r}_\sim) \ll 1$ we can to a first approximation write

$$
\left.
\begin{aligned}
(\mathbf{k} \cdot \mathbf{r}) &\cong (\mathbf{k} \cdot \mathbf{v}t), \\
\mathbf{E} &= \mathbf{E}_0 \cos[\omega_{\mathrm{pe}}t - (\mathbf{k} \cdot \mathbf{r})]. \\
m_{\mathrm{e}} \frac{d\mathbf{v}_{\mathrm{e}}}{dt} &= e\mathbf{E}_0 \cos[\omega_{\mathrm{pe}}t - (\mathbf{k} \cdot \mathbf{v})t], \\
\mathbf{v}_{\mathrm{e}} &= \mathbf{v}_\sim \sin(\omega_{\mathrm{pe}}t), \\
\mathbf{v}_\sim &\approx \frac{e\mathbf{E}_0}{m_{\mathrm{e}}\omega_{\mathrm{pe}}}, \qquad (\mathbf{k} \cdot \mathbf{v}) \ll \omega_{\mathrm{pe}}, \\
\frac{v_\sim^2}{v_{T\mathrm{e}}^2} &= \frac{E_0^2}{4\pi n_{\mathrm{e}}T_{\mathrm{e}}} = \frac{W^\ell}{n_{\mathrm{e}}T_{\mathrm{e}}}.
\end{aligned}
\right\} \tag{5.2}
$$

For ion-sound waves we have $\omega_{\mathbf{k}}^{\mathrm{s}} \ll kv_{T\mathrm{e}}$ and $v_\sim \cong eE_0/m_{\mathrm{e}}kv_{T\mathrm{e}} \; (v \simeq v_{T\mathrm{e}})$, that is

$$
\begin{aligned}
\frac{v_\sim^2}{v_{T\mathrm{e}}^2} &\cong \frac{\omega_{\mathrm{pe}}^2}{k^2 v_{T\mathrm{e}}^2} \frac{E_0^2}{4\pi n_{\mathrm{e}}T_{\mathrm{e}}} \\
&\cong \frac{E_0^2}{4\pi n_{\mathrm{e}}T_{\mathrm{e}}} \left.\frac{\partial \varepsilon_{\mathbf{k}}}{\partial \omega}\right|_{\omega=\omega_{\mathbf{k}}^{\mathrm{s}}} \cong \frac{W^{\mathrm{s}}}{n_{\mathrm{e}}T_{\mathrm{e}}},
\end{aligned} \tag{5.3}
$$

and in this case also the parameter $v_\sim^2/v_{T_e}^2$ is the same as (5.1). The non-linear interactions lead to a spread $\delta\omega_{\mathbf{k}}^{NL}$ in the frequency of the oscillations; part of this will be the reciprocal of the characteristic time of the non-linear interactions. The frequency of the collective oscillations is therefore no longer well defined. A rough estimate of the width of the "line" describing a collective wave with a frequency close to $\omega_{\mathbf{k}}$ can be found from dimensional analysis; it should be proportional to the small expansion parameter for the case of a weak non-linearity:

$$\delta\omega_{\mathbf{k}}^{NL} \approx \xi\omega_{\mathbf{k}} \frac{W}{nT} \cong \xi\omega_{\mathbf{k}} \frac{v_\sim^2}{v_{T_e}^2}, \tag{5.4}$$

where ξ is a factor of the order of unity – we shall find the sign and the actual value of ξ below. For weak non-linearities $(v_\sim^2/v_T^2 \ll 1)$ the corresponding "line" is still narrow:

$$\frac{\delta\omega_{\mathbf{k}}^{NL}}{\omega_{\mathbf{k}}} \ll 1. \tag{5.5}$$

This inequality can be considered to be a necessary criterion for a weak non-linearity as it shows that the waves are only weakly modified by the non-linear interactions. The presence of a non-linear width is very important for the non-linear interaction processes. When we speak about non-linear interactions we have in mind an interaction of two or more modes and clearly this interaction will be different in the case when the difference $\omega_{\mathbf{k}} - \omega_{\mathbf{k}'}$ between the frequencies of the interacting modes is larger than the widths $\delta\omega_{\mathbf{k}}^{NL}$, $\delta\omega_{\mathbf{k}'}^{NL}$ of the individual modes from the case when this difference is smaller than the widths. For example, if one of the two interacting waves, say, the one with the wavevector $\mathbf{k}$, is a strong one and the other, with wavevector $\mathbf{k}'$ much weaker one may expect that in the case where the frequency of the weak wave lies inside the width of the strong wave the latter will dominate the dynamics of the weak one. In particular, the phase of the weak one may become synchronised with the phase of the strong wave – a process known as *phase locking*. Another example of such kind of interaction is the interaction of a high-frequency wave with a low-frequency wave for the case when the width of the high-frequency wave is larger than the frequency of the low-frequency wave. For instance, these waves might be a Langmuir wave and an ion-sound wave. If the width $\delta\omega_{\mathbf{k}}^{NL}$ of the high-frequency wave is large one can hardly expect a low-frequency wave with not too large an amplitude to exist.

Since the non-linear frequency width increases when the amplitude of the wave increases there should be a critical energy density W_{cr} of the waves for which this width becomes equal to the frequency difference between interacting waves. For Langmuir waves, for instance, the difference between the frequencies of two waves,

$$\omega_{\mathbf{k}}^{\ell} - \omega_{\mathbf{k}'}^{\ell} \simeq \frac{3 v_{Te}^2 (k^2 - k'^2)}{2 \omega_{pe}},$$

is always small as compared to the frequency ω_{pe}. If the value of k is of the same order as that of k', the difference between the frequencies will be of the order of

$$\omega_{\mathbf{k}}^{\ell} - \omega_{\mathbf{k}'}^{\ell} \approx \omega_{pe} k^2 d_e^2 \ll \omega_{\mathbf{k}}^{\ell}. \tag{5.6}$$

Comparing this equation with Eq.(5.5) we obtain an estimate for the critical energy density in the case of Langmuir waves:

$$\frac{W_{cr}}{nT_e} \approx k^2 d_e^2 \ll 1. \tag{5.7}$$

The non-linear interactions are rather strong if the energy density of the waves is larger than the level – the thermal level – of the statistical fluctuations in the particle energy density, $W/nT \gg 1/N_d$, and a weak non-linearity corresponds to the range determined by the inequalities

$$\frac{1}{N_d} \ll \frac{W}{nT} \ll 1; \tag{5.8}$$

we can now for this range distinguish two cases:

$$\frac{1}{N_d} \ll \frac{W}{nT} \ll \frac{W_{cr}}{nT} \ll 1, \tag{5.9}$$

and

$$\frac{1}{N_d} \ll \frac{W_{cr}}{nT} \ll \frac{W}{nT} \ll 1. \tag{5.10}$$

Both cases correspond to weak non-linearities but in the first case the phase-locking effect is small, non-linear processes can randomise the phases of the interacting waves, and one can use a kinetic description of the interacting waves. On the other hand, in the second case the phase-locking effect can synchronise oscillations which have frequencies which are close to each other and can produce in that way coherent non-linear structures.

The magnitude of W_{cr} is not large for Langmuir waves. For instance, in the case when these waves are excited by a beam we have $k \approx \omega_{pe}/v_b$ and we find $W_{cr}/n_e T_e \approx v_{Te}^2/v_b^2$. If we take into account that in the case of a quasi-linear relaxation of a beam in a plasma the energy density of the Langmuir waves reaches a level $W \approx n_b m_e v_b^2$, the relation $W \ll W_{cr}$, corresponding to (5.9), gives the following inequality for the beam density:

$$\frac{n_b}{n_e} \ll \frac{v_{Te}^4}{v_b^4}. \tag{5.11}$$

This relation is often not satisfied for the powerful relativistic beams which exist, for which $v_b \approx c$. This means that the situation should be described by Eq.(5.10).

Another example is the excitation of Langmuir waves by a laser beam when the energy density of the excited waves can be of the order of the energy density W_0 of the laser while the wavelength of the excited waves is of the order of the wavelength of the laser. As the frequency of the laser radiation should be close to $\omega_{\rm pe}$ for the excitation of Langmuir waves we have $k \approx \omega_{\rm pe}/c$ so that the relation $W \ll W_{\rm cr}$ can be written in the form

$$\frac{W_0}{nT_e} \ll \frac{v_{Te}^2}{c^2}. \tag{5.12}$$

If the energy density of the laser radiation is comparable to the thermal particle energy density Eq.(5.10) is usually satisfied. We see thus that for powerful laser and particle beams the $W \gg W_{\rm cr}$ case is of most interest, but if the excitation of the waves is not too strong the $W \ll W_{\rm cr}$ case will be the more important one.

In the case of ion-sound waves the dispersion is not small so that in that case $W_{\rm cr}$ is large. In the present chapter we shall give a theory of non-linear interactions for both the $W \gg W_{\rm cr}$ and the $W \ll W_{\rm cr}$ limits. This theory uses only a single small parameter, $W/nT \ll 1$. However, the description obtained from the general theory will be different in the two limiting cases. In the $W \ll W_{\rm cr}$ case we can clearly distinguish the resonances in the non-linear interactions, whereas in the $W \gg W_{\rm cr}$ case the resonances are smeared out by the non-linear interactions.

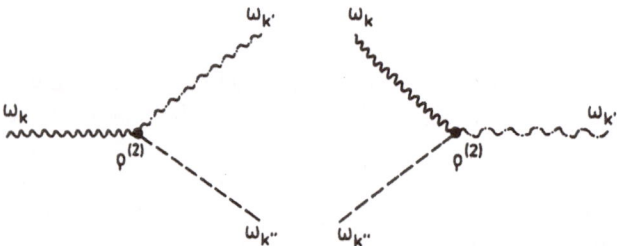

Fig. 5.1. Decay and fusion of two waves in a plasma

When $W \ll W_{\rm cr}$ we can classify the interactions by the resonances of the non-linear interactions. The simplest resonance corresponds to the excitation of a third wave by two waves which are already present in the plasma. For each of the waves the field is proportional to a factor $e^{\mp i[\omega_k t - (k \cdot r)]}$. For the first wave we put $k = k'$ and for the second one $k = k''$. The non-linear charge density $\varrho_{k,\omega}^{\rm NL}$, which is proportional to the product of these fields, will contain a factor $\exp[\mp i(\omega_{k'} \pm \omega_{k''})t \pm i(\{k' \pm k''\} \cdot r)]$. This shows that the frequency ω and the wavevector $\mathbf{k}$ of the source, which excites the third wave, are, respectively,

$$\omega = \omega_{\mathbf{k}'} \pm \omega_{\mathbf{k}''}, \qquad \mathbf{k} = \mathbf{k}' \pm \mathbf{k}''. \tag{5.13}$$

In the Poisson equation,

$$\mathbf{E}_{\mathbf{k},\omega} = \frac{\mathbf{k}}{ik^2 \varepsilon_{\mathbf{k},\omega}} \varrho_{\mathbf{k},\omega}^{NL}, \tag{5.14}$$

there will be a resonance for the field $\mathbf{E}_{\mathbf{k},\omega}$ of the third wave if the dielectric permittivity in Eq.(5.14) is close to zero, that is, if $\omega \cong \omega_{\mathbf{k}}$, or, according to Eq.(5.13)

$$\omega_{\mathbf{k}} = \omega_{\mathbf{k}'} \pm \omega_{\mathbf{k}''}, \qquad \mathbf{k} = \mathbf{k}' \pm \mathbf{k}''. \tag{5.15}$$

The equations are called the *decay equations* and the non-linear process is called a *decay process*. One has to bear in mind that in talking about a decay process, we also describe a *fusion process*, that is the fusion of two waves into a single wave (see Fig.5.1).

Non-linear interactions of higher order in the fields can lead to resonances involving more than three waves (see Fig.5.2).

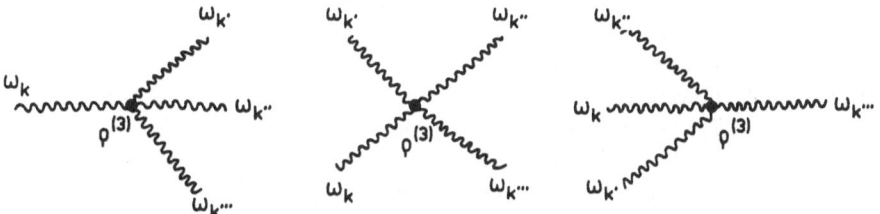

Fig. 5.2. A four-wave decay process in a plasma

As an example we shall estimate the critical energy density of Langmuir waves when the decay of a Langmuir wave into another Langmuir wave and an ion-sound wave is a resonance decay process. In this case the non-linear frequency width of the Langmuir wave must be less than the frequency of the ion-sound wave. From Eq.(5.15) we have

$$\omega_{\mathbf{k}+\mathbf{k}'}^s = |\mathbf{k}+\mathbf{k}'| v_s = \omega_{\mathbf{k}}^\ell - \omega_{\mathbf{k}'}^\ell = \frac{3 v_{Te}^2 (k^2 - k'^2)}{2\omega_{pe}},$$

and if k' is of the same order of magnitude as k, we have

$$k \approx \sqrt{\frac{m_e}{m_i}} \frac{1}{3 d_e}.$$

We thus find

$$\frac{W_{cr}}{nT_e} \approx \frac{m_e}{9m_i}.$$
(5.16)

This decay process occurs only in the $W \ll W_{cr}$ case. In deriving Eq.(5.16) we used Eq.(5.7). The quantity $m_e/9m_i$, which for hydrogen ions is approximately 10^{-4}, is rather small, but $1/N_d$ can be even much smaller and the condition $W \ll W_{cr}$ can often be met in experiments. We shall discuss decay processes in Chap.8.

Another type of resonance is resonance with particles. We showed earlier that the linear resonance with particles is described by the condition $\omega = (\mathbf{k} \cdot \mathbf{v})$. For non-linear interactions ω and $\mathbf{k}$ satisfy Eqs.(5.13), that is, non-linear resonance is described by the condition

$$\omega_\mathbf{k} \pm \omega_{\mathbf{k}'} = \{\mathbf{k} \pm \mathbf{k}'\} \cdot \mathbf{v}).$$
(5.17)

Such processes are called *stimulated scattering processes* or *stimulated emission or absorption processes* in which two waves are scattered by a plasma particle. In the case of stimulated emission we have the minus sign in Eq.(5.17) and in the case of stimulated absorption the plus sign. A condition for the occurrence of the resonances (5.17) is again that we must have $W \ll W_{cr}$. The scattering is strongly affected by the non-linear broadening as it contains the difference of the frequencies of the two waves. The value of W_{cr} for a scattering process is of the same order of magnitude as for a decay process. We shall see that in the case when $T_e = T_i$, when the decay process of a Langmuir wave into another Langmuir wave and an ion-sound wave is not allowed, W_{cr} for the scattering process is equal to (5.16), whereas in the $T_e \gg T_i$ case, when the decay process is allowed, W_{cr} for the scattering process is even smaller than (5.16) – it contains an extra factor T_i/T_e, see Chap.6. In the case when for a given wavevector both scattering and decay are possible, decay will dominate (see Chap.8) and Eq.(5.16) is therefore appropriate. We shall discuss the stimulated scattering processes in the next chapter.

In the $W \gg W_{cr}$ case the resonances are smeared out by the non-linear interactions but the physics of the non-linear interactions is similar in the sense that one of the interacting waves is modulating the plasma parameters and other waves are propagating in the medium with modulated parameters and emit or absorb waves or interact with the particles. The modulation of the plasma parameters is the main process which leads to the other ones. Since there are no resonances in the $W \gg W_{cr}$ case we shall call this limit the case of *modulational non-linear interactions*. Chapters 9 and 10 are devoted to discussing them. We shall show there that the decay processes, the stimulated scattering processes, and the modulational interaction processes are closely related to one another.

There is another kind of non-linear interactions which exists both when $W \ll W_{cr}$ and when $W \gg W_{cr}$; it is the *transitional damping* which is an amplification of non-resonant waves by waves which are in linear resonance with the particles, corresponding to a wave-particle resonance $\omega_{\mathbf{k}} = (\mathbf{k} \cdot \mathbf{v})$. This process is not determined by a frequency difference, but by the frequency itself; it is practically independent of W_{cr}, but depends on the inhomogeneities in the density of the resonant particles. It leads to a non-linear maser effect; these interactions will be considered in Chap.11.

An important problem is the effect of collisions on non-linear interactions. Collisions are related to statistical particle fluctuations. If one neglects these fluctuations we are dealing with *collisionless* non-linear interactions. The criterion for neglecting collisions when discussing non-linear interactions is

$$\omega_{\mathbf{k}} - \omega_{\mathbf{k}'} \gg \nu_{\text{eff}} \tag{5.18}$$

for the case when $\delta\omega_{\mathbf{k}}^{\text{NL}} \ll \omega_{\mathbf{k}} - \omega_{\mathbf{k}'}$, and

$$\delta\omega_{\mathbf{k}}^{\text{NL}} \gg \nu_{\text{eff}} \tag{5.19}$$

for the opposite case when $\delta\omega_{\mathbf{k}}^{\text{NL}} \gg \omega_{\mathbf{k}} - \omega_{\mathbf{k}'}$. In the case when Eq.(5.19) is not satisfied we call the non-linear interactions *modulational collision interactions*.

There are also non-linear interactions between the collective oscillations – the waves – and the fields corresponding to the statistical particle fluctuations. These processes also involve collisions and the non-linear damping of the waves due to these interactions describe *stimulated bremsstrahlung processes*. We shall discuss these processes in Chap.12 and we shall consider collective processes related to bremsstrahlung in Chap.13.

5.2 General Theory of Weak Non-linear Interactions

We shall start with a general description of the non-linear interactions of electrostatic fields and then consider in detail, as an example, the non-linear interactions of Langmuir waves. The mathematical description and physical interpretation of the non-linear interactions of Langmuir waves enables us to derive also the interactions between ion-sound waves. On the other hand, the non-linear interactions between Langmuir waves contain all the information necessary to construct a theory of the interactions between Langmuir and ion-sound waves, as even when there are no ion-sound waves present originally, the Langmuir waves can excite them. We shall thus construct a complete theory of the non-linear interactions of electrostatic oscillations in a plasma – for the case when there is no magnetic field present. The theory of non-linear interactions between electromagnetic waves or between electromagnetic and electrostatic waves is similar to the theory of non-linear interations between electrostatic waves.

The Fourier components of a Langmuir field have both a positive and a negative frequency component. Since the Langmuir field is a longitudinal

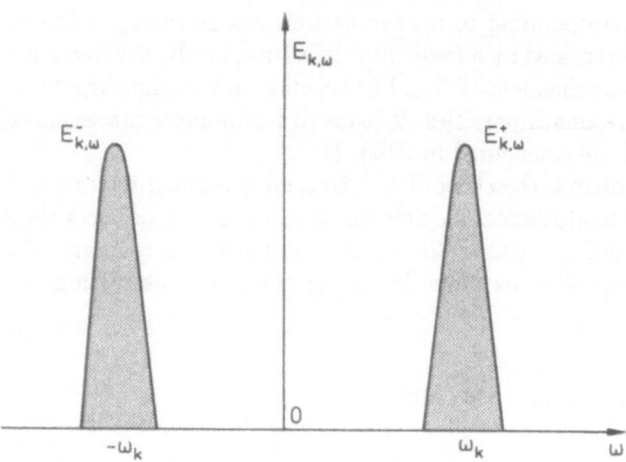

Fig. 5.3. The Fourier components of the Langmuir field in a plasma

field we have $\mathbf{E}_{\mathbf{k},\omega} = (\mathbf{k}/|\mathbf{k}|)E_{\mathbf{k},\omega}$ and we shall consider the magnitude $E_{\mathbf{k},\omega}$ of the field, which is a scalar, rather than $\mathbf{E}_{\mathbf{k},\omega}$; we shall denote the positive-frequency component of it by $E^+_{\mathbf{k},\omega}$ and the negative-frequency component by $E^-_{\mathbf{k},\omega}$. The positive-frequency component contains the positive frequency $\omega^\ell_{\mathbf{k}}$ – which is broadened by non-linear interactions – and the negative-frequency component contains the broadened negative frequency $-\omega^\ell_{\mathbf{k}}$ (see Fig.5.3).

Figure 5.3 was drawn for a given, fixed value of the wavevector **k**. Let us now take into account the fact that the frequency of a Langmuir wave is always close to $\pm\omega_{\mathrm{pe}}$ and that the dispersion corrections are small as compared to ω_{pe}. Waves with different values of **k** will have different frequencies, but they will all be close to ω_{pe}. This means that we can put all the waves with different values of **k** in a single figure and still have a relatively narrow "line". The width of that "line" will be determined both by the non-linear processes, that is, by $\delta\omega^{\mathrm{NL}}$, and by the dispersion, that is, by $\delta\omega^{\mathrm{d}}$. It is important for what follows that the total width, $\delta\omega = \delta\omega^{\mathrm{NL}} + \delta\omega^{\mathrm{d}}$ is relatively small: $\delta\omega \ll \omega_{\mathrm{pe}}$ (see Fig.5.4).

In Fig.5.4 we plot horizontally the ratio of the frequency ω to the plasma frequency ω_{pe}. We also show in that figure the so-called *virtual fields* E^{v} which appear at the doubled frequency and close to zero frequency. These are called virtual waves or virtual fields since they cannot propagate in a plasma unless there is a source present; a propagating Langmuir wave can be such a source. We have denoted these virtual fields by the superscript "v". Langmuir waves will always excite such virtual waves, but they will exist only in those locations where there are Langmuir waves and the amplitude of the virtual waves is approximately $v_\sim/v_{Te}$ times smaller than the amplitude of the Langmuir wave.

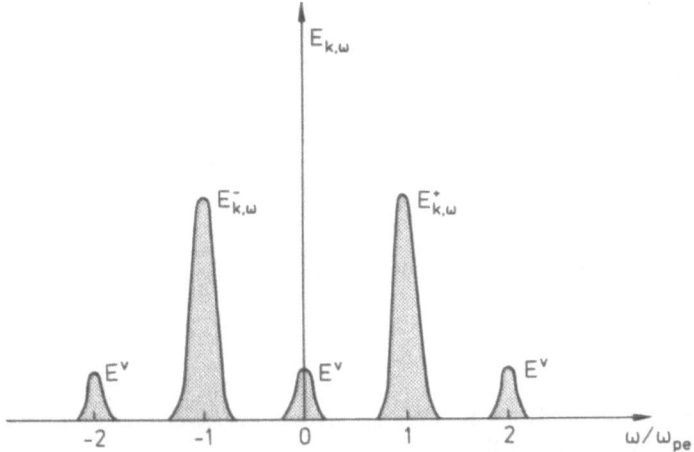

Fig. 5.4. The appearance of virtual fields when there are non-linear interactions between Langmuir waves

The plasma responds not only linearly to an applied field **E**, but also non-linearly, that is, the charge density induced by such a field will have components containing the square of the field strength (quadratic non-linearities), the cube of the field strength (cubic non-linearities), and so on. The quadratic non-linearities for Langmuir waves are special cases and are small; we shall therefore consider here both the quadratic and the cubic non-linearities. We can write down a general equation for the non-linear charge density ϱ^{NL}, or its Fourier components $\varrho^{\mathrm{NL}}_{\mathbf{k},\omega} = \varrho^{\mathrm{NL}}_{\mathbf{k}}$ ($\mathbf{k} \equiv \{\mathbf{k},\omega\}$) both for the quadratic and for the cubic non-linearities. First of all, let us remind ourselves that the Fourier component of the product AB of any two quantities A and B satisfies the well known relation

$$(AB)_{\mathbf{k},\omega} = (AB)_{\mathbf{k}}$$
$$= \int A_{\mathbf{k}_1} B_{\mathbf{k}-\mathbf{k}_1}\, d^4 k_1 = \int A_{\mathbf{k}_1,\omega_1} B_{\mathbf{k}-\mathbf{k}_1,\omega-\omega_1}\, d^3 k_1\, d\omega. \quad (5.20\mathrm{a})$$

We must stress that this is purely a mathematical relation and that it does not have any relations with decay processes, even though we can write it in the form

$$(AB)_{\mathbf{k}} = \int A_{\mathbf{k}_1} B_{\mathbf{k}_2} \delta(\mathbf{k}-\mathbf{k}_1-\mathbf{k}_2)\, d^4 k_1\, d^4 k_2$$
$$= \int A_{\mathbf{k}_1} B_{\mathbf{k}_2} \delta(\mathbf{k}-\mathbf{k}_1-\mathbf{k}_2)\delta(\omega-\omega_1-\omega_2)\, d^3 k_1\, d^3 k_2\, d\omega_1\, d\omega_2, \quad (5.20\mathrm{b})$$

which means that we have $\mathbf{k} = \mathbf{k}_1 + \mathbf{k}_2$, $\omega = \omega_1 + \omega_2$. The difference between these relations and the decay relations is that in the latter the frequencies are those of definite plasma modes which are given functions of the wavevector, whereas here the frequency is arbitrary. Using Eqs.(5.20) we can write down a general relation for the case of quadratic and cubic non-linearities:

$$
\varrho_k^{NL} = \int \varrho_{k_1,k_2}^{(2)} E_{k_1} E_{k_2} \delta(k - k_1 - k_2) \, d^4k_1 \, d^4k_2
$$

$$
+ \int \varrho_{k_1,k_2,k_3}^{(3)} E_{k_1} E_{k_2} E_{k_3} \delta(k - k_1 - k_2 - k_3) \, d^4k_1 \, d^4k_2 \, d^4k_3. \quad (5.21)
$$

These relations can be considered to be the definitions of the non-linear plasma response coefficients

$$
\varrho_{k_1,k_2}^{(2)} \quad \text{and} \quad \varrho_{k_1,k_2,k_3}^{(3)}.
$$

None of the δ-functions in Eq.(5.21) describe decay processes, even if all the fields on the right-hand side of Eq.(5.21) are fields of waves, that is, if

$$
\omega_1 = \omega_{k_1}, \qquad \omega_2 = \omega_{k_2}, \qquad \omega_3 = \omega_{k_3},
$$

since for a decay process it is necessary to have a further condition satisfied, namely, the resonance condition for the wave excited by the non-linear charge density, that is, we need to have $\omega = \omega_k$. Resonant decay processes thus will have a special form!

$$
\begin{aligned}
\omega_k &= \omega_{k_1} + \omega_{k_2}, & k &= k_1 + k_2; \\
\omega_k &= \omega_{k_1} + \omega_{k_2} + \omega_{k_3}, & k &= k_1 + k_2 + k_3,
\end{aligned}
$$

whereas the relations $\omega = \omega_1 + \omega_2$ and $\omega = \omega_1 + \omega_2 + \omega_3$ are just the consequence of the Fourier transformation of products. We shall, by the way, see later on that in these last relations virtual waves will be involved which, of course, do not have any definite frequency-wavevector dependences.

We must substitute the non-linear charge density (5.21) into the Poisson equation

$$
ik\varepsilon_k E_k = 4\pi \varrho_k^{NL}. \quad (5.22)
$$

Note that this equation is consistent: the right-hand side is a small quantity since the non-linearity is small, whereas the left-hand side is small since the solution of the equation should be close to the linear solution which means that ε_k is small. However, the question arises why we must take into account two terms of the expansion in the field strength of the plasma non-linear response, rather than just a single one – the quadratic non-linearity. To understand this, let us consider Eq.(5.21) and Fig.5.4, and try and see whether quadratic non-linearities can work for Langmuir waves. If the frequencies of both the fields in the non-linear quadratic response are close to the Langmuir frequency – close to +1 or to −1 in Fig.5.4 – we can have the combinations $+1 + 1 = 2$, $+1 - 1 = -1 + 1 = 0$, or $-1 - 1 = -2$, that is, none of the sum frequencies is the Langmuir frequency. This means that this interaction creates virtual fields. The dielectric permittivity is not small for the frequencies of the virtual field and it follows therefore from Eq.(5.22) that the strengths of these virtual fields are small. To find the amplitude of the virtual fields we need consider only the quadratic non-linearities in the Poisson equation:

$$E_k = \frac{4\pi}{ik\varepsilon_k} \int \varrho^{(2)}_{k_1,k_2} E_{k_1} E_{k_2} \delta(k - k_1 - k_2) d^4k_1 d^4k_2. \tag{5.23}$$

We now simplify our notation by putting

$$\left.\begin{array}{l} E_{k_1} \rightarrow E_1, \qquad E_{k_2} \rightarrow E_2, \qquad E_k \rightarrow E, \\[2mm] \varrho^{(2)}_{k_1,k_2} \rightarrow \varrho^{(2)}_{1,2}, \qquad \varepsilon_k = \varepsilon_{k_1+k_2} \rightarrow \varepsilon_{1+2}, \\[2mm] \delta(k - k_1 - k_2) d^4k_1 d^4k_2 \rightarrow d_{1,2}. \end{array}\right\} \tag{5.24}$$

Using this notation we write Eq.(5.23) in the form

$$E_k = \frac{4\pi}{ik} \int \frac{1}{\varepsilon_{1+2}} \varrho^{(2)}_{1,2} E_1 E_2 d_{1,2}. \tag{5.25}$$

Substituting on the right-hand side of this equation the Langmuir fields $E_1^{\pm}$ and $E_2^{\pm}$ we got the virtual fields at the second harmonic, $E^{v(\pm 2)}$, and at the zero frequency (or rather close to zero, as compared to the plasma frequency), $E^{v(0)}$, as follows:

$$E^{v(0)} = \frac{8\pi}{ik} \int \frac{1}{\varepsilon_{1+2}} \varrho^{(2)}_{1,2} E_1^+ E_2^- d_{1,2}, \tag{5.26}$$

$$E^{v(+2)} = \frac{4\pi}{ik} \int \frac{1}{\varepsilon_{1+2}} \varrho^{(2)}_{1,2} E_1^+ E_2^+ d_{1,2}, \tag{5.27}$$

$$E^{v(-2)} = \frac{4\pi}{ik} \int \frac{1}{\varepsilon_{1+2}} \varrho^{(2)}_{1,2} E_1^- E_2^- d_{1,2}. \tag{5.28}$$

We have assumed here and shall assume in what follows that the quadratic non-linear response coefficient $\varrho^{(2)}_{1,2}$ is symmetric in the indices 1 and 2:

$$\varrho^{(2)}_{1,2} = \varrho^{(2)}_{2,1}. \tag{5.29}$$

This can always be achieved by a suitable choice of notation in Eq.(5.25). This symmetry of the quadratic non-linear response led in Eq.(5.26) to the factor 8π rather than 4π because of the two possible combinations of E^+ and E^- fields.

We see that we can find the virtual fields accompanying the Langmuir field, once we know the $\varrho^{(2)}_{1,2}$ coefficient. Since ε_{1+2} is not small the ratio of the strength of one of the virtual fields to that of the Langmuir field will be of the order of $v_\sim/v_T \ll 1$. However, sometimes they can be even smaller: we shall show in a moment that the virtual fields $E^{v(\pm 2)}$ are smaller than the virtual field $E^{v(0)}$ by the small factor $v_{Te}^2/v_{ph}^2 \ll 1$. The $E^{v(\pm 3)}$ and $E^{v(\pm 4)}$

fields are even smaller as they are the next terms in an expansion in the small parameter $v_\sim / v_T$. In the case of a Langmuir field $E^{v(0)}$ is the most important virtual field.

Not only do the virtual fields accompany the Langmuir field, they, in fact, determine their non-linear interaction processes. Indeed, let us consider the quadratic non-linearity (5.25) for a Langmuir field, assuming that on the left-hand side of this equation the frequency is close to the frequency of the E^+ Langmuir field. The dimensionless frequency 1 can then be obtained only from the two combinations $1 = 1 + 0$ and $= 2 - 1$. This means, if we write it in the form of a non-linear equation involving a quadratic non-linearity, that we have

$$ik\varepsilon E^+ = 8\pi \varrho^{(2)}_{1,2} E_1^+ E_2^{v(0)} d_{1,2} + 8\pi \varrho^{(2)}_{1,2} E_1^{v(+2)} E_2^- d_{1,2}. \qquad (5.30)$$

We can write down a similar equation for E^-. We now see the reason why the quadratic non-linearities are specially small in the case of a Langmuir field: they always contain a virtual field which is at least $v_\sim / v_{Te}$ times smaller than the Langmuir field and they are therefore of the same order as the cubic non-linearities. The quadratic non-linearity is therefore effectively a cubic non-linearity. This effective cubic non-linearity can be found from Eq.(5.30) by substituting Eqs.(5.26) to (5.28) for the virtual fields. It should be taken into account together with the direct cubic non-linearities.

It is useful to simplify the notation again when we consider the cubic non-linearities and we shall write

$$\varrho^{(3)}_{k_1,k_2,k_3} \rightarrow \varrho^{(3)}_{1,2,3}, \quad \delta(k-k_1-k_2-k_3)\, d^4k_1\, d^4k_2\, d^4k_3 \rightarrow d_{1,2,3}. \qquad (5.31)$$

We shall also assume that the cubic non-linear response coefficient satisfies the symmetry relation

$$\varrho^{(3)}_{1,2,3} = \varrho^{(3)}_{1,3,2}. \qquad (5.32)$$

which can again be achieved by a suitable change in the notation for the non-linear responses. We only need the restricted symmetry (5.32) since the term where the field 1 has a negative frequency turns out to be small – of the order of the contributions from the virtual fields of the second harmonic – and it is better to keep it separate from the other terms.

We can construct the positive frequency Langmuir field corresponding to the cubic non-linear charge density from Langmuir fields only – without involving virtual fields – using the following combinations of dimensionless frequencies: $1 = 1 + 1 - 1$, $1 = 1 - 1 + 1$, and $1 = -1 + 1 + 1$. This corresponds to the following non-linear equation

$$ik\varepsilon E^+ = 8\pi \varrho^{(3)}_{1,2,3} E_1^+ E_2^+ E_3^- d_{1,2,3} + 4\pi \varrho^{(3)}_{1,2,3} E_1^- E_2^+ E_3^+ d_{1,2,3}. \qquad (5.33)$$

Combining Eqs.(5.30) and (5.33) we find the general non-linear equation

$$ik\varepsilon E^+ = 8\pi \varrho^{\text{eff}}_{1,2,3} E_1^+ E_2^+ E_3^- d_{1,2,3} + 4\pi \varrho^{\text{eff}}_{1,2,3} E_1^- E_2^+ E_3^+ d_{1,2,3}, \qquad (5.34)$$

where

$$\varrho^{\text{eff}}_{1,2,3} = \varrho^{(3)}_{1,2,3} + \frac{8\pi}{i|\mathbf{k}_2 + \mathbf{k}_3|\varepsilon_{2+3}} \varrho^{(2)}_{1,2+3}\varrho^{(2)}_{2,3}. \tag{5.35}$$

The second term on the right-hand side of Eq.(5.34) describes the processes involving the virtual field at the second harmonic and contains the last term in Eq.(5.33). The second term on the right-hand side of Eq.(5.34) turns out to be much smaller than the first term. The general equation is valid if the total width $\delta\omega$ is smaller than ω, but the frequency differences can be arbitrary as compared to the non-linear frequency broadening. The frequency difference appears in Eq.(5.34) together with the non-linear broadening, since the frequencies of the fields 2 and 3 are – to a first approximation – each other's opposite. Because there is a certain uncertainty in the frequencies one cannot draw any diagrams except symbolically, asigning to each line an ω and a $\mathbf{k}$ (see Fig.5.5 where the incident lines are, in general, not on the "mass surface" where ω is a well defined function of $\mathbf{k}$).

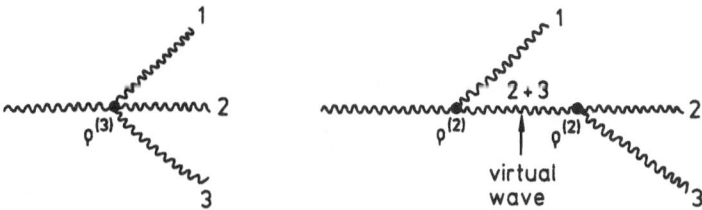

Fig. 5.5. Schematic diagrams corresponding to the direct cubic non-linearity and effective cubic non-linearity due to a contribution from the quadratic non-linearity

In this scenario the quadratic non-linearities produce an effective cubic non-linearity through the virtual waves. The Green function of these virtual waves is the same as in the case of the collision integral and we see from Eq.(5.35) that it is equal to $1/|\mathbf{k}_2 + \mathbf{k}_3|\varepsilon_{2+3}$. The conservation laws are satisfied in each of the vertices of Fig.5.5.

5.3 Non-linear Plasma Responses

One cannot always expand in the field strengths when calculating the non-linear responses. We may mention plasma responses at frequencies below the effective collision frequency in the case when the effective collisions in the system are produced by random fields – an example is provided by the collisions due to quasi-linear interactions. Bearing this in mind we shall nevertheless consider here the limit when such an expansion is possible. The expansions in the field strengths differ somewhat in the two cases of completely random

fields and of completely regular fields; however, the non-linear response coefficients turn out to be the same. In the case of regular fields we can use
Eq.(2.62). We denote the stationary distribution function by $\Phi_{\mathbf{p}}$ while now
$\delta f_{\mathbf{p}}$ denotes not the first-order correction to it, but the total deviation. We
now have

$$\frac{\partial}{\partial t} \delta f_{\mathbf{p}} + \left(\mathbf{v} \cdot \frac{\partial}{\partial \mathbf{r}} \right) \delta f_{\mathbf{p}} = -e \left(\mathbf{E} \cdot \frac{\partial}{\partial \mathbf{p}} \right) \Phi_{\mathbf{p}} - e \left(\mathbf{E} \cdot \frac{\partial}{\partial \mathbf{p}} \right) \delta f_{\mathbf{p}}. \quad (5.36)$$

For the case of random fields the same Eq.(2.62) has the form

$$\frac{\partial}{\partial t} \delta f_{\mathbf{p}} + \left(\mathbf{v} \cdot \frac{\partial}{\partial \mathbf{r}} \right) \delta f_{\mathbf{p}} = -e \left(\mathbf{E} \cdot \frac{\partial}{\partial \mathbf{p}} \right) \Phi_{\mathbf{p}}$$

$$-e \left[\left(\mathbf{E} \cdot \frac{\partial}{\partial \mathbf{p}} \right) \delta f_{\mathbf{p}} - \left\langle \left(\mathbf{E} \cdot \frac{\partial}{\partial \mathbf{p}} \right) \delta f_{\mathbf{p}} \right\rangle \right]. \quad (5.37)$$

In this last equation $\delta f_{\mathbf{p}}$ is the random part and $\Phi_{\mathbf{p}}$ the regular part of the
distribution. The quantities $\Phi_{\mathbf{p}}$ and $\delta f_{\mathbf{p}}$ have thus a different meaning in
Eqs.(5.36) and (5.37). Neverthelss, the expansion in the field strengths is
completely similar in the two cases and once we know one of the expansions
we can simply use it to find the other one. The formal difference lies solely
in that in Eq.(5.37) we subtract the average. In the case of random fields we
have instead of Eq.(5.25) the equation

$$E_{\mathbf{k}} = \frac{4\pi}{ik} \int \frac{1}{\varepsilon_{1+2}} \varrho_{1,2}^{(2)} \left[E_1 E_2 - \langle E_1 E_2 \rangle \right] d_{1,2}. \quad (5.38)$$

This equation contains the same non-linear response coefficient, $\varrho_{1,2}^{(2)}$, as the
equation for the interaction of regular fields. In the case of the cubic non-
linearities of random fields we can again use Eq.(5.34) provided we make the
substitution

$$E_1 E_2 E_3 \rightarrow E_1 E_2 E_3 - E_1 \langle E_2 E_3 \rangle - \langle E_1 E_2 E_3 \rangle. \quad (5.39)$$

Once we have seen this we can consider just Eq.(5.36) to find the non-linear
response coefficients. We shall neglect here the existence of the natural statistical particle fluctuations. If they are present the particle distribution to
zeroth order in the fields may be very non-stationary and inhomogeneous.
Here we shall assume that the distribution function when there are no fields,
$\Phi_{\mathbf{p}}$, is stationary and homogeneous – our results remain valid, of course, for
the case when this distribution varies slowly in time and space. We shall
consider the presence of statistical particle fluctuations in Chap.12. We write
$\delta f_{\mathbf{p}}$ as an expansion in the field strength, $\delta f_{\mathbf{p}} = \delta f_{\mathbf{p}}^{(1)} + \delta f_{\mathbf{p}}^{(2)} + \delta f_{\mathbf{p}}^{(3)}$, where
the superscripts indicate the powers of the field strengths. We now have the
following equations

$$\left. \begin{aligned} \frac{\partial}{\partial t} \delta f_{\mathbf{p}}^{(1)} + \left(\mathbf{v} \cdot \frac{\partial}{\partial \mathbf{r}} \right) \delta f_{\mathbf{p}}^{(1)} &= -e \left(\mathbf{E} \cdot \frac{\partial}{\partial \mathbf{p}} \right) \Phi_{\mathbf{p}} \\ \longrightarrow \mathrm{i}(\omega - (\mathbf{k} \cdot \mathbf{v})) \delta f_{\mathbf{p},\mathbf{k}}^{(1)} &= e \left(\mathbf{E_k} \cdot \frac{\partial}{\partial \mathbf{p}} \right) \Phi_{\mathbf{p}}, \end{aligned} \right\} \tag{5.40}$$

$$\left. \begin{aligned} \frac{\partial}{\partial t} \delta f_{\mathbf{p}}^{(2)} + \left(\mathbf{v} \cdot \frac{\partial}{\partial \mathbf{r}} \right) \delta f_{\mathbf{p}}^{(2)} &= -e \left(\mathbf{E} \cdot \frac{\partial}{\partial \mathbf{p}} \right) \delta f_{\mathbf{p}}^{(1)} \\ \longrightarrow \mathrm{i}(\omega - (\mathbf{k} \cdot \mathbf{v})) \delta f_{\mathbf{p},\mathbf{k}}^{(2)} &= e \int \left(\mathbf{E_{k_1}} \cdot \frac{\partial \delta f_{\mathbf{p},\mathbf{k}-\mathbf{k_1}}^{(1)}}{\partial \mathbf{p}} \right) d^4 k_1, \end{aligned} \right\} \tag{5.41}$$

$$\left. \begin{aligned} \frac{\partial}{\partial t} \delta f_{\mathbf{p}}^{(3)} + \left(\mathbf{v} \cdot \frac{\partial}{\partial \mathbf{r}} \right) \delta f_{\mathbf{p}}^{(3)} &= -e \left(\mathbf{E} \cdot \frac{\partial}{\partial \mathbf{p}} \right) \delta f_{\mathbf{p}}^{(2)} \\ \longrightarrow \mathrm{i}(\omega - (\mathbf{k} \cdot \mathbf{v})) \delta f_{\mathbf{p},\mathbf{k}}^{(3)} &= e \int \left(\mathbf{E_{k_1}} \cdot \frac{\partial \delta f_{\mathbf{p},\mathbf{k}-\mathbf{k_1}}^{(2)}}{\partial \mathbf{p}} \right) d^4 k_1. \end{aligned} \right\} \tag{5.42}$$

We have given here both the original differential equations and their Fourier transforms, where we have again used Eq.(5.20). Equations (5.40) to (5.42) can be used to find the non-linear charge densities:

$$\delta \varrho_{\mathbf{k}}^{\mathrm{NL}(2)} = e \int \delta f_{\mathbf{p},\mathbf{k}}^{(2)} \frac{d^3 \mathbf{p}}{(2\pi)^3}, \qquad \delta \varrho_{\mathbf{k}}^{\mathrm{NL}(3)} = e \int \delta f_{\mathbf{p},\mathbf{k}}^{(3)} \frac{d^3 \mathbf{p}}{(2\pi)^3}. \tag{5.43}$$

If we now take into account that the fields are longitudinal, so that we have $\mathbf{E} = E\mathbf{k}/k$ and also take into account that there are contributions from different kinds of particles, we find, symmetrising the results in the way dscribed earlier:

$$\begin{aligned} \varrho_{1,2}^{(2)} = -\sum_\alpha \frac{e_\alpha^3}{2k_1 k_2} \int \frac{1}{\omega - (\mathbf{k} \cdot \mathbf{v}) + \mathrm{i}0} \\ \times \left[\left(\mathbf{k_1} \cdot \frac{\partial}{\partial \mathbf{p}} \right) \frac{1}{\omega_2 - (\mathbf{k_2} \cdot \mathbf{v}) + \mathrm{i}0} \left(\mathbf{k_2} \cdot \frac{\partial}{\partial \mathbf{p}} \right) \right. \\ \left. + \left(\mathbf{k_2} \cdot \frac{\partial}{\partial \mathbf{p}} \right) \frac{1}{\omega_1 - (\mathbf{k_1} \cdot \mathbf{v}) + \mathrm{i}0} \left(\mathbf{k_1} \cdot \frac{\partial}{\partial \mathbf{p}} \right) \right] \Phi_{\mathbf{p}}^\alpha \frac{d^3 \mathbf{p}}{(2\pi)^3}, \end{aligned} \tag{5.44}$$

$$\begin{aligned} \varrho_{1,2,3}^{(3)} = \mathrm{i} \sum_\alpha \frac{e_\alpha^4}{2k_1 k_2 k_3} \int \frac{1}{\omega - (\mathbf{k} \cdot \mathbf{v}) + \mathrm{i}0} \\ \times \left(\mathbf{k_1} \cdot \frac{\partial}{\partial \mathbf{p}} \right) \frac{1}{\omega - \omega_1 - ([\mathbf{k} - \mathbf{k_1}] \cdot \mathbf{v}) + \mathrm{i}0} \\ \times \left[\left(\mathbf{k_2} \cdot \frac{\partial}{\partial \mathbf{p}} \right) \frac{1}{\omega_3 - (\mathbf{k_3} \cdot \mathbf{v}) + \mathrm{i}0} \left(\mathbf{k_3} \cdot \frac{\partial}{\partial \mathbf{p}} \right) \right. \\ \left. + \left(\mathbf{k_3} \cdot \frac{\partial}{\partial \mathbf{p}} \right) \frac{1}{\omega_2 - (\mathbf{k_2} \cdot \mathbf{v}) + \mathrm{i}0} \left(\mathbf{k_2} \cdot \frac{\partial}{\partial \mathbf{p}} \right) \right] \Phi_{\mathbf{p}}^\alpha \frac{d^3 \mathbf{p}}{(2\pi)^3}. \end{aligned} \tag{5.45}$$

The causality principle was used in arriving at the correct expressions for the denominators. Equations (5.44) and (5.45) solve the problem of the non-linear plasma responses.

5.4 Approximate Expressions for the Non-linear Responses for Langmuir Waves

The phase velocities of Langmuir waves are much higher than the electron thermal velocities and the main contribution to the non-linear charge densities comes from the electrons. One can easily show that the second term in Eq.(5.33) is in this case smaller than the first one by a factor of the order of $v_{Te}^2/v_{ph}^2 \ll 1$. Indeed, in this second term the field 1 has a negative frequency and in the cubic non-linear resonance only denominators with sums of frequencies enter whereas in the first term frequency differences enter. The same is true of the second term in Eq.(5.35) for the product $\varrho_{1,2+3}^{(2)} \varrho_{2,3}^{(2)}$ in the effective cubic response. We shall therefore consider only the first term in Eq.(5.34). We have the following expression for $\varrho_{2,3}^{(2)}$:

$$
\varrho_{2,3}^{(2)} = - \frac{e^0}{2k_2 k_3} \int \frac{1}{\omega_2 + \omega_3 - ([\mathbf{k}_2 + \mathbf{k}_3] \cdot \mathbf{v}) + i0}
$$
$$
\times \left[\left(\mathbf{k}_2 \cdot \frac{\partial}{\partial \mathbf{p}} \right) \frac{1}{\omega_3 - (\mathbf{k}_3 \cdot \mathbf{v}) + i0} \left(\mathbf{k}_3 \cdot \frac{\partial}{\partial \mathbf{p}} \right) \right.
$$
$$
\left. + \left(\mathbf{k}_3 \cdot \frac{\partial}{\partial \mathbf{p}} \right) \frac{1}{\omega_2 - (\mathbf{k}_2 \cdot \mathbf{v}) + i0} \left(\mathbf{k}_2 \cdot \frac{\partial}{\partial \mathbf{p}} \right) \right] \Phi_{\mathbf{p}}^{e} \frac{d^3 \mathbf{p}}{(2\pi)^3}. \quad (5.46)
$$

In this expression only the first denominator is small. However, if the first differential operator acts upon the distribution function the same kind of small expression will appear in the numerator; this does not occur, if this operator acts upon the second denominator. We can therefore write approximately

$$
\varrho_{2,3}^{(2)} \cong - \frac{e^3 (\mathbf{k}_2 \cdot \mathbf{k}_3)}{2k_2 k_3 m_e} \int \frac{1}{\omega_2 + \omega_3 - ([\mathbf{k}_2 + \mathbf{k}_3] \cdot \mathbf{v}) + i0}
$$
$$
\times \left[\frac{1}{\left(\omega_3 - (\mathbf{k}_3 \cdot \mathbf{v}) + i0 \right)^2} \left(\mathbf{k}_3 \cdot \frac{\partial}{\partial \mathbf{p}} \right) \right.
$$
$$
\left. + \frac{1}{\left(\omega_2 - (\mathbf{k}_2 \cdot \mathbf{v}) + i0 \right)^2} \left(\mathbf{k}_2 \cdot \frac{\partial}{\partial \mathbf{p}} \right) \right] \Phi_{\mathbf{p}}^{e} \frac{d^3 \mathbf{p}}{(2\pi)^3}. \quad (5.47)
$$

We can now in the last bracket neglect both the thermal corrections ($\mathbf{k}_{2,3} \cdot \mathbf{v}$) as compared to $\omega_{2,3}$ as well as the frequency spread $\delta\omega$ as compared to ω_{pe}. We can express the final result in terms of the linear electron dielectric permittivity $\varepsilon_{\mathbf{k}}^{(e)}$,

$$\varepsilon_{2+3}^{(e)} = 1 + \frac{4\pi e^2}{|\mathbf{k}_2 + \mathbf{k}_3|^2} \int \frac{1}{\omega_2 + \omega_3 - ([\mathbf{k}_2 + \mathbf{k}_3] \cdot \mathbf{v}) + i0}$$
$$\times \left([\mathbf{k}_2 + \mathbf{k}_3] \cdot \frac{\partial}{\partial \mathbf{p}}\right) \Phi_\mathbf{p}^e \frac{d^3\mathbf{p}}{(2\pi)^3}. \tag{5.48}$$

We can therefore write

$$\varrho_{2,3}^{(2)} \cong - \frac{e(\mathbf{k}_2 \cdot \mathbf{k}_3)|\mathbf{k}_2 + \mathbf{k}_3|^2}{8\pi k_2 k_3 m_e \omega_{pe}^2} \left[\varepsilon_{2+3}^{(e)} - 1\right]. \tag{5.49}$$

We now turn to the quantity $\varrho_{1,2+3}^{(2)}$. In Eq.(5.44) we can neglect the second term as it does not contain a small denominator and we can write

$$\varrho_{1,2+3}^{(2)} = - \frac{e^3}{2k_1|\mathbf{k}_2 + \mathbf{k}_3|} \int \frac{1}{\omega_- - (\mathbf{k} \cdot \mathbf{v}) + i0} \left(\mathbf{k}_1 \cdot \frac{\partial}{\partial \mathbf{p}}\right)$$
$$\times \frac{1}{\omega_2 + \omega_3 - ([\mathbf{k}_2 + \mathbf{k}_3] \cdot \mathbf{v}) + i0} \left([\mathbf{k}_2 + \mathbf{k}_3] \cdot \frac{\partial}{\partial \mathbf{p}}\right) \Phi_\mathbf{p}^e \frac{d^3\mathbf{p}}{(2\pi)^3}. \tag{5.50}$$

We can integrate by parts with respect to the first derivative and after that neglect $(\mathbf{k} \cdot \mathbf{v})$ as compared to ω and $\delta\omega$ as compared to ω_{pe}. Once again we can express the result in terms of the linear electron dielectric permittivity $\varepsilon_{2+3}^{(e)}$:

$$\varrho_{1,2+3}^{(2)} \cong \frac{e(\mathbf{k} \cdot \mathbf{k}_1)|\mathbf{k}_2 + \mathbf{k}_3|}{8\pi k_1 m_e \omega_{pe}^2} \left[\varepsilon_{2+3}^{(e)} - 1\right]. \tag{5.51}$$

Next, when we calculate the cubic response we must carry out both procedures, that is, first integrate by parts and then use the procedure we employed to calculate the first quadratic response. We find $(\mathbf{k} - \mathbf{k}_1 = \mathbf{k}_2 + \mathbf{k}_3)$

$$\varrho_{1,2,3}^{(3)} \cong -i \frac{e^2(\mathbf{k} \cdot \mathbf{k}_1)(\mathbf{k}_2 \cdot \mathbf{k}_3)|\mathbf{k}_2 + \mathbf{k}_3|^2}{8\pi k_1 k_2 k_3 m_e^2 \omega_{pe}^4} \left[\varepsilon_{2+3}^{(e)} - 1\right]^2. \tag{5.52}$$

Finally we can write down the expression for the term in Eq.(5.35) which is due to the quadratic responses:

$$-i \frac{8\pi}{|\mathbf{k}_2 + \mathbf{k}_3|\varepsilon_{2+3}} \varrho_{1,2+3}^{(2)} \varrho_{2,3}^{(2)} \cong \frac{e^2(\mathbf{k} \cdot \mathbf{k}_1)(\mathbf{k}_2 \cdot \mathbf{k}_3)|\mathbf{k}_2 + \mathbf{k}_3|^2}{8\pi k_1 k_2 k_3 m_e^2 \omega_{pe}^4 \varepsilon_{2+3}} \left[\varepsilon_{2+3}^{(e)} - 1\right]^2. \tag{5.53}$$

We can now take into account the fact that

$$\varepsilon_\mathbf{k} = \varepsilon_\mathbf{k}^{(e)} + \varepsilon_\mathbf{k}^{(i)} - 1, \tag{5.54}$$

since the vacuum value (1) of the dielectric permittivity is by definition included in both $\varepsilon_\mathbf{k}^{(e)}$ and $\varepsilon_\mathbf{k}^{(i)}$, and combine Eqs.(5.35), (5.52), and (5.53). The result is

$$\varrho_{1,2,3}^{\text{eff}} = -i\frac{e^2(\mathbf{k}\cdot\mathbf{k}_1)(\mathbf{k}_2\cdot\mathbf{k}_3)}{8\pi k_1 k_2 k_3 m_e^2 \omega_{\text{pe}}^4}\frac{\left(\varepsilon_{2+3}^{(e)}-1\right)\varepsilon_{2+3}^{(i)}}{\varepsilon_{2+3}}|\mathbf{k}_2+\mathbf{k}_3|^2. \tag{5.55}$$

This is a general expression which includes most of the non-linear plasma effects. If one can neglect the imaginary part of the electron permittivity and uses for its real part the Debye screening approximation

$$\varepsilon_{2+3}^{(e)} \cong \frac{\omega_{\text{pe}}^2}{|\mathbf{k}_2+\mathbf{k}_3|^2 v_{Te}^2} \gg 1, \tag{5.56}$$

the expression for the effective response simplifies and becomes

$$\frac{8\pi\varrho_{1,2,3}^{\text{eff}}}{ik} = -\frac{(\mathbf{k}\cdot\mathbf{k}_1)(\mathbf{k}_2\cdot\mathbf{k}_3)}{4\pi k k_1 k_2 k_3 n_e T_e}\frac{\varepsilon_{2+3}^{(i)}}{\varepsilon_{2+3}}. \tag{5.57}$$

Using Eq.(5.57) we can write the non-linear equation (5.34) in the form

$$\varepsilon E^+ = -\int \frac{(\mathbf{k}\cdot\mathbf{k}_1)(\mathbf{k}_2\cdot\mathbf{k}_3)}{k k_1 k_2 k_3}\frac{\varepsilon_{2+3}^{(i)}}{\varepsilon_{2+3}}\frac{E_1^+ E_2^+ E_3^-}{4\pi n_e T_e}\,d_{1,2,3}. \tag{5.58}$$

In the derivation of Eq.(5.58) we assumed that the following parameters were small:

$$\frac{|E|^2}{4\pi n_e T_e} \ll 1, \qquad \frac{k v_{Te}}{\omega_{\text{pe}}} \ll 1, \qquad \frac{\delta\omega}{\omega_{\text{pe}}} \ll 1. \tag{5.59}$$

In deriving Eq.(5.58) we needed to take into account in the non-linear responses $\varrho^{(2)}$ and $\varrho^{(3)}$ only the contribution from the electrons whereas in the linear response ε_{2+3} we must take into account the contributions from both the electrons and the ions. The ion non-linear responses contain m_i^2 in the denominator and they can be neglected in the case of Langmuir waves. The fact that $\varepsilon_k^{(i)}$ occurs is due to the cancellation of the cubic non-linearities by an effective contribution from the quadratic non-linearities. Formally, this fact makes it necessary to take into account the contribution from the ions in the linear response in Eq.(5.58). The physical reason for this is that the displacement of the electrons leads to ambipolar fields. When we calculate the imaginary parts of the non-linearities this cancellation is even more important and it leads to significant physical consequences which we shall discuss in the next chapter. Here we mention merely that these cancellations describe the dynamic screening of the particle fields when the Langmuir waves are scattered.

The non-linearity which we have calculated here is sometimes called the *striction non-linearity* while the non-linearity which is a correction to it in the next approximation in the small parameter $k^2 v_{Te}^2/\omega_{\text{pe}}^2$ is called the *electron non-linearity*.

5.5 Approximate Expressions
for the Non-linear Responses for Ion–Sound Waves

We shall now show that one can, in a certain approximation, use for ion-sound waves the same expression as the one we have found for Langmuir waves, provided one substitutes in it the index "i" for the index "e", that is, ions play for ion-sound waves the same rôle as electrons for Langmuir waves. To show this we must bear in mind that there were two important features of Langmuir waves which we used in calculating the non-linear responses. First of all, the main contribution to the non-linear responses comes from the electrons and, secondly, three Langmuir fields cannot satisfy the decay relations which means that virtual fields make an appearance. The ion-sound waves are also waves which have the non-decay property. We saw that it was not important to include dispersion effects in the widths and we can work just with the positive and the negative components of the ion-sound field for a given $\mathbf{k}$ – since the dispersion of the ion-sound waves is not weak – and take into account the fact that the non-linear frequency shift $\delta\omega$ is small as compared to the frequency. We must mention in passing that because of the large dispersion the critical energy for ion-sound waves is usually close to the weak non-linearity limit. The range for which phase locking occurs is very · narrow and not as important as in the case of Langmuir waves. For ion-sound waves the ions behave, indeed, like the electrons for Langmuir waves; this is because the phase velocity of the ion-sound waves is much larger than the ion thermal velocity just as the phase velocity of the Langmuir waves is much larger than the electron thermal velocity. We can therefore in the ion non-linear responses neglect the thermal motions of the ions just as we neglected the thermal motions of the electrons in the electron non-linear responses in the case of Langmuir waves. To prove our earlier statement we must now prove, first of all, that the contribution from the electrons to the non-linear responses for ion-sound waves is small. We must remind ourselves that ion-sound waves exist only provided we have

$$\frac{T_{\mathrm{i}}}{ZT_{\mathrm{e}}} \ll 1, \tag{5.60}$$

and in our estimates we must take into account that this parameter is small. We start with an estimate of $\varrho^{(3)}$ using a thermal electron distribution for which

$$\left(\mathbf{k} \cdot \frac{\partial}{\partial \mathbf{p}}\right) \varPhi_{\mathbf{p}}^{\mathrm{e}} = -\frac{(\mathbf{k} \cdot \mathbf{v})}{T_{\mathrm{e}}} \varPhi_{\mathbf{p}}^{\mathrm{e}}.$$

We shall also neglect $\delta\omega$ as compared to $(\mathbf{k} \cdot \mathbf{v})$ in all denominators. We can easily calculate the non-linear response $\varrho^{(3)}$, but – in contrast to the case of the Langmuir oscillations – the second terms on the right-hand sides of Eqs.(5.33) and (5.34) are now of the same order of magnitude as the first ones. The following combination of non-linear responses occurs in the non-linear equation (5.33):

$$\varrho^{(3)}_{1,2,3} + \varrho^{(3)}_{2,1,3} + \varrho^{(3)}_{3,1,2} = i\frac{e^4 n_e}{2k_1 k_2 k_3 T_e^3}. \tag{5.61}$$

We shall now show that this contribution is small as compared to the contributions from the ions. Before doing that we shall show that some of the other contributions are of the same order as or less than expression (5.61). It will then be possible to neglect all of those together with (5.61). The contributions from the electrons to the quadratic non-linearities in the expression for $\varrho^{\text{eff}}_{1,2,3}$ are of the same order as or less than expression (5.61). Indeed, calculations similar to our earlier ones lead to

$$\frac{8\pi}{ik} \varrho^{\text{eff (e)}}_{1,2,3} = \frac{4\pi e^4 n_e}{kk_1 k_2 k_3 T_e^3} \left[1 - \frac{\varepsilon^{(e)}_{2+3} - 1}{\varepsilon_{2+3}} - \frac{\varepsilon^{(e)}_{1+3} - 1}{\varepsilon_{1+3}} - \frac{\varepsilon^{(e)}_{1+2} - 1}{\varepsilon_{1+2}} \right]. \tag{5.62}$$

The contribution from the ions to the second terms on the right-hand sides of Eqs.(5.33) and (5.34) are also of the same order of magnitude as expression (5.61). We see this as follows. This contribution does not contain any differences of ion-sound frequencies and the phase velocity for all the fields are thus significantly higher than the ion thermal velocity. For an estimate of this term we can therefore use Eq.(5.61), substituting $m_i \omega^2/k^2$ for T_e, $n_i = n_e/Z$ for n_e, and $Z^4 e^4$ for e^4. The result,

$$\varrho^{\text{eff (i)}}_{3,1,2} \approx i\frac{Z^4 e^4 n_i k^6}{k_1 k_2 k_3 m_i^3 \omega^6} \approx i\frac{Z^3 e^4 n_e}{k_1 k_2 k_3 Z^3 T_e^3}; \qquad \frac{\omega^2}{k^2} \cong \frac{ZT_e}{m_i}, \tag{5.63}$$

is, indeed, of the same order as expression (5.61). Provided expression (5.61) is, indeed, small we can restrict our discussion to the first term in Eq.(5.33) with non-linear responses due to the ions only. In that case we can use the approximate expression which we found for the Langmuir waves and change the index "(e)" to "(i)". However, there arises a complication due to the fact that the parameter (5.60) is small.

Consider Eq.(5.47). We should not distinguish between $\left(\omega_2 - (\mathbf{k}_2 \cdot \mathbf{v})\right)^2$ and $\left(\omega_3 - (\mathbf{k}_3 \cdot \mathbf{v})\right)^2$, since their difference will be proportional to $\omega_2 + \omega_3 - ([\mathbf{k}_2 + \mathbf{k}_3] \cdot \mathbf{v})$ which cancels the small denominator. We therefore can write

$$\varrho^{(2)(i)}_{2,3} \cong - \frac{Z^3 e^3 (\mathbf{k}_2 \cdot \mathbf{k}_3)}{2k_2 k_3 m_i} \int \frac{1}{\omega_2 + \omega_3 - ([\mathbf{k}_2 + \mathbf{k}_3] \cdot \mathbf{v}) + i0}$$
$$\times \frac{1}{(\omega_2 - (\mathbf{k}_2 \cdot \mathbf{v}) + i0)^2} \left([\mathbf{k}_2 + \mathbf{k}_3] \cdot \frac{\partial}{\partial \mathbf{p}}\right) \Phi^i_{\mathbf{p}} \frac{d^3 p}{(2\pi)^3}. \tag{5.64}$$

One can expand the second denominator as follows:

$$\frac{1}{(\omega_2 - (\mathbf{k}_2 \cdot \mathbf{v}) + i0)^2} \cong \frac{1}{\omega^2} \left\{ 1 + \frac{2(\mathbf{k}_2 \cdot \mathbf{v})}{\omega_2} + \frac{3(\mathbf{k}_2 \cdot \mathbf{v})^2}{\omega_2^2} + \cdots \right\}. \tag{5.65}$$

The second term in this expansion describes the *Doppler shift* of the frequency of the field acting upon a moving ion. In principle, these corrections are always small and we neglected them for the case of the Langmuir oscillations. However, the fact that there are cancellations in the contributions from the cubic non-linearities and the effective contributions from the quadratic non-linearities raises the question whether the final result after these cancellations have been taken into account are smaller or larger than the Doppler corrections which we have just found. This question also arises for the case of Langmuir oscillations. In that case the answer is that, indeed, the Doppler corrections are in most cases of interest small as compared to the result which we found earlier and which took the cancellations into account. In the case of ion-sound oscillations the answer is the opposite, that is, the cancellations are much stronger for ion-sound waves and the Doppler corrections dominate. It turns out that their relative order of magnitude is that of the square root of the small parameter (5.60), that is, they are proportional to

$$\sqrt{\frac{T_i}{ZT_e}}. \tag{5.66}$$

We must still prove that the result of the cancellations is proportional to the small parameter (5.60) itself rather than to its square root. Indeed, if we neglect the Doppler corrections we can write Eq.(5.64) in the form

$$\varrho_{2,3}^{(2)(i)} \cong \frac{Ze(\mathbf{k}_2 \cdot \mathbf{k}_3)}{8\pi k_2 k_3 m_i \omega^2} |\mathbf{k}_2 + \mathbf{k}_3|^2 \left(\varepsilon_{2+3}^{(i)} - 1\right). \tag{5.67}$$

Still neglecting the Doppler corrections we get instead of Eqs.(5.51) and (5.52):

$$\varrho_{1,2+3}^{(2)(i)} \cong -\frac{Ze(\mathbf{k}_1 \cdot \mathbf{k})|\mathbf{k}_2 + \mathbf{k}_3|}{8\pi k_1 m_i \omega^2} \left(\varepsilon_{2+3}^{(i)} - 1\right), \tag{5.68}$$

$$\varrho_{1,2,3}^{(3)(i)} \cong -i \frac{Z^2 e^2 (\mathbf{k}_1 \cdot \mathbf{k})(\mathbf{k}_2 \cdot \mathbf{k}_3)|\mathbf{k}_2 + \mathbf{k}_3|^2}{8\pi k_1 k_2 k_3 m_i^2 \omega^2 \omega_2^2} \left(\varepsilon_{2+3}^{(i)} - 1\right). \tag{5.69}$$

This leads to the final result

$$\frac{8\pi \varrho_{1,2,3}^{\text{eff (i)}}}{ik} \cong -\frac{Z^2 e^2 (\mathbf{k}_1 \cdot \mathbf{k})(\mathbf{k}_2 \cdot \mathbf{k}_3)|\mathbf{k}_2 + \mathbf{k}_3|^2}{8\pi k k_1 k_2 k_3 m_i^2 \omega^2 \omega_2^2} \frac{\left(\varepsilon_{2+3}^{(i)} - 1\right)\varepsilon_{2+3}^{(e)}}{\varepsilon_{2+3}}. \tag{5.70}$$

The electronic part of the dielectric permittivity appears due to the above-mentioned cancellations which lead to an additional factor, as compared to expression (5.69) for the cubic non-linearity, which is equal to

$$1 - \frac{\left(\varepsilon_{2+3}^{(i)} - 1\right)}{\varepsilon_{2+3}} = \frac{\varepsilon_{2+3}^{(e)}}{\varepsilon_{2+3}}. \tag{5.71}$$

If $\omega_2 + \omega_3 \ll |\mathbf{k}_2 + \mathbf{k}_3| v_{Ti}$, the cancellation shown by Eq.(5.71) is, indeed, very severe. The physical reason for such a large cancellation is the presence of dynamic screening of the fields of the ions which are involved in the interaction; this means that the non-linearity is determined by the interaction with "dressed" ions. The cancellation gives a factor T_i/ZT_e whereas the factor determining the Doppler corrections is $(\mathbf{k} \cdot \mathbf{v})/\omega$ which is approximately equal to $\sqrt{T_i/ZT_e}$. The last point which we must take into account, as it is important, is that when $\omega_2 + \omega_3 \ll |\mathbf{k}_2 + \mathbf{k}_3| v_{Ti}$ expression (5.71) is of the same order of magnitude as the terms determined by the electrons, that is expression (5.62). We have thus proved that the electron non-linearities are negligible in this case.

For our estimates we have concentrated on the real parts of the effective non-linear responses. The discussion of the imaginary parts is simpler. The imaginary part of the electron effective response appears only in the dielectric permittivity (5.62) and it is again smaller than the imaginary part of the effective ion response by a factor equal to expression (5.60).

The final result is that we can use for the ion-sound waves the result for the Langmuir waves provided we replace the indices "e" by "i" and take into account the Doppler corrections to the frequency of the field acting on the ions. A simpler and more translucent interpretation of this result will be given when we consider the probabilities for the corresponding non-linear processes when the cancellations which we have discussed have a simple physical meaning and when there is a natural explanation for the dependence of the cross-sections of the non-linear processes on the Doppler corrections. It turns out that the probabilities depend on the square of the Doppler corrections. We shall discuss these points in the next chapter.

To conclude this section we return to the electron non-linearities for the ion-sound interactions and say a few words about the electron imaginary parts. Without carrying out any calculations we can state that in any consistent theory their relative contribution should be small. In fact, they are due to the resonance of the ion-sound waves with the electrons. However, to first order this resonance leads to Landau damping and in a consistent theory where we have the small parameter $W/nT \ll 1$ this should lead to corrections to the interactions which are small as long as this parameter is small. In the past great efforts have been necessary to show that, in fact, these corrections are small as compared to the Landau damping, and the calculations have shown that the relative order of magnitude of the corrections contains fractional powers of W/nT: they contain factors $(W/nT)^{1/3}$ or $(W/nT)^{1/2}$.

5.6 Non-linear Changes in the Particle Distributions

The problem we considered earlier is in some sense a generalisation of the Landau problem as we started from an initial distribution and tried to find not only the linear plasma responses but also the non-linear ones. We have mentioned the possibility to generalise this approach by assuming that the initial distribution varied slowly both in time and space. It is important that such an approach be self-consistent. We remind ourselves that we showed that in a quasi-linear approach the emission processes, taken into account in the linear approximation, also change the particle distribution functions. The slow variations of the distribution function in time and space should not be put in artificially but should reflect the effect of the processes responsible for the non-linear interactions on the distribution function. This shows the necessity to find a self-consistent set of equations which describe both the non-linear changes in the fields and the non-linear changes in the particle distributions. The quasi-linear interaction provides such a description in the case of random fields but it is not always acceptable for non-linear interactions since phase locking in the case of strong waves makes it impossible to describe them using a statistical approach; however, if there is no phase locking one can for random waves just use a non-linear generalisation of the quasi-linear approach. We thus need a self-consistent approach in non-linear theory for both random and regular waves. As far as random waves are concerned we can use an approach similar to the one used in the quasi-linear description. This means that we can propose a generalisation of the non-linear changes in the distribution function, taking into account processes of the same kind as those responsible for the non-linear interactions and then use a correspondence principle – as was done in the case of the quasi-linear interactions – to find the coefficients in these equations. Indeed, one can find such a self-consistent set of equations in the case of random waves and they describe the changes due to non-linear interactions in both the distribution functions of the waves and those of the particles. However, we also need a similar description for regular waves, not only for the case of non-linear interactions, but even for the simpler case of linear interactions – more precisely in the case of a generalisation of the quasi-linear interactions. We thus need an approach which is more general than the quasi-linear one and which is valid for regular waves and which in the limit of random waves goes over into the quasi-linear description. The aim of the present section is to describe a general approach of this kind, using a procedure which is similar to the one used earlier for a very simple, but not self-consistent, non-linear approach. We will then find here a set of general equations which describe the changes in the particle distribution function due to non-linear interactions with fields which are no longer assumed to be random. To a first approximation this approach is a generalisation of the quasi-linear approach. As in the earlier case we shall consider the case of a Langmuir field.

We introduce again the the fields E^+, E^-, $E^{v(0)}$, and $E^{v(\pm 2)}$ as functions of r and t which are the Fourier integrals of the components $E_{k,\omega}^{\pm}$, $E_{k,\omega}^{v(0)}$, and $E_{k,\omega}^{v(\pm 2)}$ over the k and ω ranges corresponding to the broadened lines close to the plasma frequency, the zero frequency, the doubled plasma frequency, and so on. Our new approach will lie in carrying out a similar procedure for the distribution function. We now introduce a slowly varying distribution function Φ_p which corresponds to the Fourier integral over a broadened "line" close to the zero frequency. We next introduce a component δf_p which consists of various contributions: the distribution function $\delta f_p^{\pm}$, representing the Fourier integral over the "line" close to the plasma frequency, the distribution function $\delta f_p^{\pm 2}$ which represents the Fourier integral over the "line" close to the doubled plasma frequency, and so on. Collecting terms in the initial kinetic equation corresponding to the respective "lines" we find:

$$\frac{\partial \Phi_p}{\partial t} + \left(v \cdot \frac{\partial \Phi_p}{\partial r}\right) = -e\left(\frac{\partial}{\partial p} \cdot \left[E^{v(0)}\Phi_p + E^-\delta f_p^+ + E^+\delta f_p^- \right.\right.$$
$$\left.\left. + E^{v(-2)}\delta f_p^{(+2)} + E^{v(+2)}\delta f_p^{(-2)}\right]\right), \quad (5.72)$$

$$\frac{\partial \delta f_p^{\pm}}{\partial t} + \left(v \cdot \frac{\partial \delta f_p^{\pm}}{\partial r}\right) = -e\left(\frac{\partial}{\partial p} \cdot \left[E^{\pm}\Phi_p + E^{v(0)}\delta f_p^{+} \right.\right.$$
$$\left.\left. + E^{v(\pm 2)}\delta f_p^{\mp} + E^{\mp}\delta f_p^{(\pm 2)}\right]\right), \quad (5.73)$$

$$\frac{\partial \delta f_p^{(\pm 2)}}{\partial t} + \left(v \cdot \frac{\partial \delta f_p^{(\pm 2)}}{\partial r}\right) = -e\left(\frac{\partial}{\partial p} \cdot \left[E^{v(\pm 2)}\Phi_p \right.\right.$$
$$\left.\left. + E^{\pm}\delta f_p^{\pm} + E^{v(0)}\delta f_p^{(\pm 2)}\right]\right). \quad (5.74)$$

This approach has only one constraint: $\delta\omega \ll \omega$. The set of equations can, of course, be continued, but as in the case of the non-linear interactions which we described earlier we can restrict ourselves to the components which are nearest $E^{\pm}$. The difference between the present approach and the one described earlier is that now the linear responses are not split off but appear in the equations for each of the harmonics. This kind of description leads to a generalisation of the quasi-linear approach, if we restrict ourselves to the terms containing $\left(E^{+,-} \cdot \partial \delta f_p^{-,+}/\partial p\right)$ in the equation for Φ_p and the terms $\left(E^{\pm} \cdot \partial \Phi_p/\partial p\right)$ in the equations for $\delta f_p^{\pm}$. In the more general case this kind of description will lead to a generalisation of the non-linear description since the distribution function Φ_p in the non-linear description is no longer arbitrary, but the one which includes the changes due to the non-linear interactions.

We can make further progress if we introduce some simplifications. We remind ourselves that the strengths of the fields at the second harmonic terms are small and therefore the changes in the distribution function at these harmonics will also be small. We can therefore neglect in Eq.(5.74) the last term on the right-hand side as it is proportional to the fourth power of the Langmuir field – remember that the virtual field at the zero harmonic is proportional to the square of the Langmuir field. Similarly, we use in the second term on the right-hand side of Eq.(5.74) the linear approximation for the Langmuir field. This enables us to find the distribution function at the second harmonic up to terms proportional to the square of the Langmuir field. This is a reasonable and necessary approximation since we are dealing only with terms up to the second harmonic and we calculate it only up to terms of second order in the Langmuir field.

The final result is

$$
\delta f_{\mathbf{p},\mathbf{k}}^{(\pm 2)} = \frac{eE_{\mathbf{k}}^{\mathrm{v}(\pm 2)}}{(\omega - (\mathbf{k}\cdot\mathbf{v}) + \mathrm{i}0)k}\left(\mathbf{k}\cdot\frac{\partial\Phi_{\mathbf{p}}}{\partial\mathbf{p}}\right) - \frac{e^2}{2(\omega - (\mathbf{k}\cdot\mathbf{v}) + \mathrm{i}0)}
$$
$$
\times \int \frac{d^4k_1 d^4k_2 \delta(\mathbf{k} - \mathbf{k}_1 - \mathbf{k}_2)}{k_1 k_2} E_{\mathbf{k}_1}^{\pm} E_{\mathbf{k}_2}^{\pm}
$$
$$
\times \left[\left(\mathbf{k}_1\cdot\frac{\partial}{\partial\mathbf{p}}\right)\frac{1}{\omega_2 - (\mathbf{k}_2\cdot\mathbf{v}) + \mathrm{i}0}\left(\mathbf{k}_2\cdot\frac{\partial}{\partial\mathbf{p}}\right)\right.
$$
$$
\left. + \left(\mathbf{k}_2\cdot\frac{\partial}{\partial\mathbf{p}}\right)\frac{1}{\omega_1 - (\mathbf{k}_1\cdot\mathbf{v}) + \mathrm{i}0}\left(\mathbf{k}_1\cdot\frac{\partial}{\partial\mathbf{p}}\right)\right]\Phi_{\mathbf{p}}. \tag{5.75}
$$

The quantity $\Phi_{\mathbf{p}}$ is here a function which varies slowly in time and space: $\Phi_{\mathbf{p}} = \Phi_{\mathbf{p}}(\mathbf{r},t)$. The result we have obtained could have been obtained from the general equation by expanding in the small parameters $\delta\omega/\omega$ and $\delta k/k$, where $\delta\omega$ characterises the slow time-dependence of the distribution function and δk characterises its slow space-dependence.

We must also write down the Poisson equation for the separate harmonics:

$$
\mathrm{i}kE_{\mathbf{k}}^{\mathrm{v}(\pm 2)} = \sum_{\alpha}\int \delta f_{\mathbf{p},\mathbf{k}}^{\alpha(\pm 2)}\frac{d^3p}{(2\pi)^3}. \tag{5.76}
$$

The linear term in the field at the second harmonic on the right-hand side of Eq.(5.75) leads to the appearance of the dielectric permittivity on the left-hand side of Eq.(5.76):

$$
\mathrm{i}k\varepsilon E^{\mathrm{v}(\pm 2)} = 4\pi \int \varrho_{1,2}^{(2)} E_1^+ E_2^+ d_{1,2}. \tag{5.77}
$$

The approximate expression which we have obtained corresponds to the use of a truncation procedure; the difference with what had been done before is that $\Phi_{\mathbf{p}}$ is now a definite slowly varying distribution function satisfying Eq.(5.72). We can further simplify the equation for $\Phi_{\mathbf{p}}$ if we can use a further truncation process for calculating approximately the slowly varying fields. Bearing in mind that those fields are also to a first approximation quadratic in the Langmuir field we can use the simplified equation for $\Phi_{\mathbf{p}}$ to find them:

$$\frac{\partial \Phi_\mathbf{p}}{\partial t} + \left(\mathbf{v} \cdot \frac{\partial \Phi_\mathbf{p}}{\partial \mathbf{r}}\right) + e\left(\mathbf{E}^{\mathbf{v}(0)} \cdot \frac{\partial \Phi_\mathbf{p}}{\partial \mathbf{p}}\right)$$
$$= -e\left[\left(\mathbf{E}^+ \cdot \frac{\partial \delta f_\mathbf{p}^-}{\partial \mathbf{p}}\right) + \left(\mathbf{E}^- \cdot \frac{\partial \delta f_\mathbf{p}^+}{\partial \mathbf{p}}\right)\right]. \tag{5.78}$$

Using the expressions for the $\delta f_\mathbf{p}^\pm$ which are linear in the Langmuir field we can find an approximate expression for $\Phi_\mathbf{p}$: $\Phi_\mathbf{p} = \Phi_\mathbf{p}^{(0)} + \Phi_\mathbf{p}^{(1)}$, where $\Phi_\mathbf{p}^{(1)}$ describes the effects which are quadratic in the Langmuir fields:

$$\Phi_{\mathbf{p},\mathbf{k}}^{(1)} = \frac{eE_\mathbf{k}^{\mathbf{v}(0)}}{i(\omega - (\mathbf{k} \cdot \mathbf{v}) + i0)k}\left(\mathbf{k} \cdot \frac{\partial \Phi_\mathbf{p}^{(0)}}{\partial \mathbf{p}}\right) - \frac{e^2}{\omega - (\mathbf{k} \cdot \mathbf{v}) + i0}$$
$$\times \int \frac{d_{1,2}}{k_1 k_2} E_{\mathbf{k}_1}^+ E_{\mathbf{k}_2}^- \left[\left(\mathbf{k}_1 \cdot \frac{\partial}{\partial \mathbf{p}}\right)\frac{1}{\omega_2 - (\mathbf{k}_2 \cdot \mathbf{v}) + i0}\left(\mathbf{k}_2 \cdot \frac{\partial}{\partial \mathbf{p}}\right)\right.$$
$$\left. + \left(\mathbf{k}_2 \cdot \frac{\partial}{\partial \mathbf{p}}\right)\frac{1}{\omega_1 - (\mathbf{k}_1 \cdot \mathbf{v}) + i0}\left(\mathbf{k}_1 \cdot \frac{\partial}{\partial \mathbf{p}}\right)\right]\Phi_\mathbf{p}^{(0)}. \tag{5.79}$$

This expansion assumes that the changes in the slowly varying part of the distribution function due to the non-linear processes are small. We can now also write down a Poisson equation for the zero harmonic:

$$ik\varepsilon E^{\mathbf{v}(0)} = 8\pi \int \varrho_{1,2}^{(2)} E_1^+ E_2^- \, d_{1,2}. \tag{5.80}$$

Using this equation again implies a truncation and perturbation method approach. The difference between the approach used here and the one used earlier is that now there enters a self-consistent slowly varying part $\Phi_\mathbf{p}$ of the distribution function. Using the approximate expressions for the virtual fields we can now write down Eq.(5.73) for the distribution function at the Langmuir frequency $\delta f_\mathbf{p}^\pm$. This equation is complicated, but if the non-linear terms can be treated as perturbations we can in them substitute the linear expressions for the $\delta f_\mathbf{p}^\pm$ in terms of the Langmuir field $E^\pm$:

$$\delta f^{\pm}_{\mathbf{p},k} = \frac{eE^{\pm}_{\mathbf{k}}}{i(\omega - (\mathbf{k} \cdot \mathbf{v}) + i0)k} \left(\mathbf{k} \cdot \frac{\partial \Phi_{\mathbf{p}}}{\partial \mathbf{p}} \right) + \frac{e}{i(\omega - (\mathbf{k} \cdot \mathbf{v}) + i0)}$$

$$\times \int d_{1,2} \left(\mathbf{k}_1 \cdot \frac{\partial}{\partial \mathbf{p}} \right) \left\{ 2E^{\mathrm{v}(0)}_{\mathbf{k}_1} E^{\pm}_{\mathbf{k}_2} + 2E^{\mathrm{v}(\pm 2)}_{\mathbf{k}_1} E^{\mp}_{\mathbf{k}_2} \right\}$$

$$\times \frac{e}{ik_1 k_2 (\omega_2 - (\mathbf{k}_2 \cdot \mathbf{v}) + i0)} \left(\mathbf{k}_2 \cdot \frac{\partial \Phi_{\mathbf{p}}}{\partial \mathbf{p}} \right)$$

$$- \frac{e^3}{i(\omega - (\mathbf{k} \cdot \mathbf{v}) + i0)} \int \left[2E^{\pm}_{\mathbf{k}_1} E^{\pm}_{\mathbf{k}_2} E^{\mp}_{\mathbf{k}_3} + E^{\mp}_{\mathbf{k}_1} E^{\pm}_{\mathbf{k}_2} E^{\pm}_{\mathbf{k}_3} \right]$$

$$\times \frac{d_{1,2,3}}{k_1 k_2 k_3} \left(\mathbf{k}_1 \cdot \frac{\partial}{\partial \mathbf{p}} \right) \frac{1}{2(\omega - \omega_1 - ([\mathbf{k} - \mathbf{k}_1] \cdot \mathbf{v}) + i0)}$$

$$\times \left[\left(\mathbf{k}_2 \cdot \frac{\partial}{\partial \mathbf{p}} \right) \frac{1}{\omega_3 - (\mathbf{k}_3 \cdot \mathbf{v}) + i0} \left(\mathbf{k}_3 \cdot \frac{\partial \Phi_{\mathbf{p}}}{\partial \mathbf{p}} \right) \right.$$

$$\left. + \left(\mathbf{k}_3 \cdot \frac{\partial \Phi_{\mathbf{p}}}{\partial \mathbf{p}} \right) \frac{1}{\omega_2 - (\mathbf{k}_2 \cdot \mathbf{v}) + i0} \left(\mathbf{k}_2 \cdot \frac{\partial \Phi_{\mathbf{p}}}{\partial \mathbf{p}} \right) \right]. \tag{5.81}$$

This equation now gives us the equation for E^{+} which has the same form as Eq.(5.34) which we obtained earlier:

$$\mathbf{1}k_{t} E^{+} - 8\pi \varrho^{\mathrm{eff}}_{1,2,3} E^{+}_{1} E^{+}_{2} F^{-}_{3} d_{1,2,3} + 4\pi \varrho^{\mathrm{eff}}_{1,2,3} E^{-}_{1} E^{+}_{2} E^{+}_{3} d_{1,2,3}. \tag{5.82}$$

The difference between Eq.(5.82) and Eq.(5.34) is that in Eq.(5.82) the distribution function $\Phi_{\mathbf{p}}$ is self-consistently determined. To find the equation for $\Phi_{\mathbf{p}}$ one must in Eq.(5.72) substitute expression (5.81) for $\delta f^{\pm}_{\mathbf{p}}$, expression (5.75) for $\delta f^{(\pm 2)}_{\mathbf{p}}$, expression (5.76) for $E^{\mathrm{v}(\pm 2)}$, and expression (5.80) for $E^{\mathrm{v}(0)}$. We shall not write down the final equation as it is a very complicated one and in what follows we shall need only some limiting cases of it which we shall write down later on. It is important to note that we have now developed a method which is a consistent procedure for finding self-consistent equations for both the fields and the distribution functions and which enables us to estimate what terms are neglected in the truncation and in the approximations used. One can, if necessary, use a different truncation method; for instance, one can use as the zeroth approximation not the linear approximation, but a generalisation of the quasi-linear equation. This last equation can be found from Eqs.(5.72) and (5.73) by retaining only the terms which are linear in the Langmuir fields and neglecting all virtual fields. In the chapters that follow we shall use both approaches: a generalisation of the quasi-linear equation and a generalisation which is a self-consistent non-linear approach.

5.7 Non-linear Changes in the Particle Distributions for the Case of Random Fields

We shall finish this discussion with a few comments on the non-linear changes in the particle distributions in random fields. We have already mentioned that in this case we should properly separate the random and the regular components. We can start with the equation for the regular component:

$$\frac{\partial \Phi_{\mathbf{p}}}{\partial t} + \left(\mathbf{v} \cdot \frac{\partial \Phi_{\mathbf{p}}}{\partial \mathbf{r}} \right) = -e \frac{\partial}{\partial \mathbf{p}} \langle \mathbf{E} \, \delta f_{\mathbf{p}} \rangle, \tag{5.83}$$

and use an expansion of the random component in the field strengths in the case when all fields are random:

$$\delta f_{\mathbf{p}} = \delta f_{\mathbf{p}}^{(0)} + \delta f_{\mathbf{p}}^{(1)} + \delta f_{\mathbf{p}}^{(2)} + \delta f_{\mathbf{p}}^{(3)}. \tag{5.84}$$

The term $\delta f_{\mathbf{p}}^{(0)}$ corresponds to the natural particle statistical fluctuations and the upper indices correspond to the power of the field strengths. The field is the sum of the wave field and that of the particle statistical fluctuations. An important point is that these fields can be expressed in terms of quadratic combinations of those components because of the non-linear responses; of course, the non-linear processes must be taken into account both in the wave fields and in the fields of the particle fluctuations as well as in the interactions of the wave fields with the fields of the particle fluctuations. This means that the equation

$$ik\varepsilon E = 4\pi \int \varrho_{1,2}^{(2)} \left(E_1 E_2 - \langle E_1 E_2 \rangle \right) d_{1,2} \tag{5.85}$$

can be used to express the average of the product of three random components in Eq.(5.83), for example, in the term $\langle E \delta f_{\mathbf{p}}^{(2)} \rangle$, in terms of the averages of four random components which in the case of weak correlations can be approximated by sums of products of averages of two random components. We have already given earlier the actual expressions for the $\delta f_{\mathbf{p}}^{(1,2,3)}$. The results obtained for the average distribution function can be used to describe various effects:

1. The changes in the distribution function due to stimulated scattering. These will be found when we neglect all fields due to the natural particle statistical fluctuations and assume that the fields of the collective oscillations are rather strong. We shall not write down the necessary expressions since the simplest way to obtain them explicitly is by using the correspondence principle and the probabilities for stimulated scattering; we shall do that in the next chapter. An independent way to derive these expressions is to use a balance equation involving scattering probabilities, using a procedure similar to the quasi-linear approach but for non-linear interactions. We shall give the probabilities for the non-linear interactions in the next chapter.

2. Non-linear plasma maser interactions in the case when one of the waves is in resonance with the particles while the other one is not in resonance (see Chap.11).

3. The non-linear interactions of random waves with the random particle fluctuations (Chap.12) which lead to bremsstrahlung absorption (Chap.13).

We cannot use the statistical approach in the case when $W \gg W_{cr}$; in that case the modulational interactions of the waves are important (Chap.9). In that case we must use for the self-consistent treatment the approach described in earlier sections of the present chapter.

Problems

1. Find the change in the particle distribution in a regular monochromatic field for the case (a) when it is decaying in time with a damping rate much smaller than the field frequency and (b) when it is growing in time with a growth rate much smaller than the field frequency.
2. Calculate the non-linear frequency shift of a monochromatic Langmuir wave.
3. Calculate the non-linear frequency shift due to the electron and ion non-linearities of an ion-sound wave with a given wavevector k_0.
4. Calculate the amplitude of the virtual fields produced by a monochromatic Langmuir wave at the zero frequency and at the doubled frequencies.
5. Repeat the calculations of Prob.4 for the case of a random Langmuir wave.
6. Repeat the calculations of Prob.4 for the case of a regular ion-sound wave and the case of a random ion-sound wave.
7. Write down an equation for the slowly varying component of the particle distribution in a regular Langmuir field, neglecting all virtual fields.

6 Stimulated Scattering of Waves by Particles

6.1 A Simple Physical Picture
of the Scattering Process

Let us remind ourselves of the well known picture of the scattering of waves by individual particles when there is no plasma. One must distinguish between *spontaneous and stimulated scattering*. In the case of spontaneous scattering there is only a single wave which forces the particle to oscillate in its field; in the simplest case the oscillating particle, if it is charged, will emit dipole radiation with an intensity which is proportional to that of the incident wave. In the case of stimulated scattering there must be at least two waves present. The first wave forces the particle to oscillate and the second wave has a frequency and wavenumber corresponding to the waves which are emitted by the oscillating particle. The stimulated emission by the particle is the stimulated scattering which has an intensity proportional to the intensities of both incident waves. One must, of course, take into account also the process where the second wave causes the particle to oscillate and the first wave leads to the stimulated emission.

In the presence of other particles – in particular, when there is a plasma present – the scattering mechanism changes in a fundamental way and the usual picture of the scattering which we have just described appears to be completely wrong. Of course, the concept of stimulated scattering remains, but the mechanism for the emission of waves by another wave forcing the charge to oscillate can no longer be applied. Moreover, it appears that scattering is possible even in the case when the particle is not oscillating, or when it is very heavy – even when it has an infinitely large mass. In contrast to the case of a vacuum, longitudinal waves can exist in a plasma. This possibility not only changes the scattering processes, but more importantly, there appears a new scattering mechanism. We shall here be mainly concerned with the spontaneous and stimulated scattering processes for longitudinal waves. As the stimulated scattering intensity is proportional to the square of the intensity of the plasma waves stimulated scattering describes their non-linear interaction and that is just the reason why we are here interested in the stimulated scattering process.

When describing collisions in a plasma we found that it was necessary to introduce the concept of "dressed" particles and the dynamical screening of the field of the colliding particles. We shall start the discussion of scattering processes in a plasma by accepting the hypothesis that the scattering can be qualitatively described by a "dressed" particle model. The longitudinal waves exist only for wavelengths which are much longer than the screening length of the "cloud" of opposite charges which "dresses" the particle. Not only the central particle, but also the screening shell – the "cloud" – will produce scattering. As the size of the shell is much smaller than the wavelength the shell to a first approximation will act as a radiating point charge. The total charge of the shell is equal to that of the central particle. It follows that for the case when the central particle is an electron and the shell is made up of electrons – to be more precise, its positive charge is produced by the absence of electrons, but the charge to mass ratio is the same as for electrons – the scattering should be zero: the shell and the central charge oscillate with opposite phases in the field of the incident wave and together they produce a zero dipole moment. In other words, an electron surrounded by an electron shell does not have a net charge and should not give rise to scattering. The result is that in a plasma electrons will not give rise to scattering whereas they do in a vacuum. We see here a radical change in the scattering picture. Moreover, ions do not scatter in a vacuum, because of their large mass, but they can scatter appreciably in a plasma. Indeed, in this case the central charge will perform hardly any oscillations in the field of the incident wave but if the central charge is screened by electrons the shell will oscillate, as it has the charge to mass ratio of an electron, and will emit radiation. This will lead to a scattering cross-section of the order of that for an electron in vacuo.

Slightly exaggerating we can describe the situation as follows. The whole picture of the scattering is completely reversed in the case of a plasma. Of course, this strong statement needs corrections because the screening shell does not consist only of electrons – or more precisely of enhancements or depletions of the electron density close to the central charge – since the ions can also participate in the screening as we have explained earlier in great detail. Another reason for corrections is that the screening should be dynamical and should depend on the particle velocity. Finally, we must take into account that although the wavelength is much larger than the size of the screening shell, their ratio is finite. We must bear in ind that the screening can be produced by charges of any sign, but the enhancement or depletion of the plasma particle densities in the neighbourhood of the screened charge will always be relatively small.

We shall now estimate the scattering more quantitatively, taking into account the corrections we have just mentioned. It is clear that because of these corrections the scattering by electrons will not be zero, but it still will be much weaker than in a vacuum. Let us first of all take into account the possible presence of an ion screening cloud around an electron. We have seen in earlier chapters that when the electron is moving fast the ion screening cloud cannot follow the electron and the ion screening will be negligible. However, in the case of a slowly moving electron or an electron at rest ion screening will be important. The difference between fast and slow electrons is due to the fact that the screening is dynamical. The screening by ions is approximately static screening only for electron velocities much slower than the ion-sound velocity v_s. This means that the ions play an important rôle in the screening of an electron charge when $v \ll v_s$, but not when $v \gg v_s$. Since the ions in the shell are heavy they do not contribute to the scattering. The screening of a charge at rest is determined by the Debye radius d to which both ions and electrons contribute:

$$\frac{1}{d^2} = \frac{1}{d_e^2} + \frac{1}{d_i^2}. \tag{6.1}$$

We therefore find that the effective charge of the screening shell will be

$$e_{\text{eff}} = -e \frac{\frac{1}{d_e^2}}{\frac{1}{d_e^2} + \frac{1}{d_i^2}} = -e \frac{T_i}{ZT_e + T_i}. \tag{6.2}$$

This charge is much smaller than the electron charge when $ZT_e \gg T_i$ and it cannot completely cancel the charge of the central electron. Let us denote the cross-section for scattering by a "bare" electron by σ_0. This cross-section can be found by proceeding in the manner described at the start of the present section and formally considering a longitudinal wave in vacuo and assuming that the scattered wave is also longitudinal – of course, longitudinal waves cannot propagate in vacuo and this is the reason for using the qualification "formally". This cross-section σ_0 will be of the order of the usual Thomson cross-section and will differ from it only because of the polarisation properties of the longitudinal waves. The cross-section for scattering by an electron at rest in a plasma will be different. The total charge which will scatter the wave will be the sum of the charge of the central electron and that of its electron shell:

$$e + e_{\text{eff}} = e \frac{ZT_e}{ZT_e + T_i} = eZ_{\text{eff}}^{(e)}. \tag{6.3}$$

The scattering cross-section is proportional to the square of this charge so that we have

$$\sigma = \sigma_0 \left(\frac{ZT_e}{ZT_e + T_i} \right)^2. \tag{6.4}$$

This cross-section is of the order of σ_0. An important point is now that the total number of electrons for which $v \ll v_s$, for which the scattering cross-section is large, is very small. Their relative number is determined by the ratio of the phase volume for electrons with velocities less than the ion-sound speed, v_s, to the phase volume of electrons with average thermal velocities, v_{Te}:

$$\frac{\delta n_e}{n_e} \approx \frac{v_\mathrm{s}^3}{v_{Te}^3} = \left(\frac{m_e}{m_\mathrm{i}}\right)^{3/2}. \tag{6.5}$$

If we are interested in scattering by some volume of plasma we need to know the scattering cross-section related to all the electrons in that volume. However, if $v \gg v_\mathrm{s}$ the ions do not participate in the screening and the total charge of the central electron together with the electron shell will be close to zero, as we mentioned earlier when we made a rough estimate. The strong statement which we made earlier thus refers to the main part of the electron distribution: the electrons for which $v \gg v_\mathrm{s}$ will contribute practically nothing to the scattering. The average scattering cross-section per electron will therefore be smaller than expression (6.4) by the factor (6.5). The statement that electrons for which $v \gg v_\mathrm{s}$ do not scatter at all is not completely accurate, if one takes into account the finite size of the screening shell as compared to the wavelength. Quadrupole emission by the shell is possible; its intensity is smaller than that of dipole emission by a factor equal to the square of the ratio of the size of the shell to the wavelength. This last factor is of the order of magnitude $k^2 d_e^2 \ll 1$, which means that we have the following estimate for the cross-section:

$$\sigma \approx \sigma_0 k^2 d_e^2 \ll \sigma_0. \tag{6.6}$$

In fact, the average cross-section will be determined by the larger of the two small factors which we have just mentioned. We therefore reach the conclusion that the cross-section for scattering by electrons in a plasma is much smaller than that in vacuo for wavelengths which are much longer than the screening length. This statement is obviously true for waves with any polarisation, but for longitudinal waves it will be true for all wavelengths, since the wavelengths of longitudinal waves are necessarily longer than the screening length. The cross-sections for scattering by electrons are particularly small for waves with long wavelengths since the cross-section is in that case determined by the small factor (6.5). The stimulated scattering of waves by electrons in a plasma is usually neglected as a possible mechanism for nonlinear interactions. Electromagnetic waves have wavelengths which are longer than the size of the screening shell for frequencies less than $\omega_{pe} c / v_{Te}$ and all effects which we have just described occur also for those waves in that range of frequencies.

Scattering by ions is changed much less by these corrections since their velocities are usually much smaller than the ion-sound velocity. The screening is almost static. Scattering of Langmuir waves by the central ion is negligibly small and the screening is in that case due to the electron shell. For ions the scattering charge will thus differ from (6.2) by a factor $-Z$, since the

screening charge has the same absolute magnitude and the opposite charge
of those of the central charge:

$$e_{\text{eff}}^{(i)} = Ze\frac{T_i}{T_i + ZT_e} = eZ_{\text{eff}}^{(i)}. \tag{6.7}$$

We get thus the following estimate for the cross-section for scattering by ions
in a plasma:

$$\sigma = \sigma_0 \left(\frac{ZT_i}{ZT_e + T_i}\right)^2, \tag{6.8}$$

which is, indeed, of the order of the cross-section for scattering by electrons
in a vacuum. This discussion confirms the earlier statement that the cross-
sections are changed radically in a plasma. We can remind ourselves that
in a vacuum the cross-section for scattering by ions is smaller than that for
scattering by electrons, σ_0 by a factor $(m_e/m_i)^2$ whereas for a plasma, in the
case when (6.5) is valid, the scattering by electrons is less than the scattering
by ions by a factor $(m_e/m_i)^{3/2}$: the change could be six to seven orders of
magnitude. This is, of course, a much more drastic change than in the case
of particle collisions when there was a change by a factor which appeared
under the logarithm sign. We must mention that this qualitative change of
the cross-section is independent of the plasma density. Decreasing the density
only increases the size of the screening shell, but the plasma waves always
have wavelengths which are longer than the size of the screening shell.

We can estimate the scattering cross-sections for ion-sound waves in the
same qualitative manner. The difference with the case of Langmuir waves is
that scattering by ions becomes dominant, but the effect of the screening shell
is also very important. The domination of the ions in the scattering is due
to the low values of the frequencies of the ion-sound waves which therefore
can move the ions over appreciable distances during their long period. The
easiest way to prove this statement is to consider the differential scattering
probability. To estimate it one must divide the intensity of the emission in
the scattering process both by the phase volume, which is proportional to ω^3
and by the frequency, ω; this means that the probability is proportional to
$1/\omega^4$ and for ions it is also proportional to $1/m_i^2$. For ion-sound waves the
m_i-dependence drops out of the product of these two factors. The charge of
the central ion is $-Ze$ and that of the screening shell is

$$e_{\text{eff}} = Ze\frac{ZT_e}{ZT_e + T_i} \approx Ze. \tag{6.9}$$

In the case when $ZT_e \gg T_i$, which is when ion-sound waves exist, it almost
cancels the central charge. However, if the ion is moving with a velocity $\mathbf{v}$,
the field will be acting on the ion with a frequency $\omega - (\mathbf{k}\cdot\mathbf{v})$ and the relative
Doppler corrections for the scattering by the central charge will be determined
by $(\mathbf{k}\cdot\mathbf{v})/\omega$. On the other hand, the screening shell is produced statistically

by various particles moving in all possible directions and therefore in the approximation which is linear in $(\mathbf{k} \cdot \mathbf{v})/\omega$ the Doppler effect will vanish, that is, the screening charge can still be described by Eq.(6.9) as long as we neglect the higher-order Doppler corrections. The Doppler corrections for the central charge are of the order of $\sqrt{T_i/ZT_e}$ which is much larger than the corrections of order T_i/ZT_e to the effective charge. Since the scattering cross-section is proportional to the square of the amplitude of the oscillating charge dipole we find

$$\sigma^{(i)} \sim \sigma_0^{(i)} \left(\frac{2(\mathbf{k} \cdot \mathbf{v})}{\omega} \right)^2 , \tag{6.10}$$

where $\sigma_0^{(i)}$ is the cross-section for the scattering of ion-sound waves by a "bare" ion. This process is thus completely determined by the Doppler corrections to the frequency of the field acting upon the ion. This is, of course, a consequence of the efficient screening by the shell. There is a very important relation between our present qualitative considerations and our discussion of the non-linear interactions of ion-sound waves which also depended, as we proved, on these Doppler corrections to the frequency. We shall, indeed, show that the reason for this relation is that the non-linear interactions are nothing but the processes of stimulated scattering of ion-sound waves by ions.

The whole picture sketched here refers to all the plasma particles. The question again arises how these particles can be the central particles for the scattering and at the same time produce the screening shells for other particles. The answer is the same as the one we gave when we were discussing collisions. The most important factor here are the fluctuations since the particles participating in fluctuations can screen the fields of other particles. The total number of particles in the plasma remains unchanged. We have assumed here that the picture we found for the case of particle collisions in a theory of fluctuations remains valid also for the case of scattering processes. We should prove this for the scattering processes using fluctuation theory. The easiest way to do this is to consider stimulated scattering processes. We shall show that these correspond to non-linear interactions of random waves and their theory we have already constructed earlier. We shall thus turn again to the problem of non-linear interactions of random waves.

6.2 Description of the Stimulated Scattering Processes of Waves in a Plasma. Non-linear Landau Damping

We start by defining the scattering probability. Let $w_{\mathbf{p}}(\mathbf{k}, \mathbf{k}')$ be the probability per unit time that a wave with wavevector (momentum) $\mathbf{k}$ will be scattered by a particle with momentum $\mathbf{p}$ and become a wave with wavevector (momentum) $\mathbf{k}'$. We shall indicate by an upper index (α) the kind of particle–electron (e) or ion (i)–that is scattering the waves. The types of the incident and scattered waves will be denoted, respectively, by the upper indices σ' and σ. If the type of wave is changed in the process, we shall still call it a scattering process–for instance, when Langmuir waves are scattered and produce Langmuir waves we use the two upper indices "ℓ, ℓ", and when Langmuir waves are scattered and produce ion-sound waves we use the two upper indices "ℓ, s". The probability is normalised to one incident and one scattered wave so that the summing over these waves corresponds to multiplying by the numbers of waves N_k and $N_{k'}$ and integration over the phase volumes $d^3k/(2\pi)^3$ and $d^3k'/(2\pi)^3$.

To find the energy Q^{sc} scattered by a unit plasma volume we must multiply the probability by the number N_k of incident waves per unit volume, by the number of particles of momentum $\mathbf{p}$, that is, by $\Phi_{\mathbf{p}}$, and integrate over the phase volumes of the waves and the particles:

$$Q^{sc} = \int \omega_k w_{\mathbf{p}}(\mathbf{k}, \mathbf{k}') N_{k'} \Phi_{\mathbf{p}} \frac{d^3k' \, d^3p \, d^3k}{(2\pi)^9} = \int Q_k^{sc} \frac{d^3k}{(2\pi)^3}. \tag{6.11}$$

The energy Q_k^{sc} scattered by a single particle is given by the equation

$$Q_k^{sc} = Q_{\mathbf{p},k}^{sc} \Phi_{\mathbf{p}} \frac{d^3p}{(2\pi)^3}; \qquad Q_{\mathbf{p},k}^{sc} = \omega_k \int w_{\mathbf{p}}(\mathbf{k}, \mathbf{k}') N_{k'} \frac{d^3k'}{(2\pi)^3}. \tag{6.12}$$

One could use these equations as a definition of the scattering probability since in scattering the power of the scattered waves is proportional to the intensity of the incident wave. The scattering probability is just a coefficient in this equation, normalised in the way indicated by Eq.(6.11) or Eq.(6.12). The introduction of a scattering probability is very useful for the discussion of stimulated scattering processes just as the introduction of the Vavilov-Cherenkov emission probability was useful for the discussion of Landau damping. We must take into account when considering stimulated processes that the probability for stimulated emission differs from that for spontaneous emission by a factor N_k.

We shall now construct a balance equation, taking into account both spontaneous and stimulated scattering processes. We bear in mind that in the scattering process described by the probability $w_{\mathbf{p}}(\mathbf{k}, \mathbf{k}')$ a wave with momentum $\mathbf{k}$ is absorbed and a wave with momentum $\mathbf{k}'$ is emitted so that the increase per unit time of the number N_k is given by the equation

$$\left(\frac{dN_k}{dt}\right)_+ = \int (N_k + 1) N_{k'} w_{\mathbf{p}}(\mathbf{k}, \mathbf{k}') \Phi_{\mathbf{p}} \frac{d^3k' \, d^3p}{(2\pi)^6}. \tag{6.13}$$

The decrease in the number of waves is determined by the same probability and by the number of particles in the state after absorption of an amount of momentum $\mathbf{k}'$ and the emission of an amount of momentum $\mathbf{k}$, that is, by $\Phi_{\mathbf{p}+\mathbf{k}'-\mathbf{k}}$:

$$\left(\frac{dN_{\mathbf{k}}}{dt}\right)_{-} = \int N_{\mathbf{k}}(N_{\mathbf{k'}}+1)w_{\mathbf{p}}(\mathbf{k},\mathbf{k'})\Phi_{\mathbf{p+k'-k}}\frac{d^3k'\,d^3p}{(2\pi)^6}. \qquad (6.14)$$

We shall assume that the momentum transferred to the particles in the scattering process, $\mathbf{k'}-\mathbf{k}$, is small so that we can expand in terms of it. We then get the following balance equation:

$$\frac{dN_{\mathbf{k}}}{dt} = \left(\frac{dN_{\mathbf{k}}}{dt}\right)_{+} + \left(\frac{dN_{\mathbf{k}}}{dt}\right)_{-}$$

$$= N_{\mathbf{k}} \int N_{\mathbf{k'}}w_{\mathbf{p}}(\mathbf{k},\mathbf{k'})\left([\mathbf{k}-\mathbf{k'}]\cdot\frac{\partial\Phi_{\mathbf{p}}}{\partial\mathbf{p}}\right)\frac{d^3k'\,d^3p}{(2\pi)^6}$$

$$+ \int N_{\mathbf{k'}}w_{\mathbf{p}}(\mathbf{k},\mathbf{k'})\Phi_{\mathbf{p}}\frac{d^3k'\,d^3p}{(2\pi)^6}$$

$$- N_{\mathbf{k}} \int w_{\mathbf{p}}(\mathbf{k},\mathbf{k'})\Phi_{\mathbf{p}}\frac{d^3k'\,d^3p}{(2\pi)^6}, \qquad (6.15)$$

where the first term on the right-hand side describes the stimulated scattering, the second term describes the spontaneous scattering of incident waves, and the third term describes the rate at which waves with momentum $\mathbf{k}$ are depleted due to spontaneous scattering leading to waves with momentum $\mathbf{k'}$. If we consider only the stimulated processes we have

$$\frac{dN_{\mathbf{k}}}{dt} = N_{\mathbf{k}} \int N_{\mathbf{k'}}w_{\mathbf{p}}(\mathbf{k},\mathbf{k'})\left([\mathbf{k}-\mathbf{k'}]\cdot\frac{\partial\Phi_{\mathbf{p}}}{\partial\mathbf{p}}\right)\frac{d^3k'\,d^3p}{(2\pi)^6}$$

$$= 2\gamma_{\mathbf{k}}^{NL}N_{\mathbf{k}}. \qquad (6.16)$$

The quantity $\gamma_{\mathbf{k}}^{NL}$ is called the *non-linear Landau damping* rate and is given by the equation

$$\gamma_{\mathbf{k}}^{NL} = \tfrac{1}{2} \int N_{\mathbf{k'}}w_{\mathbf{p}}(\mathbf{k},\mathbf{k'})\left([\mathbf{k}-\mathbf{k'}]\cdot\frac{\partial\Phi_{\mathbf{p}}}{\partial\mathbf{p}}\right)\frac{d^3k'\,d^3p}{(2\pi)^6}. \qquad (6.17)$$

This name is not a very appropriate one, although it is widely used in the literature: it does not describe damping and Landau did not consider stimulated scattering processes. The name is given because of the analogy between expression (6.17) and that for the linear Landau damping, the expression for which can be written in the form

$$\gamma_{\mathbf{k}} = \tfrac{1}{2} \int w_{\mathbf{p}}(\mathbf{k})\left(\mathbf{k}\cdot\frac{\partial\Phi_{\mathbf{p}}}{\partial\mathbf{p}}\right)\frac{d^3p}{(2\pi)^3}. \qquad (6.18)$$

Both expressions contain a derivative of the distribution function and the transferred momentum. However, the expression for the stimulated scattering does not describe any damping (absorption) of waves. It follows from Eq.(6.16) that the total number of waves is conserved:

$$\frac{dN}{dt} = \int \frac{dN_\mathbf{k}}{dt} \frac{d^3\mathbf{k}}{(2\pi)^3}$$

$$= \int N_\mathbf{k} N_{\mathbf{k}'} w_\mathbf{p}(\mathbf{k}, \mathbf{k}') \left([\mathbf{k} - \mathbf{k}'] \cdot \frac{\partial \Phi_\mathbf{p}}{\partial \mathbf{p}} \right) \frac{d^3\mathbf{k}\, d^3\mathbf{k}'\, d^3\mathbf{p}}{(2\pi)^9} = 0. \quad (6.19)$$

The last equality is a consequence of the symmetry of the probablity under an interchange of $\mathbf{k}$ and $\mathbf{k}'$. The conservation of the number of waves has a simple physical meaning: in a scattering process an emitted wave appears for every wave which is absorbed. The conservation of the number of waves is valid only for processes in which there is no change in the type of waves; for instance it is true when scattering of a Langmuir wave produces another Langmuir wave or the scattering of an ion-sound wave produces another ion-sound wave. However, this conservation law is not valid if, for instance, the scattering of a Langmuir wave produces an ion-sound wave.

We showed already in the preceding section that scattering by ions is the most important scattering process in a plasma. It is useful for what follows to express scattering processes in the form of "reactions". For instance, in the case of scattering of Langmuir waves or ion-sound waves by ions we write

$$\ell + i \leftrightarrows \ell' + i', \qquad s + i \leftrightarrows s' + i', \tag{6.20}$$

for scattering of Langmuir waves by electrons we have

$$\ell + e \leftrightarrows \ell' + e', \tag{6.21}$$

and for scattering by electrons where a Langmuir wave is replaced by an ion-sound wave

$$\ell + e \leftrightarrows s + e'. \tag{6.22}$$

The primes in these formulæ indicate that we have the same kind of wave or particle, but in a different state. The number of quanta is conserved in the processes described by Eqs. (6.20) and (6.21), but not in the process described by Eq. (6.22).

One can easily show that the wave frequency decreases in stimulated scattering processes if the particles have a thermal distribution. In other words, the waves are *red-shifted* in this process. To understand this red-shift we first must show that energy is transferred from the waves to the particles, or that energy lost by the waves is gained by the particles. The distribution function is changed in this scattering process and the derivation of the equation describing these changes is just a repeat of the derivation of the quasi-linear equation, with a difference in that the momentum transferred is not $\mathbf{k}$, but $\mathbf{k} - \mathbf{k}'$ and that the probability must be multiplied by $N_{\mathbf{k}'}\, d^3\mathbf{k}'/(2\pi)^3$ and integrated over $\mathbf{k}'$. In the balance equation for particles with a momentum $\mathbf{p}$ we must take into account that there is a loss of particles in the state $\mathbf{p}$ because they make a transition to the states $\mathbf{p} + \mathbf{k} - \mathbf{k}'$ and $\mathbf{p} + \mathbf{k}' - \mathbf{k}$ and a gain of particles with momentum $\mathbf{p}$ due to the inverse processes. If we expand in terms of the transferred momentum we find for the stimulated processes:

$$\frac{d\Phi_{\mathbf{p}}}{dt} = \frac{\partial}{\partial p_i} D^{sc}_{ij} \frac{\partial}{\partial p_j} \Phi_{\mathbf{p}},$$

$$D^{sc}_{ij} = \int (k_i - k'_i)(k_j - k'_j) w_{\mathbf{p}}(\mathbf{k}, \mathbf{k}') N_{\mathbf{k}} N_{\mathbf{k}'} \frac{d^3 k \, d^3 k'}{(2\pi)^6}. \tag{6.23}$$

For comparison we write down again the quasi-linear diffusion coefficient:

$$D^{QL}_{ij} = \int k_i k_j w_{\mathbf{p}}(\mathbf{k}) N_{\mathbf{k}} \frac{d^3 k}{(2\pi)^3}. \tag{6.24}$$

The analogy here is the same as between the linear and the non-linear Landau damping. Equation (6.23) can be obtained using the non-linear fluctuation theory given in the last section of Chap.5. The equation for the non-linear Landau damping can be derived using the non-linear fluctuation theory given in the preceding chapter. We leave the derivation of Eq.(6.23) as an exercise and mention merely that this derivation is similar to the one given in the case of the quasi-linear theory. A very important point is that both the equation for the particle distribution and that for the wave distribution can be derived directly from the existing fluctuation theory. The only restriction is that it should be possible to describe the waves by their distribution function. The waves should be random and phase locking should be unimportant.

Finally we note that Eq.(6.23) satisfies the conservation law for the total number of particles:

$$\frac{dn}{dt} = \frac{d}{dt} \int \Phi_{\mathbf{p}} \frac{d^3 p}{(2\pi)^3} = 0. \tag{6.25}$$

6.3 Conservation Laws in an Elementary Scattering Process

The main problem in the case of conservation laws in an elementary scattering process is whether we should be considering the values of the energy and the momentum of the central charge or those for part of the central charge or those of the shell, and so on. For instance, there is the difficulty that in the case of scattering by ions the emission of the scattered wave is due to the electrons in the shell and one would expect that the recoil of this emission will affect the shell but not the central ion. The question arises whether in this picture of "dressed" particles in a scattering process the dressed particle enters as a "rigid" object and whether the momentum transferred to the shell is transferred to the central charge. Doubts arise because the shell is formed statistically due to fluctuations and various particles with different momenta produce the screening. The picture of the shell is a varying construction rather than a rigid one. In a quantum picture of the scattering a momentum $\mathbf{k}'$ is absorbed and a momentum $\mathbf{k}$ is lost by the scatterer, which means that the particle gains a momentum $\mathbf{k}' - \mathbf{k}$. In the case of scattering of Langmuir waves by ions it appears that the electron shell gains this momentum, rather than the central ion. However, this is not the correct answer and, in fact, the central particle should gain this momentum $\mathbf{k}' - \mathbf{k}$.

The main point is that before the scattering, that is, before the absorption and subsequent emission processes, as well as after the scattering the particles are free and move with a constant velocity. Why is this important? A particle moving with a constant velocity in a plasma has a well defined shell which is fixed when the velocity of the particle is fixed. This can be seen in any description of a test particle in a plasma. The particle will not emit any waves and will be able to move with a constant velocity, only if the shell is fixed. Therefore, if the shell receives the recoil momentum in the scattering process, the shell will no longer correspond to the shell of a stationary particle and it will continue to emit waves until the momentum has been transferred to the central charge. In this sense we can say that the emission is produced by the dressed particle as a whole and we can always refer the momentum and the energy to the central charge. The statistically produced shell thus appears to be even more rigid than a quantum shell of bound electrons. Of course, the picture we have just painted is the result of exact calculations using fluctuation theory and is, in fact, an interpretation of these results. However, the physical picture is so important that we wanted to give it first, before going over to a mathematical treatment; the meaning of the result of such a treatment is usually not emphasised and the physical picture is often hidden by the complicated calculations of fluctuation theory.

Having the picture we have just drawn in mind we shall write down the energy conservation law for scattering by a "particle" with initial momentum $\mathbf{p}$ and final momentum $\mathbf{p} + \mathbf{k}' - \mathbf{k}$:

$$\varepsilon_{\mathbf{p}} + \omega_{\mathbf{k}'} = \varepsilon_{\mathbf{p}+\mathbf{k}'-\mathbf{k}} + \omega_{\mathbf{k}}. \tag{6.26}$$

Expanding in terms of the transferred momentum we find

$$\omega_{\mathbf{k}'} - \omega_{\mathbf{k}} = \left([\mathbf{k} - \mathbf{k}'] \cdot \mathbf{v}\right). \tag{6.27}$$

We can use this conservation law to find the conservation law for the total sum of the energies of the wave and the particle. The proof is similar to that given for the case of the quasi-linear equation. The only differences lie in that we must use the non-linear damping rate rather than the linear one, the diffusion equation for scattering, which we found a moment ago, instead of the quasi-linear equation, and Eq.(6.27) instead of the resonance condition for Vavilov-Cherenkov emission. We can similarly prove a conservation law for the total sum of the momenta of the particles and the waves. We find thus that the "dressed" particles which occur in the conservation laws are well defined objects.

We can also use Eq.(6.27) to prove that the waves are red-shifted if they are scattered by particles in thermal equilibrium. This means that in the scattering energy is transferred from the waves to the particles – bear in mind that the number of waves is conserved. Indeed, for particles in thermal equilibrium we have

$$\left([\mathbf{k} - \mathbf{k}'] \cdot \frac{\partial \Phi_\mathbf{p}}{\partial \mathbf{p}} \right) = \frac{([\mathbf{k} - \mathbf{k}'] \cdot \mathbf{v})}{T} \Phi_\mathbf{p} = \frac{\omega_\mathbf{k} - \omega_{\mathbf{k}'}}{T} \Phi_\mathbf{p}, \tag{6.28}$$

and thus

$$\gamma_\mathbf{k}^{NL} = \frac{1}{2T} \int N_{\mathbf{k}'}(\omega_{\mathbf{k}'} - \omega_\mathbf{k}) w_\mathbf{p}(\mathbf{k}, \mathbf{k}') \Phi_\mathbf{p} \frac{d^3 k' \, d^3 p}{(2\pi)^6}. \tag{6.29}$$

For the case of a narrow wavepacket with $\mathbf{k}' \approx \mathbf{k}_0$ waves for which $\omega_\mathbf{k} < \omega_{\mathbf{k}_0}$ will be amplified, whereas waves for which $\omega_\mathbf{k} > \omega_{\mathbf{k}_0}$ will be damped. In the general case in a wavepacket waves with higher than average frequencies will be damped and waves with lower than average frequencies will be amplified. For Langmuir waves the wavenumber dependence of the frequency is described by a small term and the conservation of the total number of quanta (waves) means that whereas the energy of the wave will not change much, there will be a considerable change in the wavenumbers. The latter effect is due to the fact that in Eq.(6.29) there occurs a frequency difference which means that the plasma frequency drops out of the integrand. We find thus that in the case of stimulated scattering of Langmuir waves by particles in thermal equilibrium there is a transfer of waves in the direction of smaller wavenumbers, that is, in the direction of larger scales.

In the case of ion-sound waves there is also a transfer in the direction of larger scales but now there is an appreciable decrease in energy due to the scattering process.

This process should lead to a concentration of waves in the region of large scales. In the case of Langmuir waves this concentration is known as the *Langmuir condensate*. We can estimate the value of the wavenumbers at which the Langmuir condensate is formed and also the maximum amplitude of its field. First of all we note that the scattering by ions should take place in small steps. In each of the steps the absolute value of the wavenumber only changes a little. Indeed, let us consider the case when as the result of the scattering the wavevector is changed considerably in direction but not in absolute magnitude. This kind of scattering might almost be called elastic scattering of the waves since the momentum of the waves is changed in direction, but their energy remains unchanged. The reason why to a first approximation the scattering is elastic is that the ion thermal velocity is small and the frequency change in the scattering is small. If k' is of the same order as k, the quantity $|\mathbf{k} - \mathbf{k}'|$ will also be of the same order as k and we have thus

$$\omega_\mathbf{k} - \omega_{\mathbf{k}'} = \frac{3v_{Te}^2}{2\omega_{pe}}\left(k^2 - k'^2\right) \approx \frac{3v_{Te}^2}{\omega_{pe}}\,k\,\Delta k \approx kv_{Ti}, \quad \Delta k = k - k'. \tag{6.30}$$

Let us denote the change in absolute magnitude of k in a single scattering by k_*. From Eq.(6.30) we find for $\Delta k = k_*$:

$$k_* = \frac{v_{Ti}}{3v_{Te}^2}\,\omega_{pe} = \frac{v_{Ti}}{3v_{Te}}\frac{1}{d_e} = \sqrt{\frac{m_e}{9m_i}\frac{T_i}{T_e}}\frac{1}{d_e}. \tag{6.31}$$

We see thus that k_* is, indeed, small. After the energy is transferred in the first step, energy will be transferred in the next step. The energy thus cascades down to smaller wavenumbers and larger scales. The last step corresponds to $k \approx k_*$; this gives us an estimate of the wavenumber $\Delta k \approx k_*$ of the Langmuir condensate where the Langmuir waves will concentrate. The next question is: how much energy can be accumulated in the condensate? We found earlier the existence of a critical energy density W_{cr}. For energy densities larger than W_{cr} the nature of the non-linear interactions is changed – the interactions become modulational kind of interactions – and one may expect that no more energy will be accumulated in the condensate. We can find an estimate for the critical energy density from the relation

$$\frac{W_{cr}}{nT_e} \approx (\Delta k)^2 d_e^2. \tag{6.32}$$

After cascading the initial energy W_0 of the Langmuir oscillations will be concentrated in the condensate without any large change in its total magnitude W_0. This energy will be transferred to the wavenumber region $\Delta k \approx k_*$. Substituting this value into Eq.(6.32) we find

$$\frac{W_0}{nT_e} > k_*^2 d_e^2 = \frac{m_e}{9m_i}\frac{T_i}{T_e}. \tag{6.33}$$

This value differs only by the factor T_i/T_e from the estimate (5.16) which gave the threshold above which in a process where a Langmuir wave decayed into another Langmuir wave and an ion-sound wave the non-linear broadening of the Langmuir waves would be larger than the frequency of the ion-sound waves.

Scattering of ion-sound waves by ions is also a cascade process. but the size of each step in the cascade gets smaller as the wavenumber decreases. Again making an assumption that the scattering of ion-sound waves by ions is almost elastic we find

$$\omega_\mathbf{k} - \omega_{\mathbf{k}'} \approx \Delta k v_s; \qquad \frac{\Delta k}{k} = \sqrt{\frac{T_i}{ZT_e}}. \tag{6.34}$$

Although the energy of the ion-sound waves due to the decrease in the steps in the cascade is also accumulating at small k values the large dispersion of the ion-sound waves for $k \ll 1/d_e$ makes it difficult to exceed the threshold for modulational interactions in the case when $k \ll 1/d_e$. The modulational instability is possible for ion oscillations with $\omega \approx \omega_{pi}$.

Scattering is not the only process which can lead to non-linear interactions of random waves at energies below the critical value. There are also simultaneous emission or absorption processes involving two waves. After absorbing two waves the particle has changed its energy from ε_p to $\varepsilon_{p+k+k'}$ so that the conservation of energy in the elementary two-wave emission process is of the form

$$\varepsilon_p + \omega_k + \omega_{k'} = \varepsilon_{p+k+k'} \longrightarrow \omega_k + \omega_{k'} = ([k + k'] \cdot v). \tag{6.35}$$

This process is forbidden for two Langmuir waves or for two ion-sound waves but the simultaneous emission of one Langmuir wave and one ion-sound wave in scattering by electrons is possible:

$$e \leftrightarrows s + \ell + e'.$$

This process can play a very important rôle, and one can make big mistakes if one neglects it – as was done in a number of early papers.

6.4 Scattering Probabilities for Langmuir Waves

We shall now prove that fluctuation theory leads to a "dressed" particle model, we shall find the exact values of the effective screening charges, and we shall show that these effective charges depend, in general, on the wavenumbers of the incident and the scattered waves, but that in some limiting cases these effective screening charges lead to the simple expressions which we found earlier using some simple physical assumptions. We must find an explicit expression for the scattering probability from the general non-linear equations and show that this expression describes, indeed, the non-linear Landau damping (stimulated scattering). Let us remind ourselves that the total number of screened particles is equal to the total number of particles and that screening is produced by the fluctuations. If there were no fluctuations a homogeneous system would not scatter, a fact well known for scattering in any medium. Spontaneous scattering is due to the statistical particle fluctuations, whereas stimulated scattering is due to the fluctuations produced by random collective waves. The non-linear equations which we derived earlier are valid for fields which are well above the level of statistical fluctuations. This means that we can find only the stimulated scattering – or non-linear Landau damping – from these equations. It is thus necessary to average the general non-linear equations, assuming that the collective waves are random.

We shall use Eq.(5.35) to consider random Langmuir waves, neglecting the second term on the right-hand side – we showed that it is smaller than the first term by a factor which is the square of the ratio of the thermal

velocity to the phase velocity – and we shall use the approximate expression
(5.55) for the non-linear response together with the substitution (5.39) for
random waves

$$\varepsilon E^+ = -\frac{e^2}{m_e^2 \omega_{pe}^4} \int \frac{(\mathbf{k} \cdot \mathbf{k}_1)(\mathbf{k}_2 \cdot \mathbf{k}_3)}{k k_1 k_2 k_3} |\mathbf{k}_2 + \mathbf{k}_3|^2 d_{1,2,3}$$

$$\times \frac{(\varepsilon_{2+3}^{(e)} - 1)\varepsilon_{2+3}^{(i)}}{\varepsilon_{2+3}} \left[E_1^+ E_2^+ E_3^- - E_1^+ \langle E_2^+ E_3^- \rangle - \langle E_1^+ E_2^+ E_3^- \rangle \right]. \quad (6.36)$$

For the averaging we shall use Eqs.(3.16), (3.20), and (3.26) as well as the
relations

$$\mathbf{E}_{\mathbf{k}',\omega'} = \mathbf{E}_{\mathbf{k}'} = E_{\mathbf{k}'} \frac{\mathbf{k}'}{k'} = -E_{\mathbf{k}'} \frac{\mathbf{k}}{k},$$

$$\langle E_{\mathbf{k}}^+ E_{\mathbf{k}'}^- \rangle = -|E^+|_{\mathbf{k}}^2 \delta(k + k'), \qquad k = \{\mathbf{k}, \omega\}. \quad (6.37)$$

We have here taken into account that a positive-frequency component can be
correlated only with a negative-frequency component. An important feature
of the discussion which follows is that we shall neglect here the non-linear
broadening as compared to the dispersion broadening, that is, we shall assume
that the energy of the waves is much smaller than the critical energy. This
is the only case when we can use Eqs.(3.20) and (3.26). We now write down
the appropriate relation for the correlation function of the positive-frequency
component:

$$|E^+|_{\mathbf{k}}^2 = |E|_{\mathbf{k}}^2 \delta(\omega - \omega_{\mathbf{k}}) = \frac{N_{\mathbf{k}}}{2\pi^2 (\partial \varepsilon / \partial \omega)|_{\omega=\omega_{\mathbf{k}}}} \delta(\omega - \omega_{\mathbf{k}}). \quad (6.38)$$

The equation we are looking for can be obtained by multiplying Eq.(6.36)
by $E_{\mathbf{k}'}^-$, averaging the result, and integrating it over $\mathbf{k}'$. On the right-hand
side we must take into account that the last term vanishes since $\langle E \rangle = 0$,
that the average value of four random fields can be written approximately as
a sum of products of average values of two fields, and that the average value
of the product of two fields vanishes unless the two fields have frequencies
of opposite sign. We are thus left with only a single combination in which
$k_3 = -k_1$, $k_2 = -k' = k$:

$$\varepsilon_{\mathbf{k}} |E^+|_{\mathbf{k}}^2 = -|E^+|_{\mathbf{k}}^2 \frac{e^2}{m_e^2 \omega_{pe}^4}$$

$$\times \int \frac{(\mathbf{k} \cdot \mathbf{k}_1)^2}{k^2 k_1^2} \frac{(\varepsilon_{\mathbf{k}-\mathbf{k}_1}^{(e)} - 1)\varepsilon_{\mathbf{k}-\mathbf{k}_1}^{(i)}}{\varepsilon_{\mathbf{k}-\mathbf{k}_1}} |\mathbf{k} - \mathbf{k}_1|^2 |E^+|_{\mathbf{k}_1}^2 d^4 k_1. \quad (6.39)$$

We can write this equation in the form

$$\left(\varepsilon_{\mathbf{k}} + \varepsilon_{\mathbf{k}}^{NL} \right) |E_{\mathbf{k}}^+|^2 = 0, \quad (6.40)$$

where $\varepsilon_{\mathbf{k}}^{NL}$ is the non-linear plasma permittivity

$$\varepsilon_{\mathbf{k}}^{\mathrm{NL}} = \frac{e^2}{m_e^2 \omega_{pe}^4} \int \frac{(\mathbf{k} \cdot \mathbf{k}')^2}{k^2 k'^2} \frac{(\varepsilon_{\mathbf{k}-\mathbf{k}'}^{(e)} - 1)\varepsilon_{\mathbf{k}-\mathbf{k}'}^{(i)}}{\varepsilon_{\mathbf{k}-\mathbf{k}'}} |\mathbf{k} - \mathbf{k}'|^2 |E^+|_{\mathbf{k}'}^2 d^4 k'. \quad (6.41)$$

We have written here $\mathbf{k}'$ for $\mathbf{k}_1$. We have emphasised that we are considering the stimulated scattering for the case where the non-linear broadening is small; we can therefore also neglect the non-linear frequency shift when calculating the non-linear growth rate and write

$$\gamma_{\mathbf{k}}^{\mathrm{NL}} = -\left. \frac{\mathrm{Im}\{\varepsilon_{\mathbf{k}}^{\mathrm{NL}}\}}{\partial \mathrm{Re}\{\varepsilon_{\mathbf{k}}\}/\partial\omega} \right|_{\omega=\omega_{\mathbf{k}}}. \quad (6.42)$$

We neglected in $\mathrm{Re}\{\varepsilon_{\mathbf{k}}\}$ the contribution from the non-linear real part because the non-linear frequency shift is small. We shall omit in what follows the Re sign since the effect is determined by $\mathrm{Im}\{\varepsilon\}$ which is much smaller than $\mathrm{Re}\{\varepsilon\}$ because the total ε to a first approximation is the same as its real part. Using now Eq.(6.38) and neglecting the non-linear frequency broadening in the imaginary part of the non-linear permittivity as compared to the dispersion broadening we find

$$\gamma_{\mathbf{k}}^{\mathrm{NL}} = -\frac{4\pi e^2}{m_e^2 \omega_{pe}^4} \int N_{\mathbf{k}'} \frac{d^3 k' (\mathbf{k} \cdot \mathbf{k}')^2 |\mathbf{k} - \mathbf{k}'|^2}{(2\pi)^0 k^2 k'^2} \left[\frac{\partial \varepsilon_{\mathbf{k}}}{\partial\omega} \frac{\partial \varepsilon_{\mathbf{k}'}}{\partial\omega'} \right]^{-1}_{\omega=\omega_{\mathbf{k}};\omega'=\omega_{\mathbf{k}'}}$$

$$\times \mathrm{Im} \left\{ \frac{\left(\varepsilon_{\mathbf{k}-\mathbf{k}',\omega_{\mathbf{k}}-\omega_{\mathbf{k}'}}^{(e)} - 1\right) \varepsilon_{\mathbf{k}-\mathbf{k}',\omega_{\mathbf{k}}-\omega_{\mathbf{k}'}}^{(i)}}{\varepsilon_{\mathbf{k}-\mathbf{k}',\omega_{\mathbf{k}}-\omega_{\mathbf{k}'}}} \right\}. \quad (6.43)$$

This equation is already very nearly the same as the one we found by introducing the scattering probability. The imaginary part on the right-hand side of Eq.(6.43) contains the imaginary parts of both the electron and the ion permittivities and describes the stimulated scattering both by electrons and by ions. We shall use the relations

$$\varepsilon = \varepsilon^{(e)} + \varepsilon^{(i)} - 1,$$

$$\mathrm{Im}\left\{ \frac{(\varepsilon^{(e)} - 1)\varepsilon^{(i)}}{\varepsilon} \right\} = \left| \frac{\varepsilon^{(e)} - 1}{\varepsilon} \right|^2 \mathrm{Im}\{\varepsilon^{(i)}\} + \left| \frac{\varepsilon^{(i)}}{\varepsilon} \right|^2 \mathrm{Im}\{\varepsilon^{(e)}\}, \quad (6.44)$$

where we have omitted the lower indices, and

$$\mathrm{Im}\left\{ \varepsilon_{\mathbf{k}-\mathbf{k}',\omega_{\mathbf{k}}-\omega_{\mathbf{k}'}}^{(i)} \right\} = -\frac{4\pi^2 Z^2 e^2}{|\mathbf{k} - \mathbf{k}'|^2} \int \left([\mathbf{k} - \mathbf{k}'] \cdot \frac{\partial \Phi_{\mathbf{p}}^i}{\partial \mathbf{p}} \right)$$

$$\times \delta\left(\omega_{\mathbf{k}} - \omega_{\mathbf{k}'} - ([\mathbf{k} - \mathbf{k}'] \cdot \mathbf{v})\right) \frac{d^3 p}{(2\pi)^3}. \quad (6.45)$$

The first term in Eq.(6.44) describes the scattering by ions:

$$\gamma_{\mathbf{k}}^{NL} = -\frac{2Z^2 e^4}{m_e^2 \omega_{pe}^4} \int N_{\mathbf{k}'} \frac{(\mathbf{k} \cdot \mathbf{k}')^2}{k^2 k'^2} \left| \frac{\varepsilon_{\mathbf{k}-\mathbf{k}',\omega_{\mathbf{k}}-\omega_{\mathbf{k}'}}^{(e)} - 1}{\varepsilon_{\mathbf{k}-\mathbf{k}',\omega_{\mathbf{k}}-\omega_{\mathbf{k}'}}} \right|^2 \left[\frac{\partial \varepsilon_{\mathbf{k}}}{\partial \omega} \frac{\partial \varepsilon_{\mathbf{k}'}}{\partial \omega'} \right]^{-1}_{\substack{\omega=\omega_{\mathbf{k}} \\ \omega'=\omega_{\mathbf{k}'}}}$$

$$\times \left([\mathbf{k} - \mathbf{k}'] \cdot \frac{\partial \Phi_{\mathbf{p}}^i}{\partial \mathbf{p}} \right) \delta(\omega_{\mathbf{k}} - \omega_{\mathbf{k}'} - ([\mathbf{k} - \mathbf{k}'] \cdot \mathbf{v})) \frac{d^3 k' d^3 p}{(2\pi)^3}. \quad (6.46)$$

Within the approximation we are working in we have for Langmuir waves

$$\left. \frac{\partial \varepsilon_{\mathbf{k}}}{\partial \omega} \right|_{\omega=\omega_{\mathbf{k}}} \approx \left. \frac{\partial}{\partial \omega} \left[1 - \frac{\omega_{pe}^2}{\omega^2} \right] \right|_{\omega=\omega_{pe}} = \frac{2}{\omega_{pe}}. \quad (6.47)$$

Equation (6.46) must be compared with Eq.(6.17) for the non-linear Landau damping which, indicating the scattering process which we are considering at the moment, can be written in the form

$$\gamma_{\mathbf{k}}^{NL} = \frac{1}{2} \int N_{\mathbf{k}'} w_{\mathbf{p}}^{\ell,\ell'(i)}(\mathbf{k}, \mathbf{k}') \left([\mathbf{k} - \mathbf{k}'] \cdot \frac{\partial \Phi_{\mathbf{p}}^i}{\partial \mathbf{p}} \right) \frac{d^3 k' d^3 p}{(2\pi)^6}. \quad (6.48)$$

We thus find the following expression for the probability of scattering of a Langmuir wave by ions producing another Langmuir wave, that is, for the $\ell + i \leftrightarrows \ell' + i'$ process:

$$w_{\mathbf{p}}^{\ell,\ell'(i)}(\mathbf{k}, \mathbf{k}') = \frac{Z^2 e^4 (2\pi)^3}{m_e^2 \omega_{pe}^2} \frac{(\mathbf{k} \cdot \mathbf{k}')^2}{k^2 k'^2}$$

$$\times \left| \frac{\varepsilon_{\mathbf{k}-\mathbf{k}',\omega_{\mathbf{k}}-\omega_{\mathbf{k}'}}^{(e)} - 1}{\varepsilon_{\mathbf{k}-\mathbf{k}',\omega_{\mathbf{k}}-\omega_{\mathbf{k}'}}} \right|^2 \delta(\omega_{\mathbf{k}} - \omega_{\mathbf{k}'} - ([\mathbf{k} - \mathbf{k}'] \cdot \mathbf{v})). \quad (6.49)$$

We obtain from the second term in Eq.(6.44) in a similar way the probability of scattering of a Langmuir wave by electrons producing another Langmuir wave, that is, for the $\ell + e \leftrightarrows \ell' + e'$ process:

$$w_{\mathbf{p}}^{\ell,\ell'(e)}(\mathbf{k}, \mathbf{k}') = \frac{(2\pi)^3 e^4}{m_e^2 \omega_{pe}^2} \frac{(\mathbf{k} \cdot \mathbf{k}')^2}{k^2 k'^2}$$

$$\times \left| \frac{\varepsilon_{\mathbf{k}-\mathbf{k}',\omega_{\mathbf{k}}-\omega_{\mathbf{k}'}}^{(i)}}{\varepsilon_{\mathbf{k}-\mathbf{k}',\omega_{\mathbf{k}}-\omega_{\mathbf{k}'}}} \right|^2 \delta(\omega_{\mathbf{k}} - \omega_{\mathbf{k}'} - ([\mathbf{k} - \mathbf{k}'] \cdot \mathbf{v})). \quad (6.50)$$

The two probabilities can be written in the same form if we introduce an effective charge, in units of the electron charge, for the two cases:

$$w_{\mathbf{p}}^{\ell,\ell'(e,i)}(\mathbf{k}, \mathbf{k}') = \frac{(2\pi)^3 e^4}{m_e^2 \omega_{pe}^2} \frac{(\mathbf{k} \cdot \mathbf{k}')^2}{k^2 k'^2}$$

$$\times \left| Z_{\text{eff}}^{(e,i)} \right|^2 \delta(\omega_{\mathbf{k}} - \omega_{\mathbf{k}'} - ([\mathbf{k} - \mathbf{k}'] \cdot \mathbf{v})), \quad (6.51)$$

with

$$Z_{\text{eff}}^{(i)} = Z \frac{\varepsilon_{k-k',\omega_k-\omega_{k'}}^{(e)} - 1}{\varepsilon_{k-k',\omega_k-\omega_{k'}}}, \qquad Z_{\text{eff}}^{(e)} = Z \frac{\varepsilon_{k-k',\omega_k-\omega_{k'}}^{(i)}}{\varepsilon_{k-k',\omega_k-\omega_{k'}}}. \tag{6.52}$$

These expressions include a minus sign – because the electron charge is negative – to show that in both cases the charges are negative in the limit when the Debye screening approximation can be used for the permittivities. In that case we have

$$Z_{\text{eff}}^{(i)} = Z \frac{T_i}{T_i + ZT_e}, \qquad Z_{\text{eff}}^{(e)} = Z \frac{T_e}{T_i + ZT_e}, \tag{6.53}$$

which is the same as the approximate expressions (6.7) and (6.3) which we used earlier in our simple physical picture of the screening shells. We have thus proved, using fluctuation theory, that, indeed, the scattering in a plasma is due to particles surrounded by their polarisation shells. The dipole moment which is responsible for the scattering of Langmuir waves both by electrons and by ions is produced only by the electrons and the electron component of the shell; it is proportional to eZ_{eff}. The charge to mass ratio of the particles which determines the time dependence of the dipole moment is equal to e/m_e and the scattering intensity is thus proportional to $Z_{\text{eff}}^2 e^2 (e/m_e)^2$.

The well known cross-section for the normal Thomson scattering of electromagnetic waves in a vacuum can be found in textbooks. One can obviously rewrite it in the form of a probability for the scattering of an electromagnetic wave by an electron producing another electromagnetic wave – in the symbolic notation we use this corresponds to the reaction $t + e \leftrightarrows t' + e'$. We leave the derivation of the probability of this process as an exercise. The result is that it differs from expression (6.51) by a polarisation factor and by an effective charge $Z_{\text{eff}} = 1$. Due to the two possible polarisations of the transverse electromagnetic waves the factor $(k \cdot k')^2 / k^2 k'^2$ in Eq.(6.51) is changed to $\frac{1}{2}(1 + (k \cdot k')^2 / k^2 k'^2)$ which is nothing but the average of two transverse polarisations. For an electron at rest we then find

$$w_p^{t,t'(e)}(k, k') = \frac{(2\pi)^3 e^4}{2m_e^2 \omega_k^2} \left[1 + \frac{(k \cdot k')^2}{k^2 k'^2} \right] \delta(\omega_k^t - \omega_{k'}^t). \tag{6.55}$$

We can, indeed, find from Eq.(6.55) the well known cross-section for Thomson scattering. Using Eqs.(6.11) and (6.12) we find for the intensity of the $t + e \leftrightarrows t' + e'$ scattering process the equation

$$\begin{aligned} Q_k^{\text{sc}} &= \int \omega_{k'}^t N_{k'}^t w_p^{t,t'(e)}(k, k') \frac{d^3k \, d^3k'}{(2\pi)^6} \\ &= \int \frac{e^4}{2m_e^2 \omega_k^2} \left[1 + \frac{(k \cdot k')^2}{k^2 k'^2} \right] \delta(\omega_k^t - \omega_{k'}^t) W_{k'}^t \, d^3k \, d^3k', \end{aligned} \tag{6.56}$$

with

$$W_{\mathbf{k}'}^{t} = \frac{\omega_{\mathbf{k}'}^{t} N_{\mathbf{k}'}^{t}}{(2\pi)^3}.$$

Note that this expression does not contain the light velocity and this is natural since the electron is assumed to be non-relativistic and the probability is determined by the electron alone. The velocity of light enters, indeed, when we take the dispersion properties of the electromagnetic waves into account. In vacuo we have $\omega_{\mathbf{k}}^{t} = kc$ and it is easy to evaluate the integral in Eq.(6.56); the angular averaging of the polarisation factor gives a factor $\frac{4}{3}$ and we can put $4\pi k^2\, dk = 4\pi(\omega_{\mathbf{k}}^{t})^2\, d\omega_{\mathbf{k}}^{t}/c^3$. We then get

$$Q_{\mathrm{p}}^{sc,t,t'\,(e)} = \frac{8\pi e^4}{3c^3 m_{\mathrm{e}}^2} \int W_{\mathbf{k}'}^{t}\, d^3\mathbf{k}', \qquad k = k'. \tag{6.57}$$

This is the well known result for Thomson scattering. We have now seen that in a plasma the expression for the scattering of longitudinal waves contains a different polarisation factor, namely, the square of the scalar product, $(\mathbf{k} \cdot \mathbf{k}')^2/k^2 k'^2$, of the polarisation vectors of the two longitudinal waves involved. However, the most important feature is the radical change in the cross-sections described by Z_{eff}.

In order to compare the result for scattering in a plasma with expression (6.57) we write down the power of the spontaneous scattering for Langmuir waves. From Eq.(6.51) we get

$$Q_{\mathrm{p}}^{sc,\ell,\ell'} = \frac{4\pi e^4 |Z_{\mathrm{eff}}|^2}{9 m_{\mathrm{e}}^2 v_{Te}^2 \omega_{\mathrm{pe}}} \int k' W_{\mathbf{k}'}^{\ell}\, d^3\mathbf{k}', \tag{6.58}$$

For scattering by electrons we substitute the electron effective charge, $Z_{\mathrm{eff}}^{(e)}$, and for scattering by ions we substitute the ion effective charge, $Z_{\mathrm{eff}}^{(i)}$, which are given by Eqs.(6.53). Expression (6.58) does not contain the velocity of light and it contains the electron mass both in the case of scattering by electrons and in the case of scattering by ions, whereas in the vacuum case the ion mass will enter in the latter case. It is important that we have found not only an expression for the effective charge which is valid in the Debye screening approximation , but also an expression for the effective charge which depends, in general, on the wavevectors $\mathbf{k}$ and $\mathbf{k}'$, that is, on the velocity of the charge: this means that it describes dynamic screening in the scattering process. Let us illustrate this fact and find expressions which cannot be obtained from the Debye screening approximation. The change in frequency in the scattering process is determined by the conservation laws in the elementary scattering process, which means that it is determined by the particle velocity. If we are dealing with scattering over an angle of the order of unity we find that $|\mathbf{k} - \mathbf{k}'|$ is of the order of k and the frequency changes by an amount of the order of $k v_T$. From this it follows that the ion part of the dielectric permittivity is for $v \gg v_s$ much smaller than the electron part and in that case the effective charge of the electron is, indeed, inversely proportional to its velocity:

$$Z_{\text{eff}}^{(i)} \approx Z, \qquad Z_{\text{eff}}^{(e)} \approx \frac{v_s^2}{v^2} \quad \text{and} \quad Z_{\text{eff}}^{(e)} \approx \frac{m_e}{m_i} \quad \text{for } v \approx v_{Te}. \qquad (6.59)$$

It is clear from this estimate that due to the presence of the plasma the rôle of the electrons and the ions is, so to say, interchanged, as the cross-section for scattering by electrons will be inversely proportional to the square of the ion mass. Moreover, because of the factor Z, which might be large, the scattering by ions in a plasma is even stronger than the scattering by electrons in vacuo.

The effect of the other particles in the plasma thus changes the order of magnitude of the scattering cross-sections: the cross-section for scattering by electrons may decrease by a factor $(m_e/m_i)^2$ whereas that for scattering by ions may increase by a factor $(m_i/m_e)^2$. These changes are, indeed, very important.

6.5 Scattering Probabilities for Ion–Sound Waves

We shall now consider scattering of an ion-sound wave leading to either another ion-sound wave or to a Langmuir wave. We showed that the first process occurs for scattering by ions and because there is a cancellation of the amplitude of the scattering by the central ion by the amplitude of the scattering by the screening shell the scattering process will be determined by the Doppler corrections to the frequency of the wave field acting on the ion. We found this from qualitative considerations and will now prove it using fluctuation theory. We consider the stimulated scattering process, calculating the non-linear Landau damping using fluctuation theory. In Eq.(5.45) for the effective non-linear response we consider only the imaginary part of the denominators, which lead to a conservation law for the scattering:

$$\varrho_{1,2,3}^{\text{eff (i)}} = -\frac{Z^4 e^4 (\mathbf{k} \cdot \mathbf{k}_1)(\mathbf{k}_2 \cdot \mathbf{k}_3)}{2 k_1 k_2 k_3 m_i^2} \int \left\{ \frac{1}{(\omega - (\mathbf{k} \cdot \mathbf{v}))^2 (\omega_1 - (\mathbf{k}_1 \cdot \mathbf{v}))^2} \right.$$
$$\left. - \frac{1}{\omega^2 (\omega_1 - (\mathbf{k}_1 \cdot \mathbf{v}))^2} + \frac{1}{\omega^2 \omega_1^2} - \frac{1}{\omega_1^2 (\omega - (\mathbf{k} \cdot \mathbf{v}))^2} \right\}$$
$$\times \delta(\omega - \omega_1 - ([\mathbf{k} - \mathbf{k}_1] \cdot \mathbf{v})) \left([\mathbf{k} - \mathbf{k}_1] \cdot \frac{\partial \Phi_p^i}{\partial \mathbf{p}} \right) \frac{d^3 \mathbf{p}}{(2\pi)^3}. \qquad (6.60)$$

After expanding this expression in terms of the small parameter $(\mathbf{k} \cdot \mathbf{v})/\omega$ we get

$$\varrho_{1,2,3}^{\text{eff (i)}} = -\frac{2\pi Z^4 e^4 (\mathbf{k} \cdot \mathbf{k}_1)(\mathbf{k}_2 \cdot \mathbf{k}_3)}{k_1 k_2 k_3 m_i^2 \omega^3 \omega_1^3} \int \left([\mathbf{k} - \mathbf{k}_1] \cdot \frac{\partial \Phi_p^i}{\partial \mathbf{p}} \right) \frac{d^3 \mathbf{p}}{(2\pi)^3}$$
$$\times (\mathbf{k} \cdot \mathbf{v})(\mathbf{k}_1 \cdot \mathbf{v}) \delta(\omega - \omega_1 - ([\mathbf{k} - \mathbf{k}_1] \cdot \mathbf{v})). \qquad (6.61)$$

We have shown earlier that when $T_i/ZT_e \ll 1$ only the first term in Eq.(5.34) is important and we have thus

$$\varepsilon E^+ = -\mathrm{i}\frac{8\pi}{k}\, \varrho_{1,2,3}^{\mathrm{eff}\ (\mathrm{i})} \left(E_1^+ E_2^+ E_3^- - E_1^+ \langle E_2^+ E_3^-\rangle\right) d_{1,2,3}. \tag{6.62}$$

We can now find an equation for random ion-sound waves by the same method we used to derive the equation for Langmuir waves. We multiply Eq.(6.62) by $E_{\mathbf{k}'}^-$ and average over a statistical ensemble. In this way we find the following equation for $\mathrm{Im}\{\varepsilon_{\mathbf{k}}^{\mathrm{NL}}\}$:

$$\mathrm{Im}\left\{\varepsilon_{\mathbf{k}}^{\mathrm{NL}}\right\} = -\frac{16\pi^2 Z^4 e^4}{m_i^2 \omega^3} \int |E^+|_{\mathbf{k}'}^2 \frac{(\mathbf{k}\cdot\mathbf{k}')^2}{k^2 k'^2 \omega'^3}\, d^4 k' \frac{d^3\mathbf{p}}{(2\pi)^3}\left([\mathbf{k}-\mathbf{k}']\cdot\frac{\partial \Phi_{\mathbf{p}}^i}{\partial \mathbf{p}}\right)$$
$$\times\ (\mathbf{k}\cdot\mathbf{v})^2 \delta\big(\omega - \omega' - ([\mathbf{k}-\mathbf{k}']\cdot\mathbf{v})\big). \tag{6.63}$$

We must express the correlation function in terms of the number of waves, the distribution function, use the relation

$$\frac{\partial \varepsilon_{\mathbf{k}}}{\partial \omega} \approx \frac{2\omega_{\mathrm{pi}}^2}{\omega^3},$$

and find the non-linear growth rate $\gamma_{\mathbf{k}}^{\mathrm{NL}}$ and compare it with the general expression for the non-linear Landau damping in order to find the probability for the $\mathrm{s}+\mathrm{i} \leftrightarrows \mathrm{s}'+\mathrm{i}'$ scattering process. We have

$$w_{\mathbf{p}}^{\mathrm{s},\mathrm{s}'\,(\mathrm{i})} = \frac{4(2\pi)^3 Z^4 e^4 (\mathbf{k}\cdot\mathbf{v})^2}{m_i^2 \omega_{\mathrm{pi}}^4} \frac{(\mathbf{k}\cdot\mathbf{k}')^2}{k^2 k'^2}\, \delta\big(\omega_{\mathbf{k}}^{\mathrm{s}} - \omega_{\mathbf{k}'}^{\mathrm{s}} - ([\mathbf{k}-\mathbf{k}']\cdot\mathbf{v})\big). \tag{6.64}$$

This expression which shows that the scattering probability increases with increasing ion velocity is valid when the change in frequency in the scattering is small in which case we have $(\mathbf{k}\cdot\mathbf{v}) \approx (\mathbf{k}'\cdot\mathbf{v})$. This expression for the scattering probability is no longer valid for velocities much higher than the thermal ion velocity.

The other processes in which ion-sound waves are involved are the scattering of an ion-sound wave with the production of a Langmuir wave and the scattering involving the simultaneous absorption or emission of two waves, one of which is an ion-sound wave and the other a Langmuir wave In all these processes the scattering is only by electrons and they are described by the reaction formulæ

$$\mathrm{e}+\ell \leftrightarrows \mathrm{e}'\pm\mathrm{s}.$$

Usually both processes are possible for the same initial conditions since we can if we neglect the ion-sound frequency as compared to the Langmuir frequency and the Langmuir wavenumber as compared to the ion-sound wavenumber – the necessity for this second neglect can be seen from the conservation law – write the conservation laws for these processes in the following approximate form:

$$\omega_{\mathbf{k}}^{\ell} \mp \omega_{\mathbf{k}'}^{\mathbf{s}} = ([\mathbf{k} \mp \mathbf{k}'] \cdot \mathbf{v}) \longrightarrow \omega_{\mathbf{k}}^{\ell} \approx \mp(\mathbf{k}' \cdot \mathbf{v}), \qquad k' > \frac{\omega_{pe}}{v_{Te}}. \quad (6.65)$$

The last inequality follows, indeed, from the first Eq.(6.65) and confirms that, as we assumed, we have $k \ll k'$. By changing the direction of propagation of the ion-sound waves we change one kind of interaction – for instance, scattering – into the other one – emission or absorption of two waves, for instance – which shows that both processes should be possible at the same time.

It follows from Eq.(6.65) that in those processes only ion oscillations with a frequency close to ω_{pi} can take part. For these waves decay processes are forbidden – like they were forbidden in the case when only ion-sound waves or only Langmuir waves are involved. In fact, conservation of momentum in decay processes involving two Langmuir waves and a single ion-sound wave cannot be satisfied as the ion-sound wave will have a momentum which is much larger than any momentum of a possible Langmuir wave. The absence of decay means that in the quadratic interactions between ion-sound and Langmuir waves the third wave must be a virtual wave. We shall again denote the positive- and negative-frequency components of the Langmuir field by E^+ and E^- and we shall denote the ion-sound field by E^s, not splitting it into positive- and negative-frequency parts. We shall split the virtual field into positive- and negative-frequency parts, E^{+v} and E^{-v}, since its frequency will be close to either $+\omega_{pe}$ or $-\omega_{pe}$, as we shall see in a moment; the wavenumbers of the virtual field will lie outside the range of the Langmuir branch. We have therefore

$$ik\varepsilon E^+ = 8\pi \int \varrho_{1,1'}^{(2)} \left[E_1^s E_{1'}^{+v} - \langle E_1^s E_{1'}^{+v} \rangle \right] d_{1,1'}. \quad (6.66)$$

Using the same quadratic non-linearity we can express the virtual field $E_{1'}^{+v}$ in terms of the Langmuir field:

$$ik_{1'}\varepsilon_{1'} E_{1'}^{+v} = 8\pi \int \varrho_{2,3}^{(2)} \left[E_2^+ E_3^s - \langle E_2^+ E_3^s \rangle \right] d_{2,3}. \quad (6.67)$$

Taking into account the cubic non-linearity we get a general equation for the interaction between Langmuir waves and ion-sound waves, similar to the non-linear equations we obtained earlier:

$$ik\varepsilon E^+ = 8\pi \int \varrho_{1,2,3}^{\mathrm{eff}} E_1^s E_2^+ E_3^s \, d_{1,2,3};$$

$$\varrho_{1,2,3}^{\mathrm{eff}} = \varrho_{1,2,3}^{(3)} + \frac{8\pi \varrho_{1,2,3}^{(2)} \varrho_{2,3}^{(2)}}{i|\mathbf{k}_2 + \mathbf{k}_3|\varepsilon_{2+3}}. \quad (6.68)$$

We have here taken into account that the average value of the product of an ion-sound field and a Langmuir field is equal to zero to first approximation. By now multiplying Eq.(6.68) by $E_{\mathbf{k}'}^-$ and averaging over a statistical ensemble we obtain the non-linear permittivity for Langmuir waves which is due to ion-sound waves:

$$\varepsilon_k^{NL} = \frac{8\pi}{ik} \int \varrho_{k',k,k'}^{eff} |E^s|_{k'}^2 \, d^4k'. \tag{6.69}$$

In order to obtain the probability for the processes we are interested in we need to know the imaginary part of the non-linear permittivity. For the imaginary part produced by the cubic non-linearity we have

$$\begin{aligned}
\mathrm{Im}\left\{ \frac{8\pi}{ik} \varrho_{k',k,k'}^{(3)} \right\} &= -\frac{4\pi^2 e^4}{k^2 k'^2} \int \frac{1}{\omega - (\mathbf{k}\cdot\mathbf{v})} \left(\mathbf{k}' \cdot \frac{\partial}{\partial \mathbf{p}} \right) \\
&\quad \times \delta\big(\omega - \omega' - ([\mathbf{k}-\mathbf{k}']\cdot\mathbf{v})\big) \frac{d^3p}{(2\pi)^3} \\
&\quad \times \left[\left(\mathbf{k}\cdot\frac{\partial}{\partial\mathbf{p}} \right) \frac{1}{\omega' - (\mathbf{k}'\cdot\mathbf{v})} \left(\mathbf{k}'\cdot\frac{\partial}{\partial\mathbf{p}} \right) \right. \\
&\quad \left. - \left(\mathbf{k}'\cdot\frac{\partial}{\partial\mathbf{p}} \right) \frac{1}{\omega - (\mathbf{k}\cdot\mathbf{v})} \left(\mathbf{k}\cdot\frac{\partial}{\partial\mathbf{p}} \right) \right] \Phi_p^e \\
&\approx -\frac{4\pi^2 e^4 (\mathbf{k}\cdot\mathbf{k}')^2}{\omega_{pe}^4 k'^2 k^2 m_e^2} \int \delta\big(\omega - \omega' - ([\mathbf{k}-\mathbf{k}']\cdot\mathbf{v})\big) \\
&\quad \times \left([\mathbf{k}-\mathbf{k}'] \cdot \frac{\partial \Phi_p^e}{\partial\mathbf{p}} \right) \frac{d^3p}{(2\pi)^3}. \tag{6.70}
\end{aligned}$$

The contribution to ϱ^{eff} from the quadratic non-linearity is smaller by a factor $k/k' \ll 1$. To find the probabilities from Eq.(6.70) we use the fact that for ion-sound waves we have $\partial \varepsilon / \partial \omega \approx 2/\omega_{pi}$ and that

$$|E^s|_{k'}^2 \approx \frac{\omega_{pi}}{4\pi^2} \left[N_{k'}^s \delta(\omega' - \omega_{k'}^s) + N_{-k'}^s \delta(\omega' + \omega_{k'}^s) \right]. \tag{6.71}$$

The first term in Eq.(6.71) leads to a scattering probability – which we denote by $w_p^{\ell,s\,(e)}$ – and the second term of Eq.(6.71) leads to the probability for the emission of two waves – which we denote by $w_p^{\ell,-s\,(e)}$:

$$w_p^{\ell,\pm s\,(e)}(\mathbf{k},\mathbf{k}') = \frac{(2\pi)^3 e^4 \omega_{pi}}{m_e^2 \omega_{pe}^3} \frac{(\mathbf{k}\cdot\mathbf{k}')^2}{k^2 k'^2} \delta\big(\omega_k^\ell \mp \omega_{k'}^s - ([\mathbf{k}\mp\mathbf{k}']\cdot\mathbf{v})\big). \tag{6.72}$$

It follows from the earlier general discussion of scattering that Eq.(6.72) will always lead to non-linear damping of waves if the frequency of the scattered wave is lower than that of the incident wave. Since the ion-sound waves always satisfy this condition they will produce non-linear damping of the Langmuir waves. In calculating the damping rate we shall take both processes into account:

$$\gamma_{\mathbf{k}}^{\ell,\mathrm{NL}} = \frac{1}{2} \int N_{\mathbf{k}'}^{s} \left[w_{\mathbf{p}}^{\ell,+s\,(e)}(\mathbf{k},\mathbf{k}') \left([\mathbf{k}-\mathbf{k}'] \cdot \frac{\partial \Phi_{\mathbf{p}}^{e}}{\partial \mathbf{p}} \right) \right.$$

$$\left. + w_{\mathbf{p}}^{\ell,-s\,(e)}(\mathbf{k},\mathbf{k}') \left([\mathbf{k}+\mathbf{k}'] \cdot \frac{\partial \Phi_{\mathbf{p}}^{e}}{\partial \mathbf{p}} \right) \right] \frac{d^3p}{(2\pi)^3}$$

$$\approx -\sqrt{\frac{\pi}{8}}\,\omega_{\mathrm{pe}} \int \frac{W_{\mathbf{k}'}^{s}\omega_{\mathrm{pe}}}{k'v_{Te}n_eT_e} \frac{(\mathbf{k}\cdot\mathbf{k}')^2}{k^2k'^2} \exp\left[-\frac{\omega_{\mathrm{pe}}^2}{2k'^2v_{Te}^2} \right] d^3k'. \quad (6.73)$$

Here $W_{\mathbf{k}'}^{s}\,d^3\mathbf{k}'$ is the ion-sound wave energy density in the range $d^3\mathbf{k}'$:

$$W_{\mathbf{k}'}^{s} = \omega_{\mathrm{pi}} \frac{N_{\mathbf{k}'}^{s}}{(2\pi)^3}. \quad (6.74)$$

The damping rate given by Eq.(6.73) is rather large since, as we shall show in the next chapter, the ion-sound wave energy density excited by an external electric field is smaller than n_eT_e only by a factor T_i/ZT_e.

We now turn to a discussion of the problem of the damping – or amplification – of ion-sound waves produced by strong Langmuir waves. So far we discussed the balance equation considering only scattering processes. The balance equations for the case when we consider the emission or absorption of two waves can be obtained in a similar way and the term corresponding to those processes enters into the expression for the damping – or growth – rate with the opposite sign:

$$\gamma_{\mathbf{k}'}^{s,\mathrm{NL}} = \frac{1}{2} \int N_{\mathbf{k}}^{\ell} \left[w_{\mathbf{p}}^{\ell,+s\,(e)}(\mathbf{k},\mathbf{k}') \left([\mathbf{k}-\mathbf{k}'] \cdot \frac{\partial \Phi_{\mathbf{p}}^{e}}{\partial \mathbf{p}} \right) \right.$$

$$\left. - w_{\mathbf{p}}^{\ell,-s\,(e)}(\mathbf{k},\mathbf{k}') \left([\mathbf{k}+\mathbf{k}'] \cdot \frac{\partial \Phi_{\mathbf{p}}^{e}}{\partial \mathbf{p}} \right) \right] \frac{d^3p}{(2\pi)^3}$$

$$\approx -\frac{1}{2}\sqrt{\frac{\pi}{8}} \int \frac{W_{\mathbf{k}}^{\ell}\omega_{\mathrm{pi}}}{n_eT_e} \frac{(\mathbf{k}\cdot\mathbf{k}')^2}{k^2k'^2}$$

$$\times \left\{ \frac{\omega_{\mathbf{k}}^{\ell} - \omega_{\mathbf{k}'}^{s}}{|\mathbf{k}-\mathbf{k}'|v_{Te}} \exp\left[-\frac{(\omega_{\mathbf{k}}^{\ell} - \omega_{\mathbf{k}'}^{s})^2}{2|\mathbf{k}-\mathbf{k}'|^2v_{Te}^2} \right] \right.$$

$$\left. - \frac{\omega_{\mathbf{k}}^{\ell} + \omega_{\mathbf{k}'}^{s}}{|\mathbf{k}+\mathbf{k}'|v_{Te}} \exp\left[-\frac{(\omega_{\mathbf{k}}^{\ell} + \omega_{\mathbf{k}'}^{s})^2}{2|\mathbf{k}+\mathbf{k}'|^2v_{Te}^2} \right] \right\} d^3k. \quad (6.75)$$

We have assumed here that the electron distribution was a thermal one and $W_{\mathbf{k}}^{\ell}$ is the Langmuir wave energy density:

$$W_{\mathbf{k}}^{\ell} \approx \omega_{\mathrm{pe}} \frac{N_{\mathbf{k}}^{\ell}}{(2\pi)^3}. \quad (6.76)$$

It follows from eq.(6.75) that for

$$\frac{k}{k'} \ll \sqrt{\frac{m_e}{m_i}} \quad (6.77)$$

expression (6.75) is positive; this does not mean, however, that the ion-sound waves are amplified since expression (6.75) is always smaller than the non-linear Landau damping:

$$
\gamma_{\mathbf{k}'}^{\mathrm{s,NL}} = \sqrt{\frac{\pi m_{\mathrm{e}}}{8 m_{\mathrm{i}}}}\, \omega_{\mathbf{k}'}^{\mathrm{s}}\, \frac{\omega_{\mathrm{pe}}}{k' v_{Te}} \left[1 - \frac{\omega_{\mathrm{pe}}^2}{k'^2 v_{Te}^2} \right]
$$

$$
\times \exp\left[-\frac{\omega_{\mathrm{pe}}^2}{2 k'^2 v_{Te}^2} \right] \int \frac{W_{\mathbf{k}}^{\ell} (\mathbf{k}\cdot\mathbf{k}')^2}{n_{\mathrm{e}} T_{\mathrm{e}} k^2 k'^2}\, d^3\mathbf{k}. \tag{6.78}
$$

If the inequality $k/k' \gg \sqrt{m_{\mathrm{e}}/m_{\mathrm{i}}}$, which is the opposite of inequality (6.77), holds we have

$$
\gamma_{\mathbf{k}'}^{\mathrm{s}} = \sqrt{\frac{\pi}{8}}\, \omega_{\mathrm{pi}}\, \frac{\omega_{\mathrm{pe}}}{k' v_{Te}} \left[1 - \frac{\omega_{\mathrm{pe}}^2}{k'^2 v_{Te}^2} \right]
$$

$$
\times \exp\left[-\frac{\omega_{\mathrm{pe}}^2}{2 k'^2 v_{Te}^2} \right] \int \frac{W_{\mathbf{k}}^{\ell}}{n_{\mathrm{e}} T_{\mathrm{e}}}\, \frac{k}{k'}\, \frac{(\mathbf{k}\cdot\mathbf{k}')^3}{k^3 k'^3}\, d^3\mathbf{k}; \tag{6.79}
$$

the sign of $\gamma_{\mathbf{k}'}^{\mathrm{s}}$ depends here on the degree of anisotropy of the Langmuir wave distribution and in the case of $k \sim k'$ and $W^{\ell}/n_{\mathrm{e}} T_{\mathrm{e}} > \sqrt{m_{\mathrm{e}}/m_{\mathrm{i}}}$ the damping rate given by Eq.(6.79) can be larger than the Landau damping.

Finally we can write down a general equation for ion-sound waves in the presence of Langmuir waves:

$$
i\varepsilon E^{\mathrm{s}} = 8\pi \int \varrho_{1,2,3}^{\mathrm{eff}} \left(E_1^+ E_2^{\mathrm{s}} E_3^- + E_1^- E_2^{\mathrm{s}} E_3^+ \right) d_{1,2,3}, \tag{6.80}
$$

and an expression for the non-linear permittivity for random Langmuir waves:

$$
\varepsilon_{\mathbf{k}'}^{\mathrm{NL}} = \int \frac{8\pi \varrho_{k,k',-k}^{\mathrm{eff}}}{i k'} \left[|E^+|_{\mathbf{k}}^2 + |E^+|_{-\mathbf{k}}^2 \right] d^4\mathbf{k}. \tag{6.81}
$$

The total correlator is equal to

$$
|E^{\ell}|^2 = |E^+|_{\mathbf{k}}^2 + |E^-|_{\mathbf{k}}^2; \qquad |E^-|_{\mathbf{k}}^2 = |E^+|_{-\mathbf{k}}^2.
$$

For a given ion-sound wave frequency both processes considered contribute to expression (6.81). The new result described by Eq.(6.81) is a change in the dispersion of ion-sound waves in the presence of Langmuir waves.

6.6 Transition Scattering

We have in previous sections several times illustrated the fundamental rôle played by polarisation shells in scattering processes. One can only talk about polarisation shells, and then approximately, in the case when the non-linear plasma response can be expressed in terms of the linear permittivity. In the more general case there is a new scattering mechanism, *transition scattering*, when the incident wave modulates the dielectric permittivity and the charge moves relative to these modulations. This can happen either because the modulations are propagating and are incident upon a charge at rest, or because the charge is moving through static modulations, or because the charge and the modulations are moving with different velocities. The modulations can be considered to be modulations of the dielectric permittivity and described by an addition, $\widehat{\delta \varepsilon}_k$, to the homogeneous ε_k. In the case when the incident wave is a Langmuir wave, $E_{k'}^\ell$, the additional contribution to the dielectric permittivity from the modulations can be found from the quadratic non-linear response,

$$(\varepsilon_k + \widehat{\delta\varepsilon}_k)E_k = \varepsilon_k E_k - \frac{8\pi}{ik} \int \varrho^{(2)}_{k-k',k'} E_{k'}^\ell E_{k-k'}\, d^4k' = 0. \qquad (6.82)$$

The simplest case of a *dielectric permittivity wave* is a plasma density wave,

$$n = n^{(0)} + n^{(1)} \cos(\omega_0 t - (\mathbf{k}_0 \cdot \mathbf{r})); \qquad (6.83)$$

this is appropriate only for small values of ω_0 and k_0 when this wave can be treated in the adiabatic approximation and need not be described by an operator. In the general case the modulations must be described by an integral operator as in Eq.(6.82). For a Langmuir wave a density variation always accompanies the wave, but in general it is not necessary that the plasma density in the wave is changed in order to have a permittivity wave. The reason is that in a scattering process the difference between the frequencies of the incident and the scattered waves is usually small due to the conservation laws to be obeyed in the scattering process. In that case the adiabatic approximation is not valid and transition scattering of, for instance, transverse waves, which do not produce density modulations, will be appreciable. Using the quadratic plasma response we can evaluate a general expression for transition scattering. It will describe only part of the total scattering since it neglects scattering by the central charge; however, if the emission by the central charge is negligible, it will be the general expression. We will use here the concept of *test particles* bearing in mind that in a non-equilibrium distribution any plasma partcle can be considered to be a test particle. We assume that the charge of this test particle is e_α and that its velocity is constant and equal to $\mathbf{v}$. The charge density of this charge will then be

$$\varrho_k^\alpha = \frac{e_\alpha}{(2\pi)^3} \delta(\omega - (\mathbf{k} \cdot \mathbf{v})). \qquad (6.84)$$

The field created by this charge will be determined by the equation

$$E_{\mathbf{k}} - \frac{8\pi}{ik\varepsilon_{\mathbf{k}}} \int \varrho^{(2)}_{\mathbf{k}',\mathbf{k}-\mathbf{k}'} E^{\ell}_{\mathbf{k}'} E_{\mathbf{k}-\mathbf{k}'} \, d^4k' = \frac{4\pi e_\alpha}{ik\varepsilon_{\mathbf{k}}(2\pi)^3} \delta(\omega - (\mathbf{k} \cdot \mathbf{v})), \quad (6.85)$$

where we have taken into account the modulation $\widehat{\delta\varepsilon}_{\mathbf{k}}$ of the dielectric permittivity. To first approximation we can neglect in Eq.(6.85) the term with the modulation of the permittivity and we then find the self-field of the particle:

$$E^\alpha_{\mathbf{k}} = \frac{4\pi e_\alpha}{ik\varepsilon_{\mathbf{k}}(2\pi)^3} \delta(\omega - (\mathbf{k} \cdot \mathbf{v})). \tag{6.86}$$

In the next approximation we find

$$\delta E_{\mathbf{k}} = \frac{8\pi}{ik\varepsilon_{\mathbf{k}}} \int \varrho^{(2)}_{\mathbf{k}',\mathbf{k}-\mathbf{k}'} E^{\ell}_{\mathbf{k}'} E^{\alpha}_{\mathbf{k}-\mathbf{k}'} \, d^4k'. \tag{6.87}$$

This field contains not only the scattered field but also the field which must be added to expression (6.86) and which is transferred together with the charge α. The charge density exciting the field (6.87) is given by the equation

$$\delta\varrho_{\mathbf{k}} = 2 \int \varrho^{(2)}_{\mathbf{k}',\mathbf{k}-\mathbf{k}'} E^{\ell}_{\mathbf{k}'} E^{\alpha}_{\mathbf{k}-\mathbf{k}'} \, d^4k'. \tag{6.88}$$

As the self-field will not produce any work when acting on the charge, the total power corresponding to the work done by the field (6.87) on the charge density (6.88) will be due to the scattered field. The scattered power $Q^{\ell'}_{\mathbf{p}}$ will be equal to this power with a minus sign:

$$Q^{\ell'}_{\mathbf{p}} = - \int \langle (\delta\mathbf{j} \cdot \delta\mathbf{E}) \rangle \, d^3r$$

$$= (2\pi)^3 \int \langle (\mathbf{k} \cdot \delta\mathbf{j}_{\mathbf{k},\omega_1}) \delta E_{-\mathbf{k},\omega_1} \rangle \, e^{-i\omega t - i\omega_1 t} \frac{d^4k}{k} \, d\omega_1. \tag{6.89}$$

Using the continuity equation in the form $(\mathbf{k} \cdot \delta\mathbf{j}_{\mathbf{k},\omega}) = \omega \, \delta\varrho_{\mathbf{k},\omega}$ and introducing a four-vector $k_1 \equiv \{-\mathbf{k}, \omega\}$ we have

$$Q^{\ell'}_{\mathbf{p}} = 2(2\pi)^3 \int \frac{\omega}{k} \varrho^{(2)}_{\mathbf{k}',\mathbf{k}-\mathbf{k}'} \frac{8\pi}{ik\varepsilon_{k_1}} \varrho^{(2)}_{\mathbf{k}'',k_1-\mathbf{k}''} \, d^4k \, d^4k' \, d^4k'' \, d\omega_1$$

$$\times \langle E^{\ell}_{\mathbf{k}'} E^{\ell}_{\mathbf{k}''} \rangle E^{\alpha}_{\mathbf{k}-\mathbf{k}'} E^{\alpha}_{\mathbf{k}-\mathbf{k}''}. \tag{6.90}$$

The presence of the correlation function of the Langmuir waves means that we have $\mathbf{k}'' = -\mathbf{k}'$ and from the occurrence of the delta-function in Eq.(6.86) for the self-field it follows that $\omega_1 = -\omega$; hence the exponential factor becomes unity. Since we have

$$\varepsilon_{\mathbf{k}} = \varepsilon^*_{-\mathbf{k}} \quad \text{and} \quad \delta\varrho^{(2)}_{\mathbf{k}'',\mathbf{k}-\mathbf{k}''} = \delta\varrho^{(2)}_{-\mathbf{k}',-\mathbf{k}+\mathbf{k}'} = \delta\varrho^{(2)*}_{\mathbf{k}',\mathbf{k}-\mathbf{k}'}, \tag{6.91}$$

only squares of absolute values of various quantities will occur in Eq.(6.90); by making the substitution $\mathbf{k} \to -\mathbf{k}$ we see that only the imaginary part of $1/\varepsilon_{-\mathbf{k}}$ will make a contribution. From the causality principle it follows that

$$\text{Im}\left\{\frac{1}{\varepsilon_{-k}}\right\} = \pi \frac{\omega}{|\omega|}\delta(\text{Re}\{\varepsilon_k\}). \tag{6.92}$$

This formula can be used only in the transparency region where the wave is emitted. We have thus

$$Q_{\text{p}}^{\ell'} = 16\pi^2(2\pi)^3 \int \frac{|\omega|}{k^2}\delta(\varepsilon_k)\left|\varrho_{k',k-k'}^{(2)}\right|^2 \frac{e_\alpha^2|E|_{k'}^2}{(2\pi^2)^2|k-k'|^2}d^4k\,d^4k'$$

$$\times \delta(\omega - \omega' - ([k-k']\cdot v)). \tag{6.93}$$

To find the probability for the transition scattering it is now only left to us to express the correlation function in terms of the distribution function and to use the relation

$$\delta(\varepsilon_k) = \{\delta(\omega - \omega_k) + \delta(\omega + \omega_k)\}\left.\frac{\partial\varepsilon_k}{\partial\omega}\right|_{\omega=\omega_k},$$

and compare the result with the power of the spontaneous scattering. The probability can be expressed in the non-linear plasma response coefficient alone:

$$w_{\text{p}}^{s,\ell'(\omega)}(k,k') = \frac{4e_\alpha^2(2\pi)^3(8\pi)^2\left|\varrho_{k',\omega_{k'};k-k',\omega_k-\omega_{k'}}^{(2)}\right|^2}{k^2|k-k'|^2\left|\varepsilon_{k-k',\omega_k-\omega_{k'}}\right|^2\left(\frac{\partial\varepsilon_k}{\partial\omega}\right)_{\omega=\omega_k}\left(\frac{\partial\varepsilon_{k'}}{\partial\omega'}\right)_{\omega'=\omega_{k'}}}$$

$$\times \delta(\omega_k - \omega_{k'} - ([k-k']\cdot v)). \tag{6.94}$$

The transition scattering process is shown in Fig.6.1.

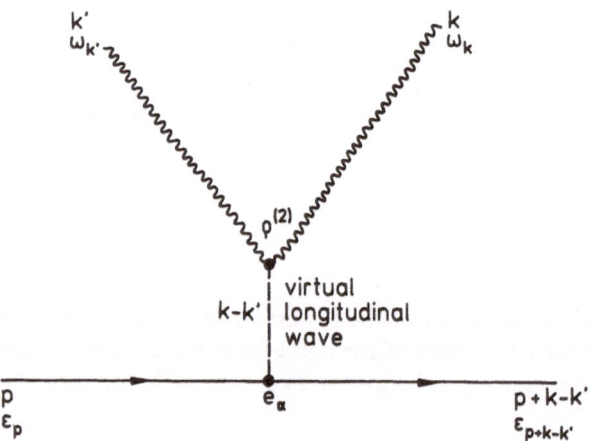

Fig. 6.1. Diagram of a scattering process involving a virtual longitudinal wave

The probability (6.94) can be expressed in terms of the scattering matrix element $M_{k,k'}^{\text{tr}}$ as follows:

$$w_{\mathbf{p}}^{\ell,\ell'\,(\alpha)}(\mathbf{k},\mathbf{k}') = \frac{2\pi \left|M_{\mathbf{k},\mathbf{k}'}^{\mathrm{tr}}\right|^2 \delta\big(\omega_{\mathbf{k}} - \omega_{\mathbf{k}'} - ([\mathbf{k}-\mathbf{k}']\cdot\mathbf{v})\big)}{(\partial\varepsilon_{\mathbf{k}}/\partial\omega)_{\omega=\omega_{\mathbf{k}}} (\partial\varepsilon_{\mathbf{k}'}/\partial\omega')_{\omega'=\omega_{\mathbf{k}'}}}. \tag{6.95}$$

where

$$M_{\mathbf{k},\mathbf{k}'}^{\mathrm{tr}} = \frac{8(2\pi)^2 e_\alpha}{k|\mathbf{k}-\mathbf{k}'|} \frac{\varrho_{\mathbf{k}',\omega_{\mathbf{k}'};\mathbf{k}-\mathbf{k}',\omega_{\mathbf{k}}-\omega_{\mathbf{k}'}}^{(2)}}{\varepsilon_{\mathbf{k}-\mathbf{k}',\omega_{\mathbf{k}}-\omega_{\mathbf{k}'}}}. \tag{6.96}$$

This probability takes into account only the transition scattering; it neglects the scattering due to the oscillation of the particle in the incident field.

6.7 Change in the Particle Distribution Due to Stimulated Wave Scattering

We shall again consider Langmuir waves; we have shown that they are accompanied by virtual fields, the largest of which is that at low frequencies – the zero-frequency virtual field. Let us discuss Langmuir waves which are such that they do not have direct resonance with particles. On the other hand, the virtual fields accompanying the Langmuir field can be in resonance with most of the particles. We shall write down the quasi-linear equation for particle diffusion in the "zero"-frequency virtual fields:

$$\frac{d\Phi_{\mathbf{p}}^\alpha}{dt} = \pi e_\alpha^2 \int \frac{d^4k}{k^2} \left|E^{\mathrm{v}\,(0)}\right|_{\mathbf{k}}^2 \left(\mathbf{k}\cdot\frac{\partial}{\partial\mathbf{p}}\right) \delta(\omega-(\mathbf{k}\cdot\mathbf{v})) \left(\mathbf{k}\cdot\frac{\partial}{\partial\mathbf{p}}\right)\Phi_{\mathbf{p}}^\alpha. \tag{6.97}$$

We must now use Eq.(5.26) for the virtual field in terms of the Langmuir field which we write in the following form:

$$E_{\mathbf{k}}^{\mathrm{v}(0)} = \frac{8\pi}{ik\varepsilon_{\mathbf{k}}} \int \varrho_{\mathbf{k}_1,\mathbf{k}_2}^{(2)} \left[E_{\mathbf{k}_1}^+ E_{\mathbf{k}_2}^- - \langle E_{\mathbf{k}_1}^+ E_{\mathbf{k}_2}^-\rangle\right] \delta(\mathbf{k}-\mathbf{k}_1-\mathbf{k}_2)\, d^4k_1\, d^4k_2. \tag{6.98}$$

To find the correlation function of the virtual field which occurs in Eq.(6.97) we multiply Eq.(6.98) by a similar expression for $E_{\mathbf{k}'}^{\mathrm{v}(0)}$ and use the fact that

$$\left.\begin{aligned}
\left\langle E_{\mathbf{k}}^{\mathrm{v}(0)} E_{\mathbf{k}'}^{\mathrm{v}(0)}\right\rangle &= -\left|E^{\mathrm{v}(0)}\right|_{\mathbf{k}}^2 \delta(\mathbf{k}+\mathbf{k}'), \\
\left\langle E_{\mathbf{k}}^+ E_{\mathbf{k}'}^-\right\rangle &= -\left|E^+\right|_{\mathbf{k}}^2 \delta(\mathbf{k}+\mathbf{k}'), \\
\left\langle E_{\mathbf{k}}^- E_{\mathbf{k}'}^+\right\rangle &= -\left|E^-\right|_{\mathbf{k}}^2 \delta(\mathbf{k}+\mathbf{k}'), \\
\left|E^-\right|_{\mathbf{k}}^2 &= \left|E^+\right|_{-\mathbf{k}}^2, \quad \varepsilon_{-\mathbf{k}} = \varepsilon_{\mathbf{k}}^*, \quad \varrho_{-\mathbf{k}_1,-\mathbf{k}_2}^{(2)} = \varrho_{\mathbf{k}_1,\mathbf{k}_2}^{(2)\,*}.
\end{aligned}\right\} \tag{6.99}$$

The result is

$$\left|E^{v(0)}\right|^2_k = \frac{(8\pi)^2}{k^2|\varepsilon_k|^2} \int \left|\varrho^{(2)}_{k_1,k_2}\right|^2 \left|E^+\right|^2_{k_1} \left|E^-\right|^2_{k_2}$$
$$\times \; \delta(k - k_1 - k_2) \, d^4k_1 \, d^4k_2. \tag{6.100}$$

If we now use the fact that

$$\left|E^+\right|^2 = N^\ell_k \frac{\delta(\omega - \omega^\ell_k)}{2\pi^2} \left[\frac{\partial\varepsilon_k}{\partial\omega}\right]^{-1}_{\omega=\omega_k},$$

$$\left|E^-\right|^2 = N^\ell_{-k} \frac{\delta(\omega + \omega^\ell_k)}{2\pi^2} \left[\frac{\partial\varepsilon_k}{\partial\omega}\right]^{-1}_{\omega=\omega_k},$$

we get finally

$$\frac{d\Phi^\alpha_P}{dt} = \frac{e^2_\alpha}{2\pi^3} \int N^\ell_k N^\ell_{k'} \frac{(8\pi)^2 \, d^3k \, d^3k' \left|\varrho^{(2)}_{k,\omega_k;-k',-\omega_{k'}}\right|^2}{|k - k'|^4 \left|\varepsilon_{k-k',\omega_k-\omega_{k'}}\right|^2}$$

$$\times \left[\frac{\partial\varepsilon_k}{\partial\omega}\Big|_{\omega=\omega_k} \frac{\partial\varepsilon_{k'}}{\partial\omega'}\Big|_{\omega'=\omega_{k'}}\right]^{-1} \left\{\left([k - k'] \cdot \frac{\partial}{\partial p}\right)\right.$$

$$\left. \times \; \delta\big(\omega_k - \omega_{k'} - ([k - k'] \cdot v)\big) \left([k - k'] \cdot \frac{\partial}{\partial p}\right)\right\} \Phi^\alpha_P. \tag{0.101}$$

Comparing this with Eq.(6.23) we find for the scattering probability involving virtual fields:

$$w^{\ell,\ell'\,(\alpha)}_p(k, k') = \frac{4e^2_\alpha(2\pi)^3(8\pi)^2 \left|\varrho^{(2)}_{k,\omega_k;-k',-\omega_{k'}}\right|^2}{|k - k'|^4 \left|\varepsilon_{k-k',\omega_k-\omega_{k'}}\right|^2 \left(\frac{\partial\varepsilon_k}{\partial\omega}\right)_{\omega=\omega_k} \left(\frac{\partial\varepsilon_{k'}}{\partial\omega'}\right)_{\omega'=\omega_{k'}}}$$

$$\times \; \delta\big(\omega_k - \omega_{k'} - ([k - k'] \cdot v)\big). \tag{6.102}$$

If we use the following relation – we leave the proof of it as an exercise –

$$\frac{1}{k} \varrho^{(2)}_{k',\omega_{k'};k-k',\omega_k-\omega_{k'}} = \frac{1}{|k - k'|} \varrho^{(2)}_{k,\omega_k;-k',-\omega_{k'}}, \tag{6.103}$$

we find that the probability given by Eq.(6.102) is exactly the same as the transition scattering probability (6.94). We have thus obtained the following important result: the quasi-linear diffusion caused by the virtual fields describes the change in the particle distributions occurring in the stimulated transition scattering processes. From the general expression for the transition probability we find the approximate expressions which we used earlier. For instance, using the approximate expression (5.49) for the non-linear response,

$$\varrho^{(2)}_{k,\omega_k;-k',-\omega_{k'}} = \frac{e(k \cdot k')|k - k'|^2}{8\pi kk'm_e\omega^2_{pe}} \left[\varepsilon^{(e)}_{k-k',\omega_k-\omega_{k'}} - 1\right], \tag{6.104}$$

we get for the case of scattering by ions, when we have $e_\alpha = Ze$:

$$w_{\mathbf{p}}^{\ell,\ell'\,(i)}(\mathbf{k},\mathbf{k}') = \frac{Z^2 e^4 (2\pi)^3}{m_e^2 \omega_{pe}^4} \frac{(\mathbf{k}\cdot\mathbf{k}')^2}{k^2 k'^2} \left| \frac{\varepsilon_{\mathbf{k}-\mathbf{k}',\omega_{\mathbf{k}}-\omega_{\mathbf{k}'}}^{(e)} - 1}{\varepsilon_{\mathbf{k}-\mathbf{k}',\omega_{\mathbf{k}}-\omega_{\mathbf{k}'}}} \right|^2$$

$$\times\ \delta\big(\omega_{\mathbf{k}} - \omega_{\mathbf{k}'} - ([\mathbf{k}-\mathbf{k}']\cdot\mathbf{v})\big), \tag{6.105}$$

which is the same as Eq. (6.49).

Let us now consider scattering due to the particle oscillations in the field of the incident wave and find the changes in the particle distributions due to it. We show this scattering process in Fig. 6.2. These effects are described by the perturbations of the particle distribution functions which are quadratic and cubic in the strengths of the Langmuir fields. We can use the general equations from §§ 5.6 and 5.7 for this purpose:

$$\frac{d\Phi_{\mathbf{p}}}{dt} = -e\left(\frac{\partial}{\partial\mathbf{p}}\cdot\left[\left\langle \mathbf{E}^{\mathrm{v}(0)}\delta f_{\mathbf{p}}^{(1),\mathrm{v}(0)}\right\rangle + \left\langle \mathbf{E}^{\mathrm{v}(0)}\delta f_{\mathbf{p}}^{(2),+,-}\right\rangle \right.\right.$$

$$\left.\left. + \left\langle \mathbf{E}^{+}\delta f_{\mathbf{p}}^{(2),\mathrm{v}(0),-}\right\rangle + \left\langle \mathbf{E}^{-}\delta f_{\mathbf{p}}^{(2),\mathrm{v}(0),+}\right\rangle + \left\langle \mathbf{E}\,\delta f_{\mathbf{p}}^{(3)}\right\rangle\right]\right). \tag{6.106}$$

We have used here a notation where the first superscript indicates the power of the field strengths and the subsequent superscripts indicate the kind of fields which are appearing, except for the last term on the right-hand side of Eq. (6.106) since it is obvious that only Langmuir fields can occur there. The first term describes the effect of particle diffusion caused by the virtual fields which we have already discussed earlier in the present section. The last term on the right-hand side of Eq. (6.106) does not contain the virtual fields and describes the scattering by a "bare" charge. The other terms describe the interference terms of these two kinds of scattering. We shall consider the last term which describes the scattering due to the particle oscillations in the incident field. To do this we write down the perturbation of the particle distribution function which is cubic in the fields (see Eq. (5.42)):

$$\delta f_{\mathbf{p},\mathbf{k}}^{(3)} = -\frac{e^3}{2i}\int \frac{d_{1,2,3}}{k_1 k_2 k_3}\left(\mathbf{k}_1\cdot\frac{\partial}{\partial\mathbf{p}}\right)\frac{1}{\omega_2 + \omega_3 - ([\mathbf{k}_2+\mathbf{k}_3]\cdot\mathbf{v}) + i0}$$

$$\times\left\{\left(\mathbf{k}_2\cdot\frac{\partial}{\partial\mathbf{p}}\right)\frac{1}{\omega_3 - (\mathbf{k}_3\cdot\mathbf{v}) + i0}\left(\mathbf{k}_2\cdot\frac{\partial}{\partial\mathbf{p}}\right)\right.$$

$$\left. + \left(\mathbf{k}_3\cdot\frac{\partial}{\partial\mathbf{p}}\right)\frac{1}{\omega - (\mathbf{k}\cdot\mathbf{v}) + i0}\left(\mathbf{k}_3\cdot\frac{\partial}{\partial\mathbf{p}}\right)\right\}\Phi_{\mathbf{p}}^{\alpha}$$

$$\times\ [E_{\mathbf{k}_1}E_{\mathbf{k}_2}E_{\mathbf{k}_3} - E_{\mathbf{k}_1}\langle E_{\mathbf{k}_2}E_{\mathbf{k}_3}\rangle - \langle E_{\mathbf{k}_1}E_{\mathbf{k}_2}E_{\mathbf{k}_3}\rangle]. \tag{6.107}$$

After substituting this expression into Eq. (6.106) we find

$$\frac{d\Phi_{\mathbf{p}}^{\alpha}}{dt} = \pi e_{\alpha}^4 \int |E|_{\mathbf{k}}^2\langle E|_{\mathbf{k}'}^2 \frac{d^4k\,d^4k'}{k^2 k'^2}\left\{\left(\mathbf{k}\cdot\frac{\partial}{\partial\mathbf{p}}\right)\frac{1}{\omega - (\mathbf{k}\cdot\mathbf{v})}\left(\mathbf{k}'\cdot\frac{\partial}{\partial\mathbf{p}}\right)\right.$$

$$\times\ \delta\big(\omega - \omega' - ([\mathbf{k}-\mathbf{k}']\cdot\mathbf{v})\big)\left[\left(\mathbf{k}\cdot\frac{\partial}{\partial\mathbf{p}}\right)\frac{1}{\omega' - (\mathbf{k}'\cdot\mathbf{v})}\left(\mathbf{k}'\cdot\frac{\partial}{\partial\mathbf{p}}\right)\right.$$

$$\left.\left. - \left(\mathbf{k}'\cdot\frac{\partial}{\partial\mathbf{p}}\right)\frac{1}{\omega - (\mathbf{k}\cdot\mathbf{v})}\left(\mathbf{k}\cdot\frac{\partial}{\partial\mathbf{p}}\right)\right]\right\}\Phi_{\mathbf{p}}^{\alpha}. \tag{6.108}$$

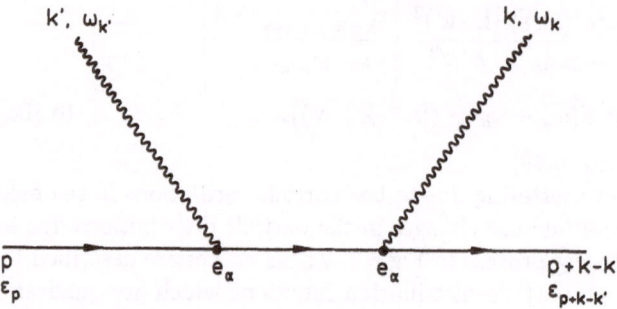

Fig. 6.2. Diagram of the scattering process due to a charge oscillating in the incident wave

Although it looks as if this equation contains third and fourth derivatives with respect to the momenta one can easily change it into a diffusion equation. Indeed, because of the presence of the δ-function the first derivatives in the two terms occurring within the square brackets must act upon the denominators as the results of them acting upon the distribution function cancel each other. Using the presence of the δ-function we can, in fact, reduce the expression within the square brackets to become

$$\frac{1}{2m_\alpha}(\mathbf{k}\cdot\mathbf{k}')\left[\frac{1}{(\omega-(\mathbf{k}\cdot\mathbf{v}))^2}+\frac{1}{(\omega'-(\mathbf{k}'\cdot\mathbf{v}))^2}\right]\left(\left[\mathbf{k}-\mathbf{k}'\right]\cdot\frac{\partial\Phi^\alpha}{\partial\mathbf{p}}\right).$$

After symmetrising with respect to k and k′, using the δ-function, we find that only the terms with the second derivatives are left; this means that we have reduced the equation to a diffusion equation:

$$\frac{d\Phi^\alpha_\mathbf{p}}{dt}=\frac{\pi e^4_\alpha}{2m_\alpha}\int\frac{(\mathbf{k}\cdot\mathbf{k}')^2}{k^2k'^2}|E|^2_\mathbf{k}|E|^2_{\mathbf{k}'}\,d^4k\,d^4k'$$
$$\times\left\{\left(\left[\mathbf{k}-\mathbf{k}'\right]\cdot\frac{\partial}{\partial\mathbf{p}}\right)\frac{1}{(\omega-(\mathbf{k}\cdot\mathbf{v}))^4}\right.$$
$$\left.\times\,\delta\bigl(\omega-\omega'-(\left[\mathbf{k}-\mathbf{k}'\right]\cdot\mathbf{v})\bigr)\left(\left[\mathbf{k}-\mathbf{k}'\right]\cdot\frac{\partial\Phi^\alpha_\mathbf{p}}{\partial\mathbf{p}}\right)\right\}. \tag{6.109}$$

From this equation we find the scattering probability which reduces to the same form as Eq.(6.95):

$$w^{\ell,\ell'\,(\alpha)}_\mathbf{p}(\mathbf{k},\mathbf{k}')=\frac{2\pi\left|M^{\text{osc}}_{\mathbf{k},\mathbf{k}'}\right|^2\delta\bigl(\omega_\mathbf{k}-\omega_{\mathbf{k}'}-(\left[\mathbf{k}-\mathbf{k}'\right]\cdot\mathbf{v})\bigr)}{(\partial\varepsilon_\mathbf{k}/\partial\omega)_{\omega=\omega_\mathbf{k}}(\partial\varepsilon_{\mathbf{k}'}/\partial\omega')_{\omega'=\omega_{\mathbf{k}'}}}, \tag{6.110}$$

where

$$M^{\text{osc}}_{\mathbf{k},\mathbf{k}'}=\frac{4\pi e^2_\alpha}{m_\alpha}\frac{(\mathbf{k}\cdot\mathbf{k}')}{kk'}\frac{1}{(\omega-(\mathbf{k}\cdot\mathbf{v}))^2}. \tag{6.111}$$

However, this probability also does not describe the complete scattering process. We shall leave as an exercise the evaluation of all the interference terms and just mention that the total probability is described by the sum $M_{\mathbf{k},\mathbf{k}'}$ of the two matrix elements which we have found:

$$M_{\mathbf{k},\mathbf{k}'} = M_{\mathbf{k},\mathbf{k}'}^{\mathrm{tr}} + M_{\mathbf{k},\mathbf{k}'}^{\mathrm{osc}}. \tag{6.112}$$

We have thus found a general expression for the scattering probabilities in a plasma. The result (6.111) shows that the Doppler corrections to the frequency, indeed, appear only for the scattering due to the particle oscillations in the incident wave. We have therefore found an exact proof of the earlier statement when we considered scattering of ion-sound waves by ions.

Appendix: Scattering Processes Involving Transverse Waves

When we are interested in scattering by non-relativistic particles the change in frequency as the result of the scattering process will be small. This means that a scattering processes involving the change of a Langmuir wave into an electromagnetic wave are possible if the frequency of the electromagnetic wave is close to the plasma frequency. We can neglect the Lorentz force in the kinetic equation for non-relativistic particles. The difference between the process considered now and the one involving two Langmuir waves lies only in the polarisation of the waves. For transverse waves we have $\mathbf{E}_{\mathbf{k}}^{t} = \mathbf{e}_{\mathbf{k}}^{t} E_{\mathbf{k}}^{t}$ where $\mathbf{e}_{\mathbf{k}}^{t}$ is one of the two polarisation unit vectors. The sum of the two polarisations will lead to the vector product of the wavevector of the transverse waves and that of the Langmuir waves. We have thus:

$$w_{\mathrm{p}}^{\ell,\mathrm{t}\,(i)}(\mathbf{k},\mathbf{k}') = \frac{Z^2 e^4 (2\pi)^3}{m_e^2 \omega_{\mathrm{pe}}^2} \frac{[\mathbf{k} \wedge \mathbf{k}']^2}{k^2 k'^2} \left| \frac{\varepsilon_{\mathbf{k}-\mathbf{k}',\omega_{\mathbf{k}}^{\ell}-\omega_{\mathbf{k}'}^{t}}^{(e)} - 1}{\varepsilon_{\mathbf{k}-\mathbf{k}',\omega_{\mathbf{k}}^{\ell}-\omega_{\mathbf{k}'}^{t}}} \right|^2$$

$$\times \ \delta\left(\omega_{\mathbf{k}}^{\ell} - \omega_{\mathbf{k}'}^{t} - ([\mathbf{k}-\mathbf{k}'] \cdot \mathbf{v})\right). \tag{6.113}$$

and

$$w_{\mathrm{p}}^{\ell,\mathrm{t}\,(e)}(\mathbf{k},\mathbf{k}') = \frac{(2\pi)^3 e^4}{m_e^2 \omega_{\mathrm{pe}}^2} \frac{[\mathbf{k} \wedge \mathbf{k}']^2}{k^2 k'^2} \left| \frac{\varepsilon_{\mathbf{k}-\mathbf{k}',\omega_{\mathbf{k}}^{\ell}-\omega_{\mathbf{k}'}^{t}}^{(i)}}{\varepsilon_{\mathbf{k}-\mathbf{k}',\omega_{\mathbf{k}}^{\ell}-\omega_{\mathbf{k}'}^{t}}} \right|^2$$

$$\times \ \delta\left(\omega_{\mathbf{k}}^{\ell} - \omega_{\mathbf{k}'}^{t} - ([\mathbf{k}-\mathbf{k}'] \cdot \mathbf{v})\right). \tag{6.114}$$

The probability for the scattering involving two transverse waves will for the case when the difference in their frequencies is small and the wavelength much larger than the Debye screening length differ from the above expressions in that one must replace the polarisation factor $[\mathbf{k} \wedge \mathbf{k}']^2/k^2 k'^2$ by the factor

$\frac{1}{2}[1 + (\mathbf{k} \cdot \mathbf{k}')^2/k^2 k'^2]$ and the square of the plasma frequency by the product of the frequencies of the two transverse waves involved in the scattering:

$$w_p^{t,t'(e)}(\mathbf{k}, \mathbf{k}') = \frac{(2\pi)^3 e^4}{2m_e^2 \omega_\mathbf{k}^t \omega_{\mathbf{k}'}^t} \left[1 + \frac{(\mathbf{k} \cdot \mathbf{k}')^2}{k^2 k'^2}\right] \left|\frac{\varepsilon_{\mathbf{k}-\mathbf{k}',\omega_\mathbf{k}^t-\omega_{\mathbf{k}'}^t}^{(i)}}{\varepsilon_{\mathbf{k}-\mathbf{k}',\omega_\mathbf{k}^t-\omega_{\mathbf{k}'}^t}}\right|^2$$
$$\times \delta(\omega_\mathbf{k}^t - \omega_{\mathbf{k}'}^t - ([\mathbf{k} - \mathbf{k}'] \cdot \mathbf{v})), \tag{6.115}$$

$$w_p^{t,t'(i)}(\mathbf{k}, \mathbf{k}') = \frac{Z^2 e^4 (2\pi)^3}{2m_e^2 \omega_\mathbf{k}^t \omega_{\mathbf{k}'}^t} \left[1 + \frac{(\mathbf{k} \cdot \mathbf{k}')^2}{k^2 k'^2}\right] \left|\frac{\varepsilon_{\mathbf{k}-\mathbf{k}',\omega_\mathbf{k}^t-\omega_{\mathbf{k}'}^t}^{(e)} - 1}{\varepsilon_{\mathbf{k}-\mathbf{k}',\omega_\mathbf{k}^t-\omega_{\mathbf{k}'}^t}}\right|^2$$
$$\times \delta(\omega_\mathbf{k}^t - \omega_{\mathbf{k}'}^t - ([\mathbf{k} - \mathbf{k}'] \cdot \mathbf{v})). \tag{6.116}$$

There are several important points to be made in connection with these results. The first one is that for electromagnetic waves, even if the change in frequency in the scattering process is small, the scattering cross-section depends strongly on the distribution functions of all particles, that is, *the collective effects in the scattering are very strong*. The scattering cross-section becomes the usual Thomson cross-section only for particles with high velocities when the change in frequency becomes large and for wavelengths larger than the screening length. The second point is that the scattering involving two electromagnetic waves is, in fact, used in many plasma laser experiments for plasma temperature measurements. One must clearly take into account that the dielectric permittivity occurs in the scattering cross-sections; it is, indeed, taken into account since the theory of the measurements uses fluctuation theory. In the standard fluctuation theory the scattering depends on the dielectric permittivity and this seems to be natural; however, very little attention is paid to this in the interpretation of the results and to the fact that the functional dependence of the cross-sections describes nothing but scattering by the dynamical shells of the particles. It can be shown that the actual fluctuation theory used to interpret the experiments is correct and gives exactly the same results as can be obtained from the probabilities (6.115) and (6.116) for spontaneous scattering by electrons and ions. We think that the interpretation which we have given here is very important. For example, the presence of appreciable scattering inside the ion line – that is, in the case when the change in frequency is of the order of the Doppler shift due to thermal ions – can formally be a mystery when one uses the standard approach but it is a trivial result if one bears in mind that it is nothing but the scattering by ions produced by their electron shells.

Finally we must point out that in our discussions we restricted ourselves solely to scattering processes where the virtual wave involved was longitudinal. This is valid only for non-relativistic particles and waves with non-relativistic phase velocities. In the case of non-relativistic particles with relativistic phase velocities there may be circumstances when one needs to take into account scattering processes involving transverse virtual waves. This process is shown in Fig.6.3.

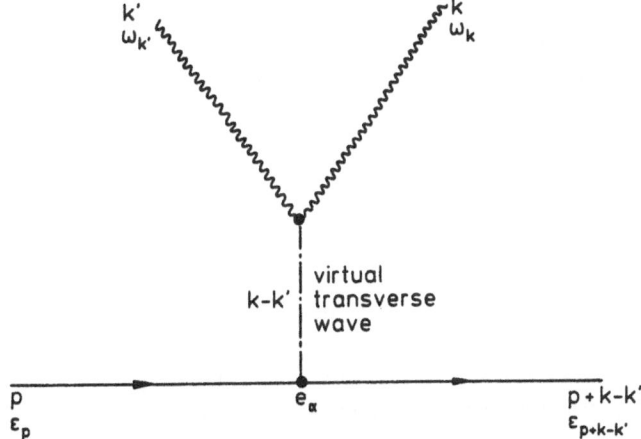

Fig. 6.3. Diagram of a scattering process involving a virtual transverse wave

The complete expression for the scattering cross-section should contain the sum of all three scattering amplitudes:

$$M_{k,k'} = M_{k,k'}^{osc} + M_{k,k'}^{tr(\ell)} + M_{k,k'}^{tr(t)}.$$

Problems

1. Find the terms describing the interference of the scattering amplitudes, which are due to the oscillations of the particles in the field of the incident wave and those corresponding to transition scattering, in the equation describing the changes in the particle distribution functions in the scattering process.
2. Prove Eq.(6.103) for the quadratic non-linear plasma response coefficients in the case when there is no direct wave-particle resonance.
3. Find the probability for the $\ell + e \rightleftarrows \ell' + e'$ scattering process involving a longitudinal virtual wave and taking into account the Doppler corrections to to the frequency of the wave acting on the electron. Estimate for what phase velocities the Doppler corrections will determine the scattering cross-section.
4. Find the probability for the $\ell + e \rightleftarrows t + e'$ scattering process involving a longitudinal virtual wave.
5. Find the probability for the $\ell + i \rightleftarrows t + i'$ scattering process involving a longitudinal virtual wave.
6. Find the probability for the $t + e \rightleftarrows t' + e'$ scattering process involving a longitudinal virtual wave.
7. Find the probability for the $t + i \rightleftarrows t' + i'$ scattering process involving a longitudinal virtual wave.

8. Find the probability for the $\ell + e \rightleftarrows \ell' + e'$ scattering process involving a transverse virtual wave.
9. Find the probability for the $\ell + i \rightleftarrows \ell' + i'$ scattering process involving a transverse virtual wave.
10. Find the probabilities for the $\ell + e \rightleftarrows t + e'$ and $\ell + i \rightleftarrows t + i'$ scattering processes involving a transverse virtual wave.
11. Find the probabilities for the $t + e \rightleftarrows t' + e'$ and $t + i \rightleftarrows t' + i'$ scattering processes involving a transverse virtual wave.
12. Calculate the scattering probability for highly relativistic particles.
13. Write down the equations for the frequency degradation of high-frequency ($\omega \gg \omega_{pe}$) electromagnetic waves scattered by relativistic particles.
14. Find the change in the distribution function of relativistic particles due to the stimulated scattering of high-frequency electromagnetic waves.

7 Plasma Turbulence

7.1 The General Concept of Turbulence

Turbulence, non-linearities, and randomness are closely related to one another. We shall discuss various aspects of turbulence step by step after first considering the corresponding non-linear processes; we shall therefore start here with turbulence problems related to stimulated scattering.

The first thing we have to do is to define what we mean when we talk about a turbulent plasma state and what are its main features.

Turbulence was first found in studies of hydrodynamic motions. It was discovered long ago that under certain circumstances for large velocities flow in gases or liquids becomes random and strong vortices of all sizes, superposed upon one another, are excited.

The simplest example is the flow of a liquid through a pipe; when the average velocity of the liquid is low the flow is regular – this is called *laminar* flow. However, when the velocity increases the flow becomes random, that is, turbulent. These days the term *turbulence* is used in quite a general sense, but a few general features are common to all turbulent motions. We shall use liquid turbulence to illustrate these. Any motion in an incompressible liquid for velocities less than the velocity of sound can be considered to the motion of a set of vortices. The vortices are the collective modes of the liquid, while other media such as plasmas can also have other collective modes. The appearance of turbulence is related to the excitation of random collective modes. In the case of a plasma we shall be interested in such collective modes as the Langmuir and ion-sound waves and we shall talk about *Langmuir turbulence* and *ion-sound turbulence*.

We shall start with a qualitative description of the appearance of hydrodynamic turbulence in flow through a pipe. One can observe the appearance of random liquid motions for large flow velocities, for instance, in water flowing from a pipe or water flowing in rivers. This random motion appears if the flow velocity exceeds some critical value determined by the critical value of the so-called *Reynolds number*. The meaning of this number can be illustrated by describing liquid flow by the *Navier-Stokes equation* which takes into account the kinematic viscosity coefficient μ of the liquid:

$$\frac{\partial \mathbf{v}}{\partial t} + (\mathbf{v} \cdot \nabla)\,\mathbf{v} = -\mu\nabla^2\mathbf{v} - \nabla p. \tag{7.1}$$

A stationary flow corresponds to a balance of the effects of the viscosity, the pressure p, and the dynamic liquid pressure. Let us consider a pipe with a circular cross-section (see Fig.7.1). The velocity of the liquid at the walls of the pipe is zero and in the case of laminar flow it increases towards the centre of the pipe. One might say that laminar flow consists of a set of cylindrical liquid layers each with its own velocity which increases towards the centre. Due to the viscosity each layer suffers friction from the layer with a smaller radius which tries to increase its velocity and from the layer with a larger radius which tries to decrease its velocity. This momentum transfer through friction may become unstable if the velocity gradient becomes large. The critical value of the velocity above which the instability occurs will be determined by the ratio of the non-linear term to the viscosity term in Eq.(7.1). This ratio will be of the order of the Reynolds number Re which in this case is given by the equation

$$\mathrm{Re} = \frac{av}{\mu}, \tag{7.2}$$

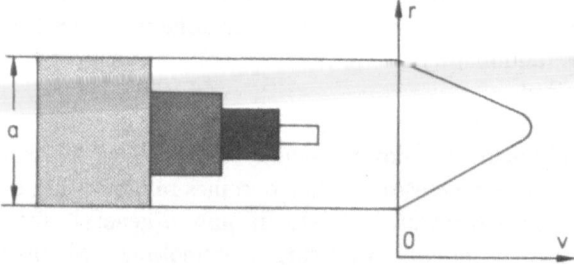

Fig. 7.1 Laminar flow of a liquid through a pipe. On the right we show the distribution of the velocity of the liquid inside the pipe

where a is the diameter of the pipe and v some typical value of the velocity. At some value of the Reynolds number the viscosity becomes insufficient to transfer the momentum from one layer to another and the liquid produces a vortex of size a which directly transfers the momentum from the wall to the centre. This instability is the first stage in the excitation of turbulence. A further increase in the velocity of the flow leads to secondary processes producing vortices of ever smaller sizes and at some critical value of the Reynolds number the motion becomes random. We should mention that the dissipation due to the viscosity plays a stabilising rôle and the random motions appear when the viscosity is no longer able to stabilise the flow and keep it laminar. This comment is made here to remind ourselves that the collective plasma modes are collisionless and that in their case the dissipation due to collisions is usually extremely small.

The turbulent state in liquids has the feature of intense random motions in vortices of many different scales. It should be possible to give a more quantitative description. First of all, the term "random" means that the motion is not reproducible; this is a charcteristic of dynamic chaos. We shall not go into details but just mention that nowadays it is possible to characterise the degree of randomness and to investigate its development during the excitation of turbulence. Secondly, the term "intense" means that the non-linear interactions between the vortices of different sizes are very important. These lead to energy transfer between vortices of different sizes and to the creation of an energy flux through the various sizes of vortices. The presence of such an energy flux is an important quantitative characteristic of hydrodynamic turbulence. This process is called an *energy cascade*. A.N.Kolmogorov was the first to propose, in 1941, such a cascade to describe the distribution of energy in the various vortex scales in a turbulent state. He started with several assumptions:

1. The turbulence of the vortices is isotropic.
2. Energy is transferred in a cascade from larger to smaller scales.
3. A single step in the energy cascade corresponds to $\Delta k \approx k$, where k is the reciprocal of the vortex size l: $k = 2\pi/l$.
4. The energy flux is conserved in the cascade.

In hydrodynamic turbulence there is no small parameter – this is in contrast to the case of plasma turbulence, as we shall see later on. It is therefore simplest to use dimensional arguments, as was done by Kolmogorov, to find the energy distribution in the turbulent state. Such a distribution is called the *turbulence spectrum* and it gives the energy distribution as a function of the absolute magnitude of the wavevector, the wavenumber $k = 2\pi/l$ where l is the vortex size. We express the energy density of the turbulent state, $W = \frac{1}{2}\varrho v^2$, where ϱ is the density of the liquid and v its velocity, in terms of a spectral distribution W_k of the turbulence through the relation

$$W = \int W_k \, dk. \qquad (7.3)$$

In turbulent spectra the following three regions are distinguished:

1. The *energy containing region* where vortices with a size of the order of a are excited which are subsequently transformed into vortices of smaller scales – in this region a stationary spectrum is established through a balance between the excitation of the vortices and the subsequent energy transfer to regions of larger k.
2. The *inertial range* in which Kolmogorov's hypotheses are valid – in this region a stationary spectrum for a given value of k is established through a balance of energy input from smaller values of k and a transfer of energy to larger values of k.
3. The *dissipation region* where the viscosity becomes important – in this region a stationary spectrum is established through a balance of an energy flux from smaller values of k and dissipation due to viscosity.

These three regions are shown in Fig.7.2.

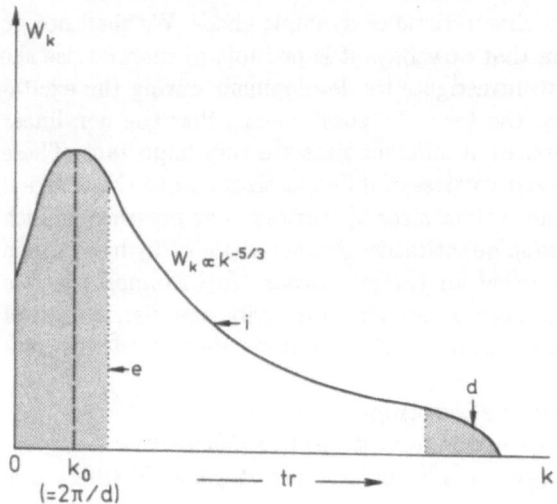

Fig. 7.2 The energy spectrum of liquid turbulence; e indicates the energy containing region, i the inertial range, d the dissipation region, and tr the direction of energy transfer

One can find the *Kolmogorov spectrum* W_k from a dimensional analysis, writing down the conservation of energy flux. If we transfer energy to a wavenumber which is larger by Δk which is of order of k, the energy at scales corresponding to k will be $W_k k \sim v^2$. The flux Φ is the change in energy per unit time so that we have $\Phi \sim W_k k \omega$ where ω is a characteristic frequency. From a dimensional analysis it follows that such a characteristic frequency will be kv which is approximately equal to $k(W_k k)^{1/2}$. From the condition that the flux be constant,

$$\Phi \sim W_k k \cdot k(W_k k)^{1/2} = \text{const},$$

it then follows that we have

$$W_k \propto \frac{1}{k^{5/3}}. \tag{7.4}$$

There are many features of hydrodynamic turbulence as we have described it here which are common to all kinds of turbulence. The randomness, the strong non-linear interactions, and the presence of energy fluxes are necessary features of any kind of turbulence, but the collective modes will vary from one kind to another and not all turbulence is isotropic. Moreover, cascades do not necessarily always go from smaller to larger values of k and energy transfer may occur involving a very large change in wavenumber, that is, there may occur an integral energy transfer superimposed upon the cascade. We shall call a state turbulent if (i) the strong collective modes are randomised, (ii) the level of their energy is appreciably higher than that corresponding to thermal fluctuations, and (iii) there appears an energy flux along the scale of the parameters characterising the collective modes. Plasma turbulence has all these features as far as the collective plasma modes are concerned. An important feature of the various kinds of plasma turbulence, such as Langmuir turbulence or ion-sound turbulence, is the presence of a small parameter – the ratio of a plasma oscillation period to the characteristic time for energy transfer. Such a small parameter does not exist in the case of hydrodynamic turbulence. This makes it possible to construct an analytic theory of plasma turbulence. The successes of theories of plasma turbulence have also made it easier to develop the appropriate, complicated theory of the Kolmogorov spectrum in liquids. In the case of plasma turbulence one can write the small parameter as the ratio of the non-linear width $\Delta\omega^{\mathrm{NL}}$ which we introduced earlier to the frequency of the oscillations:

$$\frac{\Delta\omega^{\mathrm{NL}}}{\omega_{\mathbf{k}}} \ll 1, \tag{7.5}$$

which is the same as the condition for a weak non-linearity. This means that all our earlier discussions of non-linear interactions of random waves are relevant to the problems of plasma turbulence. We start our discussion with the $W \ll W_{\mathrm{cr}}$ case, when the main non-linear interaction processes for random waves are either their stimulated scattering or the decay processes. This is the case of *weak plasma turbulence*. Later on we shall discuss the opposite case when we have $W \gg W_{\mathrm{cr}}$; in that case the main non-linear interactions are the modulational interactions and we are dealing with *strong plasma turbulence*. In the present chapter we shall discuss the rôle played by the stimulated scattering processes in the formation of the spectrum of weak plasma turbulence.

We introduce the *plasma turbulence spectrum* $W_{\mathbf{k}}$; the energy density of modes with wavevectors within the range $d^3\mathbf{k}$ is the $W_{\mathbf{k}} d^3\mathbf{k}$ and we have

$$W = \int W_{\mathbf{k}} d^3\mathbf{k}. \tag{7.6}$$

If we can neglect the non-linear frequency broadening we can express $W_{\mathbf{k}}$ in terms of the distribution function of the waves:

$$W_{\mathbf{k}} = \frac{\omega_{\mathbf{k}} N_{\mathbf{k}}}{(2\pi)^3}. \tag{7.7}$$

In the case of isotropic turbulence we look at the energy density W_k corresponding to a range dk of wavenumbers ($|\mathbf{k}| = k$):

$$W = \int W_k \, dk, \qquad W_k = \frac{\omega_k N_k k^2}{2\pi^2}. \tag{7.8}$$

If the spectrum is a power-law spectrum it is called a *Kolmogorov spectrum of plasma turbulence*. This name is as bad as the name non-linear Landau damping. The best way of naming this kind of spectrum would be to use the name of the physicist who first obtained such a spectrum in the case of plasma turbulence, but for reasons which will become obvious we shall do this only exceptionally. The name Kolmogorov spectrum, used sometimes for plasma turbulence spectra, is bad also in that Kolmogorov only used dimensional arguments to derive it while it needed enormous efforts to prove it analytically, whereas in the case of plasma turbulence an analytic theory was proposed and exact results obtained. We shall write the power-law spectrum in the form

$$W_k \propto \frac{1}{k^\nu},\tag{7.9}$$

where ν is called the *spectral index* of the spectrum.

7.2 Langmuir Turbulence

In the case of Langmuir waves the dissipation rate is smaller than the plasma frequency by a factor N_{de} and it is unlikely that the dissipation will be able to support a regular – laminar – behaviour of the plasma waves if they are strongly excited. This means that in a plasma one usually meets with a state of Langmuir turbulence. In the $W \ll W_{cr}$ case the energy transfer is due to stimulated scattering and it is in a direction opposite to the one in the case of hydrodynamic turbulence – in the direction of larger scales, that is, smaller values of k. As in liquids we can define an excitation region which is this case is not the energy-containing region. Under stationary conditions the turbulence spectrum will again be determined by the balance between excitation and energy transfer, this time to smaller k values. As before, we can define an inertial range where now under stationary conditions the spectrum is determined by energy transfer from smaller to larger scales, that is, from larger to smaller k values. This means that in this case the energy transfer is not towards the dissipation region as in liquids but towards the region where the dissipation is smallest. The oscillations will be accumulated in the region of small k values where they form the so-called *Langmuir condensate*. This serves as the energy-containing region for Langmuir turbulence. Once the energy accumulated in the condensate exceeds the critical value we found earlier, stimulated scattering as the non-linear energy transfer mechanism will no longer survive. We shall show in Chap.9 that then the direction in which the energy is transferred changes and energy will begin to flow from larger to smaller scales, as in liquids. The turbulence then becomes strong. The limit of applicability of the energy transfer by stimulated scattering is thus determined by the time needed to accumulate the critical energy density and by the value of the energy input. The latter may be either insufficient to reach the critical energy density or so large that it can create the critical energy density in the first stage. If the excitation is sufficiently weak – so that in the initial stage the energy level is less than the critical one – and the energy source continues to operate for a long time, the weak turbulent spectrum will continue to exist for a period which is determined by the time needed to reach the critical energy density in the condensate. We have earlier estimated the critical energy. We see thus that if the excitation of the Langmuir waves is not too strong weak turbulence always describes the initial stage of the development of turbulence. If the energy input is not very large and the total energy of the condensate does not exceed its critical value weak turbulence will provide the complete description of the turbulence. We show in Fig.7.3 qualitatively the distribution of Langmuir waves in the case of weak Langmuir turbulence excited by a beam.

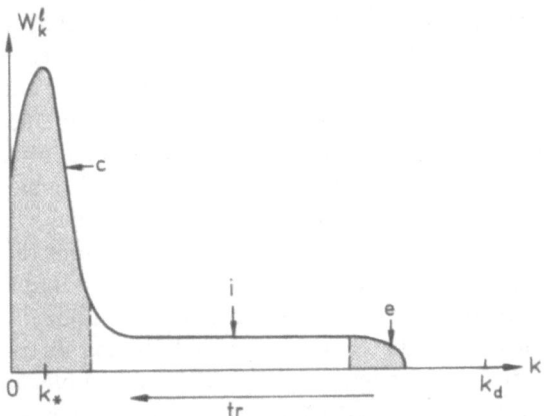

Fig. 7.3 The energy spectrum of Langmuir turbulence; e indicates the excitation region, i the inertial range, c the Langmuir condensate, and tr the direction of energy transfer

We can use the general equation for a turbulent spectrum,

$$\frac{dW_k^\ell}{dt} = 2[\gamma_{\mathbf{k}} + \gamma_{\mathbf{k}}^{\mathrm{NL}}] W_{\mathbf{k}}^\ell,$$

to find the stationary spectrum in the case of Langmuir turbulence. In that case we must have

$$\gamma_{\mathbf{k}} + \gamma_{\mathbf{k}}^{\mathrm{NL}} = 0. \tag{7.10}$$

If the phase velocities are much larger than v_{Te} the linear damping is practically completely due to collisions. In the excitation region we must take the growth rate into account. We shall assume for the moment that there are no ion-sound waves and consider only stimulated scattering of Langmuir waves by ions so that we get

$$\gamma_{\mathbf{k}} + \tfrac{1}{2} \int W_{\mathbf{k}'}^{\ell} \frac{d^3 \mathbf{k}'}{\omega_{\mathbf{k}'}} w_{\mathbf{p}}^{\ell, \ell'(i)}(\mathbf{k}, \mathbf{k}') \left([\mathbf{k} - \mathbf{k}'] \cdot \frac{\partial \Phi_{\mathbf{p}}^{i}}{\partial \mathbf{p}} \right) \frac{d^3 \mathbf{p}}{(2\pi)^3} = 0. \quad (7.11)$$

In the general case we should add to this equation an equation describing the change in the ion distribution function.

We shall try to give here a rough qualitative theoretical description of Langmuir turbulence even though it is nowadays possible to give a refined analytical description of many of its details. There is a certain amount of justification for the simplifications we shall introduce even though they make it impossible for us to describe all the details of the turbulence spectrum. When we list the simplifications we shall discuss why it is reasonable to introduce them.

The first simplification is that we assume isotropy. We found earlier that in a scattering process the wavevector can easily change its direction, whereas its absolute magnitude does not change much. Therefore after several scattering processes the waves, although excited anisotropically, will have acquired a rather isotropic distribution. A certain degree of anisotropy will remain, but nonetheless the distribution of the waves will be close to an isotropic one. Numerical calculations confirm that this assumption is not a bad one, even though some degree of anisotropy persists in the whole of the inertial range.

The second simplification is the assumption that the ion distribution will not change appreciably during the energy transfer and the formation of the turbulence spectrum. We shall assume that the ion distribution is given and that it is a thermal distribution. If the ion distribution is not thermal, the whole problem has to be reconsidered. The assumption that the ion distribution cannot change greatly during the energy transfer is not a bad one as the number of waves in conserved in the scattering process and the energy of the waves can be changed only by an amount equal to their dispersion corrections. Only that part of the energy can be transferred to the ions. We can make the following estimate for the ratio of the total change in the Langmuir wave energy when they are transferred to the condensate region to the thermal ion energy:

$$\frac{W^{\ell}}{n_i T_i} \frac{v_{Te}^2}{v_{\mathrm{ph}}^{(0)2}}, \qquad (7.12)$$

where $v_{\mathrm{ph}}^{(0)}$ is the initial phase velocity of the oscillations in the generation region – we have $v_{\mathrm{ph}}^{(0)} \cong v_b$ if we have excitation by a beam. We shall therefore neglect changes in the ion distribution of relative order (7.12).

Our third simplification is the use of an approximate expression for the scattering probability. We showed that the scattering matrix element was proportional to

$$\frac{\varepsilon_{\mathbf{k}-\mathbf{k}', \omega_{\mathbf{k}}-\omega_{\mathbf{k}'}}^{(e)} - 1}{\varepsilon_{\mathbf{k}-\mathbf{k}', \omega_{\mathbf{k}}-\omega_{\mathbf{k}'}}} \approx \frac{T_i}{Z T_e + T_i}. \qquad (7.13)$$

The last approximation in Eq.(7.13) holds in the case where the change in wavenumber, Δk, in the scattering is small as compared to a typical wavenumber, k_*, of the condensate region, $\Delta k \ll k_*$. In the case when $\Delta k \gg k_*$ this factor is close to unity. But even expression (7.13) cannot differ much from unity, because in the case when we have $ZT_e \gg T_i$ the main non-linear process becomes the decay process and we can no longer apply the considerations given here (see Chap.8). In fact, we need to know the factor (7.13) for the case when $\Delta k \approx k_*$. We shall therefore approximate the factor (7.13) by unity when deriving the spectrum. These assumption enable us to simplify Eq.(7.11) as follows:

$$\gamma_{\mathbf{k}}^{\mathrm{NL}} = \frac{Z^2 e^4}{2m_e^2 \omega_{\mathrm{pe}}^3} \int \frac{(\mathbf{k} \cdot \mathbf{k}')^2 ([\mathbf{k} - \mathbf{k}'] \cdot \mathbf{v})}{k^2 k'^2 T_i}$$
$$\times \, \delta\big(\omega_{\mathbf{k}} - \omega_{\mathbf{k}'} - ([\mathbf{k} - \mathbf{k}'] \cdot \mathbf{v})\big) \, W_{k'}^{\ell} \, dk' \, \Phi_{\mathbf{p}}^{\mathrm{i}} \, d^3\mathbf{p}. \qquad (7.14)$$

Although this has not been written out explicitly we have assumed that in this last equation we have averaged over the directions of the oscillations.

The last and most important simplification is that the change in wavenumber in each energy transfer step is a physically infinitesimal quantity. This means that he turbulent distribution we are interested in will be one which is averaged over $\Delta k \gg k_*$. The concept of a physically infinitesimal quantity is often used in disciplines such as macroscopic electrodynamics and hydrodynamics. In the latter, for instance, the particle mean free path is a physically infinitesimal quantity and one considers quantities averaged over the particle mean free path. We shall consider the quantity k_* for Langmuir turbulence to be similar to the mean free path in hydrodynamics. This means that by considering k_* to be a physically infinitesimal quantity we can hope to find only some rough characteristics of the spectrum averaged over k values larger than k_*. The roughness of the approximation is determined by the fact that k_* is in actual fact not a very small quantity. Indeed, even in the case when excitation occurs for a k value, $k \approx k_0$, comparable with the Debye wavenumber k_d it will need only $\sqrt{9m_i T_e / m_e T_i}$ steps to reach the condensate, and this will in practice be only approximately 100 steps.

Formally, if one neglects $([\mathbf{k} - \mathbf{k}'] \cdot \mathbf{v})$ in the argument of the δ function in Eq.(7.14), one can consider the step of a single spectral transfer to be zero. However, this will give a zero result in the case of a thermal distribution. It is thus necessary to take into account the next term in the expansion of the δ function:

$$\delta\big(\omega_{\mathbf{k}} - \omega_{\mathbf{k}'} - ([\mathbf{k} - \mathbf{k}'] \cdot \mathbf{v})\big) \approx \delta(\omega_{\mathbf{k}} - \omega_{\mathbf{k}'}) - ([\mathbf{k} - \mathbf{k}'] \cdot \mathbf{v})\delta'(\omega_{\mathbf{k}} - \omega_{\mathbf{k}'}), \, (7.15)$$

where δ' denotes the derivative of the δ function with respect to its argument. Substituting Eq.(7.15) into Eq.(7.14) we find

$$\gamma_{\mathbf{k}} + \frac{Z^2 e^4}{2m_e^2 \omega_{\mathrm{pe}}^3 T_i} \int \frac{(\mathbf{k} \cdot \mathbf{k}')^2}{k^2 k'^2} ([\mathbf{k} - \mathbf{k}'] \cdot \mathbf{v})^2$$
$$\times \, \delta'(\omega_{\mathbf{k}} - \omega_{\mathbf{k}'}) \, W_{k'}^{\ell} \, dk' \, \Phi_{\mathbf{p}}^{\mathrm{i}} \, d^3\mathbf{p} = 0. \qquad (7.16)$$

Integrating $([\mathbf{k} - \mathbf{k}'] \cdot \mathbf{v})^2$ over the ion distribution will give $|\mathbf{k} - \mathbf{k}'|^2 v_{Ti}^2 (2\pi)^3 n_i$ –
if we bear in mind that the distribution is normalised when integrated over
$d^3 p / (2\pi)^3$. The average of the angular factor in Eq.(7.16) gives $\frac{1}{3}$ and from
the average of $|\mathbf{k} - \mathbf{k}'|^2$ the term $2(\mathbf{k} \cdot \mathbf{k}')$ gives a zero contribution. We thus
get from Eq.(7.16)

$$\gamma_k + \frac{\pi \omega_{pe}}{12 m_i n_i} \int (k^2 + k'^2) \, \delta'(\omega_k - \omega_{k'}) \, W_{k'}^\ell \, dk' = 0. \tag{7.17}$$

Remembering that

$$\omega_{k'} = \omega_{pe} + \frac{3 k'^2 v_{Te}^2}{2 \omega_{pe}},$$

we have

$$\delta'(\omega_k - \omega_{k'}) = -\frac{d}{d\omega_{k'}} \delta(\omega_k - \omega_{k'}) = -\frac{\omega_{pe}}{3 v_{Te}^2 k'} \frac{d}{dk'} \delta(\omega_k - \omega_{k'}),$$

$$\left. \frac{d}{dk'} \frac{k^2 + k'^2}{k'} \right|_{k'=k} = 0, \qquad \delta(\omega_k - \omega_{k'}) = \frac{\omega_{pe}}{3 v_{Te}^2 k} \delta(k - k').$$

Integrating by parts, we get from Eq.(7.16)

$$\gamma_k + \gamma_k^{NL} = \gamma_k + \frac{\pi \omega_{pe}^3}{54 n_i m_i v_{Te}^4} \frac{\partial W_k^\ell}{\partial k} = 0, \tag{7.18}$$

where γ_k is the linear growth rate or the linear damping rate. In the case of
non-stationary and inhomogeneous turbulence we must use the equation

$$\frac{\partial W_k^\ell}{\partial t} + \left(\mathbf{v}_{gr} \cdot \frac{\partial W_k^\ell}{\partial \mathbf{r}} \right) = (2\gamma_k + 2\gamma_k^{NL}) W_k^\ell, \tag{7.19}$$

which for the isotropic case in which we are interested becomes

$$\frac{\partial W_k^\ell}{\partial t} + \left(\mathbf{v}_{gr} \cdot \frac{\partial W_k^\ell}{\partial \mathbf{r}} \right) = 2\gamma_k W_k^\ell + \frac{\pi \omega_{pe}^3}{27 n_i m_i v_{Te}^4} W_k^\ell \frac{\partial W_k^\ell}{\partial k}$$

$$= 2\gamma_k W_k^\ell + \alpha W_k^\ell \frac{\partial W_k^\ell}{\partial k}, \tag{7.20}$$

with

$$\alpha = \frac{\pi \omega_{pe}^3}{27 n_i m_i v_{Te}^4}.$$

We can find from Eq.(7.20) the stationary turbulence spectrum in the
inertial range when there is neither a damping nor a growth rate:

$$\gamma_k = 0: \qquad W_k^\ell = \text{const} = \frac{W_0}{k^\nu}, \qquad \nu = 0. \tag{7.21}$$

We have thus found that there is a Kolmogorov spectrum in the inertial range with a spectral index equal to zero. The plateau corresponding to this spectrum is shown in Fig.7.3. The reason for the appearance of a plateau in the spectrum is the particular property of the interactions which shows that the rate at which energy is transferred from smaller scales is the same as the rate at which it is transferred to larger scales so that there is no accumulation of energy during the transfer processes.

If one takes into account the fact that there is collisional damping of the Langmuir waves at a rate equal to $\gamma_k = -\nu_{ei}$, the spectrum W_k will show a small slope:

$$W_k = W_0 - \frac{\nu_{ei}}{\alpha}(k_0 - k), \tag{7.22}$$

where α is the same as in Eq.(7.20) and k_0 is a value of k lying in the generation region. There will be no creation of a condensate if

$$W_0 < \frac{\nu_{ei} k_0}{\alpha}, \tag{7.23}$$

as in that case the oscillations will be damped before they reach the condensate region. We can estimate the value of W_0 by considering the spectrum in the generation region. Assume that $\gamma_k - \gamma_0 > 0$ in a region $\Delta k_g = k_g - k_0 \ll k_0$. In the generation region the equation for the spectrum will gives us

$$W = W_0 - \frac{\gamma_0 - \nu_{ei}}{\alpha}(k - k_0). \tag{7.24}$$

From the condition that $W_k = 0$ for $k > k_g$ we find

$$W_0 = \frac{\gamma_0 - \nu_{ei}}{\alpha}\Delta k_g, \tag{7.25}$$

and we can write the criterion (7.23) in the form

$$\frac{\gamma_0 - \nu_{ei}}{\nu_{ei}} < \frac{k_0}{\Delta k_g}. \tag{7.26}$$

If $k_0 > \Delta k_g$ this condition can be satisfied even for a growth rate far from the instability threshold. If Eq.(7.26) is not satisfied a condensate will be formed. It begins to grow once the energy created at $k \approx k_g$ has been transferred to $k \approx k_*$. If $k_0 \gg k_*$ we can estimate the characteristic time it takes for this transfer by the equation

$$\frac{1}{\tau^{NL}} = \gamma^{NL} \approx \frac{\alpha W_0}{k_0} \approx (\gamma_0 - \nu_{ei})\frac{\Delta k_g}{k_0}; \tag{7.27}$$

it is thus larger than the time to generate the energy in the $k \approx k_0$ region by a factor $k_0/\Delta k_g$. If $W_0 \Delta k_g > n T_i m_e/9 m_i = W_{cr}$ the time τ^{NL} will be sufficient for a new type of non-linear interactions – the modulational interactions – to appear; W_{cr} is the critical energy of the condensate. If $W_0 \Delta k_g < W_{cr}$ it takes a time which is larger than τ^{NL} by a factor $W_{cr}/W_0 \Delta k_g$ to get a transfer to the modulational interactions.

7.3 Non-linear Stabilisation of a Beam-Plasma Instability

We have earlier considered the quasi-linear beam-plasma relaxation which showed that a considerable part of the initial energy of a beam is converted into plasma oscillations. However, at that time we neglected the non-linear processes. We shall now take them into account making the same assumptions as in our quasi-linear discussion – a weak beam with a velocity larger than the thermal velocities and with a considerable spread in velocities. One would expect that the non-linear effects will reduce the efficiency of the beam-plasma interactions, but this occurs when the threshold conditions are satisfied. The threshold is determined by assuming approximate equality of the quasi-linear and the non-linear effects. If the non-linear effects dominate one can easily see why the efficiency of the interactions with the plasma is reduced as follows. How much energy is transferred from the particles in the beam to plasma oscillations depends on how long the oscillations excited by the beam are acted upon in resonance by the beam particles. In the quasi-linear approach the wavenumbers of the oscillations remain unchanged and they stay a long time in resonance with the particles in the beam. In the non-linear stimulated scattering processes the wavenumbers of the oscillations do change and they can be transferred out of resonance. They therefore have less time to change the energy of the beam particles and less energy is put into oscillations. If the stimulated scattering dominates almost all the waves which are excited by the beam will be transferred away from the resonance region. Only a small part of the oscillations will survive in the resonance region and be able to support the non-linear interactions. It will take this small number of oscillations much longer to spread the beam distribution. The energy lost by the beam particles will be equal to the energy of the oscillations which are excited but the time needed for the relaxation of the beam will in the case when the non-linear effects dominate be much longer than the time of the quasi-linear relaxation. This effect is called the *non-linear stabilisation of the beam-plasma instability*. The stimulated scattering by ions provides an efficient conversion of waves from the resonance region to, for example, a region where the phase velocities are larger than the speed of light and in that case the transferred waves can never interact with resonant particles.

We shall now try to obtain an estimate for the threshold of the non-linear stabilisation of the beam-plasma instability. We shall show that under the assumptions which we have made this threshold only determines a critical value for the beam velocities – for velocities larger than the critical value the non-linear processes dominate – but it does not determine the beam density. We shall assume that the beam has already in the initial stage created enough energy in the waves for spreading the beam velocities by an amount of the order of the original spread δv_b. The energy of the excited oscillations will then be of the order of $W^\ell \approx n_b m_e v_b \delta v_b$. It is distributed over an interval δk which is determined by the resonance condition $v_b = \omega_{pe}/k$ which means that $\delta v_b = \omega_{pe} \delta k / k^2$ or $\delta k = \omega_{pe} \delta v_b / v_b^2$. We get thus the following estimate for the spectral energy density of the turbulence in the interval δk:

$$W_k^\ell \approx \frac{W^\ell}{\delta k} \approx \frac{n_b m_e v_b^3}{\omega_{pe}}, \qquad \frac{\partial W_k}{\partial k} \approx \frac{W_k}{\delta k} \approx \frac{n_b m_e v_b^5}{\omega_{pe}^2 \delta v_b},$$

and from this we can estimate the non-linear growth rate:

$$\gamma^{NL} \approx \alpha \frac{\partial W_k}{\partial k} \approx \omega_{pe} \frac{n_b}{n_i} \frac{m_e}{m_i} \frac{v_b^5}{v_{Te}^4 \delta v_b}. \tag{7.28}$$

We have omitted here a numerical factor $\pi/27$ as in the final results it will appear to the $\frac{1}{4}$ power and our estimates do not have that accuracy. The non-linear growth rate should be compared with the linear growth rate which is determined by the spread δv_b:

$$\gamma_k \approx \omega_{pe} \frac{n_b}{n_i} \frac{v_b^2}{\delta v_b^2}. \tag{7.29}$$

Comparing Eqs.(7.28) and (7.29) we find the value of the critical velocity:

$$\frac{v_b^{cr}}{v_{Te}} \gg \left[\frac{m_i}{m_e}\right]^{1/4} \left[\frac{v_b}{\delta v_b}\right]^{1/4}. \tag{7.30}$$

We should now use Eq.(7.30) to write down the criterion (5.11) that the non-linear interactions are the stimulated scattering processes:

$$\frac{n_b}{n_e} \ll \frac{v_{Te}^4}{v_b^4} \ll \frac{m_e}{m_i} \frac{\delta v_b}{v_b}. \tag{7.31}$$

The beam density therefore also enters indirectly in the stabilisation condition. The velocities satisfying Eq.(7.30) are rather high, but they still can lie in the spectral transfer range, that is, in the wavenumber range outside the condensate region. The condition that the beam excites oscillations inside the condensate and that there does not exist an inertial range is $k < k_* = \omega_{pe} v_{Ti}/3 v_{Te}^2$, or

$$v_b > \frac{3 v_{Te}^2}{v_{Ti}} = 3 v_{Te} \sqrt{\frac{m_e T_i}{m_i T_e}}, \tag{7.32}$$

which corresponds to velocities which are higher than those given by Eq. (7.30).

7.4 Solar Radio Bursts

The different types of *solar radio burst emission* were classified at a time
when the understanding of the physical nature of the bursts was very poor.
The classification was based, in particular, on the magnitude of the frequency
drift. Type III bursts have large drift velocities: the agent exciting them moves
through the whole of the solar atmosphere in a few seconds. According to the
present interpretation – first proposed by V.L.Ginzburg – such an agent is a
beam of fast electrons. It is produced in an active process on the Sun and
is often related to solar flares. The frequency spread of the emission at a
given time is small and the frequency itself corresponds to the local plasma
frequency. The frequency drift is due to the beam travelling through different
parts of the solar atmosphere with the density decreasing as it travels. Since
the electron density distribution is known from independent measurements
and theoretical models it is possible to use the observed frequency drift of
the bursts to determine the change in the beam particle velocities during
the propagation of the beam through the solar atmosphere. The puzzling
result is that the beam velocity does not change appreciably during its propagation through the solar atmosphere. On average the beam velocities are
close to $0.3c \approx 10^{10}$ cm s^{-1}. Velocities below $0.2c \approx 6 \times 10^9$ cm s^{-1} have
not been observed. The average thermal electron velocity is of the order of
5×10^8 cm s^{-1}, that is, the minimum beam velocities are approximately 10
to 15 times larger than the thermal velocity. This is exactly what one would
expect from the criterion for the non-linear stabilisation. A natural explanation of the existing observations would thus be that beams with velocities
less than the critical one are quickly relaxed and are unable to penetrate
over distances like the one which are observed, whereas beams with velocities
higher than the critical one can penetrate through the whole of the solar
atmosphere and can create Langmuir turbulence with a level which is insufficient to change the beam particle velocity but which is essential for the
generation of electromagnetic radiation through the conversion of Langmuir
waves into electromagnetic waves. The mechanism for this conversion is the
scattering by ions: $\ell + i \rightleftarrows t + i'$.

Before the non-linear stabilisation mechanism was proposed as an explanation for the observations, the idea that a beam was responsible for the
radio bursts ran into difficulties because the quasi-linear relaxation distance
was several orders of magnitude smaller than the distance actually found
from the observations. Nowadays it is known that the beam particles not
only penetrate the solar atmosphere but that they traverse the distance from
the Sun to the Earth and are found by satellite measurements. Some satellites observe not only the beam particles but also the plasma oscillations.
According to what is known at present, the total number of fast electrons,
N, creating the bursts lies in the range from 10^{30} to 10^{33}, the probable size
of the beam is of the order of 2×10^9 cm, the density of the beam lies in the

range from 10^2 cm^{-3} to 10^5 cm^{-3}, the plasma density lies in the range from 3×10^8 cm^{-3} to 10^7 cm^{-3}; for those parameter values the criterion (7.31) is satisfied for most of the bursts. For the densest beams it is perhaps necessary to look for other non-linear mechanisms for the beam stabilisation. The stabilisation may in that case be due to the modulational non-linear interactions. The data which we have given here make it possible to estimate the time for quasi-linear relaxation. One must take into account that this time is longer than the time for the last stage of the quasi-linear relaxation determined by $\gamma_k \approx \omega_{pe} n_b / n_e$, by a factor $\ln(W^{\ell(\infty)}/W^{\ell(0)})$. Here $W^{\ell(0)}$ is the initial fluctuation energy density of the Langmuir waves and $W^{\ell(\infty)}$ is that energy density in the last stage. If we put $W^{\ell(0)} \approx n_e T_e / N_{de}$, $W^{\ell(\infty)} \approx n_b m_e v_b^2$, we find

$$\ln \frac{W^{\ell(\infty)}}{W^{\ell(0)}} \approx \ln \left\{ \frac{v_b^2}{v_{Te}^2} \frac{n_b}{n_e} N_{de} \right\} \approx 20 \text{ to } 25.$$

Taking into account that three-dimensional relaxation is 2 to 3 times longer we have $\tau_{ql} \approx 10^2 n_e / n_b \omega_{pe}$ which ranges from 10^{-4} s to 0.3 s with an average value of 10^{-2} s which is two orders of magnitude smaller than the observed times of several seconds.

The non-linear stabilisation is thus a necessary process for an explanation of the observations on solar radio bursts. It looks as if the observations may correspond to values close to the boundary between weak and strong turbulence. This is quite natural since the modulational interactions are sufficiently strong to relax the parameters to the threshold between strong and weak turbulence.

7.5 Ion–Sound Turbulence

The turbulence spectrum of ion-sound oscillations is determined, on the one hand, by stimulated scattering by ions and, on the other hand, by either Landau damping by electrons if the electrons are in thermal equilibrium or by amplification by electrons if they are not in equilibrium. We shall consider here the ion-sound turbulence spectrum for $k \ll 1/d_e$. We can write the equation for a stationary ion-sound turbulence spectrum in the form

$$\gamma_k^s + \frac{1}{2} \int \frac{W_{k'}^s}{\omega_{k'}^s} w_p^{s,s'(i)}(k, k') \left([k - k'] \cdot \frac{\partial \Phi_p^i}{\partial p} \right) \frac{d^3 p}{(2\pi)^3} d^3 k' = 0. \quad (7.33)$$

We shall now introduce assumptions similar to those made earlier in the case of Langmuir turbulence. First of all, we assume that the ion distribution is given and is a thermal one; secondly, we assume that the turbulence is isotropic; and thirdly, we assume that the steps in the cascade are physically infinitesimal.

The last assumption can be used to simplify the probability (6.64). The expression $(\mathbf{k} \cdot \mathbf{v})$ in that expression can be transformed by introducing the components of the vector $\mathbf{v}$ parallel and perpendicular to $\mathbf{k} - \mathbf{k}'$. As the ion distribution is assumed to be isotropic the square of $\mathbf{k}$ which occurs in the expression for the probability will in Eq.(7.33) become the sum of the squares of expressions containing the parallel and the perpendicular components. The square containing the parallel component can be expressed in terms of the square of the difference in the frequencies of the two interacting waves:

$$(\mathbf{k} \cdot \mathbf{v}_\parallel)^2 \to \frac{(\mathbf{k} \cdot [\mathbf{k}-\mathbf{k}'])^2([\mathbf{k}-\mathbf{k}'] \cdot \mathbf{v})^2}{|\mathbf{k} - \mathbf{k}'|^4} = \frac{(\mathbf{k} \cdot [\mathbf{k}-\mathbf{k}'])^2}{|\mathbf{k} - \mathbf{k}'|^4} \left(\omega_\mathbf{k} - \omega_{\mathbf{k}'}\right)^2. \quad (7.34)$$

Since the steps in the cascade are assumed to be infinetesimal, the frequency difference is small and expression (7.34) should be neglected. Formally one can obtain that result by taking into account that to a first approximation the δ-function in the probability becomes

$$\delta\left(\omega_\mathbf{k} - \omega_{\mathbf{k}'}\right) - ([\mathbf{k} - \mathbf{k}'] \cdot \mathbf{v})\delta'(\omega_\mathbf{k} - \omega_{\mathbf{k}'}). \quad (7.35)$$

Either term of expression (7.35) gives zero when multiplied by expression (7.34). The square of the perpendicular component occurs in Eq.(7.33) only together with a thermal ion distribution and, after averaging over this thermal distribution, will contain the thermal ion velocity:

$$(\mathbf{k} \cdot \mathbf{v}_\perp)^2 \to \frac{[\mathbf{k} \wedge \{\mathbf{k} - \mathbf{k}'\}]^2}{|\mathbf{k} - \mathbf{k}'|^2} v_{Ti}^2 = \frac{[\mathbf{k} \wedge \mathbf{k}']^2}{|\mathbf{k} - \mathbf{k}'|^2} v_{Ti}^2. \quad (7.36)$$

In the probability (6.64) only the second term of (7.35) is important:

$$w_\mathrm{p}^{\mathrm{s},\mathrm{s}'(\mathrm{i})}(\mathbf{k}, \mathbf{k}') \to -\frac{4Z^4 e^4 [\mathbf{k} \wedge \mathbf{k}']^2 (\mathbf{k} \cdot \mathbf{k}')^2 ([\mathbf{k}-\mathbf{k}'] \cdot \mathbf{v})}{m_\mathrm{i}^3 \omega_\mathrm{pi}^4 k^2 k'^2 |\mathbf{k} - \mathbf{k}'|^2} \delta'\left(\omega_\mathbf{k} - \omega_{\mathbf{k}'}\right) T_\mathrm{i}. \quad (7.37)$$

The derivative of the ion distribution,

$$\left([\mathbf{k} - \mathbf{k}'] \cdot \frac{\partial \Phi_\mathrm{p}^\mathrm{i}}{\partial \mathbf{p}}\right) = -\frac{([\mathbf{k} - \mathbf{k}'] \cdot \mathbf{v})}{T_\mathrm{i}} \Phi_\mathrm{p}^\mathrm{i}, \quad (7.38)$$

contains, like expression (7.37), the first power of the parallel component and we have

$$([\mathbf{k} - \mathbf{k}'] \cdot \mathbf{v})^2 \to |\mathbf{k} - \mathbf{k}'|^2 v_{Ti}^2. \quad (7.39)$$

Because we have assumed that the turbulence is isotropic we can also average over the angular variables and this gives

$$\frac{[\mathbf{k} \wedge \mathbf{k}']^2 (\mathbf{k} \cdot \mathbf{k}')^2}{k^4 k'^4} \to \frac{2}{15}. \quad (7.40)$$

Taking all these substitutions into account we can rewrite the non-linear equation for ion-sound turbulence in the form:

$$\gamma_k^s + \frac{2\pi Z^2 v_{Ti}^2}{15 m_i n_i} \int \frac{W_{k'}^s}{\omega_{k'}^s} k^2 k'^2 \delta'\left(\omega_k^s - \omega_{k'}^s\right) dk' = 0. \qquad (7.41)$$

Using the equation

$$\delta'\left(\omega_k^s - \omega_{k'}^s\right) = -\frac{\partial}{\partial \omega_{k'}} \delta\left(\omega_k^s - \omega_{k'}^s\right) = -\left[\frac{d\omega_{k'}}{dk'}\right]^{-1} \frac{\partial}{\partial k'} \left[\frac{\delta(k - k')}{d\omega_k/dk}\right], (7.42)$$

we transform Eq.(7.41) to a differential form:

$$\gamma_k^s + \frac{2\pi Z^2 v_{Ti}^2}{15 m_i n_i} k^2 \left[\frac{d\omega_k^s}{dk}\right]^{-1} \frac{\partial}{\partial k} \left[\frac{W^s}{\omega_k^s} \frac{k^2}{d\omega_k^s/dk}\right] = \gamma_k^s + \gamma_k^{NL} = 0. \quad (7.43)$$

There is no inertial range in the case of ion-sound turbulence since Landau damping occurs for all values of k. However, the decrease in the energy of the ion-sound oscillations occurs due not only to absorption by electrons, but also to non-linear transfer by ions since in the scattering process an appreciable amount of energy is transferred to the ions. The region is therefore dissipative and the stationary turbulence which is described by Eq.(7.43) can be established only if there is an energy source present outside the region described by Eq.(7.43). Therefore there is a possibility that the excitation and dissipation regions are separated – or not – but there is no inertial range. In the general case of nonstationary and inhomogeneous turbulence we have the equation

$$\frac{\partial W_k^s}{\partial t} + \left(\mathbf{v}_{gr} \cdot \frac{\partial W_k^s}{\partial \mathbf{r}}\right) = 2\left(\gamma_k^s + \gamma_k^{NL}\right) W_k^s. \qquad (7.44)$$

One can solve Eq.(7.43) in the case when the Landau damping is determined by thermal electrons:

$$\gamma_k^s = -\sqrt{\frac{\pi}{8}} \frac{\omega_k^s}{v_{Te}} \frac{d\omega_k^s}{dk}. \qquad (7.45)$$

In the $k \ll 1/d_e$ region Eq.(7.43) has the form

$$-\sqrt{\frac{\pi}{8}} \frac{v_s}{k v_{Te}} + \frac{2\pi T_i}{15 T_e} \frac{\partial}{\partial k} \frac{k W_k^s}{n_i T_e} = 0, \qquad (7.46)$$

and its solution will be:

$$W_k^s = \frac{15}{4\sqrt{2\pi}} \frac{T_e v_s}{T_i v_{Te}} n_i T_e \frac{1}{k} \ln \frac{k}{k_0}. \qquad (7.47)$$

The constant k_0 is determined by the level of energy of the source of the turbulence outside the region we are considering at the moment – where the oscillations are damped and transferred to lower values of k. The spectrum is equal to zero at $k = k_0$, and the level of the source determines the minimum value k_0 to which the oscillations can be transferred. The larger the source, the lower the value of k_0. The ion-sound turbulence spectrum is shown in Fig. 7.4.

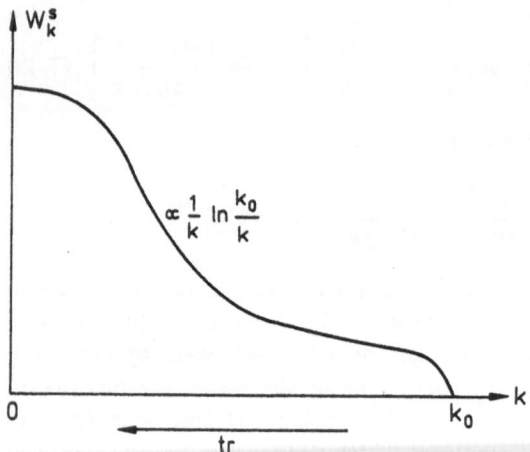

Fig. 7.4. The energy spectrum of ion-sound turbulence; tr indicates the direction of energy transfer

For $k \gg k_0$ the spectrum is close to a Kolmogorov spectrum with a spectral index

$$\nu \approx 1. \tag{7.48}$$

In the $k \gg 1/d_e$ case Eq.(7.43) differs in form for the case when the wave dispersion is determined by the electrons from when it is determined by the ions. The electrons determine the dispersion when

$$1 \ll kd_e \ll \left(\frac{ZT_e}{T_i}\right)^{1/4}; \tag{7.49}$$

in that case we have

$$\omega_k^s = \frac{kv_s}{\sqrt{1 + k^2 d_e^2}} \approx \omega_{pi} \left[1 - \frac{1}{k^2 d_e^2}\right]. \tag{7.50}$$

Equation (7.43) for the ion-sound spectrum can then be written in the form

$$\sqrt{\frac{\pi}{8}} \frac{v_s}{v_{Te}} \frac{1}{k^7 d_e^7} + \frac{2\pi T_i}{15 T_e} k \frac{\partial}{\partial k} \left[\frac{k W_k}{n_i T_e} k^4 d_e^4\right] = 0, \tag{7.51}$$

with the solution

$$W_k = \frac{15v_s}{28\sqrt{2\pi}} \frac{T_e^2 n_i}{v_{Te} T_i} \frac{1}{k} \frac{1}{k^4 d_e^4} \left\{ W_0 + \frac{1}{k^7 d_e^7} \right\}. \tag{7.52}$$

Here W_0 is a constant determined by the strength of the source outside the region where the solution is valid. If $W_0 \ll 1$ the spectrum will correspond to a spectral index

$$\nu = 12, \tag{7.53}$$

whereas for $W_0 \gg 1$ we have

$$\nu = 5. \tag{7.54}$$

The spectrum (7.53) decreases rapidly with increasing k and it soon changes to the form (7.54).

Another kind of spectrum appears when the ion-sound waves are excited by Langmuir turbulence. According to Eq.(6.79) we have in that case approximately

$$\gamma_k^s \approx \omega_{pi} \frac{1}{k^2 d_e^2} \int \frac{W_{k'}^\ell k' d_e}{n_e T_e} dk' \propto \frac{1}{k^2}. \tag{7.55}$$

As a balance between excitation and non-linear transfer a stationary spectrum is established. We find from Eq.(7.43) that now

$$\nu = 11. \tag{7.56}$$

The ion-sound wave dispersion is determined by the ions if

$$\left[\frac{ZT_e}{T_i}\right]^{1/4} \ll kd_e \ll \left[\frac{ZT_e}{T_i}\right]^{1/2}. \tag{7.57}$$

In that case we have

$$\omega_k^s = \omega_{pi} + \frac{3k^2 v_{Ti}^2}{2\omega_{pi}}, \tag{7.58}$$

and we find from Eq.(7.43) that now

$$\nu = 4. \tag{7.59}$$

7.6 Anomalous Plasma Resistivity

The anomalous plasma resistivity is one of the important manifestations of collective plasma processes. The physics of this effect is very simple. If there are no collective oscillations the only forces acting on the electrons are friction from the ions and any external force which may be present. This means that their drift velocity – and thus the current density and the electrical conductivity of the plasma – is determined by the balance between the acceleration due to the external force and the deceleration due to the friction. One can find the exact expression for the conductivity from the Landau collision integral. We shall here only give a well known qualitative estimate for the conductivity.

We can write the balance between the friction and the electric field in the form

$$\nu_{ei} m_e u = eE,$$

where u is the *drift velocity* and E the electric field strength. From this equation we find for the drift velocity and the current density j:

$$u = \frac{eE}{m_e \nu_{ei}}, \qquad j = e n_e u = \frac{e^2 n_e}{m_e \nu_{ei}} E = \sigma E, \tag{7.60}$$

where σ is the electrical conductivity which is thus given by the equation

$$\sigma = \frac{\omega_{pe}^2}{4\pi \nu_{ei}} \approx \omega_{pe} \frac{N_{de}}{4\pi}. \tag{7.61}$$

We should emphasise that the usual electrical conductivity which is determined by the electron-ion binary collisions contains a very large number – the number of particles in a Debye sphere, N_{de}. Formally, the effect of the anomalous resistivity shows up in the disappearance of the large factor N_{de} from the expression for the electrical conductivity. In fact, we shall see that the decrease in the electrical conductivity is somewhat less but the largest factor is here N_{de}. Since N_{de} is usually many orders of magnitude larger than unity the effect of changing the plasma to a state where the resistivity is anomalous drastically alters all plasma properties. For instance, in the normal collisional state the conductivity is so high that to a good approximation the plasma can be considered to be a superconductor, whereas in a turbulent state it conducts currents very badly.

The physics of the appearance of such a large resistivity lies in that for a certain value of the drift velocity exceeding some critical value the ion-sound instability becomes excited and the electrons will collide more frequently with the ion-sound waves than with the ions. Their effective collision frequency will be determined by the level of ion-sound turbulence which is established and which is determined by the balance between the damping of the excitation and the non-linear transfer due to scattering by ions. One can find the growth rate for the excitation of ion-sound waves by the drifting electrons for the case where the electrons have a thermal distribution with a drift velocity **u**:

$$\Phi_p^e = \frac{(2\pi)^{3/2}}{m_e^3 v_{Te}^3} \exp\left(-\frac{(\mathbf{v} - \mathbf{u})^2}{2v_{Te}^2}\right). \tag{7.62}$$

The result is

$$\gamma_{\mathbf{k}}^s = \sqrt{\frac{\pi}{8}} \frac{((\mathbf{k} \cdot \mathbf{u}) - \omega_{\mathbf{k}}^s)v_s}{(1 + k^2 d_e^2)^{3/2} v_{Te}} \exp\left[-\frac{(\omega_{\mathbf{k}}^s - (\mathbf{k} \cdot \mathbf{u}))^2}{2k^2 v_{Te}^2}\right]. \tag{7.63}$$

This expression differs from the expression for the Landau damping only in that the ion-sound wave frequency $\omega_{\mathbf{k}}^s$ is replaced by the Doppler shifted frequency $\omega_{\mathbf{k}}^s - (\mathbf{k} \cdot \mathbf{u})$. It is clear from Eq.(7.63) that an instability occurs when the Cherenkov condition

$$u > \frac{\omega_{\mathbf{k}}^s}{k} = \frac{v_s}{(1 + k^2 d_e^2)^{1/2}} \tag{7.64}$$

is satisfied. This means that the instability starts when the drift velocity exceeds the ion thermal velocity. If $kd_e \gg 1$ we have $\omega_{\mathbf{k}}^s/k \approx v_s/kd_e > v_s d_i/d_e = v_{Ti}$. If $u < v_s$ only the ion oscillations are excited whereas if $u > v_s$ the whole spectrum of ion sound waves is excited. If the drift velocity is only just larger than the ion-sound velocity waves moving close to the direction of the drift are excited. However, if the drift velocity is much larger than the ion-sound velocity ion-sound waves are excited at angles to the drift velocity which can be large, up to values close to $\frac{1}{2}\pi$. It follows from Eq.(7.63) that the growth rate is linearly proportional to k up to $k \approx 1/d_e$, and when k increases further it decreases as $1/k^2$. The maximum growth rate corresponds to $k \approx 1/d_e$. The use of Eq.(7.63) is limited to velocities lower than the electron thermal velocity since for drift velocities close to the electron thermal velocity the growth rate will be of the order of the frequency. The criterion for the applicability of Eq.(7.63) will thus be $u \ll v_{Te}$.

We shall now give some estimates of the ion-sound turbulent spectrum and of the anomalous resistivity under the following two assumptions: (i) the drift velocity is much larger than the ion-sound velocity: $v_s \ll u \ll v_{Te}$, and (ii) the isotropisation of the ion-sound waves through their scattering by ions is rather efficient so that we can use the approxmation that the ion-sound waves are excited isotropically. The excitation rate for $k \ll 1/d_e$ will be estimated to be given by

$$\gamma_k^s \approx \frac{kuv_s}{v_{Te}}. \tag{7.65}$$

By dropping factors of the order of unity in the $k \ll 1/d_e$ region we find the equation

$$\frac{u}{kv_{Te}} + \frac{T_i}{T_e} \frac{\partial}{\partial k} \frac{W_k^s k}{n_i T_e} = 0, \tag{7.66}$$

whence we get the spectrum which in the literature is known as the *Kadomt-sev spectrum*

$$\frac{W_k^s}{n_i T_e} = \frac{u T_e}{v_{Te} T_i} \frac{1}{k} \ln \frac{k_0}{k}. \tag{7.67}$$

This spectrum becomes zero, not for small values of k – as in the case of damping – but for large values of k. Since the growth rate decreases rapidly for $k \gg 1/d_e$ and since k_0 occurs only under the logarithm sign we can approximately put $k_0 = 1/d_e$. The total energy density of the ion-sound turbulence for such a spectrum is rather large. Integrating expression (7.67) over k up to its maximum value $k \approx 1/d_e$ we find the following estimate:

$$\frac{W^s}{nT} \approx \frac{T_e}{T_i} \frac{u}{v_{Te}}. \tag{7.68}$$

From this expression we can already see that in the framework of weak non-linearities a value of the drift velocity $u \approx v_{Te}$ is not acceptable.

The effective frequency of collisions between ion-sound waves and electrons can be found from a quasi-linear equation, since for such a high energy level of oscillations the collision integral can be neglected:

$$\frac{d\Phi_p^e}{dt} = 4\pi^2 e^2 \int \frac{k^2 d_e^2}{1 + k^2 d_e^2} W_k^s \left(\frac{k}{k} \cdot \frac{\partial}{\partial p}\right)$$

$$\times \delta\left(\omega_k^s - (k \cdot v)\right) \left(\frac{k}{k} \cdot \frac{\partial}{\partial p}\right) \Phi_p^e d^3 k. \tag{7.69}$$

We have used here the relation

$$W_k^s = \frac{|E^s|_k^2}{4\pi} \omega_k^s \left.\frac{\partial \varepsilon_k}{\partial \omega}\right|_{\omega = \omega_k^s} = \frac{1 + k^2 d_e^2}{k^2 d_e^2} \frac{|E_k^s|^2}{2\pi}. \tag{7.70}$$

For estimates one can in Eq.(7.69) neglect the ion-sound frequency in comparison with $(k \cdot v)$ under the sign of the δ-function and one must take into account that the order of magnitude of a δ-function is the reciprocal of its argument which means, for instance, that in Eq.(6.69) it is of the order of $1/kv_{Te}$. The spectrum is approximately proportional to $1/k$ and the quasi-linear integral is determined by the maximum possible value of k which $1/d_e$. Neglecting numerical factors of the order of unity we then find the following estimate:

$$\nu_{eff} \approx \omega_{pe} \frac{W^s}{n_e T_e} \approx \frac{T_e}{T_i} \frac{u}{v_{Te}} \omega_{pe}, \tag{7.71}$$

where we have used Eq.(7.68). From this result follows immediately the validity of the statement made earlier that the large factor N_{de} drops out of the effective electrical conductivity when the conductivity is determined by the turbulence. Although the factor multiplying ω_{pe} in Eq.(7.71) is not equal to unity, its minimum is the product of a large factor, T_e/T_i, and a small factor, $\sqrt{m_e/m_i}$, and this factor is clearly not less than about 0.1 and for the maximum possible drift velocities it is even larger than unity. However, in our estimates we have dropped numerical factors of order unity and there may be a certain accumulation of numerical factors in the final result. This is, indeed, what is happening. It has taken about twenty years to solve exactly the complicated non-integral equations to find the actual angular dependences and the numerical factor in the effective collision frequency. It turns out that it is approximately equal to 0.01. Therefore we should instead of Eq.(7.71) write:

$$\nu_{\text{eff}} \approx \frac{\omega_{\text{pe}}}{100} \frac{T_e}{T_i} \frac{u}{v_{Te}}. \tag{7.72}$$

This last expression is in good agreement with the results of numerous experimental studies. The same order of magnitude for the effective collision frequency can also be obtained for the frequency broadening due to non-linear interactions. This frequency broadening may be due to decay processes (see Chap.8). We can get from Eq.(7.72) an estimate for the anomalous electrical conductivity:

$$\sigma_{\text{anom}} = \frac{\omega_{\text{pe}}^2}{4\pi\nu_{\text{eff}}} \approx \frac{100}{4\pi} \omega_{\text{pe}} \frac{T_i}{T_e} \frac{v_{Te}}{u}. \tag{7.73}$$

R.Z.Sagdeev was the first to obtain this *Ohm law*. Since the anomalous electrical conductivity is inversely proportional to the drift velocity, we find

$$j \propto \frac{E}{u} \propto \frac{E}{j}, \quad \text{or} \quad j^2 \propto E,$$

and hence Ohm's law in a plasma is:

$$j \propto \sqrt{E}. \tag{7.74}$$

This law is valid for $u \gg v_s$. For drift velocities close to v_s both non-linear and quasi-linear effects are important.

The anomalous electrical resistivity has been measured in many experiments including tokamaks with a turbulent plasma and in small experimental set-ups where ion-sound turbulence is excited by external electric fields.

7.7 Turbulent Shock Waves and Turbulent Magnetic Reconnection

Dissipation is needed for the existence of shock waves since at the front of a shock wave one needs the irreversible particle heating process as well as an increase in the entropy. This is known from hydrodynamics for the case when all characteristic scales of the various processes are larger than the collisional

mean free path. In the hydrodynamic regime the irreversibility at the front
of the shock wave is produced by such dissipative processes as viscosity or
thermal conductivity. In the collisionless regime, in which we are interested,
the dissipation problem is not so easily solved since it is impossible that the
particle distributions on both sides of the shock wave can be thermal distributions. However, we can consider irreversible energy transfer from one
distribution to another. The irreversibility at the front of the shock wave
may be produced in the collisionless regime by turbulence. Such shock waves
are called *turbulent shock waves*. An example of this kind of shock wave is a
shock wave propagating across a strong magnetic field. In such a shock wave
the magnetic field strength changes its value from one side of the shock to
the other. However, a spatial change in the magnetic field strength means,
according to Ampère's law, the presence of a current. In the case of a shock
wave of a sufficiently large amplitude the value of the drift velocity in this
current will be larger than the ion-sound velocity and the electron temperature will be much higher than the ion temperature. All necessary conditions
for the excitation of ion-sound turbulence are thus present. The anomalous
resistivity will serve as the dissipative mechanism which is needed for the
formation of shock waves. Such collisionless shock waves have been observed
in many laboratory experiments. Using laser scattering one was able to measure the ion-sound turbulence spectrum which was excited and to observe
excellent agreement with theoretical predictions. Another example of a collisionless shock is the *bow shock* in the neighbourhood of the Earth which
separates the solar wind from the magnetosphere. Collisionless shocks also
appear on the Sun, in the solar wind, and in the interstellar medium; they are
widely used in the interpretation of astrophysical data. Another type of such
shock waves are produced in active cosmic experiments such as the injection
of heavy-ion clouds in the magnetosphere.

Fig. 7.5 Sketch of the magnetic field reconnection process; the arrows show the
direction of the magnetic field lines after the reconnection

Another process in which the anomalous resistivity plays an important
rôle is the turbulent magnetic field line reconnection process. As in the case of
shock waves this process can also occur in a collision-dominated regime. When
two magnetic field lines which have opposite directions come close together in
a dissipative medium reconnection is possible (see Fig.7.5). As in the case of
a shock wave the fact that the magnetic field depends on position means the
presence of a current; its field changes the magnetic field from a value **H** on

one side of the layer where the reconnection takes place to $-\mathbf{H}$ on the other side. If the drift velocity of this current is smaller than the ion-sound velocity the classical collision-dominated electrical conductivity provides a very slow rate of magnetic reconnection. However, if the drift velocity is larger than the ion-sound velocity the resistivity becomes anomalous and this leads to a very fast magnetic reconnection process. The rate of magnetic reconnection may increase by several orders of magnitude. The reconnection may occur not only for field lines which are exactly directed in opposite directions, but also in the case when only some of the components of the magnetic field are directed in opposite directions.

Magnetic reconnection processes determine many important processes in a plasma, including the disruptive instabilities in tokamaks and solar flares. Since there is tension along magnetic field lines this tends to decrease the length of the magnetic field lines after the reconnection. This leads to an expulsion of the plasma from the reconnection region. We show in Fig.7.5 the direction of such plasma motions. In tokamaks this leads to an expulsion of the plasma towards the walls and to the disruptive instabilities of tokamaks. The same process leads on the Sun to the explosions known as *solar flares*. The important feature here is not the reconnection itself but the time needed for this process. The disruptive instabilities in tokamaks and the solar flares are such fast processes that they can only be explained if the anomalous resistivity is responsible for the observed phenomenon. Such reconnections are usually called *turbulent magnetic reconnections*.

There are many other interesting applications of turbulent magnetic reconnections; we may mention here as an example the explanation by S.B.Pikelner of the spectacular phenomenon of *spicules* on the Sun's surface. These are narrow plasma columns which appear on the Sun's surface and which lift the plasma from the surface with velocities in the range of 20 to 25 km/s. The density of the plasma in these columns is higher than that of the surrounding plasma and is of the order of 10^{11} to 10^{12} cm^{-3} while the height of these columns is of the order of 10^3 km. The puzzling feature of the spicules is that if one considers the observed velocities to be produced by some active process occurring on the Sun the height to which the plasma can be lifted by inertia will be determined by the gravitational field and would lie in the range of 100 to 200 km, that is, an order of magnitude smaller than the observed heights. The only possible explanation can be that the magnetic stress forces are larger than the gravitational forces. There exists on the Sun a network of magnetic field lines created in the convection process and field lines in opposite directions can approach one another closely when the magnetic field line network changes with time. Turbulent reconnection will then create the lifting magnetic forces. The lifting velocity v_B will be determined by the rate of diffusion of the magnetic field lines and we can give the following estimate:

$$v_B \approx \frac{c^2}{4\pi\sigma_{\mathrm{anom}}h},$$ (7.75)

where h is the width of the magnetic reconnection region; the value of h can be estimated from the magnetic field equation:

$$\frac{B}{h} \approx \frac{4\pi}{c}j = \frac{4\pi}{c}en_e v_s \frac{u}{v_s}.$$ (7.76)

Since, on the other hand, there must be a balance between the magnetic and the plasma pressures,

$$B^2 = 8\pi n_e T_e,$$ (7.77)

we can express h solely in terms of the drift velocity:

$$h = \sqrt{8\pi n_e T_e}\,\frac{c}{4\pi n_e e}\sqrt{\frac{m_i}{T_e}}\frac{v_s}{u} = \frac{c}{\omega_{\mathrm{pe}}}\sqrt{\frac{2m_i}{m_e}}\frac{v_s}{u}.$$ (7.78)

If we substitute this expression and expression (7.73) for the anomalous electrical resistivity into Eq.(7.75) we find:

$$v_b \approx \frac{c}{100\sqrt{2}}\frac{m_e}{m_i}\frac{T_e}{T_i}\frac{u^2}{v_s^2}.$$ (7.79)

If the anomalous resistivity exists for a long time we would expect that the free parameters will relax to their values close to the threshold values $u \approx 2v_s$, $T_e \approx 4T_i$. For those parameter values Eq.(7.79) gives a value of 15 km/s for v_B which agrees with observations. The lifting velocities due to collisional reconnection are of the order of a few cm/s and cannot explain the observations. There are many other examples, both in laboratory experiments and in astrophysics, where one can apply the phenomenon of anomalous electrical resistivity.

Problems

1. Estimate the magnitude of the anomalous electrical conductivity for the case where the drift velocity is smaller than the ion-sound velocity, $u < v_s$, which is when ion oscillations are excited.
2. Find the non-linear stimulated scattering processes, taking into account transverse waves with frequencies close to the plasma frequency, $\omega_k^t \approx \omega_{\mathrm{pe}} + k^2 c^2 / 2\omega_{\mathrm{pe}}$. Use the same assumptions which were made in the case of Langmuir waves. Show that in the case of isotropic distributions the spectral transfer is described by the equations

$$\frac{dW_k^t}{dt} = -\nu_{\text{ei}} W_k^t + \alpha W_k^t \frac{1}{k} \frac{\partial}{\partial k} \left(k W_k^\ell \right),$$

$$\frac{dW_k^\ell}{dt} = \gamma_k W_k^\ell + \alpha W_k^\ell \left[\frac{\partial W_k^\ell}{\partial k} + k \frac{\partial}{\partial k} \frac{W_k^t}{k} \right],$$

where α is given by Eq.(7.20).

3. Find the energy of the Langmuir and the transverse waves in the case of a non-linear stabilisation of the beam-plasma instability.

$$\frac{dN}{E}$$

$$\left[\frac{1}{3!}\frac{6}{88} + \frac{3}{24}\frac{9+2}{48}\right]$$

$$\frac{dM}{6}$$

where

6. The ... effect of CO_2 behavior and the transport ... on the rate of ... and the ... of the two-phase instability.

8 Non-linear Decay Interactions

8.1 General Concepts

The simplest decay process is a wave-wave emission process, that is, a non-linear interaction in which a wave of a particular type decays and produces another wave of the same type as well as a wave of a different type. An example is the decay of a Langmuir wave into another Langmuir wave and an ion-sound wave. This process can be considered to be the emission of an ion-sound wave by a Langmuir wave. Symbolically we can write this process in the form

$$\ell \rightleftarrows \ell' + s. \tag{8.1}$$

This process is shown in Fig.8.1.

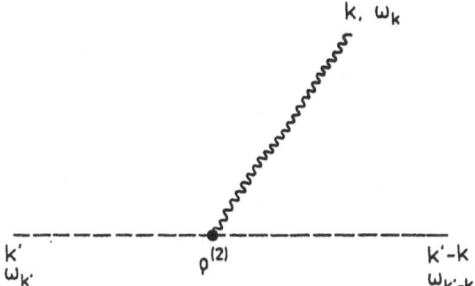

Fig. 8.1. A wave-wave emission process

Let the initial momentum of the Langmuir wave be $\mathbf{k}'$, the momentum of the emitted ion-sound wave be $\mathbf{k}$, so that the final momentum of the Langmuir wave is $\mathbf{k}' - \mathbf{k}$. The energy conservation law for this wave-wave emission process is:

$$\omega_{\mathbf{k}'}^{\ell} = \omega_{\mathbf{k}'-\mathbf{k}}^{\ell} + \omega_{\mathbf{k}}^{s}. \tag{8.2}$$

This is similar to the equation when a wave is emitted by a particle; in particular, the emission of ion-sound waves by particles is described by the equation

$$\varepsilon_{\mathbf{p}} = \varepsilon_{\mathbf{p-k}} + \omega_{\mathbf{k}}^{s}. \tag{8.3}$$

In both cases we have for the case when the momentum of the emitted wave is small as compared to the initial momentum of the incoming source either the condition for Vavilov-Cherenkov emission,

$$\omega_{\mathbf{k}}^{s} = (\mathbf{k} \cdot \mathbf{v}), \tag{8.4}$$

or the condition for *group wave resonance*,

$$\omega_{\mathbf{k}}^{s} = (\mathbf{k} \cdot \mathbf{v_{gr}}), \tag{8.5}$$

where here $\mathbf{v_{gr}}$ is the group velocity of the Langmuir waves

$$\mathbf{v_{gr}} = \frac{d\omega_{\mathbf{k'}}^{\ell}}{d\mathbf{k'}}. \tag{8.6}$$

One might say that in some sense Eq.(8.5) is the Cherenkov condition for wave-wave emission, showing that if the process is the one where an ion-sound wave is emitted by Langmuir waves the group velocity of the Langmuir wave responsible for the emission should be larger than the phase velocity of the ion-sound waves. If we proceed as in the case of the emission of waves by particles where we introduced a probability $w_{\mathbf{p}}(\mathbf{k})$ and now introduce a probability $w_{\mathbf{k'}}(\mathbf{k})$ for the emission of ion-sound waves, we can write down an equation similar to the quasi-linear equation which takes into account both the spontaneous and the stimulated processes. In some approximations this equation differs from the quasi-linear equation in the notation used and, of course, in the probability which occurs in it. It looks as if the whole problem consists in finding an expression for the probability for the emission of waves by waves. However, the problem is not as simple as that. In the case of the emission of waves by particles the momentum of the wave is practically always smaller than that of the particle, but this is not so in the case of the emission of waves by waves when the momentum of the emitted wave can be comparable with the momentum of the source wave. This is, for instance, the case for the emission of ion-sound waves by Langmuir waves. The probability for the emission of an ion-sound wave is the largest if in the final state the Langmuir wave propagates in a direction opposite to that of its initial direction with the same absolute magnitude of its momentum. The emitted ion-sound wave then has a momentum equal to twice that of the initial momentum of the Langmuir wave. There are also differences in the statistical properties of the source of the emission between the emission of waves by waves and the emission of waves by particles.

Among all possible decay processes we have not only the emission of waves, but also more complicated processes, for instance, the decay of a single wave into more than two waves. Another possibility is that two waves are changed into two other waves, and so on. Examples of those processes are the processes:

$$\ell \leftrightarrows \ell' + s + s' \quad \text{and} \quad \ell + \ell' \leftrightarrows \ell'' + \ell'''.$$

When we write down the decay process symbolically as we have just done and as we did in Eq.(8.1), we have used – and we shall continue to use – two arrows indicating that the process can go both ways. Sometimes the decay processes in which three waves are involved are called *three-wave decay processes* and those involving four waves *four-wave decay processes*. Sometimes we use the term *plasmon* instead of wave: three-plasmon and four-plasmon processes. This term is especially appropriate for random waves. The conservation laws in the decay processes – such as the one described by Eq.(8.2) – are called the *resonance decay conditions*. This term is very appropriate in the case of regular waves since the non-linear responses due to the incident wave will efficiently excite a new wave only under resonance conditions.

Our aim in the present chapter is to show that the general non-linear equation will contain the decay processes even in the case when the wave which is excited in the decay process was initially absent. The equation containing only the fields of Langmuir waves describes also the decay of Langmuir waves into ion-sound waves. This statement is true only when it is possible for ion-sound waves to exist, that is, under the condition that $ZT_e \gg T_i$. Of course, in the case when the amplitude of the ion-sound wave field is large we must use a general non-linear equation containing not only the amplitudes of the Langmuir field but also the amplitude of the ion-sound field. We have given such equations earlier. However, in order to find the decay probability we need to know only the rate at which ion-sound waves are generated by Langmuir waves for the case when initially there were no ion-sound waves. This will be the easiest way to obtain the probability.

An important problem is finding the relations between stimulated scattering processes and decay processes. We shall show that both processes describe one and the same non-linear interaction process and in the case when for a given wavenumber the conservation laws for decay and for scattering can be satisfied at the same time, the stimulated scattering process can be neglected. The reason is that decay processes can be considered as resonance scattering processes in which the virtual wave becomes a real wave and, clearly, the resonance value is the maximum one so that under resonance conditions the scattering is changed into a decay process. On the other hand, one might say that scattering is decay at the tail of a resonance curve. However, one should take into account the following important points. First of all, in many cases the decay process may be forbidden – for instance if the ion and electron temperatures are the same ion-sound waves cannot exist and any decay process involving them is impossible. Secondly, if the decay process is not forbidden, the conservation laws for the decay can be satisfied only in a restricted region of the phase space of the waves – we are thinking here of the wavevector phase space for the waves. As a rule the stimulated scattering processes are allowed in a much larger phase space volume where they may be the only non-linear processes. We shall prove all these statements in the present chapter.

We shall restrict ourselves here to the three-wave decay processes for random waves when we can introduce a decay probability. We shall show, for example, how to calculate the probability for the decay of a Langmuir wave into another Langmuir wave and an ion-sound wave from the general non-linear equation for Langmuir waves. Similar calculations can also be carried out for other kinds of decay. We shall also give a physical interpretation of the process which in many cases will make it possible to find the probabilities from simple considerations.

In the general case the wave decay process describes the decay of a σ' wave into a σ'' and a σ wave:

$$\sigma' \rightleftarrows \sigma'' + \sigma. \tag{8.7}$$

The energy and momentum conservation laws for this decay process are

$$\mathbf{k'} = \mathbf{k''} + \mathbf{k}, \tag{8.8}$$

$$\omega_{\mathbf{k'}}^{\sigma'} = \omega_{\mathbf{k''}}^{\sigma''} + \omega_{\mathbf{k}}^{\sigma}. \tag{8.9}$$

As in earlier discussions we shall mainly consider electrostatic waves and if there are no external magnetic fields these will be Langmuir and ion-sound waves. The process in which all three waves are Langmuir waves is obviously forbidden as twice the plasma frequency cannot be the frequency of a Langmuir wave. It turns out that three-wave processes in which all three waves are ion-sound waves are also forbidden. This is obvious when the frequencies of the ion-sound waves lie in the range of the ion plasma frequency. On the other hand, for ion-sound waves with $\omega_{\mathbf{k}}^{s} = kv_s$ and all three waves propagating in the same direction it looks as if conservation of momentum also guarantees conservation of energy, just by multiplying the first conservation law by v_s. However, we must take into account the dispersion corrections, even if the frequencies of the three ion-sound waves lie in the $kd_e \ll 1$ region. In that case we have

$$\omega_{\mathbf{k}}^{s} = \frac{kv_s}{\sqrt{1 + k^2 d_e^2}} \simeq kv_s \left(1 - \tfrac{1}{2} k^2 d_e^2\right),$$

and if we use this expression for the frequencies it follows that one cannot satisfy the two conservation laws simultaneously. This means that the only three-wave decay process allowed for electrostatic waves is the $\ell \rightleftarrows \ell' + s$ process.

8.2 The Balance Equation; the Conservation Laws

We now introduce the probability per unit time for the decay process (8.7):

$$w_{\sigma',\mathbf{k'}}^{\sigma'',\sigma}(\mathbf{k}), \tag{8.10}$$

with the wavevector $\mathbf{k}$ of the σ wave and the wavevector $\mathbf{k}'$ of the σ' wave lying, respectively, within the phase volumes $d^3\mathbf{k}/(2\pi)^3$ and $d^3\mathbf{k}'/(2\pi)^3$. When describing the stimulated processes we must take into account that if we are dealing with an emitted wave we must add the spontaneous and the stimulated processes, leading to a factor $N_\mathbf{k}+1$ whereas only the stimulated process is involved if we are dealing with an absorbed wave and the relevant factor is $N_\mathbf{k}$. The probability (8.10) describes a process where a σ' wave is absorbed and σ'' and σ waves are emitted. The rate at which the number of σ' waves decreases due to that process is given by the equation

$$\left[\frac{dN_{\mathbf{k}'}^{\sigma'}}{dt}\right]_{-} = - \int w_{\sigma',\mathbf{k}'}^{\sigma'';\sigma}(\mathbf{k})\, N_{\mathbf{k}'}^{\sigma'}\, (N_{\mathbf{k}'-\mathbf{k}}^{\sigma''}+1)\,(N_\mathbf{k}^\sigma+1)\,\frac{d^3\mathbf{k}}{(2\pi)^3}, \qquad (8.11)$$

whereas for the reversed process the rate at which $N_{\mathbf{k}'}^{\sigma'}$ increases is given by

$$\left[\frac{dN_{\mathbf{k}'}^{\sigma'}}{dt}\right]_{+} = \int w_{\sigma',\mathbf{k}'}^{\sigma'';\sigma}(\mathbf{k})\,(N_{\mathbf{k}'}^{\sigma'}+1)\, N_{\mathbf{k}'-\mathbf{k}}^{\sigma''}\, N_\mathbf{k}^\sigma\,\frac{d^3\mathbf{k}}{(2\pi)^3}. \qquad (8.12)$$

The balance equation thus has the form

$$\frac{dN_{\mathbf{k}'}^{\sigma'}}{dt} = \int w_{\sigma',\mathbf{k}'}^{\sigma'';\sigma}(\mathbf{k})\,\left(N_\mathbf{k}^\sigma N_{\mathbf{k}'-\mathbf{k}}^{\sigma''} - N_\mathbf{k}^\sigma N_{\mathbf{k}'}^{\sigma'} - N_{\mathbf{k}'-\mathbf{k}}^{\sigma''}N_{\mathbf{k}'}^{\sigma'}\right)\,\frac{d^3\mathbf{k}}{(2\pi)^3}. \qquad (8.13)$$

We have neglected here the term which is linear in the wave distribution functions as it is assumed to be small. We can similarly obtain balance equations for the σ and σ'' waves:

$$\frac{dN_\mathbf{k}^\sigma}{dt} = \int w_{\sigma',\mathbf{k}'}^{\sigma'';\sigma}(\mathbf{k})\,\left(N_\mathbf{k}^\sigma N_{\mathbf{k}'}^{\sigma'} + N_{\mathbf{k}'-\mathbf{k}}^{\sigma''}N_{\mathbf{k}'}^{\sigma'} - N_\mathbf{k}^\sigma N_{\mathbf{k}'-\mathbf{k}}^{\sigma''}\right)\,\frac{d^3\mathbf{k}'}{(2\pi)^3}. \qquad (8.14)$$

$$\frac{dN_{\mathbf{k}''}^{\sigma''}}{dt} = \int w_{\sigma',\mathbf{k}''+\mathbf{k}}^{\sigma'';\sigma}(\mathbf{k})$$
$$\times \left(N_\mathbf{k}^\sigma N_{\mathbf{k}''+\mathbf{k}}^{\sigma'} + N_{\mathbf{k}''}^{\sigma''}N_{\mathbf{k}''+\mathbf{k}}^{\sigma'} - N_\mathbf{k}^\sigma N_{\mathbf{k}''}^{\sigma''}\right)\,\frac{d^3\mathbf{k}}{(2\pi)^3}. \qquad (8.15)$$

In the last equation we replaced $\mathbf{k}'$ by $\mathbf{k}''+\mathbf{k}$. We now define the energy W^σ and momentum $\mathbf{P}^\sigma$ densities of the waves through the equations

$$W^\sigma = \int \omega_\mathbf{k}^\sigma N_\mathbf{k}^\sigma\,\frac{d^3\mathbf{k}}{(2\pi)^3}, \qquad \mathbf{P}^\sigma = \int \mathbf{k} N_\mathbf{k}^\sigma\,\frac{d^3\mathbf{k}}{(2\pi)^3}, \qquad (8.16)$$

and we then find the following conservation laws for the decay processes:

$$\frac{d}{dt}\left[W^\sigma + W^{\sigma'} + W^{\sigma''}\right] = 0, \qquad \frac{d}{dt}\left[\mathbf{P}^\sigma + \mathbf{P}^{\sigma'} + \mathbf{P}^{\sigma''}\right] = 0. \qquad (8.17)$$

In the case when $\sigma' = \sigma''$ we obtain the complete equation for the σ' waves by summing Eqs.(8.13) and (8.15):

$$\frac{dN_{\mathbf{k'}}^{\sigma'}}{dt} = \int \left\{ w_{\sigma',\mathbf{k'}}^{\sigma',\sigma}(\mathbf{k}) \left(N_{\mathbf{k}}^{\sigma} N_{\mathbf{k'}-\mathbf{k}}^{\sigma'} - N_{\mathbf{k}}^{\sigma} N_{\mathbf{k'}}^{\sigma'} - N_{\mathbf{k'}-\mathbf{k}}^{\sigma'} N_{\mathbf{k'}}^{\sigma'} \right) \right.$$

$$\left. + w_{\sigma',\mathbf{k'}+\mathbf{k}}^{\sigma',\sigma}(\mathbf{k}) \left(N_{\mathbf{k}}^{\sigma} N_{\mathbf{k'}+\mathbf{k}}^{\sigma'} + N_{\mathbf{k'}}^{\sigma'} N_{\mathbf{k'}+\mathbf{k}}^{\sigma'} - N_{\mathbf{k}}^{\sigma} N_{\mathbf{k'}}^{\sigma'} \right) \right\} \frac{d^3\mathbf{k}}{(2\pi)^3}. \quad (8.18)$$

In particular, this equation describes the decay interactions between Langmuir and ion-sound waves in the case where $\sigma' = \ell$ and $\sigma = $ s. In the limit where the intensity of the σ waves is zero, Eq.(8.18) describes the non-linear growth rate of the σ' waves:

$$\frac{dN_{\mathbf{k'}}^{\sigma'}}{dt} = 2\gamma_{\mathbf{k'}}^{\text{NL decay}} N_{\mathbf{k'}}^{\sigma'}, \quad (8.19)$$

where

$$\gamma_{\mathbf{k'}}^{\text{NL decay}} = \frac{1}{2} \int \left[w_{\sigma',\mathbf{k'}+\mathbf{k}}^{\sigma',\sigma}(\mathbf{k}) N_{\mathbf{k'}+\mathbf{k}}^{\sigma'} - w_{\sigma',\mathbf{k'}}^{\sigma',\sigma}(\mathbf{k}) N_{\mathbf{k'}-\mathbf{k}}^{\sigma'} \right] \frac{d^3\mathbf{k}}{(2\pi)^3}. \quad (8.20)$$

This growth rate describes the instability of σ' waves in the initial stage of the decay process when the intensity of the σ waves is small. This means that if there are no ion-sound waves Langmuir waves are unstable against the excitation of ion-sound waves. This instability is sometimes called a *parametric instability*. However, this term is not a very good one and we prefer to use the term *decay instability*. If one considers it one must bear in mind that it only describes the initial stages of a much more complicated non-linear decay process in which the energy can also flow back from the ion-sound waves to the Langmuir waves.

In the case of Langmuir waves it is more convenient to write Eq.(8.20) in the form where $N_{\mathbf{k'}}^{\ell}$ is factored out:

$$\frac{dN_{\mathbf{k}}^{\ell}}{dt} = 2\gamma_{\mathbf{k}}^{\text{NL decay}} N_{\mathbf{k}}^{\ell},$$

$$\gamma_{\mathbf{k}}^{\text{NL decay}} = \frac{1}{2} \int \left[w_{\ell,\mathbf{k'}}^{\ell,\text{s}}(\mathbf{k'}-\mathbf{k}) - w_{\ell,\mathbf{k}}^{\ell,\text{s}}(\mathbf{k}-\mathbf{k'}) \right] N_{\mathbf{k'}}^{\ell} \frac{d^3\mathbf{k'}}{(2\pi)^3}. \quad (8.21)$$

8.3 Probabilities of the Decay Interactions

We shall now consider the probability for the decay of a Langmuir wave into another Langmuir wave and an ion-sound wave. Since the maximum momentum of the ion-sound wave is twice the momentum of the initial Langmuir wave and since the wavenumbers of Langmuir waves are always less than the reciprocal of the Debye length the ion-sound wave should correspond to the sound part of their spectrum: $\omega_{\mathbf{k}}^{\text{s}} \simeq k v_{\text{s}}$. We must bear in mind that the stimulated scattering of an ion-sound wave involving a Langmuir wave is possible only for the other part of the ion-sound spectrum, corresponding to the ion oscillations. The decay considered now therefore does not compete with $\ell \rightleftarrows$ s scattering although is may compete with $\ell \rightleftarrows \ell'$ scattering, as we shall see later on.

In the preceding chapter we constructed a general non-linear theory for Langmuir waves. We can use this theory in the case when there are no waves except Langmuir waves. However, the theory should contain the decay instability (8.21) since the latter depends only on the intensity of the Langmuir waves. We have earlier obtained Eq.(6.43) for the non-linear growth rate when there are only Langmuir waves present. We shall write it down here with a small change in notation:

$$\gamma_{\mathbf{k}}^{\mathrm{NL}} = -\frac{4\pi e^2}{m_e^2 \omega_{\mathrm{pe}}^4} \int N_{\mathbf{k}'}^{\ell} \frac{d^3\mathbf{k}'(\mathbf{k}\cdot\mathbf{k}')^2|\mathbf{k}-\mathbf{k}'|^2}{(2\pi)^3 k^2 k'^2} \left[\frac{\partial \varepsilon_{\mathbf{k}}}{\partial \omega}\frac{\partial \varepsilon_{\mathbf{k}'}}{\partial \omega'}\right]_{\omega=\omega_{\mathbf{k}}^{\ell};\omega'=\omega_{\mathbf{k}'}^{\ell}}^{-1}$$

$$\times \mathrm{Im}\left\{\frac{\left(\varepsilon_{\mathbf{k}-\mathbf{k}'}^{(e)}-1\right)\varepsilon_{\mathbf{k}-\mathbf{k}'}^{(i)}}{\varepsilon_{\mathbf{k}-\mathbf{k}'}}\right\}_{\omega=\omega_{\mathbf{k}}^{\ell};\omega'=\omega_{\mathbf{k}'}^{\ell}} . \tag{8.22}$$

This expression was used in calculations relating to stimulated scattering processes by taking into account the imaginary part of the dielectric permittivities which occur on the right-hand side of Eq.(8.22). We must mention that on the right-hand side of Eq.(8.22) resonance is possible when the dielectric permittivity with argument $\mathbf{k} - \mathbf{k}'$ vanishes,

$$\varepsilon_{\mathbf{k}-\mathbf{k}'} = \varepsilon_{\mathbf{k}-\mathbf{k}',\omega_{\mathbf{k}}^{\ell}-\omega_{\mathbf{k}'}^{\ell}} \simeq 0. \tag{8.23}$$

In order that the right-hand side contains a factor zero, the frequency in the last expression must be close to a plasma frequency; in the present case it can only be an ion-sound mode $\omega_{\mathbf{k}-\mathbf{k}'}^{\mathrm{s}}$ so that we get

$$\omega_{\mathbf{k}}^{\ell} = \omega_{\mathbf{k}'}^{\ell} + \omega_{\mathbf{k}-\mathbf{k}'}^{\mathrm{s}}. \tag{8.24}$$

This is nothing but the decay law. In considering the stimulated scattering laws we did not neglect it, but in the case of resonance the scattering probability has no meaning and we must in that case calculate the imaginary part of (8.23) differently – in fact, more simply. In the presence of resonance we can use the following expression for the imaginary part of $1/\varepsilon_{\mathbf{k}-\mathbf{k}'}$:

$$\mathrm{Im}\left\{\frac{1}{\varepsilon_{\mathbf{k}-\mathbf{k}',\omega_{\mathbf{k}}^{\ell}-\omega_{\mathbf{k}'}^{\ell}}}\right\} = -\pi \frac{\omega_{\mathbf{k}}^{\ell}-\omega_{\mathbf{k}'}^{\ell}}{|\omega_{\mathbf{k}}^{\ell}-\omega_{\mathbf{k}'}^{\ell}|}\delta\left(\varepsilon_{\mathbf{k}-\mathbf{k}',\omega_{\mathbf{k}}^{\ell}-\omega_{\mathbf{k}'}^{\ell}}\right)$$

$$= -\left[\delta\left(\omega_{\mathbf{k}}^{\ell}-\omega_{\mathbf{k}'}^{\ell}-\omega_{\mathbf{k}-\mathbf{k}'}^{\mathrm{s}}\right)-\delta\left(\omega_{\mathbf{k}}^{\ell}-\omega_{\mathbf{k}'}^{\ell}+\omega_{\mathbf{k}'-\mathbf{k}}^{\mathrm{s}}\right)\right]$$

$$\times \pi \left[\frac{\partial \varepsilon_{\mathbf{k}-\mathbf{k}',\omega}}{\partial \omega}\bigg|_{\omega=\omega_{\mathbf{k}-\mathbf{k}'}^{\mathrm{s}}}\right]^{-1}. \tag{8.25}$$

The two terms in this expression correspond to the two terms of Eq.(8.21). The use of Eq.(8.25) implies that one can neglect the imaginary parts of the dielectric permittivities in comparison with their real parts. One can therefore replace $\varepsilon^{(i)}_{k-k'}$ in the numerator in Eq.(8.22) by $-(\varepsilon^{(e)}_{k-k'} - 1)$. Comparing Eq.(8.22) with Eq.(8.21) we find the following expression for the probability:

$$w^{\ell,s}_{\ell,k}(k-k') = \frac{8\pi^2 e^2 |k-k'|^2 (k \cdot k')^2}{m_e^2 \omega_{pe}^4 k^2 k'^2} \delta(\omega^\ell_k - \omega^\ell_{k'} - \omega^s_{k-k'}) \left[\varepsilon^{(e)}_{k-k', \omega^\ell_k - \omega^\ell_{k'}}\right]^2$$

$$\times \left[\frac{\partial \varepsilon_k}{\partial \omega}\right]^{-1}_{\omega=\omega^\ell_k} \left[\frac{\partial \varepsilon_{k'}}{\partial \omega'}\right]^{-1}_{\omega'=\omega^\ell_{k'}} \left[\frac{\partial \varepsilon_{k-k'}}{\partial(\omega-\omega')}\right]^{-1}_{\omega-\omega'=\omega^s_{k-k'}}. \quad (8.26)$$

We can simplify this expression by using approximate expressions for the derivative of the permittivity which are $\partial\varepsilon/\partial\omega \simeq 2/\omega_{pe}$ in the case of Langmuir waves and $\partial\varepsilon/\partial\omega \simeq 2\omega_{pi}^2/(\omega^s_{k-k'})^3$ in the case of ion-sound waves and at the same time using the Debye screening approximation for the electron permittivity:

$$w^{\ell,s}_{\ell,k}(k - k') = \frac{\pi^2 e^2 \omega^s_{k-k'}}{m_e T_e} \frac{(k \cdot k')^2}{k^2 k'^2} \delta(\omega^\ell_k - \omega^\ell_{k'} - \omega^s_{k-k'}). \quad (8.27)$$

We have also used the fact here that the wavenumbers of the ion-sound waves involved in the decay process satisfy the inequality $k \ll 1/d_e$. Returning now to the notation used at the start of the present chapter when the initial Langmuir wave has a momentum k' and the emitted ion-sound wave a momentum k we have

$$w^{\ell,s}_{\ell,k'}(k) = \frac{\pi \omega_{pe}^2 \omega^s_k}{4 n_e T_e} \frac{(k' \cdot [k' - k])^2}{k'^2 |k' - k|^2} \delta(\omega^\ell_{k'} - \omega^\ell_{k'-k} - \omega^s_k). \quad (8.28)$$

The energy conservation law for an elementary decay process can be written in the form – the plasma frequencies on the two sides of the equation cancel –

$$\frac{3k'^2 v_{Te}^2}{2\omega_{pe}} = \frac{3|k' - k|^2 v_{Te}^2}{2\omega_{pe}} + kv_s, \quad (8.29)$$

or

$$\cos\theta \equiv \frac{(k \cdot k')}{kk'} = \frac{k + 2k_*}{2k'} \leqslant 1, \qquad k_* = \frac{1}{3d_e}\sqrt{\frac{m_e}{m_i}}, \quad (8.30)$$

whence follows that

$$k' \geqslant k_*. \quad (8.31)$$

The quantity k_* corresponds to the one introduced earlier as a step in the energy transfer in the stimulated scattering process if in the latter case one substitutes $T_e = T_i$. We must bear in mind that ion-sound oscillations exist only for $T_e \gg T_i$. The region where both decay and scattering processes are possible and where one should restrict oneself to the decay processes is defined by the conditions $k \ll 1/d_e$, $k_* \leqslant k' \ll 1/d_e$. However, the decay conditions are the more rigid ones and one can find regions of k and k' where only scattering is possible. Of course, scattering is the only possible process when $T_e = T_i$. If $T_e \gg T_i$, decay processes with k values for which $k \leqslant k_*$ occur with a step k_*. Indeed, we see from Eq.(8.29) that for angles of the order of unity the difference in wavenumber of the two Langmuir waves is of the order of k_*. This enables us to use the same notation for this quantity as was used in earlier chapters.

8.4 Decay Processes and the Turbulent Spectrum

Decay processes produce a connection between the two kinds of turbulence which we have discussed earlier – Langmuir turbulence and ion-sound turbulence. Let us first consider the case when there is a source exciting Langmuir turbulence. The question now is how the decay processes will alter the spectrum of the Langmuir turbulence and what will be the nature of the ion-sound turbulence accompanying the Langmuir turbulence. Let us assume that the ion-sound turbulence level is low and that the Langmuir turbulence can be described by the equation $\gamma_{\mathbf{k}}^{\mathrm{NL\,decay}} = 0$ where $\gamma_{\mathbf{k}}^{\mathrm{NL\,decay}}$ is given by Eq.(8.22). If we use the equation for the probability which we found a moment ago we can write this equation in the form

$$\gamma_{\mathbf{k}}^{\mathrm{NL\,decay}} = \frac{\pi \omega_{\mathrm{pe}}^2}{8 n_e T_e} \int \frac{(\mathbf{k} \cdot \mathbf{k}')^2}{k^2 k'^2} \, \omega_{\mathbf{k}-\mathbf{k}'}^{\mathrm{s}} N_{\mathbf{k}'}^{\ell} \, \frac{d^3 k'}{(2\pi)^3}$$
$$\times \left\{ \delta\!\left(\omega_{\mathbf{k}}^{\ell} - \omega_{\mathbf{k}'}^{\ell} + \omega_{\mathbf{k}-\mathbf{k}'}^{\mathrm{s}}\right) - \delta\!\left(\omega_{\mathbf{k}}^{\ell} - \omega_{\mathbf{k}'}^{\ell} - \omega_{\mathbf{k}-\mathbf{k}'}^{\mathrm{s}}\right) \right\}. \quad (8.32)$$

Let us now use assumptions like the ones we used in the derivation of the Langmuir spectrum for stimulated scattering processes: (1) we assume that the turbulence is isotropic, and (2) we assume that the energy transfer step k_* is infinitesimal. The calculations are then similar to those for the case of stimulated scattering. We expand the δ-functions in terms of a small quantity – the frequency of the ion-sound oscillations – and we average the result over the angle. We change the derivative of the δ-function with respect to its argument to $-(\omega_{\mathrm{pe}}/3 v_{Te}^2 k')\partial/\partial k'$ and we then find

$$\gamma_{\mathbf{k}}^{\mathrm{NL\,decay}} = -\frac{\pi \omega_{\mathrm{pe}}^2}{12 m_i n_i v_{Te}^2} \int \frac{k^2 + k'^2}{k'} \frac{\partial}{\partial k'} \delta\!\left(\frac{3 v_{Te}^2 (k'^2 - k^2)}{2 \omega_{\mathrm{pe}}} \right) W_k^{\ell} \, dk$$
$$= \frac{\pi \omega_{\mathrm{pe}}^3}{54 m_i n_i v_{Te}^4} \frac{\partial W_k^{\ell}}{\partial k} \quad (8.33)$$

Here W_k^{ℓ} is the spectral energy density of the Langmuir turbulence:

$$W^{\ell} = \int \omega_{\mathrm{pe}} N_k^{\ell} \frac{d^3 k}{(2\pi)^3} = \int W_k^{\ell} \, dk. \quad (8.34)$$

The result (8.33) is practically the same as Eq.(7.18) which we obtained earlier for stimulated scattering processes. Equation (7.18) is "more" approximate since we dropped in it a factor $(1 + ZT_e/T_i)^2$ in the denominator (see Eq.(7.13)). Let us take this factor into account when comparing scattering and decay processes. We then see that when $ZT_e \simeq T_i$ scattering processes give a result which is only four times smaller than the result for decays when $ZT_e \gg T_i$ when the scattering should be negligible which is, indeed, the case because of the additional factor $(T_i/ZT_e)^2$ in Eq.(7.18). However, in the presence of decays scattering does not have a precise meaning as it corresponds to decay on the tail of a resonance. We find thus that in both cases, when $ZT_e \gg T_i$ and when $ZT_e \simeq T_i$, the results are very close to each other; this shows that the broadening of the resonance curve does not greatly alter the final result. In fact, the result depends on an integral over the resonance curve so that it is not sensitive to its width. Even if there is no resonance the result is almost the same, as we saw earlier.

Let us now consider what will be the spectrum of the ion-sound turbulence which accompanies the Langmuir turbulence. In the equation for the ion-sound waves we take into account both the decay processes (8.14) and the Landau damping $\gamma_{\mathbf{k}}^{\mathrm{s}}$:

$$\frac{dN_{\mathbf{k}}^{\mathrm{s}}}{dt} = \gamma_{\mathbf{k}}^{\mathrm{s}} N_{\mathbf{k}}^{\mathrm{s}} + \int w_{\ell,\mathbf{k}'}^{\ell,\mathrm{s}}(\mathbf{k}) \left[N_{\mathbf{k}}^{\mathrm{s}} \left(N_{\mathbf{k}'}^{\ell} - N_{\mathbf{k}'-\mathbf{k}}^{\ell} \right) + N_{\mathbf{k}'}^{\ell} N_{\mathbf{k}'-\mathbf{k}}^{\ell} \right] \frac{d^3 k'}{(2\pi)^3}. \quad (8.35)$$

This equation can be written in the form

$$\frac{dN_{\mathbf{k}}^{\mathrm{s}}}{dt} = 2 \left(\gamma_{\mathbf{k}}^{\mathrm{s}} + \gamma_{\mathbf{k}}^{\mathrm{NL\,s,decay}} \right) N_{\mathbf{k}}^{\mathrm{s}} + Q_{\mathbf{k}}^{\mathrm{NL}}, \quad (8.36)$$

where

$$\gamma_{\mathbf{k}}^{\mathrm{NL\,s,decay}} = \tfrac{1}{2} \int w_{\ell,\mathbf{k}'}^{\ell,\mathrm{s}}(\mathbf{k}) \left[N_{\mathbf{k}'}^{\ell} - N_{\mathbf{k}'-\mathbf{k}}^{\ell} \right] \frac{d^3 k'}{(2\pi)^3}, \quad (8.37)$$

$$Q_{\mathbf{k}}^{\mathrm{NL}} = \int w_{\ell,\mathbf{k}'}^{\ell,\mathrm{s}}(\mathbf{k}) N_{\mathbf{k}'}^{\ell} N_{\mathbf{k}'-\mathbf{k}}^{\ell} \frac{d^3 k'}{(2\pi)^3}. \quad (8.38)$$

Expression (8.37) is similar to the one for Landau damping but the rôle of the particles is now played by the Langmuir waves. The quantity $Q_{\mathbf{k}}^{\mathrm{NL}}$ plays the rôle of an ion-sound turbulence source. In the case of an isotropic distribution of Langmuir waves the non-linear damping is always a damping rate, rather than a growth rate, and Eq.(8.36) therefore always has the stationary solution

$$N_{\mathbf{k}}^{\mathrm{s}} = -\frac{Q_{\mathbf{k}}^{\mathrm{NL}}}{2 \left(\gamma_{\mathbf{k}}^{\mathrm{s}} + \gamma_{\mathbf{k}}^{\mathrm{NL\,s,decay}} \right)}. \quad (8.39)$$

The energy level of the ion-sound waves in this case turns out to be much smaller than in the case when the ion-sound wave is directly excited, for instance, by a current. For comparable values of k and k' the inequality

$N_{\mathbf{k}}^{s} \ll N_{\mathbf{k}'}^{\ell}$ seems to be satisfied which confirms our initial assumptions when we neglected the ion-sound waves when considering the Langmuir wave spectrum. We leave the problem of determining the actual ion-sound turbulence spectrum as an exercise.

In the opposite case when there exists a primary source for ion-sound turbulence and the Langmuir turbulence is the one accompanying the ion-sound turbulence we must bear in mind that decay processes cannot change the total number of Langmuir waves. Therefore, the ion-sound turbulence can only redistribute the Langmuir waves which are already present – this is not the case when there is a plasma maser; see Chap.11. In the case when the ion-sound turbulence energy is concentrated in the region of small wavenumbers – bear in mind that $W_k^s \propto 1/k$ is a typical ion-sound turbulence spectrum – and the Langmuir waves have wavevectors much larger than those of the ion-sound waves we find that the equation for the Langmuir waves is a diffusion equation similar to the case of the quasi-linear equation for wave-particle interactions:

$$\frac{dN_{\mathbf{k}'}^{\ell}}{dt} = \frac{\partial}{\partial k_i'} D_{i,j}^{\mathrm{NL}} \frac{\partial}{\partial k_j'} N_{\mathbf{k}'}^{\ell}, \quad D_{i,j}^{\mathrm{NL}} = \int k_i k_j w_{\ell,\mathbf{k}'}^{\ell,s}(\mathbf{k}) N_{\mathbf{k}}^{s} \frac{d^3\mathbf{k}}{(2\pi)^3}. \tag{8.40}$$

If there is no direct Langmuir turbulence source the stationary solution of Eq.(8.40) should be one without a flux in the momentum phase space of the waves and we have thus: $N_{\mathbf{k}'}^{\ell} = $ const. In that case the Langmuir turbulence spectrum will be proportional to k^2:

$$W_k^{\ell} = \frac{\omega_{\mathrm{pe}} k^2 N_k^{\ell}}{2\pi^2} \propto k^2. \tag{8.41}$$

8.5 Decay Processes and Plasma Diagnostics

Decay processes can be used for the diagnostics of plasma parameters without disturbing the state of the plasma. This includes measurements of the correlation functions of the turbulent fields. Probe measurements usually disturb the state of the plasma in the vicinity of the probe and need a detailed theoretical interpretation whereas decay processes can be used for the case when the perturbations by an incident signal have to be rather small. Another advantage of the use of decay processes for diagnostics is that they provide the possibility to perform local measurements and during short time intervals. Even if decay processes produce a perturbation it is much easier than in the case of probe measurements to recalculate the parameters as they were before the measurement. The possibility to have almost instantaneous measurements at different positions in the plasma is important from the general point of view that a turbulent state is a chaotic state. If one uses probe measurements one needs a great many probes or one needs to assume that the state of the plasma is practically stationary.

For diagnostics one can use a decay process in which both a turbulent mode (σ) and an electromagnetic, transverse mode (t) are involved. In practice the latter can be either a radio wave or a laser wave. Laser radiation is used in laboratory and especially for measurements of turbulence in tokamaks. Radio waves are used in the diagnostics of the ionosphere and in active and passive cosmic experiments as well as in laboratory experiments. The decay processes we are considering are the following:

$$t' \rightleftarrows t + \sigma. \tag{8.42}$$

where t is the incident signal and t' the scattered signal used to find the plasma parameters. The incident can be a small-amplitude wave in order not to disturb the state of the plasma. The secondary wave will then be even much weaker, but it will be quite possible to detect it and it contains a wealth of information about the state of the plasma. In the case of Langmuir turbulence ($\sigma = \ell$) the secondary waves will have a spectrum displaced by the plasma frequency. This can be used for the local diagnostics of the plasma density. If the frequency displacement is of the order of the ion plasma frequency or less, one expects the presence of ion-sound turbulence. However, one can obtain more information since the angular distribution of the scattered signal gives information both about the dispersion curve of the emitted mode and about the distribution of the energy between the various harmonics. One can thus diagnose the type of the mode which is excited in the system and measure the turbulence spectrum.

There are many examples of this kind of measurement. Laser diagnostics has been used for measurements of turbulence spectra on the front of collisionless shock waves and it proved the existence of ion-sound turbulence with a spectrum which is the same as the Kadomtsev spectrum.

Another example is the identification of the turbulent plasma modes which are excited in the active cosmic experiment when barium clouds are injected in the magnetosphere. The measurements proved the existence of ion-sound oscillations and other types of magnetic oscillations in the magnetic cavity produced by the expanding plasma.

The third example is the diagnostics of turbulence in tokamaks. Since in the decay processes the difference of the wavevectors of the incident and the scattered waves is equal to the wavevector of the turbulent mode the scattering angle may be small if the wavevector of the excited mode is not very large. In order to resolve the scattered wave one must therefore have a frequency of the incident wave which is not much higher than the plasma frequency. This should be accompanied by a good spatial resolution and it is therefore desirable to use lasers for this purpose. These constraints necessitated the development of special low-frequency lasers for the diagnostics of turbulence in tokamaks. Diagnostics using decay processes has been strongly developed for experiments involving the interaction of powerful lasers with matter in which case strong Langmuir and ion-sound turbulence is excited.

A last example we wish to mention is the scattering of radio waves by the ionosphere. In experiments of this kind one observed not only frequency satellites displaced by the plasma frequency in the scattered signal, but also a great deal of scattering with low frequency shifts in which one could not distinguish any special plasma modes. This apparently indicated that the plasma was in a strongly turbulent state.

8.6 Scattering in a Strongly Turbulent Plasma

The methods of diagnostics described in the preceding section were based on decay processes and therefore on the idea of weak turbulence. Let us now generalise this approach to cover the case of strong turbulence when there is no definite relation between the frequency and the wavenumber or, in other words, when the non-linear frequency broadening of the turbulent modes is of the order of the frequency. We shall consider here the case when both the incident and the scattered waves are longitudinal modes – since all discussions are here restricted to longitudinal waves – and we shall assume that the frequency of the turbulent mode is much lower than that of the incident wave. In the case of scattering by the ionosphere which we mentioned earlier we are interested in the case when both the incident and the scattered waves are transverse. It is not difficult to generalise the discussion we give here to the case of transverse waves since the only point which must be taken into account is the difference in polarisation of the longitudinal and the transverse waves. The advantage of considering both the incident and the scattered wave to be longitudial is that we can use the Poisson equation to describe them. Let us to a first approximation neglect that the dielectric permittivity depends on the wavenumber and write

$$\operatorname{div}\left\{\left[1 - \frac{4\pi n_e e^2}{m_e \omega^2}\right] \mathbf{E}\right\} = 0; \tag{8.43}$$

We now substitute into that equation $n_e = n_{e0} + \delta n$, $\omega_{pe}^2 = 4\pi n_{e0}/m_e$, where δn will describe the low-frequency oscillations due to turbulence. We define the plasma frequency as the value it would have if there were no turbulence. In our discussion we assume that the detecting signal is a Langmuir wave. The frequency of the turbulent mode is much lower than the plasma frequency and the turbulence will therefore change the properties of the detecting wave adiabatically. We put the term involving δn, which describes the interaction with the turbulence, on the right-hand side and in it we substitute $\omega = \omega_{pe}$. We denote the plasma permittivity when there is no turbulence by ε_k. We put on the left-hand side the total expression for the dielectric permittivity including the thermal effects since for the incident wave the thermal effects which lead to a wavenumber dependence of the permittivity are weak. We can also neglect the thermal effects on the right-hand side of the equation. We thus write

$$\varepsilon_k E_k = \frac{1}{n_{e0}k} (\mathbf{k} \cdot [\delta n \mathbf{E}]_k) = \frac{1}{n_{e0}k} \int \delta n_{\mathbf{k}_1} (\mathbf{k} \cdot \mathbf{E}_{\mathbf{k}-\mathbf{k}_1}) \, d^4 k_1. \qquad (8.44)$$

Due to the turbulence there will appear a fluctuating part in the electric field $\mathbf{E}$ which contains information about the turbulence. In our calculations we did not assume that there was a definite relation between the frequency and the wavevector in the turbulence and δn can therefore describe strong turbulence. One can easily show that it is possible to obtain from Eq.(8.44) the probability for the decay of a Langmuir wave into another Langmuir wave and an ion-sound wave in the case when the turbulence is weak. We shall show how one can use Eq.(8.44) to find the growth rate for the case of strong turbulence. From that expression one can in the limit of weak turbulence derive the growth rate which we used earlier and therefore one can find the probability for a decay process.

We average Eq.(8.44) over the fluctuations. On the right-hand side of the averaged equation we then find only the fluctuating part δE_k of the electric field which can be found from the same equation before it is averaged, in which case the right-hand side contains the non-fluctuating part of the field:

$$\varepsilon_{k-k_1} \delta E_{k-k_1} = \frac{1}{n_{e0}|\mathbf{k} - \mathbf{k}_1|} \int \delta n_{\mathbf{k}_2} ([\mathbf{k} - \mathbf{k}_1] \cdot \mathbf{E}_{\mathbf{k}-\mathbf{k}_1-\mathbf{k}_2}) \, d^4 k_2. \qquad (8.45)$$

We shall describe the density fluctuations through the correlation $|\delta n|^2_k$, where $k \equiv \{\mathbf{k}, \omega\}$ and

$$\langle \delta n_{\mathbf{k}_1} \delta n_{\mathbf{k}_2} \rangle = |\delta n|^2_{\mathbf{k}_1} \, \delta(\mathbf{k}_1 + \mathbf{k}_2). \qquad (8.46)$$

There is here still no definite relation between the frequency and the wavevector. Substituting Eq.(8.45) into Eq.(8.44), averaging and using Eq.(8.46) we get

$$\left(\varepsilon_k + \varepsilon_k^{NL}\right) E_k = 0, \quad \varepsilon_k^{NL} = -\int \frac{(\mathbf{k} \cdot [\mathbf{k} - \mathbf{k}_1])^2}{k^2 |\mathbf{k} - \mathbf{k}_1|^2 n_{e0}^2 \varepsilon_{k-k_1}} |\delta n|^2_{\mathbf{k}_1} \, d^4 k_1. \qquad (8.47)$$

from which we find the non-linear growth rate

$$\gamma_k^{NL} = \left[\text{Im}\{\varepsilon_k^{NL}\} \Big/ \frac{\partial \varepsilon_k}{\partial \omega} \right]_{\omega=\omega_k}$$

$$= \left[\pi \int \frac{(\mathbf{k} \cdot [\mathbf{k} - \mathbf{k}_1])^2 \, d^4 k_1}{k^2 |\mathbf{k} - \mathbf{k}_1|^2 n_{e0}^2} |\delta n|^2_{\mathbf{k}_1} \delta(\varepsilon_{k-k_1}) \Big/ \frac{\partial \varepsilon_k}{\partial \omega} \right]_{\omega=\omega_k}. \qquad (8.48)$$

We obtain the weak turbulence result by substituting into Eq.(8.48) the density fluctuations due to the ion-sound waves which have a definite relation between the frequency and the wavevector. We leave the derivation of the decay probability from the weak turbulence expression as an exercise. Equation (8.48) is valid for strong turbulence. We shall widely use this kind of description in later chapters.

8.7 Decay Processes and Emission in a Turbulent Plasma. Interaction of Strong Radiation with a Plasma

Decay processes play an important rôle in the emission from a turbulent plasma. As turbulence always generates fast particles the emission from a turbulent plasma may also be due to those fast particles. Their number is relatively small and they usually produce emission at frequencies which are very different from those close to the frequencies of the turbulent modes. The decay processes are determined by the thermal particles or, more precisely, by the main part of the particle distribution – in a turbulent state the particle distribution may differ from a thermal distribution. The emission due to the decay processes is therefore usually rather strong and occurs at frequencies close to those of the turbulent modes. This means that the two kinds of turbulent emission – emission produced by the fast particles and emission produced by the turbulence itself – are usually well separated in frequency.

We shall list here the various decay processes which may be responsible for the emission from a turbulent plasma in the case when there is no external magnetic field present – the case in which we are mainly interested. This means that we are considering turbulence corresponding to Langmuir oscillations, to ion-sound oscillations, or to low-frequency strong turbulent oscillations. The emitted electromagnetic waves will have frequencies which are larger than the plasma frequency and therefore there are only two possible emission processes:

$$\ell + s \ \rightarrow \ t, \tag{8.49}$$

$$\ell + \ell' \ \rightarrow \ t. \tag{8.50}$$

The first process produces emission at frequencies close to the plasma frequency; a strong turbulent mode instead of an ion-sound wave can also contribute to this kind of emission. The second process produces emission at frequencies close to twice the plasma frequency.

The above-mentioned processes have been observed in many laboratory experiments, in beam-plasma interactions, in laser-plasma interactions, and in experiments in which there was a strong electric field present in the plasma. The emission can be used for plasma diagnostics. It has also been observed in turbulent plasma heating experiments. In the near space this kind of emission has been observed in the polar regions of the magnetosphere; it is connected with the turbulence excited by the fast particles impinging on the upper atmosphere which create the aurora. The radiation is known as the *Earth's kilometric radiation*.

In the solar atmosphere solar flares create shock waves with very turbulent fronts which emit radiation at the plasma frequency and twice the plasma frequency. This radiation shows frequency drift because the shock waves propagate into regions where the plasma has a lower density. The observed drift corresponds to propagation at a velocity several times larger than the sound velocity. This radiation is known as the *sporadic solar radio emission of type II bursts*. We discussed earlier that particle beams create type III solar bursts. These also produce emission at the plasma frequency and at twice the plasma frequency. Decay processes also produce emission from all planets with a magnetosphere. Here the turbulence excited by the interaction of the solar wind with the boundary of the magnetisphere is important. There is also a boundary between our heliosphere and the interstellar plasma which emits at twice the plasma frequency – an effect observed recently.

The interaction between strong radiation and a plasma is related to processes which are the inverse of those connected with the emission processes. Various decay processes are important when strong laser radiation interacts with a plasma or with other matter when a plasma corona is formed at the surface of the irradiated matter. We note, first of all, that the plasma corona is usually turbulent with a high level of Langmuir and ion-sound turbulence present. The mechanism for the excitation of this turbulence is described by the decay processes. In those regions where the frequency of the radiation is much higher than the plasma frequency the following processes may be responsible for the excitation:

$$t \rightarrow t' + \ell, \tag{8.51}$$

$$t \rightarrow t' + s, \tag{8.52}$$

From the discussion of solar bursts which we gave a moment ago we see the analogy between the emission of waves by particles and the emission of waves by waves; this enables us to state that the mechanism for the excitation of turbulence by a laser beam is very similar to the mechanism of the excitation of turbulence by a beam of particles. One can have a description for laser beams which is similar to the quasi-linear description of particle beams. In the case of laser beams the dynamics are determined by the probabilities for the decay processes (8.51) and (8.52). As in the case of particle beams the relaxation of laser beams can be non-linearly stabilised; here the stabilisation process may be another decay process:

$$\ell \rightleftarrows \ell' + s. \tag{8.53}$$

One can show that the effect produced by this process differs from that of stimulated scattering only in a numerical coefficient; the process (8.53) can therefore stabilise both the particle beam relaxation and the electromagnetic radiation beam relaxation. The latter are described by the resonance condition (8.5) in the case of ion-sound waves and by a similar resonance condition in the case of Langmuir waves when the frequency is equal to the Langmuir frequency; the group velocity in these relations is close to the velocity of light. These beams are therefore similar to relativistic beams and excite Langmuir waves with wavelengths close to c/ω_{pe}. The process (8.51) is allowed for frequencies of the electromagnetic waves higher than twice the plasma frequency whereas the process (8.52) can work down to frequencies close to the plasma frequency. The process

$$t \rightleftarrows \ell + \ell' \tag{8.54}$$

is responsible for the non-linear absorption of radiation close to twice the plasma frequency and the process

$$t \rightarrow \ell + s \tag{8.55}$$

is responsible for non-linear absorption at the plasma frequency. Both are efficient mechanisms for the non-linear absorption of electromagnetic radiation by a plasma. In a dense plasma a normal acoustic wave may appear in the last process; its collision frequency is larger than the frequency of the wave. Such a wave – denoted by "a" – has no Landau damping but is damped by collision processes such as thermal conductivity or viscosity.

Possible decay processes involving acoustic waves are

$$t \rightarrow t' + a, \tag{8.56}$$
$$t \rightarrow \ell + a, \tag{8.57}$$

All these effects lead to a rather high level of turbulent fluctuations in the plasma corona. The electromagnetic radiation which arrives subsequently encounters a turbulent plasma. The decay processes in the dilute turbulent plasma will scatter and reflect the incident radiation before it reaches the dense plasma. The decay processes (8.51), (8.52), and (8.55) will produce efficient *back-scattering* of the incident radiation since the probability has a maximum for back-scattering. This produces large non-linear reflection coefficients. Scattering by Langmuir turbulence is called *Brillouin scattering* and scattering by ion-sound turbulence is called *Raman scattering*. Both types of scattering can be efficient even when the frequency of the radiation is much higher than the plasma frequency.

Langmuir waves can be excited directly by the incident radiation if its frequencies lie in a broad band and the variation of the frequency can be equal to the plasma frequency. The best way to accomplish this is to have two beams of radiation with frequencies which differ by an amount close to the plasma frequency. Such plasma waves can be used for particle acceleration and such accelerators are called *laser accelerators*. These kinds of experiments are carried out in many laboratories.

Problems

1. Use Eq.(8.47) for the non-linear permittivity in the presence of low-frequency density fluctuations to obtain the probability for the decay of a Langmuir wave into another Langmuir wave and an ion-sound wave.

2. Apply the method used for the derivation of Eq.(8.47) to derive the non-linear permittivity of electromagnetic waves if there are low-frequency density fluctuations present.

3. Use Eq.(8.39) to find the spectrum of the ion-sound waves which accompany Langmuir turbulence for the case when the characteristic wavenumbers of the ion-sound waves are much larger than those of the Langmuir waves and also for the case when they are much smaller.

4. Show that the probability for the $t \rightleftarrows t' + \ell$ decay process is given by the equation

$$w_{t,\mathbf{k}'}^{t,\ell}(\mathbf{k}) = \frac{\pi \omega_{pe}^3 k^2}{8 m_e n_e \omega_{\mathbf{k}'}^t \omega_{\mathbf{k}'-\mathbf{k}}^t} \left\{ 1 + \frac{(\mathbf{k}' \cdot [\mathbf{k}'-\mathbf{k}])^2}{k'^2 |\mathbf{k}'-\mathbf{k}|^2} \right\} \delta\left(\omega_{\mathbf{k}'}^t - \omega_{\mathbf{k}'-\mathbf{k}}^t - \omega_{\mathbf{k}}^\ell\right).$$

5. Show that the probability for the $t \rightleftarrows t' + s$ decay process is given by the equation

$$w_{t,\mathbf{k}'}^{t,s}(\mathbf{k}) = \frac{\pi \omega_{pe}^4 \omega_{\mathbf{k}}^s}{8 n_e T_e \omega_{\mathbf{k}'}^t \omega_{\mathbf{k}'-\mathbf{k}}^t} \left\{ 1 + \frac{(\mathbf{k}' \cdot [\mathbf{k}'-\mathbf{k}])^2}{k'^2 |\mathbf{k}'-\mathbf{k}|^2} \right\} \delta\left(\omega_{\mathbf{k}'}^t - \omega_{\mathbf{k}'-\mathbf{k}}^t - \omega_{\mathbf{k}}^s\right).$$

6. Show that the probability for the $t \rightleftarrows \ell + s$ decay process is given by the equation

$$w_{t,\mathbf{k}'}^{\ell,s}(\mathbf{k}) = \frac{\pi \omega_{pe}^3 \omega_{\mathbf{k}}^s}{4 n_e T_e \omega_{\mathbf{k}'}^t} \frac{[\mathbf{k} \wedge \mathbf{k}']^2}{k'^2 |\mathbf{k}'-\mathbf{k}|^2} \delta\left(\omega_{\mathbf{k}'}^t - \omega_{\mathbf{k}'-\mathbf{k}}^\ell - \omega_{\mathbf{k}}^s\right).$$

7. Show that the probability for the $\ell + \ell' \rightleftarrows t$ decay–fusion–process is given by the equation

$$w_{t,\mathbf{k}'}^{\ell,\ell}(\mathbf{k}) = \frac{\pi \omega_{pe} [|\mathbf{k}'-\mathbf{k}|^2 - k^2] [\mathbf{k}' \wedge \mathbf{k}]^2}{16 m_e n_e k'^2 k^2 |\mathbf{k}'-\mathbf{k}|^2} \delta\left(\omega_{\mathbf{k}'}^t - \omega_{\mathbf{k}'-\mathbf{k}}^\ell - \omega_{\mathbf{k}}^\ell\right).$$

8. Apply the method used for the derivation of Eq.(8.47) to find the non-linear permittivity and the probability for a decay process of electrostatic oscillations for the case of an arbitrary effective cubic non-linearity.

9 Non-linear Modulational Interactions

9.1 Non-linear Frequency Shift and Modulational Interactions

Non-linear modulational interactions occur when the energy input into the system is sufficiently large and the frequency broadening due to the non-linear interactions is larger than the difference in the frequencies of the interacting waves. The physics of this interaction consists in the modulation of the plasma parameters by one of the interacting waves which leads to the above-mentioned frequency broadening and to interactions with other waves. The physics of most non-linear interactions is the same – the modulation of the plasma properties by one of the waves and the propagation of other waves through the modulated system. In this sense, the nature of the decay and scattering processes is the same as that of the modulational interactions. The important point for the modulational interactions is that they are sufficiently strong to include the other interacting waves in the non-linear broadening. Earlier, in Chap.5, we have already estimated this broadening using a dimensional analysis and found a critical density of the waves above which the stimulated scattering processes and the decay processes become modulational interactions. We shall now start directly with a calculation of this broadening. The broadening is related to the total non-linear frequency shift, including both its real and its imaginary parts. Earlier we studied the imaginary part in detail – it is due to the non-linear processes we discussed above. For the modulational interactions the real part is important which is larger than the imaginary part and which determines the non-linear broadening.If the non-linear frequency shift is smaller than the wave dispersion which determines the difference in frequency of the interacting waves the interactions are the decay and scattering processes. On the other hand, if the frequency shift is larger than the dispersion the interactions become modulational interactions. The critical energy density separating the two regimes of non-linear interactions is thus determined by the non-linear frequency shift. We shall deal here mainly with Langmuir waves since the modulational interactions are very important in their case. We shall consider the non-linear frequency shift for ion-sound waves in the Appendix to the present chapter.

We can find the non-linear frequency shift of the Langmuir waves from the general expression (6.41) for the non-linear plasma permittivity. Since in the case of a weak non-linearity, which we are considering here, the non-linear frequency shift is much smaller than the plasma frequency ω_{pe} we can in the linear part of the dielectric permittivity put

$$1 - \frac{\omega_{pe}^2}{\omega^2} \cong \frac{2\Delta\omega}{\omega_{pe}},$$

and thus we have

$$\varepsilon_k + \varepsilon_k^{NL} = 0,$$

$$\omega_k^\ell = \omega_{pe} + \frac{3k^2 v_{Te}^2}{2\omega_{pe}} + \Delta\omega_k^{NL},$$

$$\Delta\omega_k^{NL} = -\tfrac{1}{2}\omega_{pe} \operatorname{Re}\left\{\varepsilon_{k,\omega_k^\ell}^{NL}\right\},$$

$$\operatorname{Re}\varepsilon_k^{NL} = \frac{e^2}{m_e^2 \omega_{pe}^4} \int |E^+|_{k'}^2 \frac{(\mathbf{k}\cdot\mathbf{k'})^2}{k^2 k'^2} |\mathbf{k}-\mathbf{k'}|^2$$

$$\times \operatorname{Re}\left[\frac{(\varepsilon_{k-k'}^{(e)} - 1)\varepsilon_{k-k'}^{(i)}}{\varepsilon_{k-k'}}\right] d^4 k'. \qquad (9.1)$$

The largest contribution to the frequency shift for a given value of $\mathbf{k}$ comes from those $\mathbf{k'}$ which have absolute magnitudes $|\mathbf{k'}|$ close to $|\mathbf{k}|$, namely with

$$\Delta k = |\mathbf{k}-\mathbf{k'}| < k_* = \sqrt{\frac{m_e T_i}{9 m_i T_e}}.$$

In that case we can use in Eq.(9.1) the approximate Debye screening expression for the linear dielectric permittivities (we put $Z = 1$):

$$\frac{(\varepsilon_{k-k'}^{(e)} - 1)\varepsilon_{k-k'}^{(i)}}{\varepsilon_{k-k'}} \approx \frac{T_e}{T_e + T_i} \frac{\omega_{pe}^2}{|\mathbf{k}-\mathbf{k'}|^2 v_{Te}^2}, \qquad (9.2)$$

and

$$\Delta\omega_k^{NL} = -\omega_{pe} \int \frac{W_{k'}}{4 n_e (T_e + T_i)} \frac{(\mathbf{k}\cdot\mathbf{k'})^2}{k^2 k'^2} d^3 k'. \qquad (9.3)$$

Two points are important in connection with this result. First of all, it confirms the estimate used at the beginning of Chap.5 for the non-linear frequency broadening and it is, indeed, larger than the imaginary part of the non-linear frequency shift. Secondly, it is always negative and thus lowers the energy of the oscillations in those regions of space where the intensity of the oscillations is large. We can therefore conclude that the oscillations will have a tendency to accumulate in those regions of space where their amplitude is largest. This implies a *self-contraction* of the plasma oscillations similar to the self-focussing of radiation which is a well known effect in optics. The

self-contraction means that it will be energetically favourable for a plasma
to distribute the energy of its oscillations non-uniformly. There will then be
regions in the plasma where the intensity of the oscillations will be large and
regions where it will be low. This can occur only if the negative frequency shift
is larger than the dispersion effects which tend to spread out inhomogeneities
in space of the wave distributions. This criterion will determine the critical
energy density which we introduced earlier.

One point must be made clear. The total energy density of the oscillations
enters in the non-linear frequency shift (9.3) when $\delta k < k_*$. In the opposite
case it is not the total energy density of the oscillations but only that part
of it which is concentrated in a range $\Delta k < k_*$ around the given value of k
which determines the frequency shift. This statement follows from Eq.(9.1)
since for $\Delta k > k_*$ the contribution from the other oscillations will for not too
large a value of $\Delta k/k$ decrease as $1/\Delta k^2$ and then for $\Delta k \gtrsim k$ it will decrease
as $1/k^4$. Therefore, if the spectrum has no special peculiar features only the
oscillations with $\Delta k < k_*$ contribute to the frequency shift. For wavenumbers
much larger than k_* there will, roughly speaking, enter a factor of the order
of k_*/k in the expression for the frequency shift, if we denote the total energy
density of the waves by W.

The self-contraction effect is usually accompanied by phase-locking – all
the waves in the accumulation can have synchronised phases. This can lead
to coherent structures. The synchronisation of the phases does not mean that
the phases inside such structures are the same, only that there exist definite
relations between those phases. These structures usually exist in a "sea" of
random waves and the interaction of such structures with the random waves
can lead to self-organisation processes. There is also a correlation of phases
for $W \ll W_{cr}$. Without such correlations the decay and stimulated scattering
interactions would be impossible. However, in the case when $W \gg W_{cr}$ these
correlations become stronger. Even so, only waves with frequencies which are
close to each other are correlated whereas there is no such correlation for
waves with a large frequency difference. The mechanism of the interaction
between waves which are synchronised and waves which are not synchronised
is important for self-organisation problems. In the present chapter we shall
consider only the case of regular waves, assuming that there are no random
waves. In later chapters we shall consider also the properties of dissipative
structures which are energetically supported by random waves.

The initial stage of the self-contraction of waves is usually called a *modulational instability*. In some sense it is similar to the decay instability or the
non-linear Landau damping, but it occurs in the case when $W \gg W_{cr}$. We
shall now consider qualitatively how it develops. We assume that initially
the energy of the oscillations is distributed uniformly in space (see Fig.9.1).
The plasma density is also uniform. Consider now a fluctuation in the density distribution which decreases it locally. Some oscillations may get trapped
in this density depletion – an effect which is similar to total inner refraction

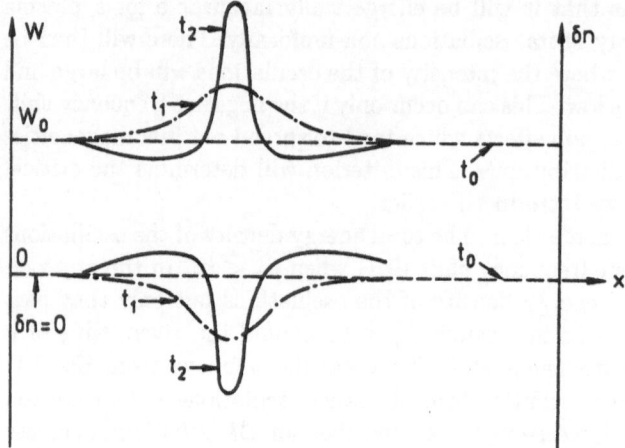

Fig. 9.1 Development of the modulational instability in time; $t_0 < t_1 < t_2$

in optics– but other oscillations will just decrease their group velocity and while they are not trapped they will nevertheless spend more time inside the region of the depletion than before it had appeared. As a result there will be a higher energy density of the oscillations in the density depletion. They will produce a striction force which will try to pull the plasma out of the depletion region. An opposite force will come from the plasma pressure and the dispersive effects. If the striction force dominates– this corresponds to an energy density of the oscillations larger than a critical value– the depletion of the density will become larger. More waves will become trapped and will accumulate in the larger density depletion, the striction force will get stronger, and so on. This is a qualitative picture of the modulational instability. The nature of the modulational instability can be clarified by considering the relation between the so-called hydrodynamic type of linear instability and the kinetic type of linear instability.

9.2 Kinetic and Hydrodynamic Linear Instabilities

The modulational instability is a non-linear instability but it is related to the stimulated scattering and decay instabilities in the same way as a hydrodynamic type of linear instability is related to a kinetic type of linear instability. We shall illustrate this question by the example of the linear beam-plasma instability the growth rate of which was given by Eq.(2.121):

$$\gamma_k^b \simeq \omega_{pe} \frac{n_b}{n_e} \frac{v_b^2}{(\delta v_b)^2}. \tag{9.4}$$

When the spread δv_b in velocities decreases the growth rate increases. To find the limits of applicability of Eq.(9.4) we must bear in mind that it was derived by using the resonance condition $\omega_k = (\mathbf{k} \cdot \mathbf{v})$. The model was a one-dimensional one and the resonance condition was $\omega_{pe} = kv$, that is, each value of the particle velocity corresponded to a definite value of the wavenumber k or to a definite phase velocity ω_{pe}/k. However, due to the instability each value of k corresponds to a growth rate γ_k^b and the resonance should therefore be broadened by an amount $\Delta\omega \simeq \gamma_k^b$. We can therefore only approximately assume that there corresponds a definite velocity to each value of k. This is possible only if the width of the resonance of a single mode is significantly smaller than the dispersion $\delta\omega_b \simeq k\delta v_b$ of the resonance frequencies, that is, we must have $\Delta\omega \ll \delta\omega_b$:

$$\gamma_k^b \ll k\delta v_b \simeq \omega_{pe} \frac{\delta v_b}{v_b}, \tag{9.5}$$

or

$$\frac{\delta v_b}{v_b} \gg \left(\frac{n_b}{n_e}\right)^{1/3}. \tag{9.6}$$

Substituting in the growth rate the minimum possible velocity dispersion we obtain an estimate of the growth rate at the limit of the applicability of Eq.(9.4):

$$\gamma_{k,max}^b \simeq \omega_{pe} \left(\frac{n_b}{n_e}\right)^{1/3}. \tag{9.7}$$

In the opposite limit to (9.6) every particle velocity within the velocity dispersion lies within a single resonance curve and all the particles excite the instability simultaneously–coherently. In the case of incoherent excitation each particle of a given velocity excites its "own" wave. It is obvious that the growth rate in the coherent case will be the maximum possible and the growth rate should therefore be independent of the spread in beam velocities if that spread is small and it should be of the order of (9.7).

In the case of a zero velocity spread the contribution from the beam to the dielectric permittivity is the normal one for a cold plasma with the Doppler effect taken into account. The dispersion equation $\varepsilon_k = 0$ then has the form

$$1 - \frac{\omega_{pe}^2}{\omega^2} - \frac{\omega_{pe}^2}{(\omega - kv_b)^2} \frac{n_b}{n_e} = 0. \tag{9.8}$$

Since $n_b/n_e \ll 1$ the last term only makes a significant contribution for kv_b close to ω which, in turn, should be close to ω_{pe}; we therefore have

$$\left.\begin{aligned} \omega &= \omega_{pe} + \delta\omega, & \delta\omega &\ll \omega_{pe}, \\ \omega &= kv_b + \delta\omega, & \delta\omega &\ll kv_b. \end{aligned}\right\} \tag{9.9}$$

This means that we are assuming that

$$\omega_{pe} \cong kv_b, \tag{9.10}$$

which is just the single-mode resonance condition. We therefore find that

$$\frac{2\delta\omega}{\omega_{pe}} + \frac{\omega_{pe}^2}{(\delta\omega)^2}\frac{n_b}{n_e} = 0. \tag{9.11}$$

This cubic equation has an unstable root:

$$\gamma_k^b = \frac{\sqrt{3}}{2^{4/3}}\omega_{pe}\left[\frac{n_b}{n_e}\right]^{1/3}. \tag{9.12}$$

Equation (9.12) just gives us the exact value of the numerical coefficient in Eq.(9.7). The instability (9.12) is called a *hydrodynamic instability* as it can be obtained from a hydrodynamical description of the beam. The instability (9.5) is called a *kinetic instability*. Equation (9.10) clearly shows that the hydrodynamic instability considers the same process as the kinetic instability, but is dealing with the opposite limit when the instability is so strong that all beam particles will lie inside a single resonance curve. In this sense all modes inside the resonance curve are excited coherently, that is, they are synchronised with one another.

Another example of the same kind of instability is the ion-sound instability in the presence of electron drift which we discussed earlier. If the drift velocity is much larger than the electron thermal velocity, the drifting electrons are cold – their velocity spread is negligible – and they move all with the same velocity u. In the frame of reference in which they are at rest the ions move with a velocity $-u$. In this frame the dispersion equation will be the same as Eq.(9.11) with the ratio of the beam density to the plasma density, n_b/n_e, replaced by the ratio of the ion plasma frequency to the electron plasma frequency, Zm_e/m_i. The growth rate will therefore be

$$\gamma_k = \frac{\sqrt{3}}{2^{4/3}}\omega_{pe}\left[\frac{Zm_e}{m_i}\right]^{1/3}. \tag{9.13}$$

This instability is the so-called *Buneman instability*. The Buneman instability and the ion-sound instability are thus the limits of the same instability of a resonance interaction of drifting electrons with electrostatic plasma perturbations.

The decay and stimulated scattering kinds of non-linear resonances play a similar rôle for the non-linear interactions as the linear resonances play for linear interactions. This is the reason why we needed to estimate the width of the non-linear resonances. The modulational interactions describe the limit of strong interactions when all interacting waves are captured within the width of the non-linear resonance. The critical energy density corresponds to the condition when all interacting waves with frequency differences due to dispersion lie within the resonance curve of the non-linear interactions.

We may expect that if the energy input into the plasma – by laser or particle beams or by other means – is large the modulational interactions will be the dominant ones. We should, however, emphasise that all interactions have the same character and change into one another when the intensity of the waves changes.

9.3 General Theory of Modulational Interactions

The general theory of the modulational interactions follows from the general non-linear equation which we derived in Chap.5. We shall use Eq.(5.55) to write down the non-linear equation:

$$\varepsilon_k E_k^+ = - \int \frac{(\mathbf{k} \cdot \mathbf{k}_1)(\mathbf{k}_2 \cdot \mathbf{k}_3)|\mathbf{k}_2 + \mathbf{k}_3|^2}{4\pi k k_1 k_2 k_3 n_e m_e \omega_{pe}^2}$$

$$\times \frac{\left(\varepsilon_{2+3}^{(e)} - 1\right) \varepsilon_{2+3}^{(i)}}{\varepsilon_{2+3}} E_1^+ E_2^+ E_3^- \, d_{1,2,3}, \tag{9.14}$$

where

$$d_{1,2,3} = d^4k_1 \, d^4k_2 \, d^4k_3 \, \delta\big(\mathbf{k} - \mathbf{k}_1 - \mathbf{k}_2 - \mathbf{k}_3\big),$$
$$E_1 = E_{\mathbf{k}_1}, \qquad E_2 = E_{\mathbf{k}_2}, \qquad E_3 = E_{\mathbf{k}_3}.$$

If we use the Debye screening approximation for ε_{2+3} we get Eq.(5.58). We shall assume here that the particle distribution function can be arbitrary. That is the reason why we use a more exact form of the non-linear equation than we did previously. The two equations give the same results for decay and stimulated scattering processes. We showed in the preceding chapter that the non-linear equation for Langmuir waves can be written in such a form that it describes their propagation in the presence of slow density variations δn (see Eq.(8.44)). In that case it was done as a particular example. Now we shall show that this can be done in a general form in the case of strong interactions when the modulational interactions dominate. To be more precise, one can find effective density variations which serve as the density variations encountered by propagating Langmuir waves. These effective density variations will depend on the wavenumbers of the interacting waves and such a dependence will especially appear in the expressions for the effective temperature. The relation we shall obtain is obviously a generalisation of Eq.(8.44) as the latter was obtained not only for a particular example, but also for the case of a quadratic non-linearity whereas Eq.(9.14) contains also the cubic non-linearity. We must mention that by changing the integration variable k_1 to $k - k_1$ we can write Eq.(8.44) in the form

$$\varepsilon_k E_k^+ = \int \frac{\delta n_{k-k_1}}{n_e} \frac{(\mathbf{k} \cdot \mathbf{k}_1)}{k k_1} E_{k_1}^+ \, d^4 k_1. \tag{9.15}$$

We shall drop the index "0" in the expression for the unperturbed plasma density in what follows. We have in Eq.(9.15) taken into account that there should be a superscript "+" on the right-hand side of Eq.(9.15) as we are dealing with the positive-frequency part of the field on the left-hand side of that equation. We can, indeed, write Eq.(9.14) in the form (9.15) if we define the effective density variations by the equation

$$
\begin{aligned}
\delta n_{\mathbf{k}-\mathbf{k}_1} = & -\int \frac{(\mathbf{k}_2 \cdot \mathbf{k}_3)|\mathbf{k}_2+\mathbf{k}_3|^2}{4\pi k_2 k_3 m_e \omega_{pe}^2} \frac{\left(\varepsilon_{\mathbf{k}_2+\mathbf{k}_3}^{(e)}-1\right)\varepsilon_{\mathbf{k}_2+\mathbf{k}_3}^{(i)}}{\varepsilon_{\mathbf{k}_2+\mathbf{k}_3}} \\
& \times E_{\mathbf{k}_2}^+ E_{\mathbf{k}_3}^- \,\delta\left(\mathbf{k}-\mathbf{k}_1-\mathbf{k}_2-\mathbf{k}_3\right) d^4 k_2\, d^4 k_3.
\end{aligned} \tag{9.16}
$$

Since $\mathbf{k}_2 + \mathbf{k}_3 = \mathbf{k} - \mathbf{k}_1$ we can take the expressions containing the dielectric permittivities outside the integral sign. Writing $\mathbf{k}' = \mathbf{k} - \mathbf{k}_1$ we find

$$
\frac{4\pi m_e \omega_{pe}^2 \varepsilon_{\mathbf{k}'} \delta n_{\mathbf{k}'}}{\left(\varepsilon_{\mathbf{k}'}^{(e)}-1\right)\varepsilon_{\mathbf{k}'}^{(i)} k'^2} = -\int \frac{(\mathbf{k}_2 \cdot \mathbf{k}_3)}{k_2 k_3} E_{\mathbf{k}_2}^+ E_{\mathbf{k}_3}^- \delta\left(\mathbf{k}'-\mathbf{k}_2-\mathbf{k}_3\right) d^4 k_2\, d^4 k_3. \tag{9.17}
$$

One sees easily that the right-hand sides of Eqs.(9.15) and (9.17) are the Fourier components of the quantities $\delta n\, \mathbf{E}^+$ and $(\mathbf{E}^+ \cdot \mathbf{E}^-)$. The set of Eqs.(9.15), (9.17) is equivalent to the initial integral equation (9.14). It is useful to introduce a complex amplitude of the Langmuir field corresponding to the envelope of the Langmuir field. This can be done by splitting off a factor with the Langmuir frequency:

$$
\mathcal{E}(\mathbf{r}, t)\, e^{-i\omega_{pe}t} = \int \frac{\mathbf{k}}{k} E_{\mathbf{k}}^+ e^{-i\omega t + i(\mathbf{k}\cdot\mathbf{r})} d^4 k. \tag{9.18}
$$

The vector $\mathcal{E}$ is a complex, rather than a real, vector describing variations in the amplitude and the phase of the Langmuir field which are slow as compared to the plasma frequency. We shall call it the *amplitude field*. If we use the relations

$$
\mathbf{E}_{-\mathbf{k}} = \mathbf{E}_{\mathbf{k}}^*, \qquad E_{-\mathbf{k}}^{+*} = -E_{\mathbf{k}}^-,
$$

we can express the complex conjugate of the amplitude field,

$$
\begin{aligned}
\mathcal{E}^* &= \int \frac{\mathbf{k}}{k} E_{\mathbf{k}}^{+*} e^{-i\omega_{pe}t + i\omega t - i(\mathbf{k}\cdot\mathbf{r})} d^4 k \\
&= -\int \frac{\mathbf{k}}{k} E_{-\mathbf{k}}^{+*} e^{-i\omega_{pe}t - i\omega t + i(\mathbf{k}\cdot\mathbf{r})} d^4 k
\end{aligned} \tag{9.19}
$$

in terms of the negative-frequency Fourier component of the Langmuir field:

$$
\mathcal{E}^* = \int \frac{\mathbf{k}}{k} E_{\mathbf{k}}^- e^{-i\omega_{pe}t - i\omega t + i(\mathbf{k}\cdot\mathbf{r})} d^4 k. \tag{9.20}
$$

If we now add on the right-hand side of Eq.(9.17) a factor $e^{-i\omega_{pe}t + i\omega_{pe}t} = 1$ we see that it contains the Fourier component of the absolute square of the amplitude field:

$$\delta n_{k'} = -\frac{k'^2 \left(\varepsilon_{k'}^{(e)} - 1\right) \varepsilon_{k'}^{(i)}}{4\pi m_e \omega_{pe}^2 \varepsilon_{k'}} \left[|\mathcal{E}|^2\right]_{k'}. \tag{9.21}$$

It is also possible to write Eq.(9.15) in a form which contains the envelope amplitude field, if we take into account that the right-hand side of Eq.(9.15) is nothing but the Fourier component of the product of δn and $\mathcal{E}$:

$$\varepsilon_{k,\omega-\omega_{pe}}(\mathbf{k}\cdot\mathcal{E}_k) = \frac{\left(\mathbf{k}\cdot[\delta n\,\mathcal{E}]_{k,\omega-\omega_{pe}}\right)}{n_e}. \tag{9.22}$$

To obtain Eq.(9.22) we have used the fact that according to Eq.(9.18) we have

$$\mathcal{E}_k = \frac{\mathbf{k}}{k} E^+_{k,\omega-\omega_{pe}}, \tag{9.23}$$

and changed ω and ω_1 to $\omega - \omega_{pe}$ and $\omega_1 - \omega_{pe}$, respectively.

So far we have only rewritten the general equation and not introduced any additional assumptions. We shall now introduce an approximation corresponding to an appropriate description of the modulational interactions. As in the hydrodynamic kind of description we shall assume that the non-linear interactions are so strong that that it is possible to restrict our considerations to the real part of the plasma response. We shall later consider the imaginary part, using perturbation theory. We remind ourselves that when we discussed decay and stimulated scattering processes we started by taking the imaginary parts into account. Considering only the real part of the responses we shall take into account the whole of the non-linear dynamics since the differences in frequency of the interacting waves can be anything and are not necessarily equal to the frequency differences due to a linear dispersion. Any non-linear broadening is thus included; the only restriction is that we must have $\Delta\omega \ll \omega_{pe}$, that is, that it should be possible to describe the situation by an envelope amplitude field. We shall thus use for the linear permittivities which occur in the non-linear response their real parts and bear in mind that ω in Eq.(9.21) corresponds to the difference in frequency of two interacting Langmuir waves (see Eq.(9.16)):

$$\varepsilon_k^{(i)} \simeq -\frac{\omega_{pi}^2}{\omega^2}, \qquad |\varepsilon_k^{(i)}| \gg 1,$$

$$\varepsilon_k^{(e)} \simeq \frac{\omega_{pe}^2}{k^2 V_{Te}^2}, \qquad |\varepsilon_k^{(e)}| \gg 1, \tag{9.24}$$

where

$$\begin{aligned}
V_{Te}^{-2} &= -\int \frac{m_e}{n_e}\frac{\partial \Phi_p^e}{\partial \varepsilon_p}\frac{d^3p}{(2\pi)^3} \\
&= -\int \frac{1}{v}\frac{\partial \Phi_p^e}{\partial v}\frac{d^3p}{(2\pi)^3} \Big/ \int \Phi_p^e \frac{d^3p}{(2\pi)^3},
\end{aligned} \tag{9.25}$$

and

$$\varepsilon_{\mathrm{k}} = \varepsilon_{\mathrm{k}}^{(e)} + \varepsilon_{\mathrm{k}}^{(i)} - 1 \simeq \frac{\omega_{\mathrm{pe}}^2}{k^2 V_{Te}^2} - \frac{\omega_{\mathrm{pi}}^2}{\omega^2}. \tag{9.26}$$

The first inequality (9.24) means that we must use a more rigid restriction: $\Delta\omega \ll \omega_{\mathrm{pi}}$. This restriction will be used only in the final simplification of the equations. If we do not assume that ε_{k} is large we must take into account all terms of the same order of magnitude and amongst them there will be terms which were neglected when we derived an approximate expression for the effective non-linear plasma responses. Following Eq.(2.85) we have here also introduced an effective temperature T_{eff} and an effective thermal velocity V_{Te} for an arbitrary, though isotropic, particle distribution. If the particle distribution is anisotropic the effective temperature will depend on k. In Eq.(9.26) we have

$$V_{Te}^2 = \frac{T_{\mathrm{eff}}}{m_{\mathrm{e}}},$$

which is the definition of T_{eff}. With these various assumptions we can write Eq.(9.21) in the form

$$(\omega^2 - k'^2 v_{\mathrm{s}}^2)\,\delta n_{\mathrm{k}'} = \frac{k'^2 v_{\mathrm{s}}^2}{4\pi m_{\mathrm{e}} V_{Te}^2}\,\left[|\mathcal{E}|^2\right]_{\mathrm{k}'} = \frac{Z k'^2}{4\pi m_{\mathrm{i}}}\,\left[|\mathcal{E}|^2\right]_{\mathrm{k}'}, \tag{9.27}$$

with

$$v_{\mathrm{s}}^2 = \frac{Z T_{\mathrm{eff}}}{m_{\mathrm{i}}}.$$

Going back to the case where we consider δn and $\mathcal{E}$ to be functions of $\mathbf{r}$ and t we find that Eq.(9.27) becomes

$$\frac{\partial^2 \delta n}{\partial t^2} - v_{\mathrm{s}}^2 \nabla^2 \delta n = \frac{Z}{4\pi m_{\mathrm{i}}} \nabla^2 |\mathcal{E}|^2. \tag{9.28}$$

In order that we can write Eq.(9.22) in the form where δn and $\mathcal{E}$ are functions of $\mathbf{r}$ and t we take into account that we have

$$\varepsilon_{\mathrm{k},\omega-\omega_{\mathrm{pe}}} \simeq 1 - \frac{\omega_{\mathrm{pe}}^2}{(\omega - \omega_{\mathrm{pe}})^2} - \frac{3k^2 v_{Te}^2}{\omega_{\mathrm{pe}}^2} \simeq \frac{2\omega}{\omega_{\mathrm{pe}}} - \frac{3k^2 v_{Te}^2}{\omega_{\mathrm{pe}}^2}, \tag{9.29}$$

and that ω and k^2 should be replaced by operators: $\omega \to i\partial/\partial t$, $k^2 \to -\nabla^2$. This leads to the equation

$$\mathrm{div}\left(i\frac{\partial}{\partial t} + \frac{3v_{Te}^2}{2\omega_{\mathrm{pe}}}\nabla^2\right)\mathcal{E} = \frac{\omega_{\mathrm{pe}}}{2n_{\mathrm{e}}}\,\mathrm{div}(\delta n\,\mathcal{E}). \tag{9.30}$$

The set of equations (9.28) and (9.30) which we have now obtained is a closed system and – within the approximations made – is equivalent to the general integral equation. For a thermal distribution we have $V_{Te}^2 = v_{Te}^2 = T_e/m_e$. Neglecting unity in comparison with the electron and ion permittivities means that we have approximate quasi-neutrality in the slow motions. The equations which we have obtained as approximations from the general non-linear equations are known as the *Zakharov equations*. The general equations contain a richer fund of information, especially as far as kinetic effects and imaginary parts are concerned. The latter are very important for dissipative coherent structures. Sometimes they can be treated using perturbation theory just as the weak damping is treated for a linear mode. The general non-linear equation also contains all corrections to Eqs.(9.28) and (9.30) of order $k^2 v_{Te}^2/\omega_{pe}^2$ and v_{Te}^2/c^2. The first of these are called *electron non-linearities* and the latter are called *relativistic non-linearities*. These non-linearities can play an important rôle; the reason is that to the first approximation – corresponding to Eq.(9.14) – the quadratic and cubic non-linearities may strongly cancel each other, because of the presence of $\varepsilon^{(i)}$ in the numerator in Eq.(9.14). This cancellation is large for fast motions. The non-linearity which is taken into account in Eqs.(9.28) and (9.30) is sometimes called the *striction non-linearity*; it can be the dominant one only for slow motions and for $W \gg W_{cr}$. Its name derives from the fact that because of the approximate quasi-neutrality any electron displacement forces an ion displacement through an ambipolar polarisation field (see next section). The physical meaning of Eq.(9.30) is clear: it describes the propagation of a Langmuir field in the presence of slowly changing density variations. The meaning of the effects described by Eq.(9.28) is also clear. Its left-hand side describes the propagation of sound waves when there is no Langmuir field present while its right-hand side describes a change in the density due to the striction force which pushes the plasma away from the regions of strong Langmuir oscillations. An important point is that because we have derived Eqs.(9.28) and (9.30) from the general non-linear equation we are able to find all necessary corrections as well as the domain of applicability of these equations.

9.4 Averaging over the Plasma Frequency

Equations (9.28) and (9.30) can be found by averaging over a period corresponding to the plasma frequency. We shall illustrate this for the case of a one-dimensional particle motion in the limit of negligible thermal motion. The Langmuir fields act mainly on the electrons; their displacement creates the ambipolar polarisation field which then acts on the ions. Both the electrons and the ions can thus be moved by an inhomogeneous Langmuir field. The motion of an electron in the field of Langmuir waves is described by the equation

$$m_e \frac{d^2 x}{dt} = -2e\mathcal{E}(x)\cos(\omega_{pe}t - kx). \tag{9.31}$$

We have here introduced an extra factor 2 in order that we have the same definition of the amplitude field as before – when it was the coefficient of the positive-frequency exponential. To a first approximation we shall neglect the amplitude δx of the electron oscillations in the Langmuir field and the Doppler frequency shift; moreover, we shall denote the initial electron position by x_0: $x = x_0 + \delta x$.

We now expand the force in terms of δx:

$$m_e \frac{d^2 \delta x}{dt} \simeq -2e\mathcal{E}(x_0) \cos(\omega_{pe}t) - 2e \frac{\partial \mathcal{E}(x_0)}{\partial x_0} \delta x \cos(\omega_{pe}t), \qquad (9.32)$$

and to a first approximation we find

$$\delta x \simeq \frac{2e\mathcal{E}(x_0) \cos(\omega_{pe}t)}{m_e \omega_{pe}^2}. \qquad (9.33)$$

Substituting this expression into the second term on the right-hand side of Eq.(9.32) and averaging over a plasma period we find

$$m_e \overline{\frac{d^2 \delta x}{dt^2}} = -\frac{2e^2}{m_e \omega_{pe}^2} \mathcal{E}(x_0) \frac{\partial \mathcal{E}(x_0)}{\partial x_0} \simeq -\frac{1}{4\pi n_e} \frac{\partial}{\partial x_0} [\mathcal{E}(x_0)]^2. \qquad (9.34)$$

For the sake of simplicity we have taken $\mathcal{E}$ to be real. Apart from the force due to the Langmuir field there is also the polarisation field, which we denote by E_p, created by the displacement of the electrons. For the electrons it cancels the striction force given by the expression on the right-hand side of Eq.(9.34) since for slow motions we can neglect the electron inertia. Dropping the index "0" we thus have

$$eE_p = -\frac{1}{4\pi n_e} \frac{\partial}{\partial x} |\mathcal{E}(x)|^2. \qquad (9.35)$$

The polarisation field is the only field acting upon the ions:

$$m_i \frac{dv_i}{dt} = ZeE_p = -\frac{Z}{4\pi n_e} \frac{\partial}{\partial x} |\mathcal{E}(x)|^2. \qquad (9.36)$$

From the equation of continuity for the ions we find to the first approximation

$$\frac{\partial \delta n_i}{\partial t} = -n_i \frac{\partial}{\partial x} v_i, \qquad (9.37)$$

where n_i is the unperturbed ion density. We have thus

$$\frac{\partial^2}{\partial t^2} \delta n = \frac{Z}{4\pi m_i} \frac{\partial^2}{\partial x^2} |\mathcal{E}|^2, \qquad (9.38)$$

that is, we have found the right-hand side of Eq.(9.28). If we take into account the term $-\nabla \delta n T_e / n_e$ which involves the pressure gradient, we get on the left-hand side of Eq.(9.38) an extra term $-v_s^2 \partial^2 \delta n / \partial x^2$.

The derivation given here makes the physics of the effects which we are taking into account in the equations much clearer, but this kind of derivation becomes very complicated, indeed, if we want to calculate the correction tems and it is completely inapplicable for a study of kinetic effects and collective dissipative effects. One of our aims was to show the various relations which are often insufficiently emphasised.

9.5 Modulational Instability

We have mentioned earlier that the initial stage of non-linear interactions, including the modulational interactions, can be described as a non-linear instability. For the sake of simplicity we shall illustrate this by considering the modulational instability for the one-dimensional case. The set of equations describing the modulational interactions is

$$\left.\begin{array}{l} \left[i\dfrac{\partial}{\partial t} + \dfrac{3v_{Te}^2}{2\omega_{pe}}\dfrac{\partial^2}{\partial x^2} \right]\mathcal{E} = \dfrac{\delta n}{2n_0}\omega_{pe}\mathcal{E}, \\[3mm] \left[\dfrac{\partial^2}{\partial t^2} - v_s^2\dfrac{\partial^2}{\partial x^2} \right]\delta n = \dfrac{Z}{4\pi m_i}\dfrac{\partial^2}{\partial x^2}\left|\mathcal{E}\right|^2. \end{array}\right\} \tag{9.39}$$

If we are considering an instability we must first find an initial equilibrium state satisfying this set of equations. As such a state we choose the amplitude field of a monochromatic wave. The term monochromatic means here that it contains only a single value of a wavenumber and thus that it has only a single frequency. However, it does not mean that there is no frequency shift. The state should satisfy Eqs.(9.39) and should therefore be a solution of the non-linear equation. We shall assume that the equilibrium values of δn and $|\mathcal{E}|^2$ are independent of x and t. We then find from the second equation of the set (9.39):

$$\delta n_0 = -\frac{|\mathcal{E}_0|^2}{4\pi T_e}, \tag{9.40}$$

where the index "0" indicates stationary equilibrium values. From the first equation of the set (9.39) we now get

$$\mathcal{E}_0 = \mathcal{E}_0^0\, e^{-i\omega_0 t + ik_0 x},$$

$$\omega_0 = \frac{3k_0^2 v_{Te}^2}{2\omega_{pe}} - \frac{|\mathcal{E}_0|^2}{8\pi n_0 T_e}\omega_{pe} = \frac{3k_0^2 v_{Te}^2}{2\omega_{pe}} - \frac{\left|\mathcal{E}_0^{(0)}\right|^2}{8\pi n_0 T_e}\omega_{pe}, \tag{9.41}$$

which means that a monochromatic wave is, indeed, an exact solution of the non-linear equations provided we take into account that the non-linear frequency shift is equal to $-|\mathcal{E}|^2/8\pi n_0 T_e$ – we must bear in mind that we must subtract ω_{pe} from the total frequency and that ω_0 only takes into account the deviation of the frequency from ω_{pe}. It turns out that this exact solution is unstable. We write

$$\delta n = \delta n_0 + \delta n', \qquad \mathcal{E} = \mathcal{E}_0 + \delta\mathcal{E}, \qquad \mathcal{E}^* = \mathcal{E}_0^* + \delta\mathcal{E}^*,$$

and linearise Eqs.(9.39) with respect to the deviations from the stationary solution:

$$\left[i\frac{\partial}{\partial t} + \frac{3v_{Te}^2}{2\omega_{pe}}\frac{\partial^2}{\partial x^2}\right]\delta\mathcal{E} = \frac{\omega_{pe}}{2n_0}\left(\delta n_0\,\delta\mathcal{E} + \mathcal{E}_0\,\delta n'\right), \tag{9.42}$$

$$\left[-i\frac{\partial}{\partial t} + \frac{3v_{Te}^2}{2\omega_{pe}}\frac{\partial^2}{\partial x^2}\right]\delta\mathcal{E}^* = \frac{\omega_{pe}}{2n_0}\left(\delta n_0\,\delta\mathcal{E}^* + \mathcal{E}_0^*\,\delta n'\right), \tag{9.43}$$

$$\left[\frac{\partial^2}{\partial t^2} - v_s^2\frac{\partial^2}{\partial x^2}\right]\delta n' = \frac{Z}{4\pi m_i}\frac{\partial^2}{\partial x^2}\left(\mathcal{E}_0^*\delta\mathcal{E} + \mathcal{E}_0\delta\mathcal{E}^*\right). \tag{9.44}$$

We shall look for a solution of these equations in the form

$$\left.\begin{array}{rcl}
\delta n' &=& \delta n'_0\,e^{-i\omega t + ikx}, \\[4pt]
\delta\mathcal{E} &=& \delta\mathcal{E}_0\,e^{-i(\omega+\omega_0)t + i(k+k_0)x}, \\[4pt]
\delta\mathcal{E}^* &=& \delta\mathcal{E}_0^*\,e^{-i(\omega-\omega_0)t + i(k-k_0)x}.
\end{array}\right\} \tag{9.45}$$

This is the only form such that the exponentials in the set of equations cancel. We now get a set of three linear equations for the amplitudes of $\delta n'$, $\delta\mathcal{E}$, and $\delta\mathcal{E}^*$ – as $\mathcal{E}$ is a complex quantity it corresponds to two independent perturbations and we have chosen those to be $\delta\mathcal{E}$ and $\delta\mathcal{E}^*$. We now find

$$\left\{\omega_0 + \omega - \frac{3(k_0+k)^2 v_{Te}^2}{2\omega_{pe}}\right\}\delta\mathcal{E}_0 = \frac{\omega_{pe}}{2n_0}\left(\delta n_0\delta\mathcal{E}_0 + \mathcal{E}_0^{(0)}\delta n'_0\right), \tag{9.46}$$

$$\left\{\omega_0 - \omega - \frac{3(k_0-k)^2 v_{Te}^2}{2\omega_{pe}}\right\}\delta\mathcal{E}_0^* = \frac{\omega_{pe}}{2n_0}\left(\delta n_0\delta\mathcal{E}_0^* + \mathcal{E}_0^{(0)*}\delta n'_0\right), \tag{9.47}$$

Substituting into these equations expression (9.41) for ω_0 and (9.40) for δn_0 we can express the perturbation of the envelope field in terms of $\delta n'_0$ and subsequently substitute them into the equation for $\delta n'_0$ (see Eq.(9.44)):

$$(\omega^2 - k^2 v_s^2)\,\delta n'_0 = \frac{Zk^2}{4\pi m_i}\left[\mathcal{E}_0^*\delta\mathcal{E}_0 + \mathcal{E}_0\delta\mathcal{E}_0^*\right], \tag{9.48}$$

and we can then find the dispersion equation for the modulational instability:

$$\omega^2 - k^2 v_s^2 = \frac{|\mathcal{E}_0|^2}{8\pi n_0 T_e} k^2 v_s^2 \omega_{pe}\left[\left\{\omega - \frac{3(k^2 + 2kk_0)v_{Te}^2}{2\omega_{pe}}\right\}^{-1}\right.$$
$$\left. - \left\{\omega + \frac{3(k^2 - 2kk_0)v_{Te}^2}{2\omega_{pe}}\right\}^{-1}\right]. \tag{9.49}$$

Since we have used a one-dimensional model in deriving this equation it means that the initial wave–sometimes called the *pump wave*–and its perturbations are parallel to one another. We shall now investigate qualitatively the kind of solutions Eq.(9.49) has. We consider the case of rather large pump amplitudes–we shall find later what is meant by this statement–when the instability growth rate–more precisely, the real and the imaginary parts of the frequency–are larger than both kv_s and $k^2 v_{Te}^2/2\omega_{pe}$. We shall also be interested in perturbations with wavelengths much shorter than the pump wavelength, that is, $k \gg k_0$. We have thus the following inequalities which have to be satisfied:

$$|\omega| \gg kv_s, \qquad |\omega| \gg \frac{3k^2 v_{Te}^2}{2\omega_{pe}}, \qquad k \gg k_0, \tag{9.50}$$

and we shall expand Eq.(9.49) in terms of the small parameters which follow from them. We find

$$\omega^4 = \frac{3|\mathcal{E}_0|^2}{8\pi n_0 T_e} k^4 v_s^2 v_{Te}^2 = k^4 v_s^2 v_{\sim}^2, \tag{9.51}$$

where

$$v_{\sim}^2 = \frac{3|\mathcal{E}_0|^2}{8\pi n_0 T_e} v_{Te}^2 = \frac{3|\mathcal{E}_0|^2}{8\pi n_0 m_e}.$$

The physical meaning of the last quantity is that it corresponds, as far as order of magnitude is concerned, to the amplitude of the oscillations of the velocity of an electron in the pump wave field. Indeed, from the electron equation of motion it follows that this amplitude is equal to $e\mathcal{E}_0/m_e\omega_{pe}$ and its square is thus $e^2|\mathcal{E}_0|^2/m_e^2\omega_{pe}^2 = |\mathcal{E}_0|^2/4\pi n_0 m_e$ which differs from expression (9.51) only by a factor $\frac{2}{3}$. Equation (9.51) has an unstable root:

$$\omega = i\gamma_k^{mod}, \qquad \gamma_k^{mod} = k\sqrt{v_s v_{\sim}}, \tag{9.52}$$

which increases with increasing wavenumber. The second inequality (9.50) determines the maximum possible value of k in Eq.(9.52):

$$k \ll k_{max} \cong \frac{\omega_{pe}}{v_{Te}} \sqrt{\frac{v_s v_{\sim}}{v_{Te}^2}} \cong \frac{\omega_{pe}}{v_{Te}} \left[\frac{Zm_e}{m_i} \frac{|\mathcal{E}_0|^2}{4\pi n_0 T_e} \right]^{1/4}. \tag{9.53}$$

We have omitted here a numerical factor of the order of unity. When $k \gg k_{max}$ the growth rate decreases when k increases. One can see that by considering the limit which is the opposite of the last inequality (9.50) in which case the dispersion equation (9.49) has the form

$$\omega^2 = v_s^2 \left\{ k^2 - \frac{|\mathcal{E}_0|^2}{24\pi n_0 T_e} \frac{\omega_{pe}^2}{v_{Te}^2} \right\}, \tag{9.54}$$

which gives an unstable root only when

$$\frac{|\mathcal{E}_0|^2}{8\pi n_0 T_e} > \frac{3k^2 v_{Te}^2}{\omega_{pe}^2},$$

that is, when

$$k < k_* = \frac{\omega_{pe}}{v_{Te}} \sqrt{\frac{|\mathcal{E}_0|^2}{24\pi n_0 T_e}}; \tag{9.55}$$

the growth rate decreases with increasing k and vanishes for $k = k_*$. The quantity k_{max} is smaller than k_* if

$$\frac{|\mathcal{E}_0|^2}{4\pi n_0 T_e} > \frac{Z m_e}{m_i}. \tag{9.56}$$

We found this criterion before when we discussed the possibility of creating a Langmuir condensate in connection with the stimulated scattering processes. We remind ourselves that this criterion was obtained from the expression for non-linear broadening – which we found again at the beginning of the present chapter when we calculated the non-linear frequency shift. Here the criterion meant the occurrence of the modulational instability – since we found it as the limit of applicability of the stimulated scattering concept in the last stage of the energy cascade. The fact that the criteria are the same only means that in the case of interest for applications when the criterion (9.56) is satisfied our analysis has found the domain of applicability. It also means that one can estimate the maximum growth rate of the modulational instability from Eq.(9.52) by substituting there $k = k_{max}$:

$$\gamma_{k,max}^{mod} \cong k_{max}\sqrt{v_s v_\sim} \cong \omega_{pe}\sqrt{\frac{Z m_e}{m_i} \frac{|\mathcal{E}_0|^2}{4\pi n_0 T_e}} \cong Z\omega_{pi}\sqrt{\frac{|\mathcal{E}_0|^2}{4\pi n_0 T_e}}. \tag{9.57}$$

We should emphasise two points in connection with this result. Firstly, it follows from Eq.(9.57) that the growth rate is smaller than the ion plasma frequency. This confirms the qualitative picture of the development of the modulational instability which we gave earlier, since the ions can be expelled from the density hole only on time scales which are longer than the ion plasma period. Secondly, the maximum growth rate corresponds to $k \gg k_0$. This means that the energy of the Langmuir oscillations is transferred from smaller k-values to larger ones. This can be considered to be the initial stage of the energy transfer. The direction of the energy transfer in the case of modulational interactions is thus the opposite of that for stimulated scattering and decay processes. The direction of the energy transfer changes when during the dynamics of the process the energy density of the oscillations crosses the critical value W_{cr}. Any difficulty connected with the accumulation of energy of oscillations in the Langmuir condensate is overcome as once the energy density in the condensate reaches its critical value the direction of the energy transfer changes and the condensate will have a tendency to disappear. The most important point in that connection is that when the energy transfer is in the opposite direction the Langmuir oscillations ultimately will be transferred to the Landau damping region. If the source right from the start creates an energy density larger than the critical one the energy will from the start flow to the Landau damping region and if the situation is turbulent, the whole picture could be similar to that pertaining to liquids. In the case of a weak source which continues to operate over a long period the development of the system may have a long stage involving weak turbulence during which the energy flows in the direction of larger scales until the energy density in the condensate will have become sufficiently large for the direction of the energy transfer to change.

9.6 Conservation Laws

The concept of the transfer of the energy of the oscillations implies that there must be some quantities which are conserved in the non-linear interaction processes. We found earlier that in the processes where Langmuir waves are undergoing stimulated scattering leading to other Langmuir waves the number of waves is conserved and that this led to an approximate energy conservation. A similar conservation law emerges when the decay process of a Langmuir wave leads to another Langmuir wave and an ion-sound wave. We shall now show that the number of waves is also conserved in the case of modulational interactions. We shall see later that the term "number of waves" has a slightly different meaning in the case of modulational interactions and that it can be introduced even for regular waves in the framework of their description by means of the simplified Eqs.(9.39). It is impossible to draw conclusions from the modulational instabilities since they describe only the initial stage of the modulational interactions. The conservation laws which we are about to derive are valid for all stages of the development of the modulational interactions. In contrast to the case of stimulated scattering, energy cannot be transferred to the particles, but some of it can be transferred to the energy of the density variations, just as some energy was transferred to the energy of the ion-sound oscillations in the case of energy transfer in decay processes. Moreover, the question of the change in energy in the energy transfer processes in the modulational interactions is more sophisticated since energy can be gained due to negative frequency shifts.

We shall here demonstrate the conservation laws in the case of modulational interactions for the one-dimensional case. We leave the three-dimensional case as an exercise.

We multiply Eq.(9.39) by $\mathcal{E}^*$ and the conjugate complex equation for $\mathcal{E}^*$ by $\mathcal{E}$ and subtract the second equation from the first one. We then find:

$$\frac{\partial}{\partial t} \int |\mathcal{E}|^2 \, dx = 0,$$

or

$$\frac{\partial N}{\partial t} = 0, \qquad N = \frac{4\pi^2}{\omega_{pe}} \int |\mathcal{E}|^2 \, dx. \tag{9.58}$$

This conservation law corresponds to the conservation of the number of waves. This conservation law is also valid for regular waves, but in that case it is difficult to see what corresponds to the number of quanta. Even in the limit of random waves this quantity differs from the quantity which we introduced earlier and called the number of quanta. Indeed, according to Eq.(3.26) we have

$$\left\langle |\mathcal{E}|^2 \right\rangle = \int |E^+|_k^2 \, dk,$$

$$N_k = 2\pi^2 \left.\frac{\partial \varepsilon_k}{\partial \omega}\right|_{\omega=\omega_k} |E^+|_k^2 \cong \frac{4\pi^2}{\omega_{pe}} \left\{1 + \frac{3k^2 v_{Te}^2}{2\omega_{pe}^2}\right\} |E^+|_k^2. \tag{9.59}$$

The presence of the extra dispersion term in Eq.(9.59) as compared to Eq.(9.58) shows the difference between these two quantities. The quantity corresponding to the number of waves introduced in Chap.3 should not be expression (9.58) but

$$\int \frac{4\pi^2}{\omega_{pe}} \left[|\mathcal{E}|^2 + \frac{3v_{Te}^2}{2\omega_{pe}^2} \left|\frac{\partial \mathcal{E}}{\partial x}\right|^2 \right] dx. \tag{9.60}$$

Of course, the difference is small and we can, incidentally, add to expression (9.60) the expression involved in the energy conservation multiplied by an arbitrary coefficient such that it is of the same order of magnitude as the small dispersion correction in expression (9.60). This means that the term "number of waves" has a somewhat different meaning in the case of the modulational interactions. The same is true for the term "energy"; we shall see that this does not contain the "main" term corresponding to the number of waves multipied by the plasma frequency. One can obviously combine the two conservation laws in such a way that one obtains the energy of the Langmuir waves as defined previously. The energy conservation law for the modulational interactions can be obtained by differentiating the first of Eqs.(9.39) with respect to x and multiplying the result by $\partial \mathcal{E}^*/\partial x$, differentiating the complex conjugate of the first of Eqs.(9.39) with respect to x and multiplying the result by $\partial \mathcal{E}/\partial x$, and subtracting the two expressions obtained in this way from one another. The result is:

$$i\frac{\partial}{\partial t} \int \left|\frac{\partial \mathcal{E}}{\partial x}\right|^2 dx = \frac{\omega_{pe}}{2n} \int \delta n \left\{ \mathcal{E}^* \frac{\partial^2 \mathcal{E}}{\partial x^2} - \mathcal{E} \frac{\partial^2 \mathcal{E}^*}{\partial x^2} \right\} dx. \tag{9.61}$$

On the other hand, we obtain directly from Eqs.(9.39):

$$i \frac{\partial}{\partial t} |\mathcal{E}|^2 + \frac{3v_{Te}^2}{2\omega_{pe}} \left\{ \mathcal{E}^* \frac{\partial^2 \mathcal{E}}{\partial x^2} - \mathcal{E} \frac{\partial^2 \mathcal{E}^*}{\partial x^2} \right\} = 0, \tag{9.62}$$

and we can thus write Eq.(9.61) in the form

$$\frac{\partial}{\partial t} \int \frac{3v_{Te}^2}{2\omega_{pe}} \left| \frac{\partial \mathcal{E}}{\partial x} \right|^2 dx + \frac{\omega_{pe}}{2n} \int \delta n \frac{\partial}{\partial t} |\mathcal{E}|^2 dx = 0, \tag{9.63}$$

or

$$\frac{\partial}{\partial t} \int dx \left[\frac{3v_{Te}^2}{2\omega_{pe}} \left| \frac{\partial \mathcal{E}}{\partial x} \right|^2 + \frac{\omega_{pe}}{2n} \delta n |\mathcal{E}|^2 \right]$$

$$= \frac{\omega_{pe}}{2n_0} \int |\mathcal{E}|^2 \frac{\partial \delta n}{\partial t} dx \equiv K. \tag{9.64}$$

To convert this equation into a conservation law we introduce the hydrodynamic velocity v due to the changes δn in density. We define it as a solution of the equation of continuity,

$$\frac{\partial \delta n}{\partial t} = -n_0 \frac{\partial v}{\partial x}. \tag{9.65}$$

This equation can be obtained from Eq.(9.36) for δn_i by dividing it by Z. The quantity δn in Eq.(9.65) is thus the change in the electron density and n_0 is the stationary value of the electron density. The quantity v can also be introduced formally as a solution of Eq.(9.65). We emphasise this point as we shall not introduce any approximations for the conservation laws of Eqs.(9.39). The arguments given in support of Eq.(9.65) are merely to help us to understand the physical meaning of the quantity v. The fact that Eq.(9.65) is the same as an approximate equation of continuity is due to the approximate nature of the set (9.39) which deals with weak non-linearities. However, in the framework of these equations we shall consider Eq.(9.65) to be an exact definition of the quantity v. Using Eq.(9.65) we can rewrite Eq.(9.64) in the form

$$K = \tfrac{1}{2}\omega_{pe} \int v \frac{\partial}{\partial x} |\mathcal{E}|^2 dx. \tag{9.66}$$

On the other hand, substituting Eq.(9.65) into the second of Eqs.(9.39) and integrating over x we find

$$n_0 \frac{\partial v}{\partial t} + v_s^2 \frac{\partial \delta n}{\partial x} = -\frac{Z}{4\pi m_i} \frac{\partial}{\partial x} |\mathcal{E}|^2, \tag{9.67}$$

and hence

$$\frac{2}{4\pi\omega_{pe}} K = \tfrac{1}{2}\omega_{pe} \int dx \left[v_s^2 \delta n \frac{\partial v}{\partial x} - n_0 v \frac{\partial v}{\partial t} \right]$$

$$= -\frac{m_i n_0}{Z} \frac{\partial}{\partial t} \int dx \left\{ \tfrac{1}{2} v^2 + \tfrac{1}{2} v_s^2 \frac{\delta n^2}{n_0^2} \right\}. \tag{9.68}$$

Using the fact that $n_0/Z = n_i$ we see that the first term on the right-hand side of Eq.(9.68) corresponds to the energy, $\frac{1}{2}n_i m_i v^2$, of the hydrodynamic motion while the second term corresponds to the energy related to the change in density during such motions. If there are no Langmuir oscillations the right-hand side corresponds to the total energy of the sound oscillations. The energy conservation can thus be written in the form:

$$\frac{\partial H}{\partial t} = 0, \qquad H = n_i m_i \int \left[\tfrac{1}{2}v^2 + v_s^2 \frac{\delta n^2}{2n_0^2} \right] dx$$

$$+ \int \left[\delta n \frac{|\mathcal{E}|^2}{4\pi n_0} + \frac{3v_{Te}^2}{4\pi \omega_{pe}^2} \left| \frac{\partial \mathcal{E}}{\partial x} \right|^2 \right] dx. \qquad (9.69)$$

The physical meaning of the various terms in this expression is clear. The last term is due to the dispersion and the penultimate term describes the non-linearities when the density variation is proportional to the square of the amplitude field. It is negative if $\delta n < 0$. This may lead to the whole energy being negative – if this term dominates. A transition to such states with $H < 0$ is possible with a gain of energy.

We can also find the conservation law for the momentum P of the system. To do this we multiply Eq.(9.59) by $\mathcal{E}^*$, Eq.(9.39) for $\mathcal{E}^*$ by $\partial \mathcal{E}/\partial x$ and subtract the results from one another. We then multiply Eq.(9.60) by $\mathcal{E}$ and Eq.(9.39) for $\mathcal{E}$ by $\partial \mathcal{E}^*/\partial x$ and subtract the results from one another. We finally subtract the two resulting equations from each other and integrate the result over x. We then find:

$$\frac{\partial}{\partial t} \frac{1}{4\pi\omega_{pe}} \int dx \left[\mathcal{E} \frac{\partial \mathcal{E}^*}{\partial x} - \mathcal{E}^* \frac{\partial \mathcal{E}}{\partial x} \right] = -\frac{1}{4\pi} \int |\mathcal{E}|^2 \frac{\partial \delta n/n_0}{\partial x} dx$$

$$= \frac{1}{4\pi} \int \frac{\delta n}{n_0} \frac{\partial}{\partial x} |\mathcal{E}|^2 dx = -\frac{m_i}{Z} \int \delta n \frac{\partial v}{\partial t} dx. \qquad (9.70)$$

We have used here Eq.(9.67) and the fact that the integral of $\delta n\, \partial \delta n/\partial x$ over x is zero. Using Eq.(9.65) we now find that the integral of $v\, \partial \delta n/\partial t$ is also zero and thus that the right-hand side of Eq.(9.70) is the time derivative of the hydrodynamic momentum. We have thus found that

$$\frac{\partial P}{\partial t} = 0, \qquad P = \frac{1}{4\pi\omega_{pe}} \int dx \left[\mathcal{E} \frac{\partial \mathcal{E}^*}{\partial x} - \mathcal{E}^* \frac{\partial \mathcal{E}}{\partial x} \right] + \frac{m_i}{Z} \int v\delta n\, dx. \quad (9.71)$$

We leave the derivation of the conservation laws in the three-dimensional case as an exercise.

9.7 Solitons. Self-contracting Cavitons

We discussed earlier the possibility of the self-contraction of wavepackets due to the modulational interactions. Self-contraction is a general property of these interactions and it is connected with the fact that energy can be gained by contracting. The question then arises of what the final result of such contractions can be. The first possibility is that the self-contraction is finally compensated by the linear dispersion to form a wavepacket which does not disperse and can exist as a localised packet for a long time. The second possibility is that the self-contraction is compensated by other non-linear non-dissipative effects. This compensation may be dynamical, that is, the final state might be one which is periodic in time or even may have a more complicated time dependence. We shall call all these coherent structures *cavitons*. The third possibility is that the self-contraction is compensated by dissipation which be either linear or non-linear. Such coherent structures we call *dissipative structures*. In the one-dimensional case the compensation of the self-contraction by linear dispersion is always possible and the localised solitary waves which are formed are called *solitons*. The relative phases of the harmonics forming the solitons are fixed with respect to one another or, in other words, synchronised. This is the reason why solitons can be considered to be coherent structures.

If we want to find the distribution of the harmonics in a soliton we can start from the one-dimensional Eqs.(9.39) and write the complex amplitude field $\mathcal{E}$ in the form $\mathcal{E} = \mathcal{E}' e^{i\phi'}$ where $\mathcal{E}'$ and ϕ' are real quantities – the amplitude and the phase of the amplitude field. We assume that for the soliton solution the amplitude $\mathcal{E}'$ moves with a constant velocity u – and also that $|\mathcal{E}|^2 = \mathcal{E}'^2$ – but that the phase ϕ' can change inside the soliton according to its non-linear properties which are governed by the non-linear equation. The problem is thus to find the relations between the phases of the coherent harmonics inside the soliton and the spatial shape of the amplitude in its own rest frame.

We assume thus that the amplitude $\mathcal{E}'$ is a function of the variable $x' = x - ut$ and that the phase ϕ' depends on both x' and t. Separating the real and imaginary parts in the first of Eqs.(9.39) we find:

$$-\mathcal{E}_0 \left[\frac{\partial \phi'}{\partial t} - u \frac{\partial \phi'}{\partial x'} \right] + \frac{3v_{Te}^2}{2\omega_{pe}} \left[\frac{\partial^2 \mathcal{E}'}{\partial x'^2} - \mathcal{E}_0 \left(\frac{\partial \phi'}{\partial x'} \right)^2 \right] = \frac{\omega_{pe}}{2n_0} \delta n\, \mathcal{E}', \quad (9.72)$$

$$-u \frac{\partial \mathcal{E}'}{\partial x'} + \frac{3v_{Te}^2}{2\omega_{pe}} \left[\mathcal{E}' \frac{\partial^2 \phi'}{\partial x'^2} + 2 \frac{\partial \mathcal{E}'}{\partial x'} \frac{\partial \phi'}{\partial x'} \right] = 0. \quad (9.73)$$

By making a change in variable: $\phi' = \phi + ux'\omega_{pe}/3v_{Te}^2$ we find from Eq.(9.73) the following equation for ϕ:

$$\frac{\partial \phi}{\partial x'} = \frac{\text{const}}{\mathcal{E}'^2}. \quad (9.74)$$

We see that Eq.(9.72) will not have a solution such that $\mathcal{E}' \to 0$ as $x \to \pm\infty$ unless the constant in Eq.(9.74) is equal to zero. We therefore conclude that $\phi = \phi(t)$ and since all terms in Eq.(9.72) except $\partial\phi'/\partial t = \partial\phi/\partial t$ are independent of t we conclude that

$$\frac{\partial\phi}{\partial t} = \text{const} \equiv -\omega_0, \qquad \phi = -\omega_0 t.$$

Equation (9.72) can then be written in the form

$$\left\{\omega_0 + \frac{\omega_{pe}u^2}{2v_{Te}^2}\right\}\mathcal{E}' + \frac{3v_{Te}^2}{2\omega_{pe}}\frac{d^2\mathcal{E}'}{dx'^2} = \frac{\omega_{pe}}{2n_0}\delta n\,\mathcal{E}' = -\frac{Z\omega_{pe}}{8\pi n_0 m_i}\frac{\mathcal{E}'^3}{v_s^2 - u^2}. \quad (9.75)$$

We have here substituted for δn from the second of Eqs.(9.39) which in the case of a density which is moving with a constant velocity u can easily be integrated:

$$\delta n = -\frac{Z}{4\pi m_i}\frac{\mathcal{E}'^2}{v_s^2 - u^2}. \qquad (9.76)$$

We notice that $\delta n < 0$ if $u < v_s$ and that $\delta n > 0$ if $u > v_s$. From the qualitative discussion of the modulational interactions in the first section of the present chapter it is clear that self-contraction is possible only when $\delta n < 0$. Therefore solitons can exist provided

$$u < v_s. \qquad (9.77)$$

The first integral of Eq.(9.75) which satisfies the conditions $d\mathcal{E}'/dx' \to 0$, $\mathcal{E}' \to 0$, as $x' \to \pm\infty$ is

$$\frac{3v_{Te}^2}{\omega_{pe}^2}\left(\frac{d\mathcal{E}'}{dx'}\right)^2 = -\left\{\frac{\omega_0}{\omega_{pe}} + \frac{u^2}{2v_{Te}^2}\right\}\mathcal{E}'^2 - \frac{Z}{16\pi n_0 m_i}\frac{\mathcal{E}'^4}{v_s^2 - u^2}. \qquad (9.78)$$

We now replace the constant ω_0 by $\mathcal{E}_0$ through the relation

$$\omega_0 = -\omega_{pe}\frac{u^2}{2v_{Te}^2} - \frac{Z\omega_{pe}}{16\pi n_0 m_i}\frac{\mathcal{E}_0^2}{v_s^2 - u^2}, \qquad (9.79)$$

and we replace the variables x' and $\mathcal{E}'(x')$ by new variables x'' and $\varepsilon(x'')$ through the relations

$$\mathcal{E}' = \mathcal{E}_0\,\varepsilon(x''), \qquad x' = x''\frac{4v_{Te}}{\omega_{pe}\mathcal{E}_0}\sqrt{3\pi n_0 m_i(v_s^2 - u^2)}. \qquad (9.80)$$

Equation (9.78) can then be written in the form

$$\left(\frac{d\varepsilon(x'')}{dx''}\right)^2 = [\varepsilon(x'')]^2\left\{1 - (\varepsilon(x''))^2\right\}. \qquad (9.81)$$

The only solution of this equation which decreases as $x'' \to \pm\infty$ is $\varepsilon = 1/\cosh x''$ and we thus find for the *soliton solution*

$$\mathcal{E} = \mathcal{E}_0 e^{-i\omega_0 t + i\omega_{\text{pe}} u(x-ut)/3v_{Te}^2}$$

$$\times \left\{ \cosh \left[\frac{\mathcal{E}_0 \omega_{\text{pe}}(x - ut)}{4v_{Te}\sqrt{3\pi n_0 m_i \left(v_s^2 - u^2\right)}} \right] \right\}^{-1}. \tag{9.82}$$

The meaning of this solution is clear. First of all, $\mathcal{E}_0$ is the soliton amplitude; secondly, the soliton width is inversely proportional to its amplitude; and thirdly, ω_0 contains a non-linear frequency shift proportional to $\mathcal{E}_0^2$ (see Eq.(9.79)) while the other terms in the phase describe the linear dispersion. Indeed, the sum of the first term in Eq.(9.79) and the last term of the phase in Eq.(9.82) gives $u^2 \omega_{\text{pe}} t / 6 v_{Te}^2 = 3 k^2 v_{Te}^2 t / 2\omega_{\text{pe}}$, where we have used the fact that u is the group velocity, $u = 3v_{Te}^2 k / \omega_{\text{pe}}$. The solitons cannot have a velocity higher than the ion-sound velocity.

For velocities much lower than the ion-sound velocity one can find a general equation for the modulational interactions. The equation for the density variations has in that case a trivial solution:

$$\delta n = -\frac{|\mathcal{E}|^2}{4\pi T_e}, \tag{9.83}$$

and hence

$$i\frac{\partial \mathcal{E}}{\partial t} + \frac{3v_{Te}^2}{2\omega_{\text{pe}}} \frac{\partial^2 \mathcal{E}}{\partial x^2} + \omega_{\text{pe}} \frac{|\mathcal{E}|^2 \mathcal{E}}{8\pi n_0 T_e} = 0. \tag{9.84}$$

This equation is known as the *non-linear Schrödinger equation*. One can prove a number of general theorems for the non-linear Schrödinger equation. It can be shown that in the one- dimensional case any initial perturbation will decay into a set of solitons. For motions with velocities comparable to or larger than the ion-sound velocity such a theorem cannot be proved. This is obvious from a physical point of view since there appears an extra degree of freedom – the ion-sound wave. The perturbations can decay by creating ion-sound waves. Ion-sound waves can also be created when solitons collide. In that case the collisions are inelastic. Solitons with velocities much smaller than the ion-sound velocity pass through one another without generating ion-sound waves. This is the reason why an initial perturbation can decay in solitons only in the case of velocities much lower than the ion-sound velocity. In real situations velocities of the order of or even higher than the ion-sound velocity are of most interest. We found earlier that the maximum growth rate of the modulational instability – in the linear stage of the modulational interactions – corresponds to velocities much higher than the ion-sound velocity.

We shall now illustrate qualitatively why a stable balance between self-contraction and linear dispersion is possible for one-dimensional motion. We shall also show that in the case of three-dimensional motion such a balance is unstable. Due to the conservation of the number of waves we have as to order of magnitude that $|\mathcal{E}|^2 \propto 1/r_0^s$ where r_0 is the characteristic size of the caviton and s is the dimensionality of the space – $s = 3$ for a caviton and $s = 1$ for a soliton. The condition for the capturing of waves in the caviton is $\delta n/n_0 \approx k^2 d^2 \sim d^2/r_0^2$, the striction force producing the self-contraction is proportional to $\partial|\mathcal{E}|^2/\partial r \propto 1/r_0^{s+1}$, and the thermal pressure force is proportional to $\partial \delta n T/\partial r \propto 1/r_0^3$.

Let us now assume that there is equilibrium between the self-contraction and the thermal expansion forces for a certain value of r_0. If we now decrease r_0 in the one-dimensional case ($s = 1$) the striction force will grow more slowly than the thermal pressure force and the caviton will expand increasing r_0 again. On the other hand, if r_0 increases the self-contraction force will decrease more slowly than the thermal pressure force and the caviton will contract. The one-dimensional equilibrium is therefore stable.

However, in the three-dimensional case a decrease of the radius of the caviton will further increase the relative magnitude of the self-contraction force and self-contraction will proceed. In the case of Langmuir wavepackets this process is called the *Langmuir collapse*. This is not a very good term as it reminds us of gravitational collapse when the self-contraction can, indeed, be infinite. In the case of modulational interactions self-contraction is a natural process but it cannot proceed without limits.

The real problem is to understand the physical mechanisms which can restrict the self-contraction. Linear and non-linear Landau damping are the natural dissipative mechanisms stopping the self-contraction. As a rough estimate one can assume that self-contraction will stop at a radius of the order of the Debye radius due to the linear Landau damping. However, this conclusion is incorrect; in fact, linear Landau damping can halt the self-contraction at a caviton radius much larger than the Debye radius. This is due to the coherence of the modes in the caviton and is called the *arrest of the collapse* (see Chap.10).

Another kind of collective damping – *transitional damping* – due to inhomogeneities in the caviton can be important in preventing the self-contraction to small sizes which are comparable to the Debye radius. Other kinds of non-linearities which were neglected in the simplified equation we derived for the modulational interactions may also be important. The reason is that the rate at which self-contraction proceeds increases during the self-contraction and for velocities much larger than the ion-sound velocity the cancellation of the cubic and the effective quadratic responses becomes rather strong and neglected non-linearities such as the electron and the relativistic non-linearities may be important. Moreover, the conclusion of a rapid self-contraction was based on the conservation of the number of waves in the caviton. However, another kind of self-contraction where waves are lost from the caviton is also possible. Different kinds of self-contraction can be described by various self-similar solutions of the non-linear equations. We leave that problem as an exercise.

We must still mention three points. First of all, it is likely that the field in the caviton is oriented in some way so that a completely spherical collapse is impossible. Usually the cavities found in computer experiments have a dipole structure. The second point is that laboratory experiments have, indeed, shown that the self-contraction is halted at sizes much larger than the Debye radius. The third point is that a single self-contraction can become unstable and develop smaller-size self-contractions inside it and the whole self-contraction picture may become random in time. We shall not discuss these points in detail as at this moment in time there are still no definite answers to the various problems involved in these scenarios.

Appendix: Non-linear Frequency Shifts Produced by Ion–Sound Waves

Ion-sound waves can produce non-linear frequency shifts both for Langmuir waves and for ion-sound waves. As an example we shall consider the non-linear frequency shift of Langmuir waves produced by ion oscillations with $k' \gg \omega_{pe}/v_{Te}$. We see from Eqs.(6.69) and (6.70) that the frequency shift is determined by the real part of the non-linear response:

$$
\begin{aligned}
\Delta\omega_k^{NL} &= -\tfrac{1}{2}\omega_{pe}\,\mathrm{Re}\left\{\varepsilon_{k,\omega_k}^{NL}\right\} \\
&= -\omega_{pe}\int |E^s|_{k'}^2\,\mathrm{Re}\left\{\frac{4\pi}{ik}\,\varrho_{k';k,\omega_k;-k'}^{eff}\right\}d^4k'.
\end{aligned}
\tag{9.85}
$$

In the case when $k' \gg \omega_{pe}/v_{Te}$, $\omega_{k'}^s \approx \omega_{pi}$ we have

$$
\begin{aligned}
\Delta\omega_k^{NL} &= -\omega_{pe}\int \frac{|E^s|_{k'}^2}{8\pi n_e T_e}\frac{(\mathbf{k}\cdot\mathbf{k}')^2}{k^2 k'^2}\frac{\omega_{pe}^2}{k'^2 v_{Te}^2}\,d^4k' \\
&= -\omega_{pe}\int \frac{W_{k'}^s}{2n_e T_e}\frac{(\mathbf{k}\cdot\mathbf{k}')^2}{k^2 k'^2}\frac{\omega_{pe}^2}{k'^2 v_{Te}^2}\,d^3k'.
\end{aligned}
\tag{9.86}
$$

This means that ion oscillations lead to a negative frequency shift for Langmuir oscillations and there is therefore a tendency for ion-sound waves to concentrate in the region of the ion oscillations. We showed earlier that these ion oscillations will produce a significant damping of the Langmuir oscillations due to the stimulated scattering by electrons.

The frequency shift of the ion-sound waves produced by ion-sound waves can be found from the approximate expressions for the non-linear responses due to electrons:

$$
\mathrm{Re}\left\{\frac{8\pi}{ik}\varrho_{-k',k,k'}^{eff\,(e)}\right\} \approx \frac{\omega_{pe}^4}{k^2 k'^2 v_{Te}^4}\frac{1}{4\pi n_e T_e},
\tag{9.87}
$$

and due to ions:

$$
\mathrm{Re}\left\{\frac{8\pi}{ik}\varrho_{k',k,-k'}^{eff\,(i)}\right\} \approx \frac{(\mathbf{k}\cdot\mathbf{k}')^2}{k^2 k'^2}\frac{\omega_{pe}^4\left(1+k^2 d_e^2\right)\left(1+k'^2 d_e^2\right)}{4\pi n_e T_e k^2 k'^2 v_{Te}^4}.
\tag{9.88}
$$

The frequency shift is determined by the equation

$$\Delta\omega_{\mathbf{k}}^{NL} \frac{2\omega_{pi}}{(\omega_{\mathbf{k}}^s)^3} = -\int \mathrm{Re}\left\{\frac{8\pi}{ik}\varrho_{\mathbf{k}',\mathbf{k},-\mathbf{k}'}^{eff}\right\} |E^s|_{\mathbf{k}'}^2 \, d^4\mathbf{k}';$$ (9.89)

it differs in sign for the electron non-linearities,

$$\Delta\omega_{\mathbf{k}}^{NL\,(e)} = \omega_{\mathbf{k}}^s \int \frac{W_{\mathbf{k}'}^s}{2n_e T_e} \frac{d^3\mathbf{k}'}{(1+k^2d_e^2)(1+k'^2d_e^2)},$$ (9.90)

and for the ion non-linearities,

$$\Delta\omega_{\mathbf{k}}^{NL\,(i)} = -\omega_{\mathbf{k}}^s \int \frac{(\mathbf{k}\cdot\mathbf{k}')^2}{k^2 k'^2} \frac{W_{\mathbf{k}'}^s d^3\mathbf{k}'}{2n_e T_e}.$$ (9.91)

According to Eq.(6.81) Langmuir oscillations can produce a non-linear frequency shift for ion-sound waves:

$$\Delta\omega_{\mathbf{k}'}^{NL} \approx -\frac{\omega_{pi}}{ik} 4\pi \int \varrho_{\mathbf{k},\mathbf{k}',\mathbf{k}}^{eff} |E^\ell|_{\mathbf{k}}^2 \, d^4\mathbf{k}$$

$$\approx \omega_{pi} \int \frac{(\mathbf{k}\cdot\mathbf{k}')^2}{k^2 k'^2} \frac{\omega_{pe}^4}{k'^4 v_{Te}^4} \frac{W_{\mathbf{k}}^\ell d^3\mathbf{k}}{2n_e T_e},$$ (9.92)

which is positive; in Eq.(9.92) we have $k' \gg 1/d_e$.

Problems

1. Find the growth rate of the modulational instability of a monochromatic pump wave in the three-dimensional case.
2. Show that the expressions for the number of waves, N, energy, H, and momentum, $\mathbf{P}$, which are conserved in the three-dimensional case, are

$$N = \frac{4\pi^2}{\omega_{pe}} \int |\mathcal{E}|^2 \, d^3\mathbf{r},$$

$$H = \int \left\{\frac{\delta n}{n_0} \frac{|\mathcal{E}|^2}{4\pi} + \frac{3v_{Te}^2}{4\pi\omega_{pe}^2}|\mathrm{div}\mathcal{E}|^2\right\} d^3\mathbf{r}$$

$$+ \int n_i m_i \left[\tfrac{1}{2}\mathbf{v}^2 + \tfrac{1}{2}\frac{\delta n}{n_0^2} v_s^2\right] d^3\mathbf{r},$$

$$\mathbf{P} = \frac{1}{4\pi\omega_{pe}} \int [\mathcal{E}\,\mathrm{div}\mathcal{E}^* - \mathcal{E}^*\,\mathrm{div}\mathcal{E}] \, d^3\mathbf{r} + \frac{m_i}{Z} \int \delta n\,\mathbf{v}\, d^3\mathbf{r},$$

where $\mathbf{v}$ is introduced through the continuity equation

$$\frac{\partial \delta n}{\partial t} = -n_0\mathrm{div}\,\mathbf{v}.$$

3. Show that the number of waves, N, the energy, H, and the momentum, P, of a Langmuir soliton are given by the equations

$$N = 2\sqrt{6}n_0 T_e \frac{(2\pi)^3 v_{Te}}{\omega_{pe}^2} \left\{ \frac{\mathcal{E}_0^2}{4\pi n_0 T_e} \right\}^{1/2},$$

$$H = -\frac{n_0 T_e}{9\sqrt{3}} \frac{v_{Te}}{\omega_{pe}} \left\{ \frac{\mathcal{E}_0^2}{4\pi n_0 T_e} \right\}^{3/2} \left[1 - \frac{3u^2}{v_s^2} \right],$$

$$P = \frac{4n_0 m_i u}{\sqrt{3}} \frac{v_{Te}}{\omega_{pe}} \left\{ \frac{\mathcal{E}_0^2}{4\pi n_0 T_e} \right\}^{3/2}.$$

4. Show that in the case of subsonic self-contraction the self-similar solutions of Eqs.(9.28) and (9.30) are

$$\mathcal{E} \propto t^{-2/3} \mathbf{f}(r_\parallel t^{-2/3}, \mathbf{r}_\perp t^{-1/3}), \qquad \delta n \propto t^{-4/3} f(r_\parallel t^{-2/3}, \mathbf{r}_\perp t^{-4/3}),$$

where $\mathbf{f}$ and f are arbitrary functions of their arguments and where $\mathbf{r} \equiv \{r_\parallel, \mathbf{r}_\perp\}$.

5. Show that in the case of supersonic self-contraction the self-similar solutions of Eqs.(9.28) and (9.30) are

$$\mathcal{E} \propto t^{-1} \mathbf{f}(rt^{-2/3}), \qquad \delta n \propto t^{-4/3} f(rt^{-2/3}),$$

where $\mathbf{f}$ and f are arbitrary functions of their arguments.

6. Show that in the case of near-sonic self-contraction the self-similar solution of Eqs.(9.28) and (9.30) is

$$\mathcal{E} \propto t^{-1} \mathbf{f}(r_\parallel t^{-1}, \mathbf{r}_\perp t^{-1/2}),$$

where $\mathbf{f}$ is an arbitrary function of its arguments and where $\mathbf{r} \equiv \{r_\parallel, \mathbf{r}_\perp\}$.

7. Evaluate the non-linear frequency shifts (9.86), (9.90), and (9.91).

10 Dissipative Structures and Strong Langmuir Turbulence

10.1 Introduction. Self-organisation and Strong Turbulence

The main problem of strong turbulence is the coexistence of regular structures and random fields which lead to the formation of self-organised dissipative structures. Many problems of strong turbulence are still being studied. However, this field of research is so important that it needs a special discussion. At the present time the majority of the important physical concepts are clear, but the details of the formation of strong turbulence and its development are still being widely investigated, often by using super-computers. It is likely that the complicated phenomena which appear in a strongly turbulent state will be studied for a long time to come. We shall consider here what we feel are the most interesting and at this time well understood new physical ideas in this field without going into a detailed discussion but paying attention mainly to the physical mechanisms of the formation of dissipative structures.

We shall start with reminding ourselves of some of the statements we made earlier in the Introduction and making some comments on the relations between turbulence and structures, since at this stage we have already acquainted ourselves with the actual processes which determine the turbulence and with the actual processes which lead to the formation of structures. The main point we made in the Introduction was that increasing the level of turbulence leads to an increase in the energy fluxes in phase space including those directed to the regions in space corresponding to regular motions and this may be the reason for the growth of regular structures when there is turbulence. This is a *self-organisation process* and the structures formed in these processes are known as *dissipative structures*. For their formation it is necessary to have both energy fluxes which can increase their amplitude as well as damping giving dissipation. They are called dissipative structures because their dissipation is balanced by an increase in energy due to the external fluxes and they therefore survive only dynamically. The problems of strong turbulence are mainly problems of the transformation of regular structures in the presence of random fields into dissipative coherent structures existing at the same time as the random fields. These self-organised structures can take energy from the random fields and may be damped. Their behaviour may be quite different from the behaviour of regular structures when there are no random fields. They may have a finite lifetime even in the one-dimensional case when a soliton can live for a long time when there is no turbulence. These self-organised dissipative structures can be considered to be the elementary "bricks" which build the strong turbulence and their properties are very similar to those of other kinds of self-organised structures in open systems such as are known in biology. The dissipative structures themselves "live" in the random fields and are open systems. They can receive energy from the random fields and they can also lose energy.

The efforts to solve the strong turbulence problem using super-computers is a developing field of research. However, one should stress that one cannot solve the whole of the strong turbulence problem by using super-computers without paying attention to the behaviour of such self-organised structures. Here we shall pay most attention to the physical processes which may determine the transformation of coherent structures into dissipative structures. The main physical processes are the linear and non-linear interactions of the structures, including their interactions with plasma particles, inhomogeneities, and random fields.

Let us first try to understand why it is possible that regular structures and random fields can coexist using as an example Langmuir fields and the coherent structures formed by this field, such as self-contracting cavitons and solitons. The fluxes in the direction of the large scales in the weak turbulence regime are an example of fluxes towards the region in phase space where regular structures can exist and can be easily formed by the modulational interactions. We must, however, mention here that in the large wavenumber – small scale – region it is difficult to satisfy the criterion $W/nT_e \gg k^2 d_e^2$ for modulational interactions – in the weak non-linearity case, $W/nT_e \ll 1$ which we are considering here – since it contains the quantity $k^2 d_e^2$ which for scales of the order of the Debye radius is of the order of unity. This means that small-scale weak turbulence may always be present even at the same time as the regular structures. Moreover, the modulational self-contraction processes lead to a decrease in the scales of the regular structures and they will closely approach the region where weak turbulence exists in the sense that the wavevectors involved will be in the region where they can capture the random weakly turbulent waves. We estimated this capture in the previous chapter and showed that it depended on the amplitude of the field and on the ratio $\Delta l/k_*$. We can therefore expect from physical estimates that regular structures may coexist with random fields of scales less than the characteristic scales of a regular structure.

A second point is that if there are regular structures present the random fields will be inhomogeneous in the neighbourhood of these structures. This seems obvious since the structures affect the turbulence and interact with it. The structures themselves are inhomogeneous and one may expect the turbulence to be inhomogeneous on the scale of the size of the regular structures.

The question now is whether or not this inhomogeneity increases the fluxes. If it increases the fluxes, there may occur some kind of self-organising instability since the increase in the fluxes will increase the amplitude of the structure, thus increasing the inhomogeneity which, in turn will further increase the fluxes, and so on.

A third point is that the distribution of the particles interacting with the structures can also have inhomogeneities of the order of the scale of the structures. A more complicated and more real problem is connected with the presence of self-organised structures of different scales which is a qualitative picture of strong turbulence. In that case we must for the larger-scale structures take into account the inhomogeneities in the particle distribution produced by the smaller-scale structures. Thus, the particles, as well as the random fields, will be distributed inhomogeneously with many inhomogeneity scales superimposed upon one another. These inhomogeneities may lead to additional damping of the structures. This additional damping is one of the important physical differences between the structures in the presence of strong turbulence and those when there is no turbulence. We shall consider this kind of damping – transition damping – in the present chapter.

The last, but most important, point is that the linear and non-linear damping of coherent structures is quite different from the damping of a single mode with a wavenumber of the order of the reciprocal of the size of the structure. Indeed, the interactions responsible for the formation of the structure – for Langmuir structures these are the modulational interactions – "tie" together all the harmonics of the wavepacket which is the structure. The structure interacts with the fields and with the particles as a single entity. For instance, the linear growth or damping through wave-particle interactions should be quite different in the case when the waves form a structure or when the waves behave independently from one another. The linear resonance condition,

$$\omega_{\mathbf{k}}^{\ell} = (\mathbf{k} \cdot \mathbf{v}), \tag{10.1}$$

means in the case of independent harmonics that some harmonic is either excited or damped. Excitation happens in the case of beam-plasma interactions whereas damping by thermal particles is Landau damping. For some of the harmonics inside the structure the resonance condition can be satisfied and for others it cannot be satisfied. However, the structure can be excited or damped only as a complete entity. This means that the damping or growth of some of the harmonics is redistributed between all the harmonics which are part of the structure. The harmonics which if there is no structure will not even be interacting with the particles will therefore in the case when there is a structure present be either damped or amplified.

This collective effect changes the linear Landau damping of the structures very significantly. It will be sufficient to have only a few harmonics which are heavily damped for the damping to be redistributed between all the harmonics in the structure. We shall consider this process in the present chapter and we shall see that it greatly restricts the self-contraction of the caviton which may be damped and vanish at a stage when its radius is much larger than the Debye radius. We shall thus show that the linear damping of dissipative structures is strong. The interaction between a beam and a structure will also be changed since the beam will excite the structure as a complete entity.

Another process is the non-linear Landau damping described by the resonance

$$\omega_{\mathbf{k}}^{\ell} - \omega_{\mathbf{k}'}^{\ell} = ([\mathbf{k} - \mathbf{k}'] \cdot \mathbf{v}). \tag{10.2}$$

We shall show that for coherent structures the non-linear Landau damping is also different from what it is in the case when there are no interactions which keep all the harmonics in the structure in phase. In particular, we shall show that the non-linear damping diminishes the velocities of coherent solitons.

Finally, the interactions between coherent structures and random fields also occur in such a way that the parameters of the whole structure are changed due to such interactions. All these effects are important when the structure is sufficiently strong so that it can keep all the harmonics in the structure synchronised; this happens when $W \gg W_{\mathrm{cr}}$.

In earlier chapters we described self-consistent kinds of interactions between particles and random waves – quasi-linear and non-linear interactions. It was possible self-consistently to describe both the particle dynamics and the random waves. In the problem of dissipative structures we must find a way to describe the parameters of both the regular structures and the particle distributions self-consistently. This will be a generalised quasi-linear or non-linear description. In a similar way we must find a self-consistent set of equations describing the changes in the random fields and in the parameters of the regular structures. This description must be part of the description of strong turbulence as a state with many simultaneously present dissipative structures.

10.2 Resonant Interactions of Strong Regular Fields with Uniformly Distributed Particles

We shall start with the problem of the interaction between strong regular fields – including the fields of a structure or of several structures – and resonant particles. In a non-self-consistent approach the particle distribution function is specified and determined by the initial particle distribution function. In a self-consistent approach, on the other hand, the particle distribution function will be modified by the field of the structures. The difference between such an approach and a quasi-linear approach will be that that the structures will change not only the particle velocity distribution, but also the particle spatial distribution since this distribution will change locally in space in the regions where the structures are localised. We start with an approach where the distribution function in the linear and non-linear responses is assumed to be given. Earlier we took into account the real parts of the dielectric responses and showed how they can determine the structure of the coherent wavepackets. Now we shall take into account the imaginary parts of the dielectric reponses. The imaginary part of the linear response describes the Landau damping of the structures while the imaginary parts of the non-linear responses describe the non-linear damping of the structures. We may remind ourselves that the frequencies of the density variations δn are equal to differences of frequencies of the Langmuir field. We shall show that the non-linear damping of the Langmuir field corresponds to the linear damping of the density variations. The general Eq.(9.14) contains all these imaginary pars. To take into account the linear damping of the Langmuir waves we must add to their frequencies ω_k^ℓ their imaginary parts $i\gamma_k$ where γ_k describes the linear Landau damping. In the Fourier representation this additional term will be:

$$i\gamma_k \mathcal{E}_k = \frac{i\gamma_k}{(2\pi)^4} \int \mathcal{E}(\mathbf{r}', t')\, e^{i\omega t' - i(\mathbf{k}\cdot\mathbf{r}')}\, d^3r'\, dt', \tag{10.3}$$

while in the equation for $\mathcal{E}(\mathbf{r}, t)$ it will have the form

$$\int \gamma_k \mathcal{E}_k\, e^{-i\omega t + i(\mathbf{k}\cdot\mathbf{r})}\, d^3k\, d\omega$$

$$= \frac{1}{(2\pi)^3} \int \gamma_k\, e^{i\left(\mathbf{k}\cdot[\mathbf{r}-\mathbf{r}']\right)}\, d^3k\, \mathcal{E}(\mathbf{r}', t)\, d^3r'. \tag{10.4}$$

The linear Landau damping can thus lead to long-range interactions between structures, since the integral over $\mathbf{r}'$ in Eq.(10.4) contains not only the given structure but also all other structures which may be present in the system. If we take into account the linear Landau damping we must change Eq.(9.28) to read as follows:

$$\mathrm{div}\left[i\frac{\partial \mathcal{E}}{\partial t} + \frac{3v_{Te}^2}{2\omega_{pe}} \nabla^2 \mathcal{E} + \frac{1}{(2\pi)^3} \int \gamma_k\, e^{i\left(\mathbf{k}\cdot[\mathbf{r}-\mathbf{r}']\right)}\, d^3k\, \mathcal{E}(\mathbf{r}', t)\, d^3r' \right]$$

$$= \frac{\omega_{pe}}{2n_0}\, \mathrm{div}(\delta n \mathcal{E}). \tag{10.5}$$

We also write down Eq.(9.21) for the density variations, changing k' to k:

$$\varepsilon_k \delta n_k = -\frac{k^2 \left(\varepsilon_k^{(e)} - 1 \right) \varepsilon_k^{(i)}}{4\pi m_e \omega_{pe}^2}\, \left[|\mathcal{E}|^2 \right]_k. \tag{10.6}$$

We shall now consider that part of the imaginary parts of the quantities in Eq.(10.6) which are due to resonance with electrons, that is, the imaginary part of $\varepsilon_k^{(e)}$ in ε_k on the left-hand side of Eq.(10.6) and the imaginary part of $\varepsilon_k^{(e)} - 1$ on the right-hand side. If Φ_p^e is given and homogeneous we have in the general case:

$$\mathrm{Im}\left\{\varepsilon_k^{(e)}\right\} = -\frac{4\pi^2 e^2}{k^2} \int \delta(\omega - (\mathbf{k}\cdot\mathbf{v})) \left(\mathbf{k}\cdot\frac{\partial \Phi_p^e}{\partial \mathbf{p}}\right) \frac{d^3p}{(2\pi)^3},$$

which for a thermal electron distribution in the case when we have $\omega \ll k v_{Te}$ becomes

$$\mathrm{Im}\left\{\varepsilon_k^{(e)}\right\} = \frac{\omega_{pe}^2}{k^2 v_{Te}^2}\sqrt{\frac{\pi}{2}}\frac{\omega}{|k| v_{Te}}. \tag{10.7}$$

If Eq.(10.7) is valid we get from Eq.(10.6):

$$\delta n_k = \frac{Zk^2}{4\pi m_i}\frac{1 + i\sqrt{\pi/2}\,(\omega/k v_{Te})}{\omega^2 - k^2 v_s^2 + i\omega^2 \sqrt{\pi/2}\,(\omega/k v_{Te})}\left[|\mathcal{E}|^2\right]_k$$

$$\approx \frac{Zk^2}{4\pi m_i}\left[\omega^2 - k^2 v_s^2 + ik^2 v_s^2 \sqrt{\frac{\pi}{2}}\frac{\omega}{k v_{Te}}\right]^{-1}\left[|\mathcal{E}|^2\right]_k. \tag{10.8}$$

The extra term in the denominator can be written as $-2i k v_s \gamma_k^s$ where γ_k^s is the linear Landau damping of the sound waves. We have thus shown that the non-linear Landau damping of the Langmuir field corresponds to the linear damping of the density variations which follow this field. Using Eq.(10.8) we get the following equation for the density variations:

$$\frac{\partial^2 \delta n}{\partial t^2} - v_s^2 \nabla^2 \delta n + \sqrt{\frac{\pi Z m_e}{2m_i}}\int \frac{k v_s}{(2\pi)^3}\,e^{i(\mathbf{k}\cdot[\mathbf{r}-\mathbf{r}'])}\,d^3k\,\frac{\partial}{\partial t}\delta n(\mathbf{r}',t)\,d^3r'$$

$$= \frac{Z}{4\pi m_i}\nabla^2 |\mathcal{E}|^2. \tag{10.9}$$

Performing the integration over $\mathbf{k}$ we have

$$\frac{\partial^2 \delta n}{\partial t^2} - v_s^2 \nabla^2 \delta n + \sqrt{\frac{Z m_e}{2\pi^3 m_i}}\,v_s \int \frac{d^3r'}{|\mathbf{r}-\mathbf{r}'|^4}\,\frac{\partial}{\partial t}\delta n(\mathbf{r}',t)$$

$$= \frac{Z}{4\pi m_i}\nabla^2 |\mathcal{E}|^2. \tag{10.10}$$

This equation shows explicitly the long-range interaction between different structures. Although this interaction may be weak if the structures are far from one another, it could be important in the case when the structures are unstable against finite perturbations.

We can also write down these equations for the case of arbitrary non-thermal particle distribution functions. In the case of the one-dimensional description it is useful to introduce the particle velocity distribution function Φ_v:

$$\int \Phi_p \frac{d^3p}{(2\pi)^3} = \int \Phi_v\,dv. \tag{10.11}$$

The imaginary part of the dielectric permittivity can be expressed in terms of the velocity distribution function Φ_v as follows:

$$\text{Im}\left\{\varepsilon_k^{(e)}\right\} = -\frac{\pi\omega_{pe}^2}{n_0 k|k|}\frac{\partial \Phi_v^e}{\partial v}\bigg|_{v=\omega/k}, \tag{10.12}$$

and the one-dimensional equivalents of Eqs.(10.5) and (10.9) have in the case of arbitrary but given particle distributions the form:

$$i\frac{\partial\mathcal{E}}{\partial t} + \frac{3v_{Te}^2}{2\omega_{pe}}\frac{\partial^2\mathcal{E}}{\partial x^2} + i\frac{\omega_{pe}^3}{4n_0}\int dx'\,\mathcal{E}(x',t)\frac{e^{ik(x-x')}}{k|k|}\frac{\partial\Phi^e}{\partial v}\bigg|_{v=\omega_{pe}/k}$$

$$= \frac{\omega_{pe}}{2n_0}\,\delta n\,\mathcal{E}; \tag{10.13}$$

$$\frac{\partial^2\delta n}{\partial t^2} - v_s^2\frac{\partial^2\delta n}{\partial x^2} + i\frac{\pi v_s^2 v_{Te}^2}{Zn_0}\int\frac{k^3\,d\omega\,dt'\,dx'}{(2\pi)^2|k|}\,\delta n(x',t')$$

$$\times\, e^{ik(x-x')-i(\omega-\omega')t}\frac{\partial\Phi_v^e}{\partial v}\bigg|_{v=\omega/k} = \frac{Z}{4\pi m_i}\frac{\partial^2|\mathcal{E}|^2}{\partial x^2}. \tag{10.14}$$

The dissipative processes are in the one-dimensional case very sensitive to the actual particle distributions which can change due to the feedback through the interaction of the structures with the particle distributions. Similar equations can be derived from Eq.(10.6) for the interactions of the structures with resonant ions. We leave that problem as an exercise.

10.3 Self-consistent Description of the Interaction Between Structures and Resonant Particles

We shall now consider a self-consistent description which takes into account the changes in the particle distribution functions of the resonant particles due to their interactions with the fields of the regular structures. We shall for the moment leave aside the problem of a self-consistent description of the non-linear resonance (10.2) and concentrate on the problem of a self-consistent description of the linear resonance (10.1). We turn to Eq.(5.70) in which we take into account only the effects which are quadratic in the $\mathbf{E}^+$ and $\mathbf{E}^-$ fields:

$$\frac{d\Phi_P^e}{dt} = \frac{\partial\Phi_P^e}{\partial t} + \left(\mathbf{v}\cdot\frac{\partial\Phi_P^e}{\partial\mathbf{r}}\right)$$

$$= -e\left(\mathbf{E}^{(0)}\cdot\frac{\partial\Phi_P^e}{\partial\mathbf{p}}\right) - e\left(\mathbf{E}^+\cdot\frac{\partial f_P^{-,e}}{\partial\mathbf{p}}\right) - e\left(\mathbf{E}^-\cdot\frac{\partial f_P^{+,e}}{\partial\mathbf{p}}\right), \tag{10.15}$$

$$\frac{\partial f_P^{\pm,e}}{\partial t} + \left(\mathbf{v}\cdot\frac{\partial f_P^{\pm,e}}{\partial\mathbf{p}}\right) = -e\left(\mathbf{E}^\pm\cdot\frac{\partial\Phi_P^e}{\partial\mathbf{p}}\right). \tag{10.16}$$

Equation (10.15) contains only the quadratic terms in $\mathbf{E}^+$ and $\mathbf{E}^-$ whereas Eq.(10.16) contains the linear terms in $\mathbf{E}^+$ and $\mathbf{E}^-$. We shall be interested in the change in the distribution of the resonant particles, while $\mathbf{E}^{(0)}$ is the virtual field which describes mainly the perturbations of the non-resonant particles. To be more precise, we can divide the whole distribution function into resonant and non-resonant parts and assume that only the resonant part which is described by $\Phi_{\mathbf{p}}^{\text{res(e)}} = \Phi_{\mathbf{p}}^{\text{res(e)}}(\mathbf{r}, t)$ is time-dependent and inhomogeneous in space. We can then neglect the virtual field in Eq.(10.15). The non-resonant particles will be described by a distribution function $\Phi^{\text{non-res}}$ and are in the zeroth approximation assumed to be distributed uniformly. The inhomogeneous perturbations of the distribution of the non-resonant particles are produced by non-linear effects, including the above-mentioned virtual field. We can therefore treat the effect of the virtual field independently of and differently from the effect of the resonant field on the distribution of the resonant particles. We leave the calculation of the effect produced by the virtual field as an exercise and shall consider here the approximation in which the virtual field in Eq.(10.15) is neglected:

$$\frac{\partial \Phi_{\mathbf{p}}^e}{\partial t} + \left(\mathbf{v} \cdot \frac{\partial \Phi_{\mathbf{p}}^e}{\partial \mathbf{r}} \right) = -e \left(\mathbf{E}^+ \cdot \frac{\partial f_{\mathbf{p}}^{-,e}}{\partial \mathbf{p}} \right) - e \left(\mathbf{E}^- \cdot \frac{\partial f_{\mathbf{p}}^{+,e}}{\partial \mathbf{p}} \right). \qquad (10.17)$$

When solving Eq.(10.16) for the resonant particles we shall assume that their distribution depends on both $\mathbf{r}$ and t:

$$\varepsilon_{\mathbf{k}} E_{\mathbf{k},\omega_{\text{pe}}+\omega}^+ = -\frac{4\pi e^2}{k} \int \frac{1}{\omega_{\text{pe}} + \omega - (\mathbf{k} \cdot \mathbf{v}) + i0}$$

$$\times \left(E_{\mathbf{k}-\mathbf{k}',\omega_{\text{pe}}+\omega-\omega'}^+ \cdot \frac{\partial \Phi_{\mathbf{k}'}^{\text{res(e)}}}{\partial \mathbf{p}} \right) d^4 k' \frac{d^3 \mathbf{p}}{(2\pi)^3}. \qquad (10.18)$$

In the denominator on the right-hand side of this equation we can to a first approximation neglect ω as compared to ω_{pe} and we can also substitute:

$$E_{\mathbf{k}-\mathbf{k}',\omega_{\text{pe}}+\omega-\omega'}^+ = \frac{1}{(2\pi)^4} \int \mathcal{E}(\mathbf{r}', t') \, e^{-i\left([\mathbf{k}-\mathbf{k}']\cdot\mathbf{r}'\right)+i(\omega-\omega')t'} \, d^3\mathbf{r}' \, dt',$$

$$\Phi_{\mathbf{p},\mathbf{k}',\omega'}^{\text{res(e)}} = \frac{1}{(2\pi)^4} \int \Phi(\mathbf{r}'', t'')_{\mathbf{p}} \, e^{-i\left(\mathbf{k}'\cdot\mathbf{r}''\right)+i\omega't''} \, d^3\mathbf{r}'' \, dt''.$$

We then obtain a generalisation of Eq.(10.5):

$$\text{div} \left[i \frac{\partial \mathcal{E}(\mathbf{r}, t)}{\partial t} + \frac{3 v_{Te}^2}{2\omega_{\text{pe}}} \nabla^2 \mathcal{E}(\mathbf{r}, t) - \frac{\omega_{\text{pe}}}{2n_0} \delta n(\mathbf{r}, t) \mathcal{E}(\mathbf{r}, t) \right]$$

$$= -i \frac{e^2 \omega_{\text{pe}}}{(2\pi)^3} \int \frac{e^{i\left(\mathbf{k}\cdot[\mathbf{r}-\mathbf{r}']\right)}}{\omega_{\text{pe}} - (\mathbf{k} \cdot \mathbf{v}) + i0}$$

$$\times \left(\mathcal{E}(\mathbf{r}', t) \cdot \frac{\partial \Phi_{\mathbf{p}}^{\text{res(e)}}(\mathbf{r}', t)}{\partial \mathbf{p}} \right) d^3\mathbf{k} \, d^3\mathbf{r}' \frac{d^3\mathbf{p}}{(2\pi)^3}. \qquad (10.19)$$

In the one-dimensional case we have:

$$i\frac{\partial \mathcal{E}(x,t)}{\partial t} + \frac{3v_{Te}^2}{2\omega_{pe}}\frac{\partial^2 \mathcal{E}(x,t)}{\partial x^2} - \frac{\omega_{pe}}{2n_0}\delta n(x,t)\mathcal{E}(x,t)$$

$$+ \frac{\omega_{pe}^3}{4n_0}\int \frac{e^{ik(x-x')}\,dk\,dv}{\pi k(\omega_{pe} - kv + i0)}$$

$$\times \mathcal{E}(x',t)\frac{\partial \Phi_v^{res(e)}(x',t)}{\partial v}\,dx' = 0, \tag{10.20}$$

which is a generalisation of Eq.(10.13).

The inhomogeneity of the distribution function introduces a new physical effect which is connected with the transition radiation of particles passing through the inhomogeneity. Formally, not only the Landau pole, but also the principal value of the integral contributes to the imaginary part of the right-hand side of Eq.(10.20). We shall illustrate this using the example of the one-dimensional case. Remembering that

$$\frac{1}{k(\omega_{pe} - kv + i0)} = \frac{1}{\omega_{pe}}\left[\frac{1}{k} + \frac{v}{\omega_{pe} - kv + i0}\right], \tag{10.21}$$

we see that the first term on the right-hand side of Eq.(10.21) does not give a contribution to the right-hand side of Eq.(10.20) and a simple calculation of the second term gives

$$\int \frac{e^{ik(x-x')}\,dk}{\omega_{pe} - kv + i0} = -\pi i\left(\frac{v}{|v|} + \frac{x-x'}{|x-x'|}\right)e^{i\omega_{pe}(x-x')/v}, \tag{10.22}$$

so that the one-dimensional equation for the field will have the form

$$i\frac{\partial \mathcal{E}(x,t)}{\partial t} = -\frac{3v_{Te}^2}{2\omega_{pe}}\frac{\partial^2 \mathcal{E}(x,t)}{\partial x^2} + \frac{\omega_{pe}}{2n_0}\delta n(x,t)\mathcal{E}(x,t)$$

$$+ i\frac{\omega_{pe}^2}{4n_0}\int \mathcal{E}(x',t)\,e^{i\omega_{pe}(x-x')/v}\,dx'\,dv$$

$$\times \left(\frac{v}{|v|} + \frac{x-x'}{|x-x'|}\right)\frac{\partial \Phi_v^{res(e)}(x',t)}{\partial v}. \tag{10.23}$$

The first term in brackets in Eq.(10.23) is the one which corresponds to the Landau pole and it was the only one which was taken into account in Eq.(10.13). This is the first difference with Eq.(10.13). The second difference is that Eq.(10.23) contains the actual distribution function at a given time, which changes in time and space due to the action of the regular fields, rather than the initial distribution function as in Eq.(10.13). We should also mention that Eq.(10.23) contains the distribution function at a position x' which differs from that of the coordinate of the field we are interested in. The equation which we have obtained is a generalisation of the quasi-linear equation and it is valid for any regular field. The appearance of the term involving $(x - x')/|x - x'|$ is related to the inhomogeneity of the particle distribution. The contribution of this term to the energy balance will be zero in the case of a uniform particle distribution. Indeed, we find from Eq.(10.23) that

$$\frac{d}{dt} \int |\mathcal{E}|^2 \, dx = \cdots$$

$$+ \frac{\omega_{pe}^2}{4n_0} \int \frac{x-x'}{|x-x'|} \, e^{i(x-x')\omega_{pe}/v} \, dv \, \mathcal{E}^*(x,t)\mathcal{E}(x',t)$$

$$\times \frac{\partial \Phi_v^{res(e)}(x',t)}{\partial v} \, dx \, dx' + \frac{\omega_{pe}^2}{4n_0} \int \frac{x-x'}{|x-x'|} \, e^{-i(x-x')\omega_{pe}/v} \, dv$$

$$\times \mathcal{E}(x,t)\mathcal{E}^*(x',t) \frac{\partial \Phi_v^{res(e)}(x',t)}{\partial v} \, dx \, dx'. \tag{10.24}$$

In the case of a uniform particle distribution we find, by the substitution $x \leftrightarrows x'$, that the two terms which are written out in detail in Eq.(10.24) cancel one another. Sometimes it is convenient to introduce the coordinate along the particle velocity, $l' = x'v/|v|$, and Eq.(10.23) becomes in that case

$$i\frac{\partial \mathcal{E}(x,t)}{\partial t} + \frac{3v_{Te}^2}{2\omega_{pe}} \frac{\partial^2 \mathcal{E}(x,t)}{\partial x^2} + \frac{\omega_{pe}}{2n_0} \delta n(x,t)\mathcal{E}(x,t) = i\frac{\omega_{pe}^2}{4n_0} \int_0^\infty dl'$$

$$\times e^{i\omega_{pe}l'/|v|} \frac{vdv}{|v|} \mathcal{E}(x',t) \frac{\partial \Phi_v^{res(e)}(x',t)}{\partial v}\bigg|_{x'=x-vl'/|v|} \tag{10.25}$$

In the three-dimensional case we can for every value of $\mathbf{v}$ introduce a coordinate, l, which is the component of $\mathbf{r}$ parallel to $\mathbf{v}$ and the component, $\mathbf{r}_\perp$ at right angles to $\mathbf{v}$:

$$l = \frac{(\mathbf{r} \cdot \mathbf{v})}{|\mathbf{v}|}, \qquad \mathbf{r} = \mathbf{r}_\perp + \frac{\mathbf{v}}{|\mathbf{v}|}l. \tag{10.26}$$

If we do this we can write the three-dimensional equation in the form

$$\mathrm{div}\left[i\frac{\partial \mathcal{E}(\mathbf{r},t)}{\partial t} + \frac{3v_{Te}^2}{2\omega_{pe}}\nabla^2\mathcal{E}(\mathbf{r},t) - \frac{\omega_{pe}}{2n_0}\delta n(\mathbf{r},t)\mathcal{E}(\mathbf{r},t)\right]$$

$$= 2\pi e^2 \omega_{pe} \int \frac{d^3p'}{(2\pi)^3} \frac{1}{|\mathbf{v}|} \int_0^\infty dl' \, e^{i\omega_{pe}l'/|\mathbf{v}|}$$

$$\times \left(\mathcal{E}(\mathbf{r}',t) \cdot \frac{\partial \Phi_{\mathbf{p}}^{res(e)}(\mathbf{r}',t)}{\partial \mathbf{p}}\right)_{\mathbf{r}'=\mathbf{r}-\mathbf{v}l'/|\mathbf{v}|} \tag{10.27}$$

We can find the equation for the distribution function by substituting the solution of Eq.(10.16) into Eq.(10.17):

$$\frac{d\Phi_{\mathbf{p}}^e}{dt} = \widehat{I}_{\mathbf{p}}^{QL}. \qquad \widehat{I}_{\mathbf{p}}^{QL} = -e\left(\mathbf{E}^+ \cdot \frac{\partial f_{\mathbf{p}}^{-,e}}{\partial \mathbf{p}}\right) - e\left(\mathbf{E}^- \cdot \frac{\partial f_{\mathbf{p}}^{+,e}}{\partial \mathbf{p}}\right), \tag{10.28}$$

$$\widehat{I}_{\mathbf{p}}^{QL} = \frac{\partial}{\partial p_i} \int D_{ij}(\mathbf{r},\mathbf{r}',t) \frac{\partial \Phi_{\mathbf{p}}^e(\mathbf{r}',t)}{\partial p_j} \, d^3r', \tag{10.29}$$

$$
D_{ij}(\mathbf{r}, \mathbf{r}', t) = -\frac{2e^2}{(2\pi)^3} \, \mathrm{Re} \left\{ \int \frac{e^{i\left(\mathbf{k}\cdot[\mathbf{r}-\mathbf{r}']\right)}}{i(\omega_{pe} - (\mathbf{k}\cdot\mathbf{v}) + i0)} \right.
$$
$$
\left. \times \, \mathcal{E}_i^*(\mathbf{r}, t)\mathcal{E}_j(\mathbf{r}', t)\, d^3\mathbf{k} \right\}. \tag{10.30}
$$

Equation (10.29) can also be written in the form

$$
\widehat{I}_p^{\mathrm{QL}} = 2e^2 \, \mathrm{Re} \left\{ \frac{\partial}{\partial p_i} \int_0^\infty dl' \, e^{i\omega_{pe} l'/|\mathbf{v}|} \right.
$$
$$
\left. \times \, \mathcal{E}_i^*(\mathbf{r}, t)\mathcal{E}_j\left(\mathbf{r} - \frac{\mathbf{v}}{|\mathbf{v}|}l', t\right) \frac{\partial \Phi_p^e(\mathbf{r}', t)}{\partial p_j}\bigg|_{\mathbf{r}'=\mathbf{r}-\mathbf{v}l'/|\mathbf{v}|} \right\}. \tag{10.31}
$$

In the one-dimensional case the coordinate x occurs, as in Eq.(10.25). The set of equations which we have obtained can be called the general equations for the resonant wave-particle interactions. They represent a generalisation of the quasi-linear approach and they are valid for the description of the interactions between regular coherent structures and resonant particles. From this set of equations we can easily obtain the conservation laws for the energy and the momentum:

$$
\frac{d}{dt}\left[\int \frac{p^2}{2m} \widehat{I}_p^{\mathrm{QL}} \frac{d^3\mathbf{p}\, d^3\mathbf{r}}{(2\pi)^3} + \int \frac{|\mathcal{E}|^2}{2\pi} d^3\mathbf{r} \right] = 0. \tag{10.32}
$$

In the case of the momentum conservation law the first term should be

$$
\int \mathbf{p}\, \widehat{I}_p^{\mathrm{QL}} \frac{d^3\mathbf{p}\, d^3\mathbf{r}}{(2\pi)^3}. \tag{10.33}
$$

The difference between the present and the quasi-linear conservation laws is that now we are integrating over the whole of space. For the structures this means that conservation of energy or of momentum holds only after we integrate over the whole volume occupied by the structures.

10.4 Excitation and Damping of Structures in Beam–Plasma Interactions

The phase relations between the harmonics in the structures are held in place by the modulational interactions. If the growth rate of the beam-plasma instability is smaller than the reciprocal of the characteristic time scale of the interactions responsible for the formation of the structure, the structure as a whole will be excited or damped. It can contain harmonics which are damped, harmonics which are growing, and even harmonics whih do not interact directly with the beam. All these harmonics will behave in the same way if the coherence between them is kept in place by the modulational interactions. If the harmonics are uncorrelated we can restrict our discussion to those which are unstable. However, in the case we are considering at the moment this cannot be done. For instance, in the velocity distribution of the beam there will be both a part for which the derivative with respect to the velocity is positive, and a part for which it is negative. In the case where there is coherence between the harmonics both parts will contribute to the balance and the structure as a whole will be either growing or decreasing in amplitude. We shall consider the interaction of a beam with a soliton for the case of a weak kinetic one-dimensional beam. We can write down the equations for the interaction of a beam with regular structures which we shall use for this problem instead of the quasi-linear equation which we used earlier:

$$
i \frac{\partial \mathcal{E}(x,t)}{\partial t} \frac{3v_{Te}^2}{2\omega_{pe}} \frac{\partial^2 \mathcal{E}(x,t)}{\partial x^2} - \frac{\omega_{pe}}{2n_0} \delta n(x,t) \mathcal{E}(x,t)
$$

$$
= i \frac{\pi e^2}{m_e} \int \mathcal{E}(x',t) e^{i\omega_{pe}(x-x')/v} \, dx' \, dv
$$

$$
\times \left(\frac{x-x'}{|x-x'|} + \frac{v}{|v|} \right) \frac{\partial \Phi_v^{res(e)}(x',t)}{\partial v}, \tag{10.34}
$$

$$
\frac{\partial \Phi_v^{res(e)}}{\partial t} + v \frac{\partial \Phi_v^{res(e)}}{\partial x} = \frac{e^2}{m_e^2} \operatorname{Re} \left\{ \frac{\partial}{\partial v} \int e^{i\omega_{pe}(x-x')/v} \, dx' \right.
$$

$$
\left. \times \mathcal{E}^*(x,t) \mathcal{E}(x',t) \frac{1}{v} \left[\frac{x-x'}{|x-x'|} + \frac{v}{|v|} \right] \frac{\partial \Phi_v^{res(e)}(x',t)}{\partial v} \right\}. \tag{10.35}
$$

We shall also write down the expression for a soliton structure when there is no beam:

$$
\mathcal{E}^{sol}(x,t) = \mathcal{E}_0 e^{-i\omega_0 t + i\omega_{pe} u(x-ut)/3v_{Te}^2}
$$

$$
\times \left\{ \cosh \left[\frac{\mathcal{E}_0 \omega_{pe}(x-ut)}{4v_{Te} \sqrt{3\pi n_0 m_i (v_s^2 - u^2)}} \right] \right\}^{-1}, \tag{10.36}
$$

$$
\omega_0 = \omega_{pe} \frac{u^2}{2v_{Te}^2} - \frac{Z\omega_{pe}}{16\pi n_0 m_i} \frac{\mathcal{E}_0^2}{v_s^2 - u^2}. \tag{10.37}
$$

If there is a beam present, the soliton parameters $\mathcal{E}_0$ and u will depend on both time and space. In the case of an initially homogeneous beam the soliton parameters will initially depend only on the time. However, through the interaction with the soliton the beam will become inhomogeneous. We shall not give a complete solution of the complicated coupled equations for the particle distribution function and the soliton parameters, but we shall illustrate here the effect of the coherence on the excitation and the damping of the soliton. We shall be interested in the rate of change in energy which can be found from either Eq.(10.34) or Eq.(10.35):

$$\frac{\partial}{\partial t} \int \frac{|\mathcal{E}(x,t)|^2}{2\pi} \, dx = \frac{e^2}{m_e} \operatorname{Re} \left\{ \int \mathcal{E}^*(x,t)\mathcal{E}(x',t)e^{i\omega_{pe}(x-x')/v} \, dx \, dx' \, dv \right.$$

$$\left. \times \left[\frac{x-x'}{|x-x'|} + \frac{v}{|v|} \right] \frac{\partial \Phi_v^{\mathrm{res}(e)}(x',t)}{\partial v} \right\}. \qquad (10.38)$$

We now substitute the soliton field (10.36) into the right-hand side of Eq.(10.38):

$$\frac{\partial}{\partial t} \int \frac{|\mathcal{E}(x,t)|^2}{2\pi} \, dx = \frac{e^2}{m_e} \operatorname{Re} \left\{ \mathcal{E}_0^2 \int \left[\frac{x-x'}{|x-x'|} + \frac{v}{|v|} \right] \frac{\partial \Phi_v^{\mathrm{res}(e)}(x'+ut,t)}{\partial v} \right.$$

$$\times \left[\cosh \frac{x\mathcal{E}_0\omega_{pe}}{4v_{Te}\sqrt{3\pi n_0 m_i (v_s^2 - u^2)}} \cosh \frac{x'\mathcal{E}_0\omega_{pe}}{4v_{Te}\sqrt{3\pi n_0 m_i (v_s^2 - u^2)}} \right]^{-1}$$

$$\left. \times \exp \left[i(x-x')\omega_{pe} \left(\frac{1}{v} - \frac{u}{3v_{Te}^2} \right) \right] dx \, dx' \, dv \right\}. \qquad (10.39)$$

Let us analyse this expression qualitatively. The values of x and x' are of the order of the soliton width. This means that the factor in the exponent is of the order of

$$\left[\frac{v_{Te}}{v} - \frac{u}{3v_{Te}} \right] \frac{\sqrt{48 n_0 m_i (v_s^2 - u^2)}}{\mathcal{E}_0} \leqslant 1. \qquad (10.40)$$

If inequality (10.40) is satisfied the exponent is less than unity; in the opposite case the fast oscillating exponential will make the interaction very inefficient. Resonance in Eq.(10.40) means that we have

$$u = \frac{3v_{Te}^2}{v}, \qquad u = v_{gr} = \frac{3v_{Te}^2 k}{\omega_{pe}} = \frac{3v_{Te}^2}{v_{ph}}, \qquad v = v_{ph}. \qquad (10.41)$$

The last equation appears to be trivial. However, our estimates are only as to order of magnitude since the soliton does not contain just a single harmonic with definite phase and group velocities. The only meaning of our estimate is that the soliton harmonic with a group velocity equal to the soliton velocity corresponds to resonance of its phase velocity with the particle velocity. In the present discussion the width of the resonance curve is due to the presence of other harmonics in the soliton structure. It is clear that if the width of the velocity distribution of the beam particles is such that

$$\frac{\delta v_b}{v_b} \leqslant \frac{v_b}{v_{Te}} \sqrt{\frac{\mathcal{E}_0^2}{48\pi n_0 m_i (v_s^2 - u^2)}} \equiv \frac{\delta v_b^*}{v_b}, \qquad (10.42)$$

all the particles are inside the resonance curve, including the particles for which the derivative of the distribution function with respect to the velocity is negative. These will contribute to the damping. The question whether the soliton will decrease or increase its amplitude depends on the energy balance between damped and amplified harmonics. We must also mention that since the soliton cannot have a velocity larger than the ion-sound velocity we have, if we combine this with the resonance condition, the inequality

$$\frac{v_b}{v_{Te}} \geqslant \frac{3v_{Te}}{v_s} = 3\sqrt{\frac{m_i}{Zm_e}}, \tag{10.43}$$

which is the same as the criterion for non-linear stabilisation found earlier. The stabilisation is here produced by the modulational interactions which are able to correlate the harmonics in the wavepacket on a time scale which is smaller than that of the growth of the harmonics due to the excitation by the beam. The non-linear stabilisation is reflected here in the slow growth of the soliton amplitude. We shall see that the time scale of the growing is larger than the reciprocal of the linear growth rate and that the amplitude grows proportional to the time to the power $\frac{1}{3}$.

We shall give an estimate of the growth rate of the soliton assuming that the change in the particle distribution due to the interaction is small. This will be valid for the initial stage of the excitation of a soliton by a beam. We shall also assume that the width of the beam distribution is much smaller than the one defined by Eq.(10.42) with the equal sign. The beam is assumed to be homogeneous in space and directed in the positive direction, $v > 0$.

By adding to the integrand of Eq.(10.39) the same expression but with the interchange $x \leftrightarrows x'$ and using the formula

$$\int_0^\infty \frac{\cos \alpha x}{\cosh x} \, dx = \frac{\pi}{2} \frac{1}{\cosh \frac{1}{2}\pi\alpha}, \tag{10.44}$$

we find

$$\frac{\partial}{\partial t} \int \frac{|\mathcal{E}|^2}{2\pi} \, dx = 12\pi^2 v_{Te}^2 m_i (v_s^2 - u^2) \int \frac{\partial \Phi_v^{res(e)}}{\partial v} \, dv$$

$$\times \left[\cosh \left\{ 2 \left(\frac{v_{Te}}{v} - \frac{u}{3v_{Te}} \right) \frac{\sqrt{3\pi n_0 m_i (v_s^2 - u^2)}}{\mathcal{E}_0} \right\} \right]^{-2}. \tag{10.45}$$

Integrating by parts in (10.45) and assuming that the beam particles form a beam with $\delta v_b \ll v_b$ we get

$$\frac{\partial}{\partial t} \int \frac{|\mathcal{E}|^2}{24\pi^3 m_i (v_s^2 - u^2)} \, dx = -\frac{2v_{Te}^3}{\mathcal{E}_0} \sqrt{3\pi n_0 m_i (v_s^2 - u^2)}$$

$$\times \int \frac{\Phi_v^{res(e)}}{v^2} \frac{\sinh \varrho}{\cosh^3 \varrho} \, dv, \tag{10.46}$$

where ϱ is the argument of the cosh in Eq.(10.45),

$$\varrho = 2\left(\frac{v_{Te}}{v} - \frac{u}{3v_{Te}}\right)\frac{\sqrt{3\pi n_0 m_i\left(v_s^2 - u^2\right)}}{\mathcal{E}_0}.$$

This argument can be rewritten if we use the definition of the critical spread δv_b^* given by Eq.(10.42) as follows:

$$\varrho = \frac{v_{res} - v}{2\delta v_b^*}, \qquad v_{res} = \frac{3v_{Te}^2}{u}. \tag{10.47}$$

In the case of a narrow beam with $\delta v_b \ll \delta v_b^*$ the sign of the change in energy is determined by $\varrho_b = (v_{res} - v_b)/2\delta v_b^*$ and it will be positive when $\varrho_b < 0$, or, when

$$u > \frac{3v_{Te}^2}{v_b}. \tag{10.48}$$

If $\varrho_b \ll 1$ and $u \ll v_s$ we find from Eq.(10.46) that

$$\frac{d}{dt}\int |\mathcal{E}|^2 \, dx = 12\pi^3 T_e \varrho_b \frac{v_{Te}^2}{\delta v_b^*} n_b = 6\pi^3 n_b T_e \frac{v_{Te}^2\left(v_b - v_{res}\right)}{(\delta v_b^*)^2}, \tag{10.49}$$

with

$$\delta v_b^* = \frac{v_b^2}{2v_{Te}\sqrt{3}}\left[\frac{\mathcal{E}_0^2}{4\pi n_0 T_e}\right]^{1/2}.$$

On the other hand, for a soliton we have

$$\int |\mathcal{E}|^2 \, dx = \frac{4\sqrt{6}\pi v_{Te}}{\omega_{pe}}\left[\frac{\mathcal{E}_0^2}{4\pi n_0 T_e}\right]^{1/2} n_0 T_e, \tag{10.50}$$

which gives the following equation for the soliton amplitude:

$$\frac{\mathcal{E}_0^2}{4\pi n_0 T_e}\frac{d}{dt}\left[\frac{\mathcal{E}_0^2}{4\pi n_0 T_e}\right]^{1/2} = 3\sqrt{6}\pi^2\frac{v_{Te}^3\left(v_b - v_{res}\right)}{v_b^4}\frac{n_b}{n_0}\omega_{pe}, \tag{10.51}$$

from which follows that

$$\mathcal{E}_0 \propto t^{1/3}. \tag{10.52}$$

The coherent excitation of a Langmuir field is thus quite different from the incoherent excitation. Equation (10.52) does not describe an exponential, but a much slower, growth which is proportional to $t^{1/3}$. Even more interesting can be the final result of this instability. In the case where there were no coherent structures there appeared in the final state of the beam a plateau in the velocity distribution at which stage the growth of the Langmuir field stopped. The coherent excitation should also lead to a change in the distribution function such that the soliton amplitude no longer increases, which means, according to Eq.(10.46) that

$$\int \frac{\Phi_v^{\text{res}(e)}}{v^2} \frac{\sinh \varrho}{\cosh^3 \varrho} \, dv = 0. \tag{10.53}$$

A soliton satisfying Eq.(10.53) will neither grow nor be damped: part of its distribution function produces growth and another part produces damping. These two parts cancel one another out and the amplitudes of all the harmonics neither increase nor decrease. Such a soliton possesses always both an energy sink and an energy source: it is an open system – a self-organised soliton. The development of the beam-plasma interaction leads thus to the formation of a self-organised structure – a self-organised soliton. We can give an estimate of the time needed for its formation. First of all, if $\delta v_b \ll v_b$, we can approximately neglect δv_b in the denominator in Eq.(10.53). We see from Eq.(10.47) that then the final distribution should be symmetric relative to the resonance velocity. Initially v_b should be larger than v_{res} and the change in the beam distribution should therefore occur in a range of velocities of the order of

$$\delta v_{\text{res}} \equiv v_b - v_{\text{res}}. \tag{10.54}$$

From Eq.(10.53) we can find an estimate for the distance δx_{res} over which this change in the distribution function will occur:

$$\frac{v_b}{\delta x_{\text{res}}} \cong \frac{e^2 \mathcal{E}_0^2 v_b}{m_e^2 (\delta v_{\text{res}})^3 \omega_{\text{pe}}} = \omega_{\text{pe}} \frac{\mathcal{E}_0^2}{4\pi n_0 T_e} \frac{v_b v_{Te}^2}{(\delta v_{\text{res}})^3}. \tag{10.55}$$

We have here taken into account that according to Eq.(10.49) and a similar equation for the particle distribution the effective range of $x' - x$ is of the order of $v_b^2/\delta v_{\text{res}}\omega_{\text{pe}}$. The total change in the beam particle energy will be given by the integral of the difference in the particle energies over the distance δx_{res}, that is, we have the following estimate for the energy given to the soliton:

$$n_b m_e v_b \delta v_{\text{res}} \delta x_{\text{res}} \cong n_b m_e \frac{v_b (\delta v_{\text{res}})^4}{v_{Te}^2 \omega_{\text{pe}}} \frac{4\pi n_0 T_e}{\mathcal{E}_0^2}. \tag{10.56}$$

From this equation and using Eq.(10.50) we find the following estimate for the amplitude $\mathcal{E}_{0*}$ of the self-organised soliton:

$$\left[\frac{\mathcal{E}_{0*}^2}{45\pi n_0 T_e} \right]^{3/2} \cong \frac{n_b}{n_0} \frac{v_b (v_{\text{res}})^4}{v_{Te}^5}. \tag{10.57}$$

The time τ_* to reach this amplitude is independent of the beam density and we get from Eq.(10.51) the following estimate for it:

$$\frac{1}{\tau_*} \cong \omega_{\text{pe}} \frac{v_{Te}^8}{v_b^8} \left(\frac{v_b}{\delta v_{\text{res}}} \right)^3 \leqslant \left(\frac{m_e}{9m_i} \right)^2 \left(\frac{v_b}{\delta v_{\text{res}}} \right)^3 \omega_{\text{pe}}. \tag{10.58}$$

The last inequality is a consequence of Eq.(10.43). The interaction with the soliton makes the beam inhomogeneous in space. The characteristic size of this inhomogeneity is determined by Eq.(10.55). Substituting in that equation the amplitude (10.57) we find

$$\delta x_{\text{res}} \cong d_e \left(\frac{n_0}{n_b} \right)^{2/3} \frac{(\delta v_{\text{res}})^{1/3} v_{Te}^{1/3}}{v_b^{2/3}}. \tag{10.59}$$

All these relations show explicitly the difference between a coherent and an incoherent excitation by a beam.

10.5 Non-linear Landau Damping of Structures

We have so far in our discussion of the beam-plasma interactions neglected the non-linear Landau damping. We shall now show that it can change the soliton velocity u and thus take the soliton out of resonance with the beam. We have shown earlier that the stimulated scattering also affects the beam-plasma interaction in the incoherent case. The change in the soliton velocity due to stimulated scattering is of interest independently of the problem of the beam-plasma interaction. One can consider the stimulated scattering processes in the case of coherent structures self-consistently by taking into account the changes in both the structure parameters and the particle distribution. We shall here give estimates for the case of stimulated scattering by electrons for a given thermal electron distribution.

We have shown earlier that the non-linear damping of a Langmuir field is described by the linear Landau damping of the variations δn in the density which accompany the Langmuir field. We shall use Eq.(10.14) which is satisfied by the variations δn_{sol} in the soliton density if we neglect the effects of the damping of the variations in the density. We study the effect of the damping using perturbation theory and assuming that

$$\delta n = \delta n_{\text{sol}} + \delta n', \tag{10.60}$$

and that the extra variations $\delta n'$ in the density are relatively small. We then get from Eq.(10.14)

$$\frac{\partial^2 \delta n'}{\partial t^2} - v_s^2 \frac{\partial^2 \delta n'}{\partial x^2} = -i \frac{\pi v_{Te}^2 v_s^2}{n_0} \int \frac{k^3 \, dk \, d\omega}{(2\pi)^2 |k|}$$
$$\times \; \delta n_{\text{sol}}(x', t') \, dx' \, dt' \, e^{ik(x-x')-i\omega(t-t')} \left. \frac{\partial \Phi_v^e}{\partial v} \right|_{v=\omega/k}. \tag{10.61}$$

To find the solution of Eq.(10.61) we Fourier transform it, divide the equation by $\omega^2 - k^2 v_s^2$ to find the solution and then perform the inverse Fourier transformation; the result is:

$$\delta n'(x,t) = i\frac{\pi v_{Te}^2 v_s^2}{n_0} \int \frac{k^3\, dk\, d\omega}{(2\pi)^2 |k|(\omega^2 - k^2 v_s^2)}$$

$$\times\, \delta n_{\mathrm{sol}}(x' - ut')\, dx'\, dt'\, e^{ik(x-x')-i\omega(t-t')} \left.\frac{\partial \Phi_v^e}{\partial v}\right|_{v=\omega/k}. \quad (10.62)$$

Changing the integration over x' to an integration over $x' - ut'$ we can easily integrate over t' and we find

$$\delta n'(x,t) = \delta n'(x - ut)$$

$$= i\frac{v_{Te}^2 v_s^2}{2n_0} \int \frac{k\, e^{ik(x-ut)-ikx'}\, dk}{|k|(u^2 - v_s^2)}\, \delta n_{\mathrm{sol}}(x')\, dx',\ \left.\frac{\partial \Phi_v^e}{\partial v}\right|_{v=u}. \quad (10.63)$$

We note that expression (10.63) is a real quantity; this is due to the appearance of $|k|$. The density variations will not change the number of waves and hence also leave the soliton amplitude unchanged. This means that Eq.(10.63) describes the change in the soliton velocity. In the case of a thermal particle distribution the solitons are decelerated, that is, their velocity decreases. If the velocity is much larger than the ion thermal velocity this deceleration is due to the interaction with the electrons. The electrons which participate in this process are those with velocity components along the propagation direction of the soliton which are much smaller than the average thermal velocity. The main feedback of the process of decreasing the soliton velocity on the particle distribution is a change of the derivative with respect to the velocity of the distribution of a small fraction of the particles (see Eq.(10.63)). The soliton can produce this effect since it needs only a small amount of energy to form a plateau for $v \sim u$. The nature of this process may be self-organising. It is thus possible that a self-organised soliton may keep its velocity constant by changing the electron distribution in the low-velocity range. When the soliton velocity is less than or of the order of the ion thermal velocity the majority of the ions can participate in the deceleration of the soliton and it is not possible to prevent this deceleration by the formation of a plateau in the ion distribution. The deceleration by ions is so fast that the soliton spends most of the period when its velocity decreases with velocities larger than the ion thermal velocity when the main deceleration effect is due to interactions with the electrons.

We shall now estimate the time for the deceleration of a soliton by electrons in the stage when the plateau in the electron distribution function has not yet been formed. We write down the equation for the field, adding to it the density perturbations (10.63) due to the non-linear Landau damping by electrons:

$$i\frac{\partial \mathcal{E}(x,t)}{\partial t} + \frac{3v_{Te}^2}{2\omega_{pe}}\frac{\partial^2 \mathcal{E}(x,t)}{\partial x^2} = \frac{\omega_{pe}}{2n_0}\left[\delta n_{\mathrm{sol}}(x,t) + \delta n'(x,t)\right]\mathcal{E}(x,t). \quad (10.64)$$

When we evaluate the change in momentum we must take the presence of $\delta n'$ into account (see Eq.(9.67)):

$$\frac{dP}{dt} = \frac{1}{4\pi} \int \frac{\delta n'(x,t)}{n_0} \frac{\partial |\mathcal{E}_{\text{sol}}(x,t)|^2}{\partial x} \, dx. \tag{10.65}$$

Since $|\mathcal{E}_{\text{sol}}|^2$ depends solely on $x - ut$ we have

$$\delta n_{\text{sol}}(x - ut) = -\frac{|\mathcal{E}_{\text{sol}}(x - ut)|^2}{4\pi m_i (v_s^2 - u^2)}, \tag{10.66}$$

while for a thermal distribution we have

$$\left.\frac{\partial \Phi_v^e}{\partial v}\right|_{v=u} = -\frac{un_0}{\sqrt{2\pi}v_{Te}^3} e^{-u^2/2v_{Te}^2} \approx -\frac{un_0}{\sqrt{2\pi}v_{Te}^3}, \tag{10.67}$$

which leads to

$$\frac{dP}{dt} = -i\frac{v_s u}{(v_s^2 - u^2)^2} \sqrt{\frac{\pi m_e}{8Zm_i}} \frac{1}{16\pi^3 n_0 m_i}$$
$$\times \int \frac{k \, dk \, dx \, dx'}{|k|} |\mathcal{E}_{\text{sol}}(x')|^2 \frac{\partial |\mathcal{E}_{\text{sol}}(x)|^2}{\partial x} e^{ik(x-x')}. \tag{10.68}$$

In evaluating the integral over k the small imaginary part must be taken into account and we get:

$$\frac{dP}{dt} = \frac{v_s u}{(v_s^2 - u^2)^2} \sqrt{\frac{\pi m_e}{8Zm_i}} \frac{1}{8\pi^3 n_0 m_i}$$
$$\times \int \frac{2 dx \, dx'}{x - x'} |\mathcal{E}_{\text{sol}}(x')|^2 \frac{\partial |\mathcal{E}_{\text{sol}}(x)|^2}{\partial x}. \tag{10.69}$$

We have written down this expression to demonstrate that there is a long-range interaction between different solitons. Indeed, we can use it not only in the case when there is only a single soliton present, but also for the case of any density perturbations moving with the same velocity u. The modulational interactions keep the wavepackets, which form the solitons, together while the relatively weaker interaction (10.69) describes their *long-range interactions*. In the one-dimensional case this interaction is inversely proportional to the distance apart of the solitons as can be seen from Eq.(10.69).

We now go back to Eq.(10.68) and take into account the symmetry properties of the integrand to write

$$\frac{dP}{dt} = \frac{2v_s u}{(v_s^2 - u^2)^2} \sqrt{\frac{\pi m_e}{8Zm_i}} \frac{1}{8\pi^3 n_0 m_i}$$
$$\times \int_0^\infty dk \int dx \, dx' \cos kx' \, |\mathcal{E}_{\text{sol}}(x')|^2 \sin kx \frac{\partial |\mathcal{E}_{\text{sol}}(x)|^2}{\partial x}$$
$$= -\frac{2v_s u}{(v_s^2 - u^2)^2} \sqrt{\frac{\pi m_e}{8Zm_i}} \frac{1}{8\pi^3 n_0 m_i}$$
$$\times \int_0^\infty k \, dk \left[\int \cos kx \, |\mathcal{E}_{\text{sol}}(x')|^2 \, dx\right]^2. \tag{10.70}$$

Substituting now the field of a single soliton

$$|\mathcal{E}_{sol}(x)|^2 = \frac{|\mathcal{E}_0|^2}{\cosh^2(x/x_0)}, \qquad x_0 = \frac{4v_{Te}\sqrt{3\pi n_0 n_i(v_s^2 - u^2)}}{\mathcal{E}_0\omega_{pe}}, \qquad (10.71)$$

and using the fact that

$$\int_0^\infty \frac{\cos kx\, dx}{\cosh^2(x/x_0)} = \frac{\pi kx^2}{2\sinh(\frac{1}{2}\pi kx_0)}, \qquad \int_0^\infty \frac{k^3\, dk}{\sinh^2 k} = \tfrac{3}{2}\zeta(3),$$

where $\zeta(3)(= 1.08)$ is the Riemann ζ function for the argument 3, we find that

$$\frac{dP}{dt} = \frac{6\zeta(3)v_s u}{\pi^5(v_s^2 - u^2)^2}\sqrt{\frac{\pi m_e}{8Zm_i}}\frac{|\mathcal{E}_0|^4}{n_0 m_i}. \qquad (10.72)$$

We find an expression for the characteristic time τ_{st} for stopping a soliton in the $u \ll v_s$ case by using the expression for the soliton momentum

$$P = \frac{4n_0 m_i u}{\sqrt{3}}\frac{v_{Te}}{\omega_{pe}}\left(\frac{\mathcal{E}_0^2}{4\pi n_0 T_e}\right)^{3/2}. \qquad (10.73)$$

The result is:

$$\frac{1}{\tau_{st}} = \frac{1}{u}\frac{du}{dt} = \frac{96\zeta(3)}{\pi^3}\frac{v_s}{x_0}\sqrt{\frac{\pi m_e}{8Zm_i}}, \qquad (10.74)$$

where x_0 is the soliton width given by Eq.(10.71). For $Z = 1$ the path length over which the stopping occurs is approximately $5\sqrt{m_i/m_e}$ times longer than the soliton width.

These results also demonstrate the coherent effects in non-linear damping.

10.6 Coherent Landau Damping. Arrest of the Self-contraction

The linear Landau damping of a caviton is also a coherent effect since all the harmonics of the caviton are damped at the same time when they are held together by the modulational interactions. This statement is correct both for one-dimensional cavitons–solitons–and for three-dimensional self-contracting cavitons. Due to the coherent effects a caviton can be damped even for sizes much larger than the Debye length. This effect is known as arresting the Langmuir collapse.

We shall first discuss the Landau damping of a one-dimensional caviton, that is, a soliton. We shall assume the electron distribution to be thermal.

We define the damping rate of the soliton structure as a whole by the obvious relation

$$2\gamma_{\text{str}} = \frac{1}{\int |\mathcal{E}|^2 \, dx} \frac{\partial}{\partial t} \int |\mathcal{E}|^2 \, dx. \tag{10.75}$$

In this expression we substitute expression (10.45) for the thermal distribution and use Eq.(10.50) for $\int |\mathcal{E}|^2 \, dx$. We then get

$$\gamma_{\text{str}} = -\frac{3\pi^{3/2} x_0}{d_e} \omega_{\text{pe}} \int_0^\infty \frac{y \, dy \, e^{-y^2/2}}{\cosh^2 (x_0/2d_e y)}, \qquad y = \frac{v}{v_{Te}}, \tag{10.76}$$

$$\frac{\mathcal{E}_0}{\sqrt{12\pi n_0 T_e}} \gg \frac{u}{3v_{Te}}. \tag{10.77}$$

Since $x_0 \gg d_e$ we can use the method of steepest descent for evaluating the integral in Eq.(10.76):

$$\gamma_{\text{str}} = -4\sqrt{6}\pi^2 \omega_{\text{pe}} \left(\frac{x_0}{d_e}\right)^{4/3} \exp\left\{-\frac{3}{2} \left(\frac{x_0}{d_e}\right)^{2/3}\right\}. \tag{10.78}$$

Apart from a large numerical factor this expression contains an important difference in its dependence on the caviton size from the simplified estimate for the damping of a single harmonic k_0 where k_0 is of the order of $2\pi/x_0$. Before discussing this point we shall consider the damping of a three-dimensional caviton. In order to evaluate the power dissipated in a three-dimensional caviton we turn to Eq.(10.19) and in it change the denominator ω_{pe} to $(\mathbf{k}\cdot\mathbf{v})$. This is possible since if we replace the factor ω_{pe} in the numerator by $\omega_{\text{pe}} - (\mathbf{k}\cdot\mathbf{v})$, the resulting expression will vanish as the integrand then contains a derivative of the particle distribution. We can now replace $\mathbf{k}$ by $-i\partial/\partial \mathbf{r}$:

$$\text{div}\left(i\frac{\partial \mathcal{E}}{\partial t} + \frac{3v_{Te}^2}{\omega_{\text{pe}}} \nabla^2 \mathcal{E}\right) = -\frac{e^2}{(2\pi)^3} \left(\frac{\partial}{\partial \mathbf{r}} \cdot \int \frac{e^{i(\mathbf{k}\cdot[\mathbf{r}-\mathbf{r}'])}}{\omega_{\text{pe}} - (\mathbf{k}\cdot\mathbf{v}) + i0}\right.$$

$$\times \left.\left(\mathbf{v}\cdot\mathcal{E}(\mathbf{r}',t)\right) \frac{\partial \Phi_{\mathbf{p}}^{\text{res}(e)}}{\partial \mathbf{p}} d^3k \, d^3r' \frac{d^3p}{(2\pi)^3}\right). \tag{10.79}$$

Multiplying this equation by ϕ^*, where ϕ is the scalar potential, $\mathcal{E} = -\nabla\phi$, and the complex conjugate equation by ϕ and subtracting the two resulting equations from one another we find the total power Q dissipated by the caviton:

$$Q = -\frac{\partial}{\partial t} \int \frac{|\mathcal{E}|^2}{2\pi} d^3r = \frac{2e^2}{(2\pi)^3} \text{Re}\left\{\int \frac{e^{i(\mathbf{k}\cdot[\mathbf{r}-\mathbf{r}'])} \left(\mathbf{v}\cdot\mathcal{E}^*(\mathbf{r},t)\right)}{i(\omega_{\text{pe}} - k + i0)}\right.$$

$$\times \left.\left(\mathcal{E}(\mathbf{r}',t)\cdot\frac{\partial}{\partial \mathbf{p}}\right) \Phi_{\mathbf{p}}^{\text{res}(e)}(\mathbf{r}',t) d^3r \, d^3r' \frac{d^3p}{(2\pi)^3}\right\}. \tag{10.80}$$

We shall consider the case of inhomogeneous particle distributions separately. We shall in the case of a uniform thermal distribution use an expansion of the field of the structure in terms of spatial Fourier harmonics:

$$\mathcal{E}(\mathbf{r}, t) = -\int i\mathbf{k}\phi_{\mathbf{k}}(t) e^{i(\mathbf{k}\cdot\mathbf{r})} d^3k. \tag{10.81}$$

we then obtain

$$Q = \frac{(2\pi)^{3/2}\omega_{pe}^4}{2v_{Te}^5} \int |\phi_{\mathbf{k}}(t)|^2 \delta(\omega_{pe} - (\mathbf{k}\cdot\mathbf{v})) e^{-v^2/2v_{Te}^2} d^3v\, d^3k, \tag{10.82}$$

or

$$Q = \frac{(2\pi)^{5/2}\omega_{pe}^4}{2v_{Te}^3} \int |\phi_{\mathbf{k}}(t)|^2 \frac{d^3k}{k} e^{-\omega_{pe}^2/2k^2 v_{Te}^2}. \tag{10.83}$$

The energy in the caviton is equal to

$$\int \frac{|\mathcal{E}(\mathbf{r}, t)|^2}{2\pi} d^3r = (2\pi)^2 \int k^2 |\phi_{\mathbf{k}}|^2 d^3k. \tag{10.84}$$

From a definition similar to Eq. (10.75), but now for the three-dimensional case, we have for γ_{str}:

$$\gamma_{str} = \left[\sqrt{\frac{\pi}{8}} \frac{\omega_{pe}^4}{v_{Te}^3} \int \frac{d^3k}{k} |\phi_{\mathbf{k}}(t)|^2 e^{-\omega_{pe}^2/2k^2 v_{Te}^2}\right] \Big/ \left[\int k^2 |\phi_{\mathbf{k}}(t)|^2 d^3k\right]$$

$$= \left[\int \gamma_{\mathbf{k}}^L k^2 |\phi_{\mathbf{k}}(t)|^2 d^3k\right] \Big/ \left[\int k^2 |\phi_{\mathbf{k}}(t)|^2 d^3k\right], \tag{10.85}$$

where $\gamma_{\mathbf{k}}^L$ is the Landau damping rate for a single harmonic. It can be seen from Eq. (10.82) that we have for the sake of simplicity neglected the thermal corrections to the frequency of the Langmuir waves. If one takes those into account the expression $-\omega_{pe}^2/2k^2 v_{Te}^2$ in the exponent in Eq. (10.85) must be changed to $-\omega_{pe}^2/2k^2 v_{Te}^2 - 3/2$, as in the expression for the linear Landau damping while the expression after the second equal sign in Eq. (10.85) is correct also when the thermal dispersion of the Landau waves is taken into account. The modulational interactions keep all the harmonics in the structure together so that γ_{str} is the damping rate for the structure as a whole. For a dipole-type structure we take

$$\phi(\mathbf{r}) = \frac{(\mathbf{d}\cdot\mathbf{r})}{1 + r^2/a^2}, \qquad \phi_{\mathbf{k}} = i\frac{(\mathbf{d}\cdot\mathbf{k})}{k^3} a^2(1 + ka) e^{-ka}, \tag{10.86}$$

where $\mathbf{d}$ is the unit vector along the direction of the dipole, and we find

$$\gamma_{str} = \frac{\sqrt{2\pi}a}{5} \frac{\omega_{pe}^4}{v_{Te}^3} \int (1 + ka)^2 \frac{dk}{k^3} e^{-2ka - \omega_{pe}^2/2k^2 v_{Te}^2}. \tag{10.87}$$

Using the steepest descent method to integrate we get

$$\gamma_{\text{str}} = \omega_{\text{pe}} \frac{\pi}{20\sqrt{3}} \left(\frac{2a}{d_e} \right)^{8/3} \exp\left[-\frac{3}{2} \left(\frac{2a}{d_e} \right)^{2/3} \right]. \tag{10.88}$$

This expression differs from Eq.(10.78) in the numerical coefficient. However, its general form is the same: both expressions give a much larger damping than one would obtain from an estimate putting in the expression for the Landau damping of a single harmonic k to be of the order of $2\pi/a$ which depends on a in the form $a^3 \exp(-2\pi^2 a^2 \omega_{\text{pe}}^2 / v_{Te}^2)$. In the case of that expression damping is important for sizes of the order of $2d_e$ or $3d_e$ whereas in Eqs.(10.78) and (10.88) damping will affect the self-contraction for sizes of the order of $10d_e$ or $20d_e$. The coherence of the structures thus increases their linear damping. The appearance of damping for such large sizes makes them dissipative structures rather soon after the self-contraction starts. We may remind ourselves that supersonic self-contraction which is the fastest corresponds to sizes of the order of $d_e\sqrt{3m_i/m_e}$ which for hydrogen is of the order of $80d_e$. This means that the inertial self-contraction affects a change in size of not more than by a factor 8 or 10 – in the case of a hydrogen plasma. This phenomenon of large coherent dissipation of a self-contracting caviton is called the *arrest of the collapse*. Almost all computer experiments on strong plasma turbulence have found that the dissipation of the cavitons occurs for sizes much larger than the Debye size – in fact, for sizes of about $10d_e$ or $15d_e$. Laboratory experiments show that the collapse is halted at about $20d_e$.

An important conclusion of these considerations is not only that the caviton needs an energy source to survive and that it in that case becomes a self-organising structure, but also that this damping will convert the energy mainly to fast particles. The appearance of these fast particles means that a superthermal tail appears in the particle distribution function for which the absorption is even larger. This leads to a further increase in the tail and increases the damping of larger size cavitons. We may therefore expect that this kind of instability will lead to a strong dissipation of cavitons of sizes of the order of c/ω_{pe}, since the particle velocity cannot exceed the speed of light. The process of the conversion of energy to fast particles will inevitably lead to their distribution being inhomogeneous in the neighbourhood of a caviton. In that case a new absorption mechanism will begin to operate – transition damping.

10.7 Coherent Transition Damping of Cavitons

The absorption of energy by fast particles produces inhomogeneities in their distribution in the neighbourhood of a caviton. These inhomogeneities may lead to additional absorption of waves and additional dissipation of the caviton. This absorption mechanism is the transition radiation which occurs when a particle travels through an inhomogeneous refractive index. We shall consider here the general case of arbitrary inhomogeneities in the resonant particle distribution. For the sake of simplicity we shall assume again that the electron velocity distribution is thermal. This means that we write

$$\Phi_p(r, t) = \Phi_p^{th} \psi(\mathbf{r}) n_0,$$

where Φ_p^{th} is the thermal distribution while $\psi(\mathbf{r})$ describes the coordinate dependence of the particle distribution inside the caviton. We shall also introduce the Fourier transform of the field, $\mathcal{E}_k = -ik\phi_k$. Instead of Eq.(10.85) we have

$$\gamma_{str} = \gamma_{str}^L + \gamma_{str}^{tr}, \tag{10.89}$$

where γ_{str}^L describes the Landau damping of the structure which is modified by the inhomogeneity,

$$\gamma_{str}^L = \frac{\int \gamma_k^L \, \mathrm{Re}\left\{(\mathcal{E}_k^* \cdot (\mathcal{E}\psi)_k)\right\} d^3k}{\int |\mathcal{E}_k|^2 \, d^3k}, \tag{10.90}$$

while γ_{str}^{tr} describes the additional, transition damping:

$$\gamma_{str}^{tr} = \left[\pi\omega_{pe}^4 \int \frac{v \, dv \, d^3k}{v_{Te}^5 k^3} \left(\ln\left|\frac{1 - kv/\omega_{pe}}{1 + kv/\omega_{pe}}\right| + \frac{2kv}{\omega_{pe}}\right)\right.$$
$$\left. \times \, \mathrm{Im}\left\{(\mathcal{E}_k^* \cdot (\mathcal{E}\psi)_k\right\} e^{-v^2/2v_{Te}^2}\right] \Big/ \left[(2\pi)^{3/2} \int |\mathcal{E}_k|^2 \, d^3k\right]. \tag{10.91}$$

In the case of a uniform particle distribution we have $\psi = 1$ and Eq. (10.90) becomes the same as Eq.(10.85) while the transition damping is zero. The two contributions are of the same order of magnitude in the case of an inhomogeneity with a size of the order of the size of a caviton. The difference between them lies in the fact that the Landau damping contains a δ-function so that only particles with velocities larger than the phase velocity of a given harmonic contribute whereas the transition damping has a smooth maximum at $\omega_{pe} = kv$. It is therefore unnecessary for the transition damping that the resonance condition is satisfied rigorously. To illustrate this we consider transition damping for the case when all the harmonics in the structure are far from resonance and have phase velocities much larger than the particle velocity. The Landau damping will be exponentially small – even though the exponent contains the cavity size only to the power $\frac{2}{3}$. On the other hand, the transition damping does not decrease exponentially. Indeed, for $k \ll \omega_{pe}/v$ we get from Eq.(10.91):

$$\gamma_{str}^{tr} = -\frac{3v_{Te}^2}{\omega_{pe}} \frac{\int \mathrm{Im}\left\{(\mathcal{E}_k^* \cdot (\mathcal{E}\psi)_k)\right\} d^3k}{\int |\mathcal{E}_k|^2 \, d^3k}$$
$$= -\frac{3v_{Te}^2}{\omega_{pe}} \frac{\int |\mathcal{E}_0(\mathbf{r})|^2 \, (\nabla\phi(\mathbf{r}) \cdot \nabla\psi(\mathbf{r})) \, d^3r}{\int |\mathcal{E}_0(\mathbf{r})|^2 \, d^3r}, \tag{10.92}$$

where we have put $\mathcal{E}(\mathbf{r}) = \mathcal{E}_0(\mathbf{r})e^{-i\phi(\mathbf{r})}$. When we transformed back to the coordinate representation we took into account that the integrand in the numerator contains the expression

$$\text{Im}\left\{\left(\frac{\partial\mathcal{E}^*(\mathbf{r})}{\partial\mathbf{r}} \cdot \frac{\partial[\mathcal{E}(\mathbf{r})\psi(\mathbf{r})]}{\partial\mathbf{r}}\right)\right\} = |\mathcal{E}_0(\mathbf{r})|^2\left(\frac{\partial\phi(\mathbf{r})}{\partial\mathbf{r}} \cdot \frac{\partial\psi(\mathbf{r})}{\partial\mathbf{r}}\right).$$

We shall see in the next chapter that if the inhomogeneity in the distribution function is connected with the Landau damping Eq.(10.92) does not describe the whole effect in the case when the gradients of the distribution function are smooth. The problem is that transition radiation and transition damping need large gradients which correspond to appreciable changes over distances at least comparable with or less than the wavelength, whereas Eq.(10.92) formally seems to be valid also for any small gradients of the distribution function. If the gradients are very smooth the properties of the caviton should change adiabatically.

We must also point out that the sign of the right-hand side of Eq.(10.92) can be both negative or positive depending on the gradients of the phase of the waves in the caviton and the gradient of the distribution function. The effect could thus result in an amplification of the waves in the caviton. We shall discuss all these problems in detail in Chap.11. Here we point out that if the self-contraction is sufficiently strong the caviton has a size of the order of several wavelengths of the waves which are trapped in it and if the inhomogeneity of the distribution function is of the order of the size of the caviton, it will be of the order of the wavelength of the waves in the caviton. Under those conditions transition radiation and transition damping will just exist, on the limit of their appearance. We must also mention that all the above comments refer to Eq.(10.92) which was obtained assuming that the gradients were smooth from the general Eq.(10.91) which is valid for any gradients.

The magnitude of the transition damping as estimated from the general expression may be important even in the early stages of the self-contraction. It can be much larger than the Landau damping, but it depends on the nature of the gradients of the distribution function. We found earlier that the phase factor of a soliton depends on x. Similarly, in the case of three-dimensional structures the phase of ϕ will depend on $\mathbf{r}$. The degree of inhomogeneity of this phase factor may be due to (1) the self-contraction process itself, (2) the inhomogeneity in the particle distribution function produced by resonant heating, and (3) external perturbations of the phase. The effect of smooth inhomogeneities due to resonant heating – Landau damping – is connected with the general problem of adiabatic invariants in non-linear interactions and we shall discuss it in Chap.11. Although in a self-contraction the phases inside the structure are synchronised they are still not rigorously homogeneous and they depend on the type of self-contraction. The last possibility, which is due to the inhomogeneity of the phases which occurs through interaction with external perturbations, is very important because it introduces the possibility of external sources changing the phases without appreciably changing the energy of the caviton. It is probable that such effects will occur when there are several self-contracting cavitons present in the system. In such circumstances self-contraction can proceed in exceptional circumstances when there are special conditions in the neigbourhood of a particular caviton. Of course, if one creates special experimental conditions one can eliminate such perturbations and observe an unperturbed self-contraction. Such an important experimental proof of the possibility of the existence of a self-contracting process exists in fact. However, in the case of a turbulent state we may expect that there are several contracting cavitons present as well as other perturbing infuences such as short-wavelength random fields. In that case the dissipation will be important for most of the cavitons and they will be dissipative structures. There is experimental and computational evidence that the energy dissipated in the final stage of the observed self-contraction of several cavitons is less than the total dissipated energy.

10.8 Interaction Between Structures and Random Fields

There are two random components interacting with coherent structures – randomly distributed particles and random waves. So far we have considered the problem of the interaction between structures and particles. We now turn to the problem of the interaction between structures and random waves. The most interesting case is when the size of a structures is much larger than the length scale of the random waves. We shall be interested in random waves with such small length scales that there are no modulational interactions between them. When we described pure regular or pure random waves we found that the only parameter determining whether or not modulational interactions are present is the ratio of the dispersion and the non-linear frequency broadening. If there are no regular fields present the random fields will be homogeneous. If the size of the structure is much larger than the characteristic scale of the turbulence the inhomogeneity introduced by the structure in the turbulence will be a slight inhomogeneity as compared to the scales of the turbulent waves. Here we shall consider that case. The turbulence will be characterised not only by the dispersion and by the non-linear broadening but also by the degree of inhomogeneity and one needs to find the rôle of the latter in the non-linear interactions. The inhomogeneity parameter may appear in the interactions between small-scale turbulence and a regular structure. On the other hand, when we are dealing with the interactions of turbulent motions with each other we can only compare the relative rôles of dispersion and non-linearity. We shall assume that the dispersion of the random waves is rather large. We also assume that the criterion for the occurrence of modulational interactions between the random waves when there

are no regular waves present is not satisfied. Indeed, it is difficult to satisfy
the criterion for modulational interactions in the case of small scales. In this
case we may expect that non-linear resonances will govern the interactions
not only between random waves and other random waves, but also between
random waves and the waves in the structure. However, the situation appears
to be more complicated and, in fact, only one term in the interactions between
random and regular waves is governed by the resonances determined by the
dispersion of the random waves.

Before giving a general description of the interactions between random
and regular waves we shall start our discussion assuming that for the interactions between random and regular waves the large dispersion of the random
waves makes it possible to treat these interactions on the basis of the non-
linear resonances. The point is that the large dispersion of the random waves
may dominate also the non-linear frequency shift of the waves in the structure. We shall show later that the assumption that the dispersion of the
random waves governs the interactions between the random and the regular
waves is correct only for part of those interactions. We shall give an estimate
of that part of the interactions in the present section.

The interaction is between very different scales–those of the random
waves and those of the waves in the structure–and this means that we can
neglect scattering by ions. Scattering by electrons leads to interactions between waves on very different scales. Since in the linear approximation the
result is independent of whether the excited waves are random or regular,
we can use for the excitation of structures the non-linear Eq.(6.50) which
we found earlier. We then find for the scattering by electrons the following
expression for the excitation of a structure by isotropic small-scale Langmuir
turbulence:

$$\gamma_k^{\mathrm{NL}} = \frac{4}{81}\sqrt{\frac{\pi}{2}} \int \frac{\omega_{\mathrm{pi}}^4}{k'^3 v_{Te}^3} W_{k'}^\ell \, dk'. \tag{10.93}$$

How to get Eq.(10.63) from Eq.(6.50) we leave as an exercise. The quantity
γ_k^{NL} is always positive, but as it is independent of k, the growth rate of the
structure is the same as that of each of its harmonics. Expression (10.93) is
not very large as it contains ω_{pi} to the fourth power. However, in principle
it enhances the self-contraction. Scattering by ions becomes important when
during the self-contraction the caviton meets the flux from the small-scale
turbulence. In that case the size of the random waves will be comparable
with the size of the structure and we shall no longer be able to give a simple
analytical description.

Another important process which works in the opposite direction is due to
the presence of small-scale ion-sound turbulence. We showed earlier that ion-
sound oscillations with wavelengths smaller than the electron Debye length
always lead to damping of Langmuir waves. This damping is enhanced by the
Landau damping of the self-contracting caviton since this process creates fast

electrons. Scattering by fast super-thermal electrons is allowed for ion-sound waves with wavelengths which are much smaller than the electron Debye length. Such waves are automatically created by the self-contracting cavitons since the Landau damping of the waves in the caviton is rather fast and after the damping an empty caviton is left. As the empty caviton does not self-contract it will decay in the background of the ion-sound waves. The process of the damping of the waves in a caviton by the scattering of ion-sound waves by fast electrons is described by the expression:

$$\gamma_k^{NL} = -\frac{1}{27} \int \gamma_{k'}^L \frac{\omega_{pe}^4}{k'^4 v_{Te}^4} \frac{W_{k'}^s}{n_0 T_e} \, dk', \tag{10.94}$$

where $\gamma_{k'}^L$ is the linear Landau damping by fast electrons. The non-linear damping (10.94) is also independent of k and is the same for all harmonics in the caviton. This damping can stop the self-contraction. We leave the derivation of Eq.(10.94) also as an exercise.

An important feature of the interaction of regular structures with random fields is that the structures will produce a change in the energy fluxes of the random–turbulent–fields. In the linear approximation this change is proportional to the field of the regular structure and it can be considered to be the linear response of the turbulence of the random fields to the regular field of the structure. The growth rate (10.93) and the damping rate (10.94) describe the imaginary parts of this response. We have thus a simple physical interpretation for the growth or the damping of regular structures which interact with random fields.

The next important point is the possibility of the appearance of inhomogeneities in the random fields due to their interactions with the fields of the regular structures. If such inhomogeneities increase the interactions between the random fields and the fields of the regular structures we have a feedback process leading to a *self-organising instability*.

10.9 General Theory of the Interaction Between Regular and Random Fields

We shall now consider the interactions between random and regular Langmuir fields in the framework of the cubic non-linearity. We shall show that due to this interaction no random field can exist without exciting regular fields and no regular field can exist without exciting random fields. We shall denote the random fields by δE and the regular fields by E. We discussed in Chap.5 the non-linear description of pure regular fields and found that they to a good approximation can be described by the first term of Eq.(5.34):

$$ik\varepsilon E^+ = 8\pi \int \varrho_{1,2,3}^{eff} E_1^+ E_2^+ E_3^- \, d_{1,2,3}. \tag{10.95}$$

We also showed that for random fields we have

$$ik\varepsilon\delta E^+ = 8\pi \int \varrho_{1,2,3}^{\text{eff}} \left(\delta E_1^+ \delta E_2^+ \delta E_3^- - \delta E_1^+ \langle \delta E_2^+ \delta E_3^- \rangle\right.$$
$$\left. - \langle \delta E_1^+ \delta E_2^+ \delta E_3^- \rangle\right) d_{1,2,3}. \tag{10.96}$$

The difference between the two equations is connected with the difference between the starting equations (5.36) and (5.37). If both kinds of fields are present there appear cross-terms describing the interactions between the random and the regular fields in both equations. The last term in Eq.(10.96) is unimportant since after multiplying by $\delta E_{k'}^-$ and subsequent averaging – to obtain an equation for the correlation – it gives a zero result. We shall drop it in what follows. The difference between the two equations is therefore the subtraction of a term in the equation for the random fields. However, this subtraction occurs only due to the way we have defined random and regular fields. We should therefore expect a corresponding term in the equation for the regular fields. Indeed, this is the case. If we use the same procedure for separating regular and random fields we shall find in the equation for the regular fields terms containing the random fields. In the equation for the regular fields we shall retain of those terms only those which are linear in the regular fields. This is reasonable since therms containing three random fields will be zero to a first approximation if we are dealing with weak turbulence. We have

$$ik\varepsilon E^+ = 8\pi \int \varrho_{1,2,3}^{\text{eff}} \left(E_1^+ E_2^+ E_3^- + \langle \delta E_1^+ E_2^+ \delta E_3^- \rangle\right.$$
$$\left. + \langle E_1^+ \delta E_2^+ \delta E_3^- \rangle\right) d_{1,2,3}. \tag{10.97}$$

We can express the second term on the right-hand side in terms of the non-linear permittivity of the random fields (see Eq.(6.41) while the third term on the right-hand side corresponds to the one we dropped in Eq.(10.96). We can therefore write Eq.(10.97) in the form

$$ik\left(\varepsilon + \varepsilon^{\text{NL}}\right)E^+ = 8\pi \int \varrho_{1,2,3}^{\text{eff}} \left(E_1^+ E_2^+ E_3^- + E_1^+ \langle \delta E_2^+ \delta E_3^- \rangle\right) d_{1,2,3}, \tag{10.98}$$

where

$$\varepsilon_k^{\text{NL}} = \frac{8\pi}{ik} \int \varrho_{k_1,k,-k_1}^{\text{eff}} |E^+|_{k_1}^2 \, d^4 k_1. \tag{10.99}$$

We have thus proved that we can in the linear approximation use the expression for the non-linear permittivity for the random waves. This is the way we obtained Eqs.(10.93) and (10.94). However, Eq.(10.96) contains the random fields also on the right-hand side. We shall show that this term is determined by the earlier mentioned inhomogeneities in the random fields which are produced by the regular fields.

To describe the interaction with the regular fields it is reasonable to leave in the equation for the random fields only terms which are linear in the random fields, which means that they are quadratic in the regular fields. Indeed, the terms linear in the regular fields will be quadratic in the random fields and after multiplication by E^- and averaging will give zero in the case of weak turbulence. For the random fields we therefore use the equation

$$ik\varepsilon\delta E^+ = 8\pi \int \varrho^{\text{eff}}_{1,2,3} \left(\delta E_1^+ \delta E_2^+ \delta E_3^- - \delta E_1^+ \langle \delta E_2^+ \delta E_3^- \rangle \right.$$
$$\left. + \delta E_1^+ E_2^+ E_3^- + E_1^+ \delta E_2^+ E_3^- \right) d_{1,2,3}. \tag{10.100}$$

10.10 Interactions Between Regular Structures and Inhomogeneous Turbulence

It follows from Eq.(10.98) that, indeed, the non-linear permittivity ε^{NL} on the left-hand side of Eq.(10.98) contains the difference between the frequencies of the regular and the random waves. The large dispersion of the random waves makes it possible to study the interaction due to this term in the framework of non-linear resonances – as was done in §10.8. In this term the inhomogeneity of the turbulence leads only to a small correction to the interaction. However, the last term in Eq.(10.98) shows a different behaviour. In the case of stationary, homogeneous turbulence the difference between the frequencies will be exactly equal to zero. This means that the main properties of this term will be determined by the inhomogeneity and non-stationarity of the turbulence. We must in that case define the correlation functions of the random fields for the non-stationary, inhomogeneous case. We shall use the following definition:

$$\int \langle \delta E^+_{k_2} \delta E^-_{k_2+k'} \rangle \, e^{i(k' \cdot r) - i\omega' t} \, d^3k' \, d\omega' = - \, |E^+|^2_{k_2}(r, t). \tag{10.101}$$

We have used here a notation which is close to the one of Eq.(9.17) since the last term in Eq.(10.98) can be reduced to an extra density variation $< \delta n(r, t) >$, produced by the turbulence. Evaluating this term repeats the calculations carried out when we derived Eqs.(9.30) and (9.28). The resulting equation will contain an extra term involving $< \delta n >$:

$$\text{div} \left[i \frac{\partial}{\partial t} - i \gamma^L_k - i \gamma^{\text{NL}}_k + \frac{3v^2_{Te}}{2\omega_{pe}} \nabla^2 \right] \mathcal{E}$$
$$= \frac{\omega_{pe}}{2n_e} \, \text{div} \left[(\delta n^{\text{reg}} + < \delta n >) \mathcal{E} \right], \tag{10.102}$$

where $< \delta n >$ is in the general case determined by the averaged Eq.(9.17). The quantity γ^{NL}_k in Eq.(10.102) is determined by the expressions we considered earlier (see Eqs.(10.93) and (10.94)). In the case when the changes in the turbulence are produced by rather strong regular fields governed by their modulational interactions we must take into account the fact that the space-time variations of the turbulence will be of the same order of magnitude as the space-time variations of the regular fields. We have shown already that the dispersion of the random fields does not appear in the expression for $< \delta n >$. In the case when we can to a first approximation neglect the imaginary parts of the responses when describing the regular fields we can also neglect the imaginary parts when describing $< \delta n >$. In that case we have

$$- \frac{\partial^2 \delta n^{\mathrm{reg}}}{\partial t^2} - v_{\mathrm{s}}^2 \nabla^2 \delta n^{\mathrm{reg}} = \frac{Z}{4\pi m_{\mathrm{i}}} \nabla^2 |\mathcal{E}(\mathbf{r}, t)|^2,$$

$$\frac{\partial^2 < \delta n >}{\partial t^2} - v_{\mathrm{s}}^2 \nabla^2 < \delta n > = \frac{Z}{4\pi m_{\mathrm{i}}} \nabla^2 \int |E^+|_{\mathbf{k}_1}^2 (\mathbf{r}, t) \, d^4 k_1. \tag{10.103}$$

In the case of externally given inhomogeneous turbulence the imaginary part of the responses describe the extra damping of the structures and their long-range interactions. We leave this problem as an exercise.

10.11 Inhomogeneity of the Turbulence Excited by Interactions with Regular Structures

Equation (10.100) for the turbulent waves contains also two terms describing the interaction between the turbulent waves and the regular fields. The last term on the right-hand side depends on the dispersion of the random waves whereas the penultimate term is independent of it. In order to derive the equation for the correlation functions for the random field we must first multiply this equation by $\delta E_{-\mathbf{k}+\kappa}^-(1/k) e^{i(\boldsymbol{\kappa} \cdot \mathbf{r}) - i\nu t}$, $\kappa \equiv \{\boldsymbol{\kappa}, \nu\}$, then subtract from the result the equation for $\delta E_{-\mathbf{k}+\kappa}^-$ multiplied by $\delta E_{\mathbf{k}}^+(1/|\mathbf{k} - \boldsymbol{\kappa}|) e^{i(\boldsymbol{\kappa} \cdot \mathbf{r}) - i\nu t}$, after that integrate the result over κ, using the definition (10.101), and finally take into account the fact that the scale of the random fields is much smaller than that of the regular fields, that is that $|\boldsymbol{\kappa}| \ll k$. The result when there are no regular fields is obvious and is the same as the one we obtained earlier in the case of weak turbulence. We shall write it down, emphasising that it is valid in the case when we use the definition (10.101):

$$\left[\frac{\partial}{\partial t} + \left(v_{\mathrm{gr}} \cdot \frac{\partial}{\partial \mathbf{r}} \right) \right] |E^+|_{\mathbf{k}}^2 (\mathbf{r}, t) = 2\gamma_{\mathbf{k}}^{\mathrm{NL}} (\mathbf{r}, t) |E^+|_{\mathbf{k}}^2 (\mathbf{r}, t), \tag{10.104}$$

where $\gamma_{\mathbf{k}}^{\mathrm{NL}}$ is the non-linear growth rate which we here write in the following form:

$$\gamma_{\mathbf{k}}^{\mathrm{NL}} = \int \frac{(\mathbf{k} \cdot \mathbf{k}')^2}{k^2 k'^2} \gamma_{\mathbf{k},\mathbf{k}'}^{\mathrm{NL}} |E^+|_{\mathbf{k}'}^2 (\mathbf{r}, t) \, d^4 k', \tag{10.105}$$

where according to Eq.(6.41) we have

$$\gamma_{\mathbf{k},\mathbf{k}'}^{\mathrm{NL}} = -\frac{e^2}{2m_e\omega_{\mathrm{pe}}^3}|\mathbf{k}-\mathbf{k}'|^2\,\mathrm{Im}\left\{\frac{(\varepsilon_{\mathbf{k}-\mathbf{k}'}^{(e)}-1)\varepsilon_{\mathbf{k}-\mathbf{k}'}^{(i)}}{\varepsilon_{\mathbf{k}-\mathbf{k}'}}\right\}_{\substack{\omega=\omega_{\mathbf{k}}\\\omega'=\omega_{\mathbf{k}'}}}. \qquad (10.106)$$

We derived Eq.(10.104) indirectly in Chap.6. We derived the left-hand side of the equation and its general form from the balance equation and we derived the non-linear growth rate from the correspondence principle. This was the simplest way to derive the result. A direct derivation of the equation follows the lines we have sketched a moment ago and we leave the details of this independent derivation as an exercise.

Equation (10.105) describes the interactions between short-wavelength random waves and short-wavelength random waves. The interactions with regular waves for the last term in Eq.(10.100) should be very similar to Eq.(10.105) except that the wavelengths of the regular waves are much longer which means that for them to a rough approximation we have $k' \cong 0$. In this limit we can write Eq.(10.106) in the form

$$\gamma_{\mathbf{k},0}^{\mathrm{NL}} = -\frac{e^2}{2m_e\omega_{\mathrm{pe}}^3}k^2\,\mathrm{Im}\left\{\frac{(\varepsilon_{\mathbf{k}}^{(e)}-1)\varepsilon_{\mathbf{k}}^{(i)}}{\varepsilon_{\mathbf{k}}}\right\}_{\omega=\omega_{\mathbf{k}}}, \qquad (10.107)$$

The reason why we have split off the angular factor in Eq.(10.105) is that this factor has no definite limit as $k' \to 0$. We can therefore write an extra term on the right-hand side of Eq.(10.104) corresponding to the last term in the general Eq.(10.100) for the case when $\gamma_{\mathbf{k}}^{\mathrm{NL}} \to \gamma_{\mathbf{k}}^{\mathrm{NL}} + \gamma_{\mathbf{k}}^{\mathrm{NL,reg}}$, where

$$\gamma_{\mathbf{k}}^{\mathrm{NL,reg}} = \gamma_{\mathbf{k},0}^{\mathrm{NL}}\frac{|(\mathbf{k}\cdot\boldsymbol{\mathcal{E}}(\mathbf{r},t))|^2}{k^2}. \qquad (10.108)$$

The penultimate term on the right-hand side of Eq.(10.100) is independent of the difference between the frequencies of the random and the regular waves and is determined by the development of only the regular fields. Indeed, this term is determined by the quantity δn^{reg}. In calculating the contribution from this term we can therefore use the simple equation

$$\left\{\omega-\omega_{\mathrm{pe}}-\frac{3k^2v_{Te}^2}{2\omega_{\mathrm{pe}}}\right\}\delta E_{\mathbf{k}}^+ = \frac{\omega_{\mathrm{pe}}}{2n_0k}\left(\mathbf{k}\cdot[\delta n^{\mathrm{reg}}\delta\mathbf{E}^+]_{\mathbf{k}}\right). \qquad (10.109)$$

Proceeding as before we find from this equation the following equation for the expression (10.101):

$$\left[\frac{\partial}{\partial t}+\left(v_{\mathrm{gr}}\cdot\frac{\partial}{\partial\mathbf{r}}\right)\right]|E^+|_{\mathbf{k}}^2(\mathbf{r},t) = \frac{\omega_{\mathrm{pe}}}{2n_0}\left(\frac{\partial}{\partial\mathbf{k}}|E^+|_{\mathbf{k}}^2(\mathbf{r},t)\cdot\frac{\partial}{\partial\mathbf{r}}\delta n^{\mathrm{reg}}(\mathbf{r},t)\right)$$
$$-\frac{\partial}{\partial\omega}|E^+|_{\mathbf{k}}^2(\mathbf{r},t)\frac{\partial}{\partial t}\delta n^{\mathrm{reg}}(\mathbf{r},t)$$
$$+2\left[\gamma_{\mathbf{k}}^{\mathrm{NL}}(\mathbf{r},t)+\gamma_{\mathbf{k}}^{\mathrm{NL,reg}}(\mathbf{r},t)\right]|E^+|_{\mathbf{k}}^2(\mathbf{r},t). \qquad (10.110)$$

We have included here on the right-hand side the terms which we found earlier.

10.12 Development of Self-organisation

We shall now as one of the limiting cases consider the interactions between random and regular fields for the case when we can neglect both the non-linear interactions of the random fields with one another as compared to the interactions between the random and the regular fields and can neglect also the interactions due to the non-linear resonances, that is, $\gamma_k^{NL,reg}$ and γ_k^{NL} as well as γ_k^L. For the sake of simplicity we shall assume that initially the turbulence was homogeneous and we shall consider a small perturbation due to the presence of a regular structure. Of course, this can describe only the initial stages of the interactions between the random fields and the structure, but it may show any tendencies which are present.

As Eq.(10.103) contains the correlation functions of the random fields integrated over the frequency we can introduce the turbulence spectrum:

$$W_k^\ell(\mathbf{r},t) = \int |E^+|_k^2(\mathbf{r},t)\frac{d\omega}{2\pi},$$

$$W_k^\ell(\mathbf{r},t) = W_k^{(0)} + \delta W_k^\ell(\mathbf{r},t), \qquad \delta W_k^\ell(\mathbf{r},t) \ll W_k^{(0)}. \tag{10.111}$$

We shall therefore use the following equations which are obtained by integrating the equations which we found earlier over ω:

$$\left[\frac{\partial}{\partial t} + \left(\mathbf{v}_{gr}(\mathbf{k}_1)\cdot\frac{\partial}{\partial\mathbf{r}}\right)\right]W_k^\ell(\mathbf{r},t)$$

$$= \frac{\omega_{pe}}{2n_0}\left(\frac{\partial}{\partial\mathbf{k}_1}W_{\mathbf{k}_1}^{(0)}\cdot\frac{\partial}{\partial\mathbf{r}}\delta n^{reg}(\mathbf{r},t)\right). \tag{10.112}$$

$$\frac{\partial^2 <\delta n>(\mathbf{r},t)}{\partial t^2} - v_s^2\nabla^2 <\delta n>(\mathbf{r},t)$$

$$= \frac{Z}{4\pi m_i}\nabla^2\int \delta W_{\mathbf{k}_1}(\mathbf{r},t)\,d^3\mathbf{k}_1. \tag{10.113}$$

This is a linear system and we can easily solve it for $\langle\delta n\rangle_\mathbf{k}$ after first Fourier transforming it:

$$\langle\delta n\rangle_\mathbf{k} = \delta n_\mathbf{k}^{reg}\frac{Z\omega_{pe}}{4n_0T_e}\int \frac{1}{\omega - (\mathbf{k}\cdot\mathbf{v}_{gr}(\mathbf{k}_1)) + i0}$$

$$\times \left(\mathbf{k}\cdot\frac{\partial}{\partial\mathbf{k}_1}\right)W_{\mathbf{k}_1}^{(0)}\left[1 - \frac{\omega^2}{k^2v_s^2}\right]^{-1}d^3\mathbf{k}_1. \tag{10.114}$$

From this equation it is clear that the response of the turbulence to a density perturbation δn^{reg} of a regular structure is very similar to the response of the particle distribution to an external perturbation – instead of the particle velocity the response contains the group velocity of the turbulent waves and instead of the derivative of the particle distribution with respect to the particle momentum the response contains the derivative of the distribution function of the random waves with respect to their momentum. Equation (10.114) contains both the real and the imaginary parts of the response. They could be of the same order of magnitude. The imaginary part describes the long-range interactions between the structures. Long-range interactions can therefore be produced not only through interaction with particles but also through interactions with random waves. The relative magnitude of the extra density variation is according to Eq.(10.114) equal to $W/n_0 T_e k_1^2 d_e^2$ which is less than unity, since we assume that the condition for modulational interactions for the random fields is not satisfied. However, this parameter could be of the order of unity at the limit of applicability of the perturbation theory which we have been using. There is no other small parameter in our considerations; this means that if the perturbations of the turbulent field are of the order of unity the contribution of the extra density variation is also of the order of unity. This means that ultimately the change in the turbulence due to the regular structure will no longer be weak. The important point is that the imaginary part of the response of the turbulence to the perturbation by the regular fields is of the same order as the real part so that the energy conversion processes are, in fact, very important. The change in the energy of the regular field is determined by Eq.(9.63) in which we must put $\delta n = \delta n^{\text{reg}} + < \delta n >$. The quantity δn^{reg} should be included in the energy of the regular field:

$$\frac{dH^{\text{reg}}}{dt} = - \int < \delta n > \frac{1}{4\pi n_0} \frac{\partial |\mathcal{E}|^2}{\partial t} d^3 r = (2\pi)^3 \int < \delta n >_{k,\omega}$$

$$\times \frac{i\omega'}{4\pi n_0} \left[|\mathcal{E}|^2 \right]_{-k,\omega'} e^{-i\omega t - i\omega' t} d^3 k \, d\omega \, d\omega'. \tag{10.115}$$

If we are interested in the total amount of energy δH^{reg} transferred to the regular field we must integrate the above expression over time and we get:

$$\delta H^{\text{reg}} = -(2\pi)^4 \int < \delta n >_k \frac{i\omega}{4\pi n_0} \left[|\mathcal{E}|^2 \right]_{-k} d^4 k$$

$$= (2\pi)^4 \int < \delta n >_k \frac{i\omega T_e}{Z n_0} \left(1 - \frac{\omega^2}{k^2 v_s^2} \right) \delta n_{-k}^{\text{reg}} d^4 k$$

$$= (2\pi)^4 \int |\delta n_k^{\text{reg}}|^2 \frac{\pi \omega \omega_{\text{pe}}}{4 n_0^2}$$

$$\times \delta\big(\omega - (k \cdot v_{\text{gr}}(k_1))\big) \left(k \cdot \frac{\partial}{\partial k_1} \right) W_{k_1}^{(0)} d^3 k_1 \, d^4 k. \tag{10.116}$$

The sign of the change in energy depends on the turbulence spectrum. In the case of isotropic turbulence we can, as usual, introduce $W_{k_1}^{(0)} = 4\pi k_1^2 W_{k_1}^{(0)}$ and by integrating over the angles confirm that δH^{reg} is negative:

$$\delta H^{\text{reg}} = -2\pi^5 \int \frac{|\delta n_{\mathbf{k}}^{\text{reg}}|^2}{n_0^2} W_{k=\omega\omega_{\text{pe}}/3kv_{T_e}^2}^{(0)} \, k \, dk. \qquad (10.117)$$

If there is no turbulence present the number of waves is conserved in the case of self-contraction and the energy H^{reg} is negative. If there is turbulence present there is still conservation of the number of waves if we neglect the effects, described earlier, of $\gamma_{\mathbf{k}}^{\text{NL}}$. It is also conserved for the processes which we are considering in the present section. The negative value of expression (10.117) means that isotropic turbulence can enhance the self-contraction, that is, the non-linear processes for the regular waves. In some sense this is a *self-organisation* since the regular motions are enhanced due to the interactions with random motions. We must emphasise that the conclusions we have reached are valid only for slight changes in the turbulence; in the non-linear stage when the perturbations of the turbulence are no longer small the answer will be different, especially if we take into account that the regular structures will force the turbulence to become anisotropic even if it were initially isotropic. The main point is that the interaction between random waves and regular structures is not weak and in principle the energy can be transferred in both directions – to the structure and away from the structure. All will depend on the actual properties of the fields of the regular structure.

We must also mention the possibility of filling empty cavities with turbulent waves where the original waves were damped by coherent Landau damping or transition damping. This occurs for wavelengths comparable to the size of the cavity; it cannot be described in the approxmation which we used to describe the interactions between the random and the regular fields. This process is known as a *nucleation process*.

10.13 Some General Statements About Strong Langmuir Turbulence

The physical processes which we have just considered describe the properties of self-organised structures; an ensemble of such structures form a strongly turbulent state. These structures have a finite life-time – they are created and damped by the various physical processes which we discussed earlier. In the case of strong turbulence there is no single process which dominates, as is the case for weak turbulence. Objects such as the self-organised structures are extremely sophisticated and interesting objects the properties of which depend strongly on the "sea" of random particles and random waves in which they are "living". It is therefore impossible to determine the turbulence spectrum in the general case, independent of the properties of this "sea". However, one can still reach a few important general conclusions:

1. The energy flux in the turbulence is directed to smaller scales.

2. The creation of fast particles due to coherent Landau damping and transition damping is the main process of dissipation by particles.

3. Strong turbulence creates weak small-scale turbulence which, together with the strong turbulence, relaxes to some kind of quasi-equilibrium, forming self-organised structures.

4. The development of strong turbulence creates in all components rather strong inhomogeneities. These inhomogeneities are present in the self-contracting fields, in the fast-particle distributions, and in the random-wave distributions. The presence of micro-inhomogeneities is one of the characteristic features of strong turbulence.

5. Self-organisation processes are very characteristic of strong turbulence. Turbulence is a property of an open system as it needs sources, dissipation, and energy fluxes. However, in the case of strong turbulence these processes are rather fast and the same is true for the self-organisation processes.

10.14 Collective Laser–Plasma Interaction Processes

In many experiments on laser–plasma interactions the parameter $W/n_0 T_e$ is of the order of unity so that all threshold conditions for the appearance of modulational interactions are violated. Numerical investigations of the interactions of strong waves and a plasma have initiated many theoretical studies of this subject. It took a long time of experimental studies in this field before it was realised that strong turbulence plays a dominant rôle in all interaction processes of this kind. It was also found experimentally that the energy flux towards the Landau damping region converts most of the initial energy of the lasers into fast particles. Only afterwards can this energy be converted into thermal energy through collisions. In a dense plasma the most important collisions take place close to the solid target which in the laser experiments usually creates the plasma. The mean free paths of the fast particles created by the strong turbulence are rather long and they can penetrate to the centre of the target if its size is small – as is usually the case in laser-fusion experiments. However, to get fusion, the laser radiation must first compress the target. The fast particles penetrate to the centre of the target, heat it, and prevent compression. We see therefore that experiments along these lines were bound to fail and the creation of strong turbulence produces a chain of events which create an obstacle for reaching high temperatures and densities in laser–plasma interactions. At the present time experiments on the heating of a plasma due to laser-plasma interactions are mainly performed in such a way as to exclude the appearance of strong turbulence. Recent numerical studies show that, indeed, most of the laser energy is converted into fast particles.

Problems

1. Show that the interaction between a regular electron distribution function and a zero-frequency virtual field is described by the equation

$$\frac{d}{dt}\Phi^e_p(\mathbf{r},t) = I^{(0)}_p = \frac{\partial}{\partial p_i}F_i(\mathbf{r},t)\Phi^e_p(\mathbf{r},t), \quad \mathbf{F}(\mathbf{r},t) = -e\mathbf{E}^{(0)}(\mathbf{r},t),$$

where in the case of interactions with non-resonant particles we have

$$\mathbf{E}^{(0)}(\mathbf{r},t) = \frac{e}{m_e\omega^2_{pe}}\frac{\partial}{\partial\mathbf{r}}|\mathcal{E}(\mathbf{r},t)|^2 = -\frac{\partial}{\partial\mathbf{r}}\phi(\mathbf{r},t),$$

$$\phi(\mathbf{r},t) = -\frac{e|\mathcal{E}(\mathbf{r},t)|^2}{m_e\omega^2_{pe}}.$$

2. Show that in the case of uniformly distributed resonant particles we have

$$\mathbf{E}^{(0)} = \frac{ev^2_{Te}}{4\pi^2 n_0}\frac{\partial}{\partial\mathbf{r}}\operatorname{Re}\left\{\int \operatorname{div}\mathcal{E}^*(\mathbf{r},t)\,\delta\big(\omega_{pe} - (\mathbf{k}\cdot\mathbf{v})\big)\right.$$
$$\left.\times\left(\mathcal{E}(\mathbf{r}',t)\cdot\frac{\partial\Phi^e_p}{\partial\mathbf{p}}\right)e^{i(\mathbf{k}\cdot[\mathbf{r}-\mathbf{r}'])}\frac{d^3p\,d^3k\,d^3r'}{(2\pi)^3}\right\}.$$

3. Show that in the one-dimensional case for inhomogeneously distributed resonant particles we have

$$E^{\mathrm{res}(0)}(x,t) = \frac{ev^2_{Te}}{n_0}\frac{\partial}{\partial x}\operatorname{Re}\left\{\int\left[\frac{v}{|v|} + \frac{x-x'}{|x-x'|}\right]\frac{e^{i\omega_{pe}(x-x')/v}}{v}\right.$$
$$\left.\times\frac{\partial\mathcal{E}^*(x,t)}{\partial x}\mathcal{E}(x',t)\frac{\partial\Phi^{\mathrm{res}(e)}_v(x',t)}{\partial v}\right\}.$$

4. Use Eq.(10.31) to show that the power Q_p transferred by an inhomogeneous caviton to a single plasma particle is given by the equation

$$Q_p = \frac{2e^2}{m_e}\operatorname{Re}\left\{\int dl'\,\frac{1}{v}\left[\delta_{ij} - \frac{v_iv_j}{v^2}\left(1 + k\frac{\partial}{\partial k}\right)\right.\right.$$
$$\left.\left.+\frac{v_i}{v}\left(\delta_{js} - \frac{v_jv_s}{v^2}\right)\frac{\partial}{\partial r_s}\right]e^{ikl'}\,\mathcal{E}^*_i\left(\mathbf{r} + \frac{\mathbf{v}}{v}l'\right)\mathcal{E}_j(\mathbf{r})\right\}.$$

5. Find the power dissipated by a dipole-type caviton with

$$\phi = \frac{\mathcal{E}_0 x}{1 + r^2/a^2},$$

where a is the radius of the caviton.

6. Prove that the non-linear growth rates for the interaction of regular structures with random fields are described by Eqs.(10.94), (10.95), and (10.100).

7. Derive Eq.(10.110) from Eq.(10.109).

11 Plasma Maser Effect and Adiabatic Invariants in Non-linear Interactions

11.1 Introduction

It is well known that wave amplification can be the result not only of a linear maser effect, but also of a non-linear maser effect. In the latter case the maser effect may be produced by the inverse population of particles which are participating in the non-linear interactions or by the inverse population of waves. We have discussed earlier that a plasma is a strongly non-linear state of matter, and that it may contain a large number of waves which can be described, like the particles, by a distribution function. Moreover, non-linear processes make it possible that waves are emitted by the particles and this kind of stimulated processes may be responsible for the non-linear amplification of waves. In some sense these are trivial non-linear amplification processes. The name *plasma maser effect* is given to a different, non-trivial wave amplification mechanism which is based upon the fact that a plasma is often in a turbulent state which results in the formation of small-scale inhomogeneities in the fields and in the particle distributions. These inhomogeneities are characteristic features of the plasma state of matter since a plasma is often turbulent.

In some sense these inhomogenities are similar to the inhomogeneities produced by particle fluctuations. We found that the latter are responsible for collective particle collisions. In collisions waves may be emitted; this is known as *bremsstrahlung*. Bremsstrahlung in a plasma differs from bremsstrahlung in "bare" particle collisions; we shall discuss this effect in later chapters. For the moment we use this analogy to emphasise that the interactions between particles and inhomogeneities created by turbulence may also produce emission similar to the normal bremsstrahlung. Of course, the scales of the inhomogeneities created by turbulence are much larger than those which produce the usual bremsstrahlung. On the other hand, their amplitudes are much larger. The normal particle fluctuations depend only on the average particle distribution whereas the inhomogeneities produced by turbulence will depend on the sources exciting the turbulence, on the degree of anisotropy of the fields, and on the inhomogeneities in the particle distribution. In the case of the usual bremsstrahlung a maser effect can be produced only by some kind of inverse particle population whereas in the case of turbulence the maser effect can, in principle, be produced by the anisotropy in the distribution of the inhomogeneities. This emission mechanism due to inhomogeneities produced by turbulence will be called *turbulent bremsstrahlung* and if this emission is amplified in the plasma the amplification will be called a *plasma maser effect*.

The process of emission by inhomogeneities is called *transition radiation*. This radiation is similar to bremsstrahlung. In the simplest picture, when a particle passes through a sharp inhomogeneity such as the boundary between a superconductor and the vacuum the emission can be treated as the instantaneous disappearance of the charge and its image at the boundary and we can describe this process using the standard formulas for bremsstrahlung. It is clear that emission is possible only provided the boundary is sufficiently sharp. For the problem we are interested in at the moment this means that the wavelength of the emitted wave should be larger than the size of the inhomogeneity. However, even if all the inhomogeneities have the same size it is obvious that this condition is satisfied for all waves with wavelengths larger than that particular size of the inhomogeneity, which means that a broadband spectrum can be emitted. Usually the inhomogeneities in a plasma will have a distribution over different sizes which increases the range of the wavelengths of the emitted radiation. Waves with wavelengths much smaller than the size of the inhomogeneity will not be emitted since particles moving through the inhomogeneity will change their field almost adiabatically.

The mechanism for emission by a particle moving, for example, through a density inhomogeneity is the following. We have shown earlier that particles in a plasma are "dressed" particles with a screening shell the size of which depends on the density. As the particle passes through the inhomogeneity its shell will change and thus produce a time-dependent dipole moment. For wavelengths smaller than the size of the inhomogeneity the particle shell has time to relax to correspond to the new density and the dipole moment which is produced is negligibly small. However, in the opposite case a significant emission is possible. Strong turbulence can produce inhomogeneities with a size of the order of the minimum possible wavelength of a collective mode. In the case of Langmuir turbulence this minimum wavelength is of the order of a few electron Debye lengths while in the case of ion-sound turbulence it is of the order of a few ion Debye lengths. We see thus that this kind of emission is possible for all frequencies to the left of the $k = k_{max} \simeq 1/r_{min}$ line in an $\omega - k$ diagram for the dispersion curves, where r_{min} is the minimum size of the inhomogeneities produced by the turbulence (see the shaded region in Fig.11.1).

The wavelengths of Langmuir and of ion-sound waves can be very long – corresponding in Fig.11.1 to $k \rightarrow 0$. This means that for practically any size of inhomogeneities in a plasma we can find Langmuir and ion-sound waves with wavelengths larger than the size of the inhomogeneiy. The important conclusion we can draw from this picture is that this process allows significant "jumps" in frequency, both up and down in frequency, with the possibility of a change in the type of the wave. Other non-linear emission processes

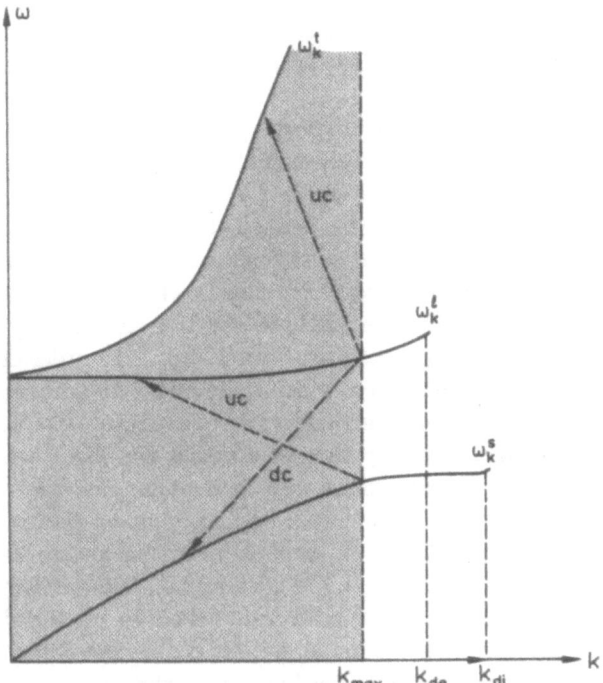

Fig. 11.1 Processes involving up- and down-conversion in frequencies. The region of wavelengths smaller than the minimum scale $l_{\min}$ of the inhomogeneities, which determines the maximum wavenumber, $k_{\max} \approx 2\pi/l_{\min}$, is shaded

which have been discussed earlier correspond to transformations of waves in a horizontal direction in Fig.11.1. We shall therefore denote the turbulent bremsstrahlung processes as *frequency up- and down-conversion processes.* For example, the inhomogeneities in ion-sound turbulence with a size of the order of the ion Debye length can lead to the emission of Langmuir waves with arbitrary wavelengths or of electromagnetic waves with frequencies lower than $\omega_{pi}c/v_{Ti} = \omega_{pe}c\sqrt{T_e}/v_{Te}\sqrt{T_i} \gg \omega_{pe}$. The inhomogeneities in Langmuir turbulence with a size of the order of the electron Debye length can lead to the emission of long-wavelength ion-sound waves, of long-wavelength Langmuir waves, and of electromagnetic waves with frequencies lower than $\omega_{pe}c/v_{Te} \gg \omega_{pe}$.

Although turbulent bremsstrahlung produces a process in which the frequency changes significantly and although the emission efficiency should decrease with increasing change in frequency, the decrease is not as drastic as, for example, in the decay processes where in the case when there is no decay resonance – or, more precisely, outside the decay-resonance curve – the emission is practically zero. In the case of turbulent bremsstrahlung we may expect a relatively weak decrease in intensity when the change in frequency increases – in fact, the decrease follows a power law. Turbulent bremsstrahlung therefore gives us conversion in the vertical direction in an $\omega - k$ plot whereas other non-linear processes mainly lead to horizontal conversions in this plot.

The up-conversion process is of interest from the point of view of interpreting many laboratory experiments and observations in space where such a conversion with a significant change in frequency can be observed. For example, there is interest in the excitation by ion-sound turbulence of the observed emission of electromagnetic waves at the plasma frequency. The modulational instability of ion oscillations with frequencies close to ω_{pi} can develop in a manner very similar to the one discussed for the case of Langmuir turbulence and it can create inhomogeneities with a size of the order of the ion Debye length. Of course, in actual cases one must take into account both vertical and horizontal frequency-conversion processes in the $\omega - k$ plot. However, the vertical conversion is the only process which is possible when one wants to produce conversion of an ion-sound wave into a Langmuir wave; this may be followed by a horizontal conversion from Langmuir to electromagnetic waves without any great change in frequency. Up-conversion to electromagnetic waves is also possible but this decreases with an increase in frequency difference while both decay and scattering processes combined with turbulent bremsstrahlung will have a maximum for conversion with a small change in frequency. This is a possible mechanism for the emission of electromagnetic waves with frequencies close to the plasma frequency by ion-sound turbulence. The up-conversion of electromagnetic waves in the case of strong Langmuir turbulence is especially important when we are interested in emission at frequencies much higher than ω_{pe}. In various experiments in which strong Langmuir turbulence was excited emission at frequencies up to $10\omega_{pe}$ or even higher has been observed.

Turbulent bremsstrahlung is a characteristic feature of strong turbulence which produces small-scale inhomogeneities. In the case of weak turbulence the quasi-linear interactions also produce inhomogeneities but since these inhomogeneities are not concentrated in structures they are smooth. The problem of the possibility of emission produced by a smooth inhomogeneity with a size larger than the wavelength is clear from the physical point of view: the emission is negligibly small and the particle fields change adiabatically. However, it is not obvious how this expectation which is physically obvious can be realised in a turbulent state since it is the result of both non-linear interactions and a change in the fields due to the inhomogeneities. The interaction between resonant and non-resonant waves is especially of interest. The resonant waves are linearly damped directly whereas the non-resonant waves are damped indirectly through their interaction with the resonant waves. On the other hand, the resonant waves also produce themselves some inhomogeneities via the quasi-linear interactions since these quasi-linear interactions lead either to a time-dependent or to a space-dependent distribution function of the resonant particles. If the non-resonant waves have wavelengths much shorter than the scale of the inhomogeneities produced by the resonant waves only the sum of the non-linear interactions and the linear interactions due to the inhomogeneities will lead to the adiabatic properties of the non-resonant waves. These properties are the so-called *adiabatic invariants* of the non-linear interactions. The adiabatic invariants can be made to disappear by destroying the balance leading to them. This may be done by external energy, particle, or momentum sources. The presence of such sources is typical for a turbulent state. We see thus that adiabatic invariants may be destroyed in an open system.

We shall in what follows consider only the problems of adiabatic invariants and wave amplification for longitudinal waves and shall not touch upon the problem of the spontaneous emission in the case of turbulent bremsstrahlung, discussing merely the possibilities of wave amplification or damping.

11.2 Damping Due to Inhomogeneities Created by External Sources

We shall first consider externally created inhomogeneities which are independent of changes in the plasma particle distribution. In this model the inhomogeneities will be produced by an external source which, in principle, can create both spatial inhomogeneities and time variations. We shall use the following equation for describing the particle distribution in the presence of a source:

$$\frac{\partial f_{\mathbf{p}}}{\partial t} + \left(\mathbf{v} \cdot \frac{\partial f_{\mathbf{p}}}{\partial \mathbf{r}} \right) + e \left(\mathbf{E} \cdot \frac{\partial f_{\mathbf{p}}}{\partial \mathbf{p}} \right) = I_{\text{source}}. \tag{11.1}$$

If there is no field the source will produce an inhomogeneous and non-stationary distribution $f_{\mathbf{p}}^{(0)}(\mathbf{r}, t)$ which satisfies the equation

$$\frac{\partial f_{\mathbf{p}}^{(0)}}{\partial t} + \left(\mathbf{v} \cdot \frac{\partial f_{\mathbf{p}}^{(0)}}{\partial \mathbf{r}} \right) = I_{\text{source}}. \tag{11.2}$$

The perturbation $f_{\mathbf{p}}^{(1)}$ of this distribution by the field $\mathbf{E}$ is found from the equation

$$\frac{\partial f_{\mathbf{p}}^{(1)}}{\partial t} + \left(\mathbf{v} \cdot \frac{\partial f_{\mathbf{p}}^{(1)}}{\partial \mathbf{r}} \right) + e \left(\mathbf{E} \cdot \frac{\partial f_{\mathbf{p}}^{(0)}}{\partial \mathbf{p}} \right) = \delta I_{\text{source}}. \tag{11.3}$$

If we assume that the source remains unchanged by the perturbations we have $\delta I_{\text{source}} = 0$. The function $f_{\mathbf{p}}^{(0)}$ is a non-stationary, inhomogeneous distribution. We shall assume that it consists of a main part, a homogeneous, stationary distribution $\Phi_{\mathbf{p}}$, and a small inhomogeneous, non-stationary part, $\delta f_{\mathbf{p}}^{(0)}$:

$$f_{\mathbf{p}}^{(0)} = \Phi_{\mathbf{p}} + \delta f_{\mathbf{p}}^{(0)}. \tag{11.4}$$

Although we assume that $\delta f_{\mathbf{p}}^{(0)} \ll \Phi_{\mathbf{p}}$, the inhomogeneous part may describe very sharp inhomogeneities. For the sake of simplicity we shall consider the one-dimensional case. Fourier transforming and putting $\mathbf{k} = \{k, \omega\}$ we have

$$i(\omega - kv) f_{p,\mathbf{k}}^{(1)} = eE_{\mathbf{k}'} \frac{\partial \Phi_p}{\partial p} + e \int E_{\mathbf{k}'} \frac{\partial \delta f_{p,\mathbf{k}-\mathbf{k}'}^{(0)}}{\partial p} d^2k', \tag{11.5}$$

$$i\varepsilon_{\mathbf{k}} k E_{\mathbf{k}} = 4\pi e^2 \int \frac{1}{i(\omega - kv + i0)} E_{\mathbf{k}'} \frac{\partial f_{p,\mathbf{k}-\mathbf{k}'}^{(0)}}{\partial p} d^2k' \frac{dp}{2\pi}, \tag{11.6}$$

where $\varepsilon_{\mathbf{k}}$ is determined by the homogeneous distribution Φ_p. Since the right-hand side of Eq.(11.6) is small the solution of the equation $\varepsilon_{\mathbf{k}} = 0$ will in first approximation be

$$\omega = \omega_{k_0} = \omega_{\mathrm{pe}} + \frac{3k_0^2 v_{Te}^2}{2\omega_{\mathrm{pe}}}.$$

For the positive-frequency part of the field we have

$$E_{\mathbf{k}}^+ = \mathcal{E}_0 \, \delta(\omega_0 - \omega_{k_0}) \, \delta(k - k_0). \tag{11.7}$$

Substituting this expression into the right-hand side of Eq.(11.6) and expanding the left-hand side around $\omega = \omega_{k_0}$ we find that

$$i(\omega - \omega_{k_0}) E_{\mathbf{k}}^+ = \frac{4\pi e^2 \mathcal{E}_0}{ik(\partial \varepsilon_{k_0} / \partial \omega_0)_{\omega_0 = \omega_{k_0}}}$$
$$\times \int \frac{1}{\omega - kv + i0} \frac{\partial}{\partial p} f_{p,\mathbf{k}-k_0,\omega-\omega_{k_0}}^{(0)} \frac{dp}{2\pi}. \tag{11.8}$$

Introducing the amplitude field $\mathcal{E}(x, t)$ through the equation

$$\mathcal{E}(x, t) = \int E_{\mathbf{k}}^+ e^{-i(\omega - \omega_{\mathrm{pe}})t + ikx} \, d\omega \, dk, \tag{11.9}$$

we find

$$e^{-ik_0 x + i(\omega_{k_0} - \omega_{\mathrm{pe}})t} \left(i \frac{\partial \mathcal{E}(x, t)}{\partial t} + \frac{3v_{Te}^2}{2\omega_{\mathrm{pe}}} \frac{\partial^2}{\partial x^2} \mathcal{E}(x, t) \right)$$
$$= -\frac{4\pi e^2 \mathcal{E}_0}{(2\pi)^3} \left[\frac{\partial \varepsilon_{k_0}}{\partial \omega_0} \right]_{\omega_0 = \omega_{k_0}}^{-1} \int \frac{dk \, dx'}{\omega_{k_0} - k_0 v + \omega - kv + i0}$$
$$\times \frac{e^{ik(x-x') - i\omega(t-t')}}{k + k_0} \frac{\partial}{\partial p} f_p^{(0)}(x', t'). \tag{11.10}$$

On the right-hand side we changed the integration over k and ω to integration over $k - k_0$ and $\omega - \omega_{k_0}$. Using Eq.(11.10) we can estimate whether external inhomogeneities or non-stationarity can act adiabatically. In the adiabatic case the amplitude field varies as $e^{ik_0 x - i\omega_{k_0} t}$ with a slowly varying factor $\mathcal{E}_0(x, t)$ in front of the exponent. We then get on the left-hand side

$$i\left(\frac{\partial}{\partial t} + v_{\text{gr}}(k_0)\frac{\partial}{\partial x}\right)\mathcal{E}_0(x, t).$$

The right-hand side of Eq.(11.10) gives a local frequency shift, which is an indication of the adiabatic change in the frequency, only if we can neglect ω as ompared to ω_{k_0} and k as compared to k_0. As ω and k on the right-hand side of Eq.(11.10) are the shifts from ω_{k_0} and k_0 they are the characteristics of the non-stationarity and the inhomogeneities and the above conditions therefore seem to be natural ones. However, an important point is that in the case of resonant waves when $\omega_{k_0} - k_0 v \simeq 0$ the adiabatic conditions are not so trivial. In the case where we have exact resonance we cannot expand in ω and k; however, for a distribution of resonant particles with a spread δv in velocities the width of the resonance is $k_0 \delta v$ and the adiabatic approximation could thus be valid for

$$\frac{k}{k_0}\frac{\omega}{\omega_{k_0}} \ll \frac{\delta v}{v}. \tag{11.11}$$

We see thus that the resonant particles play a special rôle in this interaction and for them the conservation of adiabatic invariants is not trivial. On the other hand, we may still expect than in the random phase approximation – when δv is relatively large – there will exist a broad range where we can apply the adiabatic invariants, but not for strong turbulence.

We can also show that the general equation for external inhomogeneities, obtained from Eq.(11.6) by neglecting time variations which are slow on a time scale of the order of ω_{pe}^{-1},

$$\left(i\frac{\partial}{\partial t} + \frac{3v_{T\text{e}}^2}{2\omega_{\text{pe}}}\frac{\partial^2}{\partial x^2}\right)\mathcal{E} = -\frac{e^2\omega_{\text{pe}}}{2\pi}\int\frac{e^{ik((x-x'))}\,dx'\,dk}{k(\omega_{\text{pe}} - kv + i0)}$$

$$\times\ \mathcal{E}(x't)\frac{\partial}{\partial p}f_p^{(0)}(x't), \tag{11.12}$$

is the same as the last term in Eq.(10.20) which we used to describe the change in the particle distribution function for regular waves, provided we replace $f_p^{(0)}(x', t)$ by the regular part $\Phi_p(x', t)$ of the distribution function. In Eq.(11.12) the inhomogeneities are produced by external sources whereas in Eq.(10.20) they were due to modulational interactions.

11.3 Non-linear Permittivity in the Presence of Random Resonant Waves

We shall start from the general non-linear equations (6.36) and (6.37) for random fields:

$$ik\varepsilon E^+ = 4\pi\varrho^{\mathrm{NL}} = 8\pi \int \varrho^{\mathrm{eff}}_{1,2,3}\left(E_1^+ E_2^+ E_3^- - E_1^+ \langle E_2^+ E_3^-\rangle\right)$$

$$- \langle E_1^+ E_2^+ E_3^-\rangle)d_{1,2,3} + 4\pi \int \varrho^{\mathrm{eff}}_{1,2,3}\left(E_1^- E_2^+ E_3^+\right.$$

$$- E_1^- \langle E_2^+ E_3^+\rangle - \langle E_1^- E_2^+ E_3^+\rangle)d_{1,2,3}. \tag{11.13}$$

Earlier we considered the non-linear interactions of non-resonant waves. Now we shall generalise our considerations to the case where there are resonant waves present. Let us assume that the random resonant waves are almost homogeneous and stationary. Due to the quasi-linear interactions they will produce some non-stationarity and inhomogeneities in the distribution function of the resonant particles. These inhomogeneities will affect the non-resonant waves adiabatically if condition (11.11) is satisfied. When describing the non-linear interactions of the resonant and the non resonant waves we can to a first approximation neglect the inhomogeneities and the non-stationarity and use Eq.(6.37):

$$\langle E_{\mathbf{k}}^+ E_{\mathbf{k}'}^-\rangle = -|E^+|_{\mathbf{k}}^2 \,\delta(k+k'), \qquad |E|_{\mathbf{k}}^2 = |E^+|_{\mathbf{k}}^2 + |E^-|_{\mathbf{k}}^2. \tag{11.14}$$

We shall assume one of the fields in Eq.(11.13) to be a resonant field – to be denoted as E_1^{R} – so that for that field we have $\omega_1 = (\mathbf{k}_1 \cdot \mathbf{v})$. The other field is assumed to be non-resonant – to be denoted as E^{NR} – so that in that case we have $\omega \neq (\mathbf{k}\cdot\mathbf{v})$ and the condition for scattering resonance, $\omega - \omega_1 = ([\mathbf{k}-\mathbf{k}_1]\cdot\mathbf{v})$, can therefore not be satisfied. The interaction we want to consider here is therefore supplementary to the stimulated scattering we considered in earlier chapters. By considering this interaction we complete the list of all non-linear interactions for random waves. Multiplying Eq.(11.13) by $E_{\mathbf{k}'}^-$ and averaging, using Eq.(11.14), we must take into account that the second term in this case gives the same result as the first one – in contrast to the case of stimulated scattering when the second term is negligible. We find

$$\left(\varepsilon_{\mathbf{k}} + \varepsilon_{\mathbf{k}}^{\mathrm{NL}}\right)|E^{\mathrm{NR}}|_{\mathbf{k}}^2 = 0, \qquad \varepsilon_{\mathbf{k}}^{\mathrm{NL}} = \frac{8\pi}{ik}\int \varrho^{\mathrm{eff}}_{\mathbf{k}_1,\mathbf{k},-\mathbf{k}_1}|E^{\mathrm{R}}|_{\mathbf{k}_1}^2 \, d^4\mathbf{k}_1, \tag{11.15}$$

where, according to Eq.(5.35), ϱ^{eff} contains two parts:

$$\varrho^{\mathrm{eff}}_{\mathbf{k}_1,\mathbf{k},-\mathbf{k}_1} = \varrho^{(3)}_{\mathbf{k}_1,\mathbf{k},-\mathbf{k}_1} + \frac{8\pi}{i|\mathbf{k}-\mathbf{k}_1|\varepsilon_{\mathbf{k}-\mathbf{k}_1}} \varrho^{(2)}_{\mathbf{k}_1,\mathbf{k}-\mathbf{k}_1}\varrho^{(2)}_{\mathbf{k},-\mathbf{k}_1}. \tag{11.16}$$

We shall be interested in the imaginary part of $\varepsilon_{\mathbf{k}}^{\mathrm{NL}}$. Since there is no scattering resonance the second term in Eq.(11.16) will contribute to the imaginary part of $\varepsilon_{\mathbf{k}}^{\mathrm{NL}}$ through the quadratic non-linear responses. We can show that the latter is exactly equal to zero for the interaction between resonant and non-resonant waves. We shall illustrate this for the one-dimensional case, leaving the general proof of this theorem as an exercise. For the one-dimensional case we find by partial integration

$$\varrho^{(2)}_{k_1, k-k_1} = \frac{e^3 k}{2m_e} \int \frac{dp/2\pi}{(\omega - kv)(\omega - \omega_1 - [k - k_1]v)(\omega_1 - k_1 v + \mathrm{i}0)} \frac{\partial \Phi_p}{\partial p},$$

$$\varrho^{(2)}_{k, -k_1} = \frac{e^3}{2m_e} \int \frac{(k - k_1)dp/2\pi}{(\omega - kv)(\omega - \omega_1 - [k - k_1]v)(-\omega_1 + k_1 v + \mathrm{i}0)} \frac{\partial \Phi_p}{\partial p}.$$

The imaginary part of the first expression divided by k is equal to the imaginary part of the second expression divided by $k - k_1$ whereas their real parts differ in sign so that the imaginary part of the product of these expressions is equal to zero. For the interaction we are interested in we have therefore $\varrho^{\mathrm{eff}} = \varrho^{(3)}$.

11.4 Non-linear Interactions of Resonant and Non-resonant Random Waves

The imaginary part of ϱ^{eff} is determined by the real part of $\varrho^{(3)}$:

$$\mathrm{Re}\left\{\varrho^{(3)}_{\mathbf{k}_1, \mathbf{k}, -\mathbf{k}_1}\right\} = -\frac{\pi e^4}{2k_1^2 k} \int \frac{1}{\omega - (\mathbf{k} \cdot \mathbf{v})} \left(\mathbf{k}_1 \cdot \frac{\partial}{\partial \mathbf{p}}\right)$$

$$\times \frac{1}{\omega - \omega_1 - ([\mathbf{k} - \mathbf{k}_1] \cdot \mathbf{v})} \left(\mathbf{k} \cdot \frac{\partial}{\partial \mathbf{p}}\right)$$

$$\times \delta(\omega_1 - (\mathbf{k}_1 \cdot \mathbf{v})) \left(\mathbf{k}_1 \cdot \frac{\partial}{\partial \mathbf{p}}\right) \Phi^{\mathrm{R}}_{\mathbf{P}} \frac{d^3 \mathbf{p}}{(2\pi)^3}. \qquad (11.17)$$

We can use this expression to calculate the non-linear growth rate or damping rate of non-resonant waves in the presence of resonant waves:

$$\gamma_{\mathbf{k}}^{\mathrm{NL}} = -\left[\mathrm{Im}\left\{\varepsilon_{\mathbf{k}}^{\mathrm{NL}}\right\} \left(\frac{\partial \mathrm{Re}\{\varepsilon_{\mathbf{k}}\}}{\partial \omega}\right)^{-1}\right]_{\omega = \omega_{\mathbf{k}}^{\ell}}$$

$$= -\frac{2\pi^2 e^4}{k^2} \omega_{\mathrm{pe}} \int \frac{\left|E^{\mathrm{R}}\right|^2_{\mathbf{k}_1}}{k_1^2} d^4 k_1 \frac{1}{\omega - (\mathbf{k} \cdot \mathbf{v})} \left(\mathbf{k}_1 \cdot \frac{\partial}{\partial \mathbf{p}}\right)$$

$$\times \frac{1}{\omega - \omega_1 - ([\mathbf{k} - \mathbf{k}_1] \cdot \mathbf{v})} \left(\mathbf{k} \cdot \frac{\partial}{\partial \mathbf{p}}\right)$$

$$\times \delta(\omega_1 - (\mathbf{k}_1 \cdot \mathbf{v})) \left(\mathbf{k}_1 \cdot \frac{\partial}{\partial \mathbf{p}}\right) \Phi^{\mathrm{R}}_{\mathbf{P}} \frac{d^3 \mathbf{p}}{(2\pi)^3}$$

$$
= -\frac{6\pi^2 e^4}{k^2 m_e^2} \omega_{pe} \int |E^R|^2_{k_1} d^4 k_1 \frac{k^2 (k \cdot k_1)}{(\omega_k^\ell - k)^4 k_1^2}
$$

$$
\times \delta(\omega_1 - (k_1 \cdot v)) \left(k_1 \cdot \frac{\partial}{\partial p} \right) \Phi_p^R \frac{d^3 p}{(2\pi)^3}. \tag{11.18}
$$

In the case of weak turbulence we take, as usual,

$$
|E^R|^2_{k_1} = |E^R|^2_{k_1} \delta(\omega_1 - \omega_{k_1}) + |E^R|^2_{-k_1} \delta(\omega_1 + \omega_{k_1}), \tag{11.19}
$$

and for $kv \ll \omega_{pe}$ we get

$$
\gamma_k^{NL} = \frac{3\omega_{pe}}{4n_0^2} \int |E^R|^2_{k_1} d^3 k_1 \frac{(k \cdot k_1)}{k_1^2}
$$

$$
\times \delta(\omega_{k_1} - (k_1 \cdot v)) \left(k_1 \cdot \frac{\partial}{\partial p} \right) \Phi_p^R \frac{d^3 p}{(2\pi)^3}. \tag{11.20}
$$

If the particle distribution is isotropic, $\Phi_p^R = \Phi^R(\varepsilon_p)$, where ε_p is the particle energy, we have

$$
\gamma_k^{NL} = \frac{3\pi m_e^2 \omega_{pe}^2}{2n_0^2} \int |E^R|^2_{k_1} \frac{(k \cdot k_1)}{k_1^3} \Phi^R \left(\frac{m_e \omega_{pe}^2}{2k_1^2} \right) \frac{d^3 k_1}{(2\pi)^3}. \tag{11.21}
$$

This relation shows that in the case of an isotropic particle distribution and isotropic resonant waves the growth rate is equal to zero, that is, there is no instability threshold in the case when either the turbulence or the particle distribution is anisotropic. However, we have already mentioned that in the adiabatic approximation there is no excitation of non-resonant waves. The meaning of this last statement is that the number of waves will not be changed in the adiabatic limit, but the wave parameters can be changed and therefore the wave amplitudes may be changed. The question therefore arises whether Eq.(11.21) describes a change in the number of waves or the adiabatic change of their amplitude.

11.5 Adiabatic Invariants in the Interaction of Resonant and Non-resonant Waves

We shall now take into account that the non-resonant wave is propagating in a plasma with a distribution function given by the relation

$$
\Phi_p = \Phi_p^{NR} + \Phi_p^R(r, t), \qquad \Phi_p^R(r, t) \ll \Phi_p^{NR}, \tag{11.22}
$$

where the distribution function of the non-resonant waves is stationary and homogeneous and determines the dispersion of the waves. The number of resonant particles is small and this is natural as otherwise the Langmuir waves would be heavily damped. The inhomogeneity and non-stationarity of the resonant distribution is related to the quasi-linear interactions of the resonant waves with the particles. We must emphasise that the quasi-linear equation has, as a rule, no stationary and homogeneous solutions – excepting the formation of a one-dimensional plateau which three-dimensionally is unstable. One must take into account that the non-resonant waves propagate in a non-stationary and inhomogeneous medium. The equation for them is the same as the one in the case of external sources, except that now the inhomogeneity and the non-stationarity are produced by the quasi-linear interactions rather than by external sources:

$$
\varepsilon_k E_k^{+NR} = -4\pi e^2 \int E_{k-k'}^{+NR} \frac{dk'}{k(\omega - (\mathbf{k}\cdot\mathbf{v}) + i0)|\mathbf{k}-\mathbf{k}'|}
$$

$$
\times \left([\mathbf{k}-\mathbf{k}'] \cdot \frac{\partial \Phi_{\mathbf{p},k'}^R}{\partial \mathbf{p}} \right) \frac{d^3\mathbf{p}}{(2\pi)^3} + 4\pi \varrho_k^{NL}. \tag{11.23}
$$

On the right-hand side we have added the non-linear charge density (11.13). On the right-hand side of Eq.(11.23) we consider the inhomogeneity and the non-stationarity together. In the linear relation the inhomogeneity of the resonant-particle distribution is taken into account through the k'-dependence of the Fourier components of $\Phi_{\mathbf{p}}^{NR}$ while the non-resonant homogeneous distribution contributes to ε_k The inhomogeneity also changes the non-linear interactions but we shall show that in the case of a smooth inhomogeneity there is no significant change in the non-linear growth rate and the same expression for the non-linear growth rate enters into the non-linear equation as for the homogeneous case. This equation can be found from Eq.(11.23) for the quantity $|E^+|_k(\mathbf{r},t)$ defined by Eq.(10.101) and instead of Eq.(11.14) we must use the relation

$$
\langle E_k^+ E_{-k+\kappa}^- \rangle = -\int |E^+|_k^2(\mathbf{r},t) e^{-i(\boldsymbol{\kappa}\cdot\mathbf{r})+i\nu t} d^3\mathbf{r}\,dt, \quad \kappa = \{\boldsymbol{\kappa},\nu\}, \tag{11.24}
$$

which is the Fourier transform of Eq.(10.101).

We shall now consider in the adiabatic limit all three contributions in the equation for $|E^+|_k(\mathbf{r},t)$ which come from (i) the convective term, that is, the left-hand side of Eq.(11.23), (ii) the inhomogeneity and the non-stationarity, that is, the first term on the right-hand side of Eq.(11.23), and (iii) the non-linearity, that is, the last term on the right-hand side of Eq.(11.23). Using Eq.(11.23) we construct the expression

$$
\widehat{D}_{conv} = -i \int (\varepsilon_k - \varepsilon_{k-\kappa}) \langle E_{-k+\kappa}^- E_k^+ \rangle e^{-i\nu t + i(\boldsymbol{\kappa}\cdot\mathbf{r})} d^4\kappa. \tag{11.25}
$$

We can write the equation obtained from Eq.(11.23) symbolically in the form

$$
\widehat{D}_{conv} = \widehat{D}_{NL} + D_{inh}. \tag{11.26}
$$

We shall now expand $\varepsilon_{k-\kappa}$ in terms of κ ($\equiv \{\kappa, \nu\}$). This is possible in the adiabatic approximation. We then get the convective term in the form

$$\hat{D}_{\text{conv}} = \frac{\partial \text{Re}\{\varepsilon_k\}}{\partial \omega} \frac{\partial}{\partial t} |E^+|_k(\mathbf{r}, t) - \left(\frac{\partial \text{Re}\{\varepsilon_k\}}{\partial \mathbf{k}} \cdot \frac{\partial}{\partial \mathbf{r}}\right) |E^+|_k(\mathbf{r}, t). \quad (11.27)$$

We must emphasise that we have not used the relation $\omega = \omega_k$ in Eq.(11.25) so that the correlation function can also be used in the case of strong turbulence. The non-linear term in the adiabatic approximation is very similar to the one found earlier for the case of stationary turbulence. We leave the derivation of that expression from Eq.(11.23) as an exercise as the result is rather obvious:

$$\hat{D}_{\text{NL}} = 2\gamma_k^{\text{NL}}(\mathbf{r}, t)|E^{+\text{NR}}|_{k_1}^2(\mathbf{r}, t)\frac{\partial \text{Re}\{\varepsilon_k\}}{\partial \omega}, \quad (11.28)$$

$$2\gamma_k^{\text{NL}}(\mathbf{r}, t) = -\frac{24\pi^2 e^4}{k^2 m_e^2} \int |E^{\text{R}}|_{k_1}^2(\mathbf{r}, t) \left[\frac{\partial \text{Re}\{\varepsilon_k\}}{\partial \omega}\right]^{-1} \frac{k^2 (\mathbf{k} \cdot \mathbf{k}_1) d^4 k_1}{(\omega - (\mathbf{k} \cdot \mathbf{v}))^4 k_1^2}$$

$$\times \delta(\omega_1 - (\mathbf{k}_1 \cdot \mathbf{v})) \left(\mathbf{k}_1 \cdot \frac{\partial}{\partial \mathbf{p}}\right) \Phi_{\mathbf{p}}^{\text{R}}(\mathbf{r}, t) \frac{d^3 \mathbf{p}}{(2\pi)^3}. \quad (11.29)$$

This last expression differs from Eq.(11.18) only in (i) that the local expression for the correlation functions occurs here and (ii) that it does not contain the substitution $\omega = \omega_k$.

For the inhomogeneity we first consider the contribution from the real part of the response. Retaining on the right-hand side of Eq.(11.23) only the inhomogeneous term and multiplying Eq.(11.23) after that by $E_{-k+\kappa}^-$ we get for the adiabatic case ($k' \ll k$):

$$\text{Re}\{\varepsilon_k + \delta\varepsilon_k(\mathbf{r}, t)\} |E^+|_k^2(\mathbf{r}, t) = 0, \quad (11.30)$$

where $\delta\varepsilon_k(\mathbf{r}, t)$ is the inhomogeneous local part of the dielectric permittivity due to the resonant particles:

$$\delta\varepsilon_k(\mathbf{r}, t) = \frac{4\pi e^2}{k^2} \int \frac{1}{\omega - (\mathbf{k} \cdot \mathbf{v}) + i0} \left(\mathbf{k} \cdot \frac{\partial}{\partial \mathbf{p}}\right) \Phi_{\mathbf{p}}^{\text{R}}(\mathbf{r}, t) \frac{d^3 \mathbf{p}}{(2\pi)^3}. \quad (11.31)$$

The $\mathbf{r}$- and t-dependence is determined solely by the inhomogeneity and the non-stationarity of the distribution function. The frequency of the non-resonant waves has an additional small inhomogeneous and non-stationary part:

$$\omega^{\text{NR}}(\mathbf{r}, t) = \omega_k + \delta\omega_k(\mathbf{r}, t), \quad (11.32)$$

$$\delta\omega_k(\mathbf{r}, t) = -\text{Re}\{\delta\varepsilon_k(\mathbf{r}, t)\} \left/ \frac{\partial \varepsilon_k}{\partial \omega}\right|_{\omega=\omega_k}. \quad (11.33)$$

It is very important that in the adiabatic limit the non-linear contribution $\widehat{D}_{\text{conv}}$ can be expressed in terms of $\delta\varepsilon_{\mathbf{k}}(\mathbf{r}, t)$ by using the quasi-linear equation for the distribution function of the resonant particles,

$$\frac{\partial \Phi_{\mathbf{p}}^{\text{R}}}{\partial t} + \left(\mathbf{v} \cdot \frac{\partial \Phi_{\mathbf{p}}^{\text{R}}}{\partial \mathbf{r}} \right) = \pi e^2 \int |E^{\text{R}}|_{\mathbf{k}_1}^2 \frac{d^4 k_1}{k_1^2} \left(\mathbf{k}_1 \cdot \frac{\partial}{\partial \mathbf{p}} \right)$$

$$\times \, \delta(\omega_1 - (\mathbf{k} \cdot \mathbf{v})) \left(\mathbf{k}_1 \cdot \frac{\partial}{\partial \mathbf{p}} \right) \Phi_{\mathbf{p}}^{\text{R}}. \tag{11.34}$$

Using the relation

$$\frac{6(\mathbf{k} \cdot \mathbf{k}_1) k^2}{m_{\text{e}}^2 (\omega - (\mathbf{k} \cdot \mathbf{v}) + \mathrm{i}0)^4} = \left(\mathbf{k}_1 \cdot \frac{\partial}{\partial \mathbf{p}} \right) \left(\mathbf{k} \cdot \frac{\partial}{\partial \mathbf{p}} \right) \frac{1}{(\omega - (\mathbf{k} \cdot \mathbf{v}) + \mathrm{i}0)^2},$$

integrating by parts, and also using Eq.(11.34) we can transform the non-linear contribution to the form

$$\widehat{D}_{\text{NL}} = -\frac{4\pi e^2}{k^2} |E^{+\text{NR}}|_{\mathbf{k}}^2(\mathbf{r}, t) \int \frac{1}{(\omega - (\mathbf{k} \cdot \mathbf{v}) + \mathrm{i}0)^2}$$

$$\times \left(\mathbf{k} \cdot \frac{\partial}{\partial \mathbf{p}} \right) \left[\frac{\partial}{\partial t} + \left(\mathbf{v} \cdot \frac{\partial}{\partial \mathbf{r}} \right) \right] \Phi_{\mathbf{p}}^{\text{R}}(\mathbf{r}, t)$$

$$= |E^{+\text{NR}}|_{\mathbf{k}}^2(\mathbf{r}, t) \left[\frac{\partial}{\partial t} \frac{\partial}{\partial \omega} - \left(\frac{\partial}{\partial \mathbf{k}} \cdot \frac{\partial}{\partial \mathbf{r}} \right) \right] \delta\varepsilon(\mathbf{r}, t). \tag{11.35}$$

Finally, we find the inhomogeneity contribution $\widehat{D}_{\text{inh}}$ from Eq.(11.26) by using the definition (11.25) and the inhomogeneity term in Eq.(11.23):

$$D_{\text{inh}} = \frac{4\pi \mathrm{i} e^2}{(2\pi)^8} \int \left[|E^{+\text{NR}}|_{\mathbf{k}-\mathbf{k}'}^2(\mathbf{r}', t') \frac{1}{k(\omega - (\mathbf{k} \cdot \mathbf{v}) + \mathrm{i}0)} \left(\frac{[\mathbf{k} - \mathbf{k}']}{|\mathbf{k} - \mathbf{k}'|} \cdot \frac{\partial}{\partial \mathbf{p}} \right) \right.$$

$$- |E^{+\text{NR}}|_{\mathbf{k}}^2(\mathbf{r}', t') \frac{1}{|\mathbf{k} - \boldsymbol{\kappa}|(\nu - \omega([\boldsymbol{\kappa} - \mathbf{k}] \cdot \mathbf{v}) + \mathrm{i}0)}$$

$$\times \left. \left(\frac{[\boldsymbol{\kappa} - \mathbf{k} - \mathbf{k}']}{|\boldsymbol{\kappa} - \mathbf{k} - \mathbf{k}'|} \cdot \frac{\partial}{\partial \mathbf{p}} \right) \right] \Phi_{\mathbf{p}}^{\text{R}}(\mathbf{r}'', t'') \mathrm{e}^{\mathrm{i}(\boldsymbol{\kappa} \cdot [\mathbf{r} - \mathbf{r}']) - \mathrm{i}\nu(t - t')}$$

$$\times \, \mathrm{e}^{\mathrm{i}(\mathbf{k}' \cdot [\mathbf{r}' - \mathbf{r}'']) - \mathrm{i}\omega'(t' - t'')} \, d^3 r' \, dt' \, d^3 r'' \, dt'' \, d\kappa \, dk'. \tag{11.36}$$

We can expand this expression in terms of the small parameters κ/k and k'/k. Expanding the correlation function $|E^{+\text{NR}}|_{\mathbf{k}-\mathbf{k}'}^2(\mathbf{r}', t')$ in terms of k' we can put everywhere $\kappa = 0$, except in the exponent and in the remainder we can expand in terms of κ, putting everywhere, except in the exponent, $k' = 0$. We then get

$$D_{\text{inh}} = -\left(\frac{\partial}{\partial \mathbf{k}}|E^{+\text{NR}}|_{\mathbf{k}}^2(\mathbf{r},t) \cdot \frac{\partial}{\partial \mathbf{r}}\right)\delta\varepsilon_{\mathbf{k}}(\mathbf{r},t)$$

$$+ \frac{\partial}{\partial \omega}|E^{+\text{NR}}|_{\mathbf{k}}^2(\mathbf{r},t)\frac{\partial}{\partial t}\delta\varepsilon_{\mathbf{k}}(\mathbf{r},t)$$

$$+ \left(\frac{\partial}{\partial \mathbf{r}} \cdot \left[|E^{+\text{NR}}|_{\mathbf{k}}^2(\mathbf{r},t)\frac{\partial}{\partial \mathbf{k}}\delta\varepsilon_{\mathbf{k}}(\mathbf{r},t)\right]\right)$$

$$- \frac{\partial}{\partial t}\left[|E^{+\text{NR}}|_{\mathbf{k}}^2(\mathbf{r},t)\frac{\partial}{\partial \omega}\delta\varepsilon_{\mathbf{k}}(\mathbf{r},t)\right]. \tag{11.37}$$

We see thus that in the adiabatic limit the non-linear contribution and the contribution connected with the inhomogeneities are of the same order of magnitude. Moreover, it is impossible to obtain the adiabatic invariants without taking into account the non-linear contribution. Adding Eqs.(11.35) and (11.37) we find that Eq.(11.26) takes the form:

$$\left[\frac{\partial}{\partial \omega}\varepsilon_{\mathbf{k}}(\mathbf{r},t)\right]\left[\frac{\partial}{\partial t}|E^{+\text{NR}}|_{\mathbf{k}}^2(\mathbf{r},t)\right] - \left[\frac{\partial}{\partial t}\varepsilon_{\mathbf{k}}(\mathbf{r},t)\right]\left[\frac{\partial}{\partial \omega}|E^{+\text{NR}}|_{\mathbf{k}}^2(\mathbf{r},t)\right]$$

$$- \left(\left[\frac{\partial}{\partial \mathbf{k}}\varepsilon_{\mathbf{k}}(\mathbf{r},t)\right] \cdot \left[\frac{\partial}{\partial \mathbf{r}}|E^{+\text{NR}}|_{\mathbf{k}}^2(\mathbf{r},t)\right]\right)$$

$$+ \left(\left[\frac{\partial}{\partial \mathbf{r}}\varepsilon_{\mathbf{k}}(\mathbf{r},t)\right] \cdot \left[\frac{\partial}{\partial \mathbf{k}}|E^{+\text{NR}}|_{\mathbf{k}}^2(\mathbf{r},t)\right]\right) = 0, \tag{11.38}$$

where

$$\varepsilon_{\mathbf{k}}(\mathbf{r},t) = \varepsilon_{\mathbf{k}} + \delta\varepsilon_{\mathbf{k}}(\mathbf{r},t) \tag{11.39}$$

is the total local dielectric permittivity. The equation which we have here obtained describes the conservation of an adiabatic invariant of the non-resonant waves. This adiabatic invariant is the number of non-resonant waves which for the case when there are inhomogeneities and non-stationarity present we define by the following relation:

$$N_{\mathbf{k}}^{\text{NR}}(\mathbf{r},t) = \pi^2|E^{+\text{NR}}|_{\mathbf{k}}^2(\mathbf{r},t)\left[\frac{\partial}{\partial \omega}\varepsilon_{\mathbf{k}}(\mathbf{r},t)\right]_{\omega=\omega_{\mathbf{k}}(\mathbf{r},t)}, \tag{11.40}$$

$$|E^{+\text{NR}}|_{\mathbf{k}}^2(\mathbf{r},t) = |E^{+\text{NR}}|_{\mathbf{k}}^2(\mathbf{r},t)\,\delta(\omega - \omega_{\mathbf{k}}(\mathbf{r},t)), \tag{11.41}$$

$$\omega_{\mathbf{k}}(\mathbf{r},t) = \omega_{\mathbf{k}} + \delta\omega_{\mathbf{k}}(\mathbf{r},t). \tag{11.42}$$

For the homogeneous and stationary case these equations transform into those which we have used many times earlier. Using these relations in Eq.(11.39) and integrating over ω we get finally:

$$\frac{\partial}{\partial t}N_{\mathbf{k}}^{\text{NR}}(\mathbf{r},t) + \left(\frac{\partial}{\partial \mathbf{r}} \cdot \left\{\left[\frac{\partial}{\partial \mathbf{k}}\omega_{\mathbf{k}}(\mathbf{r},t)\right]N_{\mathbf{k}}^{\text{NR}}(\mathbf{r},t)\right\}\right)$$

$$- \left(\frac{\partial}{\partial \mathbf{k}} \cdot \left\{\left[\frac{\partial}{\partial \mathbf{r}}\omega_{\mathbf{k}}(\mathbf{r},t)\right]N_{\mathbf{k}}^{\text{NR}}(\mathbf{r},t)\right\}\right) = 0. \tag{11.43}$$

One would expect that if there are inhomogeneities present the waves can propagate to regions in space where the plasma properties are different and due to that they will change their own parameters – these changes will be adiabatic in the framework of Eq.(11.43). There is also a flux in wavenumber space which is described by the last term in Eq.(11.43). We found earlier that the non-linear interactions could create such fluxes. We thus reach the conclusion that the resonant waves can via the quasi-linear interactions produce inhomogeneities which lead to fluxes in the k-space for the non-resonant waves. The conservation occurs only after integrating Eq.(11.43) both over the coordinates and the wavevectors of the non-resonant waves:

$$\frac{\partial}{\partial t} \int N_{\mathbf{k}}^{\mathrm{NR}}(\mathbf{r}, t)\, d^3\mathbf{k}\, d^3\mathbf{r} = 0.$$

11.6 The Plasma Maser Effect for Weak Turbulence

We shall call the amplification of waves by turbulent gradients a *plasma maser effect* in the case where the number of waves is not conserved. Weak turbulence does not create steep gradients. However, even quasi-linear gradients may be sufficient for the appearence of a maser effect. Indeed, we earlier considered the case when the wavelength of the non-resonant waves was much smaller than the gradients appearing in the problem. We see easily that in the opposite case the contribution from the inhomogeneity will be neglgibly for the propagation of a non-resonant wave. We shall once again write down this contribution – see Eq.(11.23) – making a slight change in the notation:

$$\left[\varepsilon_{\mathbf{k}} E_{\mathbf{k}}^{+\mathrm{NR}}\right]_{\mathrm{inh}} = -4\pi e^2 \int \frac{E^{+\mathrm{NR}}}{kk'(\omega - (\mathbf{k}\cdot\mathbf{v}) + i0)}$$
$$\times \left(\mathbf{k}' \cdot \frac{\partial \Phi_{\mathbf{p},\mathbf{k}-\mathbf{k}'}^{R}}{\partial \mathbf{p}}\right) \frac{d^4\mathbf{k}'\, d^3\mathbf{p}}{(2\pi)^3}. \tag{11.44}$$

In the case of a sharp inhomogeneity the characteristic values of k and k' of the non-resonant waves should be much smaller than the wavenumber characterising the inhomogeneity of the distribution function. However, this is impossible according to Eq.(11.44) since this wavenumber is equal to $k - k'$. The contribution from Eq.(11.44) is thus negligible and the whole effect is determined by the non-linear growth rate which we have already calculated.

We can very easily estimate the characteristic scale of the inhomogeneities produced by quasi-linear interactions and in that way find a criterion for a plasma maser effect in the case of weak turbulence:

$$k^{\mathrm{NR}} < k^{\mathrm{QL}}, \qquad k^{\mathrm{QL}} \approx \frac{\omega_{\mathrm{pe}} W^{\ell}}{n_0 m_e v_{\mathrm{R}}^3}, \tag{11.45}$$

where $k^{QL} \approx 1/L^{QL}$ with L^{QL} a characteristic scale for an inhomogeneity created by the quasi-linear interactions while v_R is a characteristic velocity of the resonant particles. Both non-resonant Langmuir waves and non-resonant electromagnetic waves may be excited. In the case of electromagnetic waves we have $\omega^t \approx kc$ and Eq.(11.45) then gives

$$\omega^t < \omega_{\text{pe}} \frac{c}{v_R} \frac{W^\ell}{n_0 m_e v_R^2}. \tag{11.46}$$

On the right-hand side we have two factors, one of which is large and the other of which is small. We see thus that a maser effect with $\omega > \omega_{\text{pe}}$ is possible for electromagnetic waves, but under restricted conditions.

Another possibility for having a maser effect is when we are dealing with an *open system*. This is possibility when there is a source present, I_{source}. Such a source can make the system stationary and homogeneous. In that case the quasi-linear interactions are balanced by the external source. If the source is regular, the only non-linear interactions left are those between resonant and non-resonant waves. If the source depends on the distribution function it may correspond to some dissipation in the open system. As an example we may consider the source term in the form

$$I_{\text{source}} = \left(\frac{\partial}{\partial \mathbf{p}} \cdot \left(\mathbf{F_p} f_{\mathbf{p}} \right) \right), \qquad \langle I_{\text{source}} \rangle = \left(\frac{\partial}{\partial \mathbf{p}} \cdot \left(\mathbf{F_p} \Phi_{\mathbf{p}} \right) \right). \tag{11.47}$$

The average value of the source can balance the quasi-linear interaction, but in its unaveraged form it will change the linear response and its fluctuating part will appear in the dielectric permittivity in the equation for the correlation function of the random fields. This additional dielectric permittivity, $\delta \varepsilon_k^F$, will be given by the equation

$$\delta \varepsilon_k^F = -\mathrm{i} \frac{4\pi e^2}{k^2} \int \frac{1}{\omega - (\mathbf{k} \cdot \mathbf{v})} \left(\frac{\partial}{\partial \mathbf{p}} \cdot \mathbf{F_p} \frac{1}{\omega - (\mathbf{k} \cdot \mathbf{v})} \right.$$
$$\left. \times \left(\mathbf{k} \cdot \frac{\partial}{\partial \mathbf{p}} \right) \Phi_{\mathbf{p}} \right) \frac{d^3 \mathbf{p}}{(2\pi)^3}. \tag{11.48}$$

We can use the identity

$$\left(\frac{\partial}{\partial \mathbf{p}} \cdot \mathbf{F_p} \frac{1}{\omega - (\mathbf{k} \cdot \mathbf{v})} \left(\mathbf{k} \cdot \frac{\partial}{\partial \mathbf{p}} \right) \Phi_{\mathbf{p}} \right)$$
$$= \frac{1}{2 (\omega - (\mathbf{k} \cdot \mathbf{v}))^2} \left(\mathbf{k} \cdot \frac{\partial}{\partial \mathbf{p}} \right) \left(\frac{\partial}{\partial \mathbf{p}} \cdot \mathbf{F_p} \Phi_{\mathbf{p}} \right)$$
$$- \frac{1}{\omega - (\mathbf{k} \cdot \mathbf{v})} \frac{\partial}{\partial p_i} \frac{\Phi_{\mathbf{p}}}{\omega - (\mathbf{k} \cdot \mathbf{v})} \left(\mathbf{k} \cdot \frac{\partial}{\partial \mathbf{p}} \right) F_{\mathbf{p},i}, \tag{11.49}$$

and the balance equation

$$\left(\frac{\partial}{\partial \mathbf{p}} \cdot \mathbf{F_p} \Phi_\mathbf{p}\right) = \pi e^2 \int \frac{d^4 k_1}{k_1^2} \left(\mathbf{k}_1 \cdot \frac{\partial}{\partial \mathbf{p}}\right)$$

$$\times \, \delta(\omega_1 - (\mathbf{k}_1 \cdot \mathbf{v})) \left(\mathbf{k}_1 \cdot \frac{\partial}{\partial \mathbf{p}}\right) \Phi_\mathbf{p}, \tag{11.50}$$

to find that the imaginary part of $\delta\varepsilon_\mathbf{k}^F$ connected with the first term on the right-hand side of Eq.(11.49) exactly cancels the imaginary part of $\varepsilon_\mathbf{k}^{NL}$. The maser effect is thus determined by the second term on the right-hand side of Eq.(11.49):

$$\gamma_\mathbf{k} = - \, \text{Im} \left\{\varepsilon_\mathbf{k}^{NL} + \delta\varepsilon_\mathbf{k}^F\right\} \bigg/ \frac{\partial \text{Re}\{\varepsilon_\mathbf{k}\}}{\partial \omega} = -\frac{2\pi e^2}{k^2} \omega_{pe} \int \frac{d^3 p}{(2\pi)^3}$$

$$\times \, \frac{1}{\omega - (\mathbf{k} \cdot \mathbf{v})} \frac{\partial}{\partial p_i} \frac{\Phi_\mathbf{p}}{\omega - (\mathbf{k} \cdot \mathbf{v})} \left(\mathbf{k} \cdot \frac{\partial}{\partial \mathbf{p}}\right) F_{\mathbf{p},i}. \tag{11.51}$$

The instability described by this last equation is called a *dissipative instability*. An open system can thus be the source of a maser instability.

11.7 The Plasma Maser Effect for Strong Turbulence

In a state of strong turbulence spatial inhomogeneity and local non-stationarity appear automatically as a result of modulational self-contraction processes. We may expect the presence of local inhomogeneities with a size of a few Debye lengths. The necessary conditions for the appearance of a plasma maser effect can therefore be satisfied over a much broader range of wavelengths. When considering a strongly turbulent state we must take into account the presence of several kinds of coherent structures including self-organised ones. It is therefore inappropriate to use the random-phase approximation when describing the various processes. To a first approximation the change in the distribution function is described by Eq.(10.30) which is valid for regular fields rather than by a quasi-linear equation. As before we shall neglect any time variations which are slow on a time scale of the order of ω_{pe}^{-1} and we shall also neglect the term describing the forces due to the virtual fields. We have thus

$$\frac{\partial \Phi_\mathbf{p}}{\partial t} + \left(\mathbf{v} \cdot \frac{\partial \Phi_\mathbf{p}}{\partial \mathbf{r}}\right) = -\frac{2e^2}{(2\pi)^3} \, \text{Re} \left\{\frac{\partial}{\partial p_i} \int \frac{d^3 k \, d^3 r'}{\mathrm{i}(\omega_{pe} - (\mathbf{k} \cdot \mathbf{v}) + \mathrm{i}0)}\right.$$

$$\left. \times \, e^{\mathrm{i}(\mathbf{k} \cdot [\mathbf{r} - \mathbf{r}'])} \mathcal{E}_i^*(\mathbf{r}, t) \mathcal{E}_j(\mathbf{r}, t) \frac{\partial \Phi_\mathbf{p}(\mathbf{r}', t)}{\partial p_j}\right\}. \tag{11.52}$$

We must mention that Eq.(11.52) contains the total field which can consist of the resonant field $\mathcal{E}^R$ and a non-resonant field $\mathcal{E}^{NR}$:

$$\mathcal{E} = \mathcal{E}^R + \mathcal{E}^{NR}. \tag{11.53}$$

The term "resonance" has a meaning provided the scale of the inhomogeneities is much larger than the wavelength of the resonant field and we shall assume that this is the case. We may expect to have a maser effect for non-resonant waves in the non-adiabatic case. This means that the most interesting case is the one for which

$$k^{\mathrm{NR}} \ll \frac{1}{L} \leqslant k^{\mathrm{R}}. \tag{11.54}$$

For coherent structures the size of the structure can be of the order of the wavelength of the waves trapped in the structure; this is reflected in the last inequality. In that case the division between resonant and non-resonant waves is somewhat vague.

We shall start by assuming that the non-resonant waves are weak whereas the resonant waves are strong and we now pose the problem of the development of the non-resonant waves and possibly of their amplification. We shall retain the name resonant waves for all the strong waves forming the structures. We found earlier that if we expand in terms of field amplitudes the maser effect is described by the cubic non-linearity in which two fields are resonant while the third field is non-resonant. In this perturbation theory the initial state corresponds to a homogeneous and stationary distribution. We now start from a state described by a strong inhomogeneous and non-stationary field distribution which satisfies Eq.(11.52). We shall call this field simply the resonant field although it may just be the field trapped in the structure. We shall consider the non-resonant field – of the field outside the structure – to be weak and we shall use a linear approximation in this field. It is of interest to mention that the expansion we used earlier will be a special case of the expansion we shall use in what follows. This expansion can be made in the framework of Eq.(11.52) since there we only neglected the virtual fields, but they do not contribute to the effect we are interested in. The equation for the state which is unperturbed by the non-resonant field is the same as Eq.(11.52) if we replace $\varPhi_{\mathrm{p}}$ and $\mathcal{E}$ by $\varPhi_{\mathrm{p}}^{\mathrm{R}}$ and $\mathcal{E}^{\mathrm{R}}$, respectively. We shall denote the perturbation of the distribution function by the non-resonant field by $\delta\varPhi_{\mathrm{p}}$. From Eq.(11.52) we get

$$\frac{\partial \delta\varPhi_{\mathrm{p}}}{\partial t} + \left(\mathbf{v} \cdot \frac{\partial \delta\varPhi_{\mathrm{p}}}{\partial \mathbf{r}}\right) = -\frac{2e^2}{(2\pi)^3} \operatorname{Re}\left\{\frac{\partial}{\partial p_i} \int \frac{e^{i(\mathbf{k}\cdot[\mathbf{r}-\mathbf{r}'])}\, d^3k\, d^3r'}{i(\omega_{\mathrm{pe}} - (\mathbf{k}\cdot\mathbf{v}) + i0)}\right.$$
$$\left. \times \left[\mathcal{E}_i^{\mathrm{R}*}(\mathbf{r},t)\mathcal{E}_j^{\mathrm{NR}}(\mathbf{r},t) + \mathcal{E}_i^{\mathrm{NR}*}(\mathbf{r},t)\mathcal{E}_j^{\mathrm{R}}(\mathbf{r},t)\right] \frac{\partial \varPhi_{\mathrm{p}}^{\mathrm{R}}(\mathbf{r}',t)}{\partial p_j}\right\}. \tag{11.55}$$

In the equation for the amplitude field,

$$\operatorname{div}\left[i\frac{\partial\mathcal{E}(\mathbf{r},t)}{\partial t} + \frac{3v_{Te}^2}{2\omega_{\mathrm{pe}}}\nabla^2\mathcal{E}(\mathbf{r},t) - \frac{\omega_{\mathrm{pe}}}{2n_0}\delta n(\mathbf{r},t)\mathcal{E}(\mathbf{r},t)\right] = -i\frac{e^2\omega_{\mathrm{pe}}}{(2\pi)^2}$$
$$\times \int \frac{e^{i(\mathbf{k}\cdot[\mathbf{r}-\mathbf{r}'])}\, d^3k\, d^3r'}{\omega_{\mathrm{pe}} - (\mathbf{k}\cdot\mathbf{v}) + i0}\left(\mathcal{E}(\mathbf{r}',t)\cdot\frac{\partial\varPhi_{\mathrm{p}}(\mathbf{r}',t)}{\partial\mathbf{p}}\right)\frac{d^3p}{(2\pi)^3}, \tag{11.56}$$

we also use an expansion in terms of the non-resonant field:

$$\mathrm{div}\left[i\frac{\partial \mathcal{E}^{NR}(\mathbf{r},t)}{\partial t} + \frac{3v_{Te}^2}{2\omega_{pe}}\nabla^2 \mathcal{E}^{NR}(\mathbf{r},t) - \frac{\omega_{pe}}{2n_0}\delta n^R(\mathbf{r},t)\mathcal{E}^{NR}(\mathbf{r},t)\right]$$

$$= -i\frac{e^2\omega_{pe}}{(2\pi)^2}\int \frac{e^{i(\mathbf{k}\cdot[\mathbf{r}-\mathbf{r}'])}}{\omega_{pe}-(\mathbf{k}\cdot\mathbf{v})+i0}\frac{d^3k\,d^3r'\,d^3p}{(2\pi)^3}$$

$$\times \left\{\left(\mathcal{E}^{NR}(\mathbf{r}',t)\cdot\frac{\partial\Phi_{\mathbf{p}}^R(\mathbf{r}',t)}{\partial \mathbf{p}}\right) + \left(\mathcal{E}^{R}(\mathbf{r}',t)\cdot\frac{\partial\delta\Phi_{\mathbf{p}}(\mathbf{r}',t)}{\partial \mathbf{p}}\right)\right\}. \quad (11.57)$$

It can clearly be seen that the first term in the curly brackets on the right-hand side of Eq.(11.57) describes the propagation of non-resonant waves through the inhomogeneities produced by the resonant waves, whereas the last term describes the non-linear effect. In the case of smooth inhomogeneities the sum of them leads to a conservation of the adiabatic invariant – the number of waves – while for sharp inhomogeneities the first term becomes negligible. The term involving δn^R on the left-hand side will not contribute to the change in the total number of non-resonant waves. As we are interested in the maser effect we can thus restrict our discussion by taking into account only the non-linear term. We shall introduce the time-dependent spatial Fourier components $\mathcal{E}_{\mathbf{k}}(t)$ of the fields. Combining all this we have:

$$i\frac{\partial \mathcal{E}_{\mathbf{k}}^{NR}}{\partial t} - \frac{3k^2v_{Te}^2}{2\omega_{pe}}\mathcal{E}_{\mathbf{k}}^{NR}(t) = -\frac{e^2\omega_{pe}}{(2\pi)^2k^2}\mathbf{k}$$

$$\times \int \frac{d^3p\,d^3k_1}{\omega_{pe}-(\mathbf{k}\cdot\mathbf{v})+i0}\left(\mathcal{E}_{\mathbf{k}_1}^R(t)\cdot\frac{\partial}{\partial \mathbf{p}}\right)\delta\Phi_{\mathbf{p},\mathbf{k}-\mathbf{k}_1}(t). \quad (11.58)$$

Since we neglect time variations which are slow on a time scale of ω_{pe}^{-1} we shall solve Eq.(11.55) neglecting the time derivative on the left-hand side:

$$\delta\Phi_{\mathbf{p},\mathbf{k}-\mathbf{k}_1}(t) = e^2\int \frac{d^3k'\,d^3k_1'}{([\mathbf{k}-\mathbf{k}_1]\cdot\mathbf{v})+i0}\left\{\left(\mathcal{E}_{\mathbf{k}'}^{NR}(t)\cdot\frac{\partial}{\partial \mathbf{p}}\right)\right.$$

$$\times \left[\frac{1}{\omega_{pe}-([\mathbf{k}-\mathbf{k}_1-\mathbf{k}']\cdot\mathbf{v})+i0} + \omega_{pe}\rightarrow-\omega_{pe}\right]$$

$$\times \left(\mathcal{E}_{\mathbf{k}_1'}^R(t)\cdot\frac{\partial}{\partial \mathbf{p}}\right) + \mathcal{E}_{\mathbf{k}'}^{NR}(t)\leftrightarrows\mathcal{E}_{\mathbf{k}_1'}^R(t)\right\}$$

$$\times \Phi_{\mathbf{p},\mathbf{k}-\mathbf{k}_1'-\mathbf{k}_1-\mathbf{k}'}^R(t). \quad (11.59)$$

After substituting this expression into Eq.(11.58) we can use perturbation theory to estimate the change in the non-resonant fields, as we did in the first section of the present chapter. To zeroth approximation we have

$$\mathcal{E}_{\mathbf{k}}^{NR}(t) = \mathcal{E}\frac{\mathbf{k}_0}{k_0}\delta(\mathbf{k}-\mathbf{k}_0)e^{-i\omega_{k_0}t}, \qquad \omega_{k_0} = \frac{3k_0^2v_{Te}^2}{2\omega_{pe}}. \quad (11.60)$$

Substituting this into the right-hand side of the non-linear term we find an estimate for the change in the energy of the non-resonant field:

$$\frac{1}{|\mathcal{E}_0|^2} \frac{\partial |\mathcal{E}^{NR}(t)|^2}{\partial t} \equiv 2\gamma_{\mathbf{k}_0}^{NL},$$

$$2\gamma_{\mathbf{k}_0}^{NL} = -\mathrm{Im}\left\{ \frac{e^4}{k_0^2} \omega_{pe} \int \frac{d^3\mathbf{k}_1\, d^3\mathbf{k}_1'}{\omega_{pe} - (\mathbf{k}_0 \cdot \mathbf{v}) + i0} \left(\mathcal{E}_{\mathbf{k}_1}^{R}(t) \cdot \frac{\partial}{\partial \mathbf{p}} \right) \right.$$

$$\times \frac{1}{([\mathbf{k}_0 - \mathbf{k}_1] \cdot \mathbf{v}) + i0} \left[\left(\mathbf{k}_0 \cdot \frac{\partial}{\partial \mathbf{p}} \right) \left[\frac{1}{\omega_{pe} + (\mathbf{k}_1 \cdot \mathbf{v}) + i0} \right. \right.$$

$$+ \omega_{pe} \to -\omega_{pe} \Bigg] \left(\mathcal{E}_{\mathbf{k}_1}^{R}(t) \cdot \frac{\partial}{\partial \mathbf{p}} \right) \left[\frac{1}{\omega_{pe} - ([\mathbf{k}_0 - \mathbf{k}_1 - \mathbf{k}_1'] \cdot \mathbf{v}} \right.$$

$$+ \omega_{pe} \to -\omega_{pe} \Bigg] \left(\mathbf{k}_0 \cdot \frac{\partial}{\partial \mathbf{p}} \right) \Bigg] \left. \Phi_{\mathbf{p}, \mathbf{k}_1 - \mathbf{k}_1'}^{R} \frac{d^3\mathbf{p}}{(2\pi)^3} \right\}. \qquad (11.61)$$

There are two kinds of inhomogeneities: inhomogeneities in the fields and inhomogeneities in the distribution function. If we can neglect the inhomogeneities in the distribution function we shall obtain a result which is very close to the one we obtained for random fields. The square of the modulus of the Fourier components of the resonant field will enter for

$$\Phi_{\mathbf{p}, -\mathbf{k}_1 - \mathbf{k}_1'}^{R} = \Phi_{\mathbf{p}}^{R} \delta(\mathbf{k}_1 + \mathbf{k}_1'), \qquad (11.62)$$

and the imaginary part appears only due to the pole in the resonant denominators:

$$\gamma_{\mathbf{k}_0}^{NL} = \frac{2\pi^2 e^4}{k_0^2} \omega_{pe} \int \frac{d^3\mathbf{k}_1 \left| \mathcal{E}_{\mathbf{k}_1}^{R}(t) \right|^2}{k_1^2 (\omega_{pe} - (\mathbf{k}_0 \cdot \mathbf{v}) + i0)} \left(\mathbf{k}_1 \cdot \frac{\partial}{\partial \mathbf{p}} \right)$$

$$\times \frac{1}{([\mathbf{k}_0 - \mathbf{k}_1] \cdot \mathbf{v}) + i0} \left(\mathbf{k}_0 \cdot \frac{\partial}{\partial \mathbf{p}} \right) \left[\delta(\omega_{pe} - (\mathbf{k}_1 \cdot \mathbf{v})) \right.$$

$$+ \delta(\omega_{pe} - (\mathbf{k}_1 \cdot \mathbf{v})) \Bigg] \left(\mathbf{k}_1 \cdot \frac{\partial}{\partial \mathbf{p}} \right) \Phi_{\mathbf{p}}^{R} \frac{d^3\mathbf{p}}{(2\pi)^3}. \qquad (11.63)$$

In the limit as $\omega_{pe} \gg k_0 v$, $k_1 \gg k_0$, we find

$$\gamma_{\mathbf{k}_0}^{NL} = \frac{3\omega_{pe}}{4n_0^2} \int \frac{(\mathbf{k}_0 \cdot \mathbf{k}_1)}{k^2} \left| \mathcal{E}_{\mathbf{k}_1}^{R}(t) \right|^2 \delta(\omega_{pe} - (\mathbf{k}_1 \cdot \mathbf{v})$$

$$\times \left(\mathbf{k}_1 \cdot \frac{\partial}{\partial \mathbf{p}} \right) \Phi_{\mathbf{p}}^{R}(t)\, d^3\mathbf{k}_1 \frac{d^3\mathbf{p}}{(2\pi)^3}. \qquad (11.64)$$

This result is very close to the one obtained for random fields and differs from the previous one in that we have here the Fourier components of the regular fields. It can be used to calculate the maser effect produced by regular structures.

In the opposite case when the inhomogeneities in the particle distribution play an important rôle it is not possible to simplify the expression which we have given already without specifying the resonant fields. However we can still make an important point. We see easily from the general expression that the small parameter k_0/k_1 will not appear and the estimate of the growth rate will be

$$\gamma_{k_0}^{NL} \cong \omega_{pe} \frac{n_R}{n_0} \frac{|\mathcal{E}^R|^2}{4\pi n_0 v_R^2}, \tag{11.65}$$

where n_R is the number of resonant particles.

The process is important for strong turbulence. Since the size of the inhomogeneity of the structures can be of the order of several Debye lengths, say md_e, the frequency of the emitted electromagnetic waves could be of the order of

$$\omega_{max} \cong \frac{c}{v_{Te}} \omega_{pe} \frac{1}{m}. \tag{11.66}$$

As c/v_{Te} is usually a very large number the emission can, even when the collapse stops at $m \cong 15$ to 20, occur at frequencies much higher than the electron plasma frequency. The appearance in experiments of emission at frequencies $\omega \gg \omega_{pe}$ may be an indication that small-scale spatial inhomogeneities have been created. In laser-plasma interaction experiments and turbulent tokamak experiments harmonics of the plasma frequency, up to the tenth, have been observed. The growth rate of electromagnetic waves decreases with frequency as $1/\omega^2$ and for sufficiently high harmonics the plasma may become optically thin and then the emissivity decreases rapidly with increasing number of the harmonic.

Problems

1. Show that the contribution from the quadratic non-linearities is exactly equal to zero in the three-dimensional case for the interaction between random resonant and non-resonant waves.
2. Find the non-linear permittivity for electromagnetic waves interacting with resonant Langmuir waves. Calculate the non-linear growth rate.
3. Find the non-linear permittivity for electromagnetic waves interacting with resonant ion-sound waves. Calculate the non-linear growth rate.
4. Find the non-linear growth rate of non-resonant Langmuir waves interacting with resonant ion-sound waves.
5. Show that in the presence of an external source, I_{source}, the change in the number of non-resonant waves which are interacting with a stationary distribution of resonant waves is determined by the imaginary part of the dielectric permittivity which is equal to

$$\text{Im}\left\{\varepsilon_{\mathbf{k},\omega}^{\text{eff}}\right\} = \frac{4\pi e^2}{k^2}$$

$$\times \int \left[\frac{1}{\omega - (\mathbf{k}\cdot\mathbf{v}) + i0}\left(\mathbf{k}\cdot\frac{\partial}{\partial\mathbf{p}}\right), \frac{1}{\omega - (\mathbf{k}\cdot\mathbf{v}) + i0}I_{\text{source}}\right]\Phi_{\mathbf{p}}^{\text{R}}\frac{d^3\mathbf{p}}{(2\pi)^3},$$

where the square brackets $\left[\widehat{A}, \widehat{B}\right]$ here indicate a *Poisson bracket*,

$$\left[\widehat{A}, \widehat{B}\right] \equiv \widehat{A}\widehat{B} - \widehat{B}\widehat{A}.$$

6. Discuss the dissipative instability of resonant electron heating due to the collisions of hot electrons with ions.

7. Derive the equation describing the change in the distribution function of resonant particles due to the interaction between resonant and non-resonant waves.

12 Non-linear Interactions
of Collective Oscillations and Waves
with Particle Fluctuations

12.1 Introduction

All non-linear interactions in a plasma occur while there are also present in the plasma natural particle fluctuations and the field produced by these fluctuations. One cannot avoid the non-linear interactions of the turbulent and regular wave fields with the fields of these fluctuations and with the fluctuations in the index of refraction produced by the particle fluctuations. To complete the discussion of the non-linear kinetics we must consider here the problems related to these non-linear interactions. We know already that the statistical particle fluctuations lead to collisions between particles and that those collisions are drastically different from the collisions between particles when there is no plasma present. It turns out that the non-linear interactions between waves and these particle fluctuations lead to the absorption or amplification of waves. However, this should then be a process connected with the emission by particles during collisions, which is nothing but bremsstrahlung. Since the cross-sections for collisions are significantly changed by collective processes in the plasma we may expect that the bremsstrahlung will also be significantly changed by collective plasma processes. A study of the non-linear interactions between the particle fluctuations and waves will thus enable us to find and investigate collective effects in bremsstrahlung. We shall see that the plasma drastically changes the bremsstrahlung processes and that processes which are very weak when there is no plasma present may become important in a plasma. The fields of the particle fluctuations are random fields. We earlier constructed a general theory for the interaction between regular and random fields. The point is, however, that we cannot use this theory here as we considered the limit when the random fields are waves and made the obvious assumption that the correlation function of the random waves and the particle distribution varied slowly on a time scale set by the period of the random waves. It is also important that we considered interactions with random waves whereas the fields of the particle fluctuations are not waves. From a physical point of view these fields of the particle fluctuations are very inhomogeneous and non- stationary. Their inhomogeneities can have sizes much smaller than the Debye length whereas the coherent fields of the waves have wavelengths much larger than the Debye length. The particle fluctuations create very small inhomogeneities in the index of refraction and may produce significant transition radiation by particles passing through these inhomogeneities. This is the reason why the bremsstrahlung processes will be significantly different when there is a plasma present since in the absence of a plasma transition radiation processes are impossible. We thus see why the previous considerations of the interactions between random and regular waves should be generalised and are formally inapplicable for the problems we shall deal with in the present section.

We showed earlier that that the zeroth order particle distribution for the statistical particle fluctuations satisfies the equation

$$\frac{\partial \delta f_{\mathbf{p}}^{\alpha(0)}}{\partial t} + \left(\mathbf{v} \cdot \frac{\partial \delta f_{\mathbf{p}}^{\alpha(0)}}{\partial \mathbf{r}} \right) = 0. \tag{12.1}$$

Although the regular part of the distribution function, $\Phi_{\mathbf{p}}^{\alpha(0)}$ also satisfies this equation, it is either constant or it varies slowly in space and time. In the case of $\Phi_{\mathbf{p}}^{\alpha(0)}$ both the time and the space derivatives are therefore small whereas for the particle fluctuations both derivatives may be very large, although they are correlated in such a way that their effects cancel one another and they satisfy Eq. (12.1). This effect of the inhomogeneities must be taken into account in and added to any discussion of the non-linear interactions. As we have mentioned already, they lead just to the bremsstrahlung absorption being modified by collective plasma effects.

The problem we are facing is thus how to take into account both the interactions between the coherent fields and the inhomogeneous fields of the particle fluctuations and the interactions between the coherent fields and the fluctuations in the distribution function, which includes the fluctuations in the index of refraction. This last mechanism corresponds to transition radiation and absorption by the inhomogeneities of the index of refraction. We shall here restrict our discussion to the limit of the linear approximation for the coherent wave fields and we shall thus neglect their non-linear interactions – that is, we shall not consider the problem of the effect of particle fluctuations on the non-linear interactions of the coherent fields. We shall also neglect the non-linear interactions of the fields of the particle fluctuations with each other. We retain only the interaction of the coherent fields with the fields of the particle fluctuations and with the fluctuations in the particle distribution – which lead to a transition type interaction. These two processes cannot be separated and must be considered together. Since we shall use the linear approximation for the waves we can without loss of generality assume them to be random since in a linear approximation the coherent and the random fields are described by the same dielectric permittivity. We shall therefore consider all fields – the fields of the particle fluctuations and the wave field – to be random fields. In that case the non-linear interactions will be described by the effect squared in the particle fluctuations and squared in the wave fields. In fact, we need to find the non-linear permittivity for any wave propagating in the fields of the natural particle fluctuations and in a plasma with a random index of refraction due to these particle fluctuations. As before we restrict ourselves to the case when the wave fields and the fields of the particle fluctuations are longitudinal.

12.2 General Relations

When we considered the interactions of strong coherent fields we had to introduce virtual fields. We now must add to them the fields $\mathbf{E}_{\mathbf{k}}^{(0)}$ of the statistical particle fluctuations. Apart from that, the particle distribution, unperturbed by the fields, will now consist of two parts: a regular part $\Phi_{\mathbf{p}}^{\alpha}$ which we assume to be constant in time and space., and a fluctuating part, $\delta f_{\mathbf{p},\mathbf{k}}^{\alpha(0)}$. Both of these parts satisfy Eq.(12.1). We gave in Chap.4 the actual expressions for the distribution $\delta f_{\mathbf{p},\mathbf{k}}^{\alpha(0)}$ and the fields $\mathbf{E}_{\mathbf{k}}^{(0)}$ which are related to it. We shall use these expressions here and shall denote the frequency and wavevector of these fluctuations by ω_0 and $\mathbf{k}_0$, respectively. Since we are not interested in the non-linear interactions of the waves with each other but only in the interactions between the waves and the particle fluctuations the frequencies and wavevectors of the virtual fields will be, respectively, $\omega_{\mathbf{k}} \pm \omega_0$ and $\mathbf{k} \pm \mathbf{k}_0$. We shall denote the virtual field again by $\mathbf{E}^{\mathbf{v}}$ and the field of the waves by $\mathbf{E}^{\sigma}$. The total field will thus be

$$\mathbf{E} = \mathbf{E}^{\sigma} + \mathbf{E}^{\mathbf{v}} + \mathbf{E}^{(0)}. \tag{12.2}$$

It is simplest to assume that the field $\mathbf{E}^{\sigma}$ is random since in the linear approximation in the fields $\mathbf{E}^{\sigma}$ the result is independent of whether the fields $\mathbf{E}^{\sigma}$ are random or regular. The zeroth order particle distribution will be

$$\Phi_{\mathbf{p}}^{\alpha} + \delta f_{\mathbf{p}}^{\alpha(0)}. \tag{12.3}$$

To construct a theory of the interactions between the particle fluctuations and the waves we use again a perturbation expansion in the fields and generalise Eq.(4.19) by taking into account that the field is given by Eq.(12.2) and the zeroth approximation of the distribution by Eq.(12.3). We now get the following equations for the first-, second- and third-order perturbations in terms of the fields:

$$\frac{\partial \delta f_{\mathbf{p}}^{\alpha(1)}}{\partial t} + \left(\mathbf{v} \cdot \frac{\partial \delta f_{\mathbf{p}}^{\alpha(1)}}{\partial \mathbf{r}} \right) = -e_{\alpha} \left(\mathbf{E} \cdot \frac{\partial}{\partial \mathbf{p}} \right) \left(\Phi_{\mathbf{p}}^{\alpha} + \delta f_{\mathbf{p}}^{\alpha(0)} \right)$$
$$+ e_{\alpha} \left\langle \left(\mathbf{E} \cdot \frac{\partial}{\partial \mathbf{p}} \right) \left(\Phi_{\mathbf{p}}^{\alpha} + \delta f_{\mathbf{p}}^{\alpha(0)} \right) \right\rangle, \tag{12.4}$$

$$\frac{\partial \delta f_{\mathbf{p}}^{\alpha(2)}}{\partial t} + \left(\mathbf{v} \cdot \frac{\partial \delta f_{\mathbf{p}}^{\alpha(2)}}{\partial \mathbf{r}} \right)$$

$$= -e_\alpha \left(\mathbf{E} \cdot \frac{\partial \delta f_{\mathbf{p}}^{\alpha(1)}}{\partial \mathbf{p}} \right) + e_\alpha \left\langle \left(\mathbf{E} \cdot \frac{\partial \delta f_{\mathbf{p}}^{\alpha(1)}}{\partial \mathbf{p}} \right) \right\rangle, \qquad (12.5)$$

$$\frac{\partial \delta f_{\mathbf{p}}^{\alpha(3)}}{\partial t} + \left(\mathbf{v} \cdot \frac{\partial \delta f_{\mathbf{p}}^{\alpha(3)}}{\partial \mathbf{r}} \right)$$

$$= -e_\alpha \left(\mathbf{E} \cdot \frac{\partial \delta f_{\mathbf{p}}^{\alpha(2)}}{\partial \mathbf{p}} \right) + e_\alpha \left\langle \left(\mathbf{E} \cdot \frac{\partial \delta f_{\mathbf{p}}^{\alpha(2)}}{\partial \mathbf{p}} \right) \right\rangle. \qquad (12.6)$$

The difference between these equations and Eqs.(5.40), (5.41), and (5.42) which we used previously lies in the presence of $\delta f_{\mathbf{p}}^{\alpha(0)}$ in Eq.(12.4). It is related to the fluctuations of the index of refraction. We shall separate from the total $\delta f_{\mathbf{p}}^{\alpha(i)}$ the part connected with the regular initial distribution function – and we shall denote it by $\delta f_{\mathbf{p}}^{\alpha(R,i)}$ – and the part connected with the initial distribution of the particle fluctuations – and denote it by $\delta f_{\mathbf{p}}^{\alpha(0,i)}$. The subsequent approximations for the latter are determined by quadratic combinations of particle fluctuations. We shall be interested only in effects not higher than cubic in the fields – as everywhere before. It is therefore sufficient to find only the contributions from the first and second approximations: $\delta f_{\mathbf{p}}^{\alpha(0,1)} + \delta f_{\mathbf{p}}^{\alpha(0,2)}$. From Eqs.(12.4) and (12.5) we get:

$$\frac{\partial \delta f_{\mathbf{p}}^{\alpha(0,1)}}{\partial t} + \left(\mathbf{v} \cdot \frac{\partial \delta f_{\mathbf{p}}^{\alpha(0,1)}}{\partial \mathbf{r}} \right)$$

$$= -e_\alpha \left(\mathbf{E} \cdot \frac{\partial \delta f_{\mathbf{p}}^{\alpha(0)}}{\partial \mathbf{p}} \right) + e_\alpha \left\langle \left(\mathbf{E} \cdot \frac{\partial \delta f_{\mathbf{p}}^{\alpha(0)}}{\partial \mathbf{p}} \right) \right\rangle, \qquad (12.7)$$

$$\frac{\partial \delta f_{\mathbf{p}}^{\alpha(0,2)}}{\partial t} + \left(\mathbf{v} \cdot \frac{\partial \delta f_{\mathbf{p}}^{\alpha(0,2)}}{\partial \mathbf{r}} \right)$$

$$= -e_\alpha \left(\mathbf{E} \cdot \frac{\partial \delta f_{\mathbf{p}}^{\alpha(0,1)}}{\partial \mathbf{p}} \right) + e_\alpha \left\langle \left(\mathbf{E} \cdot \frac{\partial \delta f_{\mathbf{p}}^{\alpha(0,1)}}{\partial \mathbf{p}} \right) \right\rangle. \qquad (12.8)$$

We shall thus use the simplest type of perturbation theory where in the "ground" state there are statistical particle fluctuations. The presence of these fluctuations must also lead to the scattering of the fluctuating fields by the wave fields and *vice versa*. These are spontaneous scattering processes. Other processes are the damping or amplification of wave fields due to the fluctuating index of refraction and fluctuating fields; together they produce the total bremsstrahlung process in a plasma. We shall see that the absorption or amplification of bremsstrahlung is closely related to the absorption due to the scattering of wave fields into particle fluctuation fields.

12.3 Interactions Unconnected with Fluctuations in the Particle Distribution

We shall start our considerations taking into account only the fields of the particle fluctuations but not the fluctuations in their distributions. We use then the non-linear equations from Chap.5 and Chap.6 for the total field (12.2):

$$ik\varepsilon_k E_k = 4\pi\varrho_k^{NL} = 4\pi \int \varrho_{1,2}^{(2)}(E_1 E_2 - \langle E_1 E_2 \rangle)\, d_{1,2}$$

$$+ 4\pi \int \varrho^{(3)}{}_{1,2,3}(E_1 E_2 E_3 - E_1\langle E_2 E_3 \rangle - \langle E_1 E_2 E_3 \rangle)\, d_{1,2,3}. \qquad (12.9)$$

We shall use this equation to investigate the properties of the collective wave field when there are particle fluctuation fields present and we shall thus assume that on the left-hand side of Eq.(12.9) the field E_k is close to the field of the collective oscillations σ. We can then directly average over the particle fluctuations, taking into account that terms linear in these fluctuations give zero and retaining only terms linear in the collective fields – as we said we would do. Under those conditions it is easy to find which parts of the total field enter into particular terms of the non-linear equation (12.9). If one of the fields is the wave field in the quadratic non-linearity the second field must be either the virtual field or the field of the particle fluctuations. However, the way we defined the virtual field it is determined by the wave field and the field of the particle fluctuations so that the terms involving the virtual field will be proportional to the square of the wave field and we are interested in terms which are linear in the wave field and we can thus neglect this term. On the other hand, the other terms which are proportional to the particle fluctuation field will also be zero since we can average over these fluctuations and the average value of these fields is equal to zero. From these considerations we see what kind of fields can enter into the quadratic non-linearities. The only possibility left is that one of the fields is the virtual field and the other one the field of the particle fluctuations. We can also find what fields enter into the cubic non-linearity. One of the fields should be the field of the wave and the other two must be the fields of the particle fluctuations. However, the wave field cannot occur as the first field E_1 since in that case averaging over the particle fluctuations would give zero due to the other term with a minus sign in Eq.(12.9). Taking into account the symmetry of the quadratic non-linearity under a $1 \leftrightarrows 2$ interchange and the symmetry of the cubic non-linearity under a $2 \leftrightarrows 3$ interchange we get

$$ik\varepsilon_k E_k = 8\pi \int \varrho_{1,1'}^{(2)} E_1^{(0)} E_{1'}^{v}\, d_{1,1'} + 8\pi \int \varrho_{1,2,3}^{(3)} E_1^{(0)} E_2^{\sigma} E_3^{(0)}\, d_{1,2,3}. \quad (12.10)$$

For the virtual field there now appear two contributions: the first – to be denoted by $E_k^{v(1)}$ – can be expressed in the standard way in terms of the regular part $\Phi_{\mathbf{p}}^{\alpha}$ of the particle distribution and the second one–denoted by $E_{\mathbf{k}}^{v(2)}$–can be expressed in terms of $\delta f_{\mathbf{p}}^{\alpha(0)}$ if we use Eq.(12.7). We shall consider the second part of the virtual field separately and consider here only the effect connected with the fields of the particle fluctuations–but not the fluctuations in their distribution. We then have

$$ik' \varepsilon_{\mathbf{k}'} E_{\mathbf{k}'}^{v(1)} = 8\pi \int \varrho_{2,3}^{(2)} E_2^{\sigma} E_3^{(0)} \, d_{2,3}.$$

(12.11)

After substituting this virtual field $E^{v(1)}$ into Eq.(12.10) we get a standard equation for the non-linear interactions (compare with Eqs.(5.34) and (5.35)):

$$ik \varepsilon_{\mathbf{k}} E_{\mathbf{k}} = 8\pi \int \varrho_{1,2,3}^{\text{eff}} E_1^{(0)} E_2^{\sigma} E_3^{(0)} \, d_{1,2,3},$$

(12.12)

where

$$\varrho_{1,2,3}^{\text{eff}} = \varrho_{1,2,3}^{(3)} + \frac{8\pi}{i|\mathbf{k}_2 + \mathbf{k}_3|\varepsilon_{2+3}} \varrho_{1,2+3}^{(2)} \varrho_{2,3}^{(2)}.$$

(12.13)

The difference between this equation and the one used previously lies in that now the positive and negative frequencies have not been separated. One should, of course, for the fields of the particle fluctuations take into account the contributions from both the positive and the negative frequency components. However, in Eq.(12.12) the frequencies of the wave fields may be of any sign. The other important difference lies in the order in which the fields occur in the equation.

We have already mentioned that we can directly average the equation over the particle fluctuations. We then find:

$$\left(\varepsilon_{\mathbf{k}} + \varepsilon_{\mathbf{k}}^{\text{NL}(1)}\right) E_{\mathbf{k}}^{\sigma} = 0, \quad \varepsilon_{\mathbf{k}}^{\text{NL}(1)} = \frac{8\pi}{ik} \int \varrho_{\mathbf{k}_1,\mathbf{k},-\mathbf{k}_1}^{\text{eff}} \left|E^{(0)}\right|_{\mathbf{k}_1}^2 \, d^4\mathbf{k}_1.$$

(12.14)

If we now compare this result with the expression for the non-linear interactions of Langmuir oscillations with each other–see Eqs.(6.41) and (5.55)– we find that the only difference is that in Eq.(12.14) the random fields of the particle fluctuations occur rather than the fields of the random Langmuir oscillations. We can thus interpret this part of the result, which is described by Eq.(12.14), as the processes of absorption and stimulated scattering of the fields of the collective particle oscillations by the statistical particle fluctuations. We remind ourselves that the correlation function $\left|E^{(0)}\right|_{\mathbf{k}}^2$ of the particle fluctuations is determined by Eq.(4.32) which we once more write down here for the sake of convenience:

$$\left\langle E_{i,\mathbf{k}'}^{(0)} E_{j,\mathbf{k}}^{(0)} \right\rangle = \frac{k_i k_j}{k^2} \left|E^{(0)}\right|_{\mathbf{k}}^2 \delta(\mathbf{k} + \mathbf{k}'),$$

(12.15)

$$\left|E^{(0)}\right|_{\mathbf{k}}^2 = \frac{16\pi^2}{k^2 |\varepsilon_{\mathbf{k}}|^2} \sum_{\beta} e_{\beta}^2 \int \delta(\omega - (\mathbf{k} \cdot \mathbf{v}')) \, \Phi_{\mathbf{p}'}^{\beta} \frac{d^3 \mathbf{p}'}{(2\pi)^6}.$$

(12.16)

We must emphasise that Eq.(12.14) does *not* describe the total effect of the interactions which we are describing here and that is the reason why we have added the superscript "(1)" to the expression for the dielectric permittivity.

12.4 Virtual Field Fluctuations

Let us now determine the additional part of the virtual fields which is described by Eq.(12.7). To find the charge density fluctuations which determine this part of the virtual fields we may assume that in Eq.(12.7) the field E is the same as the wave field E^σ since the product of this field and the particle distribution fluctuations will determine this virtual field:

$$f_{\mathbf{p}}^{\alpha(0,1,\sigma)} = \frac{e_\alpha}{i(\omega - (\mathbf{k} \cdot \mathbf{v}) + i0)} \int E_{\mathbf{k}_1}^\sigma \frac{1}{k_1} \left(\mathbf{k}_1 \cdot \frac{\partial}{\partial \mathbf{p}} \right) f_{\mathbf{p},\mathbf{k}-\mathbf{k}_1}^{\alpha(0)} \, d^4 k_1, \quad (12.17)$$

which leads to

$$E_{\mathbf{k}}^{\mathrm{v}(2)} = \sum_\alpha \frac{4\pi e_\alpha}{ik\varepsilon_{\mathbf{k}}} \int f_{\mathbf{p}}^{\alpha(0,1,\sigma)} \frac{d^3\mathbf{p}}{(2\pi)^3}$$

$$= -\sum_\alpha \frac{4\pi e_\alpha^2}{l\varepsilon_{\mathbf{k}}} \int \frac{1}{(\omega - (\mathbf{k} \cdot \mathbf{v}))k_2} E_{\mathbf{k}_2}^\sigma$$

$$\times \left(\mathbf{k}_2 \cdot \frac{\partial}{\partial \mathbf{p}} \right) f_{\mathbf{p},\mathbf{k}-\mathbf{k}_2}^{\alpha(0)} \frac{d^4 k_2 \, d^3 \mathbf{p}}{(2\pi)^3}. \quad (12.18)$$

If we take into account that – as we discussed in detail in earlier chapters – the term in the quadratic responses which is linear in the wave field is equal to zero we can write the contribution from $E_{\mathbf{k}}^{\mathrm{v}(2)}$ to the dielectric permittivity for the wave field in the form:

$$\varepsilon_{\mathbf{k}}^{\mathrm{NL}(2)} - \frac{8\pi}{ikE_{\mathbf{k}}^\sigma} \int \varrho_{\mathbf{k}_1,\mathbf{k}-\mathbf{k}_1}^{(2)} \left\langle E_{\mathbf{k}_1}^{(0)} E_{\mathbf{k}-\mathbf{k}_1}^{\mathrm{v}(2)} \right\rangle d^4 k_1, \quad (12.19)$$

or

$$\varepsilon_{\mathbf{k}}^{\mathrm{NL}(2)} = \sum_\alpha \frac{32\pi^2 e_\alpha^2}{ikE_{\mathbf{k}}^\sigma} \int \varrho_{\mathbf{k}_1,\mathbf{k}-\mathbf{k}_1}^{(2)} E^\sigma \frac{1}{|\mathbf{k}-\mathbf{k}_1|\varepsilon_{\mathbf{k}-\mathbf{k}_1} k_2}$$

$$\times \frac{1}{\omega - \omega_1 - ([\mathbf{k}-\mathbf{k}_1] \cdot \mathbf{v}) + i0} \left(\mathbf{k}_2 \cdot \frac{\partial}{\partial \mathbf{p}} \right)$$

$$\times \left\langle E_{\mathbf{k}_1}^{(0)} f_{\mathbf{p},\mathbf{k}-\mathbf{k}_1-\mathbf{k}_2}^{\alpha(0)} \right\rangle \frac{d^4 k_1 d^4 k_2 \, d^3 \mathbf{p}}{(2\pi)^3}. \quad (12.20)$$

We can carry out the averaging in Eq.(12.20) using Eq.(4.10):

$$\left\langle E_{\mathbf{k}}^{(0)} f_{\mathbf{p},\mathbf{k}'}^{\alpha(0)} \right\rangle = \frac{\Phi_{\mathbf{p}}^\alpha e_\alpha}{2\pi^2 ik\varepsilon_{\mathbf{k}}} \delta(\omega - (\mathbf{k} \cdot \mathbf{v})) \, \delta(\mathbf{k} + \mathbf{k}'). \quad (12.21)$$

We thus get

$$
\begin{aligned}
\varepsilon_{\mathbf{k}}^{NL(2)} = &-\sum_{\alpha} \frac{16e_{\alpha}^3}{k^2} \int \frac{1}{k_1|\mathbf{k}-\mathbf{k}_1|\varepsilon_{\mathbf{k}-\mathbf{k}_1}\varepsilon_{\mathbf{k}_1}} \varrho_{\mathbf{k}_1,\mathbf{k}-\mathbf{k}_1}^{(2)} \\
&\times \frac{1}{\omega - \omega - 1 - ([\mathbf{k}-\mathbf{k}_1]\cdot\mathbf{v}) + i0} \left(\mathbf{k}\cdot\frac{\partial}{\partial\mathbf{p}}\right) \\
&\times \Phi_{\mathbf{p}}^{\alpha}\,\delta(\omega_1 - (\mathbf{k}_1\cdot\mathbf{v})) \frac{d^4k_1\,d^3\mathbf{p}}{(2\pi)^3},
\end{aligned} \tag{12.22}
$$

which completes this part of our considerations.

12.5 Interactions Produced by Particle Fluctuations. Contribution from Virtual Fields

The interactions with the particle fluctuations may be due both to the linear perturbation $\delta f_{\mathbf{p}}^{\alpha(0,1)}$ in the fields – and the corresponding charge density $\delta\varrho^{(0,1)}$ – and to the quadratic perturbations $\delta f_{\mathbf{p}}^{(0,2)}$ – and the corresponding charge density $\delta\varrho^{(0,2)}$. In the first contribution one must substitute both the usual virtual field $E^{v(1)}$ (see Eq.(12.11)) and the fluctuating virtual field $E^{v(2)}$. Of course, the second, quadratic contribution does not contain the virtual fields and it describes a direct change in the particle fluctuations due both to the wave fields and to the fields of the particle fluctuations. The latter describe the above mentioned transition damping processes by a fluctuating index of refraction. We shall see that the terms with the virtual fields describe an interference due to the usual bremsstrahlung processes and the transition radiation by the fluctuating index of refraction. We shall in the present section consider the contribution from the virtual fields.

We turn to Eq.(12.7) and we point out that the field E which occurs in it cannot be either the wave field or the field of the particle fluctuations since averaging over the statistical fluctuations would give a zero result in those cases. This field can therefore only be a virtual field:

$$
f_{\mathbf{p},\mathbf{k}}^{\alpha(0,1,v)} = \frac{e_{\alpha}}{i(\omega - (\mathbf{k}\cdot\mathbf{v}) + i0)} \int E_{\mathbf{k}_1}^{v} \frac{1}{k_1}\left(\mathbf{k}_1\cdot\frac{\partial}{\partial\mathbf{p}}\right) f_{\mathbf{p},\mathbf{k}-\mathbf{k}_1}^{\alpha(0)}\,d^4k_1. \tag{12.23}
$$

Substituting here the virtual field (12.11) we find

$$
\begin{aligned}
\varrho_{\mathbf{k}}^{0,1,v(1)} = &\sum_{\alpha} e_{\alpha} \int f_{\mathbf{p},\mathbf{k}}^{\alpha(0,1,v)} \frac{d^3\mathbf{p}}{(2\pi)^3} = -\sum_{\alpha} \frac{e_{\alpha}^2}{k_1^2\varepsilon_{\mathbf{k}_1}} \frac{1}{\omega - (\mathbf{k}\cdot\mathbf{v}) + i0} \\
&\times \varrho_{\mathbf{k}_2,\mathbf{k}_1-\mathbf{k}_2}^{(2)} E_{\mathbf{k}_2}^{\sigma} E_{\mathbf{k}_1-\mathbf{k}_2}^{(0)} \left(\mathbf{k}_1\cdot\frac{\partial}{\partial\mathbf{p}}\right) f_{\mathbf{p},\mathbf{k}-\mathbf{k}_1}^{(0)} \frac{d^4k_1\,d^4k_2\,d^3\mathbf{p}}{(2\pi)^3}, \tag{12.24}
\end{aligned}
$$

and by using Eq.(12.24)

$$\varepsilon_{\mathbf{k}}^{NL(3)} = -\frac{4\pi}{ikE_{\mathbf{k}}^{\sigma}} \varrho_{\mathbf{k}}^{(0,1,\mathrm{v}(1))}$$

$$= -\sum_{\alpha} \int \frac{16e_{\alpha}^3}{k_1^2 \varepsilon_{\mathbf{k}_1} |\mathbf{k} - \mathbf{k}_1| \varepsilon_{\mathbf{k}_1-\mathbf{k}}} \varrho_{\mathbf{k},\mathbf{k}_1-\mathbf{k}}^{(2)} \frac{1}{\omega - (\mathbf{k} \cdot \mathbf{v})}$$

$$\times \left(\mathbf{k}_1 \cdot \frac{\partial}{\partial \mathbf{p}} \right) \delta\big(\omega - \omega_1 - ([\mathbf{k} - \mathbf{k}_1] \cdot \mathbf{v})\big) \Phi_{\mathbf{p}}^{\alpha} \frac{d^4 k_1 \, d^3 p}{(2\pi)^3}. \quad (12.25)$$

If we substitute into Eq.(12.23) the second virtual field (12.18) we get

$$\delta f_{\mathbf{p},\mathbf{k}}^{(0,1,\mathrm{v}(2))} = -\sum_{\beta} \frac{4\pi e_{\beta}^2 e_{\alpha}}{i(\omega - (\mathbf{k} \cdot \mathbf{v}))} \int \frac{d^4 k_1 \, d^4 k_2 \, d^3 p'}{(2\pi)^3 k_1^2 k_2 \varepsilon_{\mathbf{k}_1} (\omega_1 - (\mathbf{k}_1 \cdot \mathbf{v}'))} E_{\mathbf{k}_2}^{\sigma}$$

$$\times \left(\mathbf{k}_1 \cdot \frac{\partial}{\partial \mathbf{p}} \right) \left(\mathbf{k}_2 \cdot \frac{\partial}{\partial \mathbf{p}} \right) \left\langle \delta f_{\mathbf{p}',\mathbf{k}_1-\mathbf{k}_2}^{\beta(0)} \delta f_{\mathbf{p},\mathbf{k}-\mathbf{k}_1}^{\alpha(0)} \right\rangle. \quad (12.26)$$

Bearing in mind that

$$\left\langle \delta f_{\mathbf{p}',\mathbf{k}}^{\beta(0)} \delta f_{\mathbf{p},\mathbf{k}'}^{\alpha(0)} \right\rangle = \Phi_{\mathbf{p}}^{\alpha} \, \delta_{\alpha,\beta} \, \delta(\mathbf{k} + \mathbf{k}') \, \delta\big(\omega - (\mathbf{k} \cdot \mathbf{v})\big) \, \delta(\mathbf{p} - \mathbf{p}'), \quad (12.27)$$

we find

$$\varepsilon_{\mathbf{k}}^{NL(4)} = -\frac{4\pi}{ikE_{\mathbf{k}}^{\sigma}} \sum_{\alpha} e_{\alpha} \int \delta f_{\mathbf{p},\mathbf{k}}^{(0,1,\mathrm{v}(2))} \frac{d^3 p}{(2\pi)^3}$$

$$= -\sum_{\alpha} \frac{2e_{\alpha}^4}{\pi} \int \frac{(\mathbf{k} \cdot \mathbf{k}_1)^2}{k^2 k_1^2 \varepsilon_{\mathbf{k}}} \frac{\Phi_{\mathbf{p}}^{\alpha}}{m_{\alpha}^2} \frac{1}{(\omega - (\mathbf{k} \cdot \mathbf{v}) + i0)^4}$$

$$\times \delta\big(\omega - \omega_1 - ([\mathbf{k} - \mathbf{k}_1] \cdot \mathbf{v})\big) \frac{d^4 k_1 \, d^3 p}{(2\pi)^3}. \quad (12.28)$$

Before we used Eq.(12.27) we integrated by parts both over $\mathbf{p}$ and over $\mathbf{p}'$. From the expressions we have obtained it is clear that they contain the δ-functions which describe scattering. However, they also contain the usual imaginary parts corresponding to resonance. We shall see in what follows that the combination of resonance for scattering and the usual resonance leads to a conservation law for bremsstrahlung. We shall in what follows consider in detail the relation between spontaneous scattering and bremsstrahlung.

12.6 Direct Interactions with Particle Fluctuations

The only contribution now left is the one from $\delta f_{\mathbf{p}}^{\alpha(0,2)}$ (see Eq.(12.8)). We have

$$\delta f_{\mathbf{p},\mathbf{k}}^{\alpha(0,2)} = \frac{e_{\alpha}}{i(\omega - (\mathbf{k} \cdot \mathbf{v}))} \int \left(\mathbf{k}_1 \cdot \frac{\partial}{\partial \mathbf{p}} \right)$$

$$\times \left[E_{\mathbf{k}_1} \delta f_{\mathbf{p},\mathbf{k}-\mathbf{k}_1}^{\alpha(0,1)} - \left\langle E_{\mathbf{k}_1} \delta f_{\mathbf{p},\mathbf{k}-\mathbf{k}_1}^{\alpha(0,1)} \right\rangle \right] \frac{d^4 k_1}{k_1}. \quad (12.29)$$

We must substitute into this equation the quantity $\delta f_{p,k}^{\alpha(0,1)}$ which for the arguments of Eq.(12.29) is equal to

$$\delta f_{p,k-k_1}^{\alpha(0,1)} = \frac{e_\alpha}{i(\omega - \omega_1 - ([k - k_1] \cdot v) + i0)} \int \left(k_2 \cdot \frac{\partial}{\partial p}\right)$$

$$\times \left[E_{k_2} \delta f_{p,k-k_1-k_2}^{\alpha(0,1)} - \left\langle E_{k_2} \delta f_{p,k-k_1-k_2}^{\alpha(0,1)} \right\rangle\right] \frac{d^4 k_2}{k_2}. \tag{12.30}$$

One of the fields should be the wave field and the other one the field of the particle fluctuations. One sees easily that the field E_{k_2} cannot be the particle fluctuation field as after averaging over them expression (12.30) would be equal to zero. Using Eq.(12.20) for this averaging we now find

$$\varepsilon_k^{NL(5)} = -\frac{4\pi}{ikE_k^\sigma} \sum_\alpha e_\alpha \int \delta f_{p,k}^{\alpha(0,2)} \frac{d^3 p}{(2\pi)^3}$$

$$= -\frac{2}{\pi} \sum_\alpha e_\alpha^4 \int \frac{d^4 k_1}{k_1^2 k^2 \varepsilon_{k_1}} \frac{1}{\omega - (k \cdot v) + i0} \left(k_1 \cdot \frac{\partial}{\partial p}\right)$$

$$\times \frac{1}{\omega - \omega_1 - ([k - k_1] \cdot v) + i0} \left(k \cdot \frac{\partial}{\partial p}\right)$$

$$\times \delta(\omega_1 - (k_1 \cdot v)) \Phi_p^\alpha \frac{u^? \mu}{(2\pi)^3}. \tag{12.31}$$

This expression contains a δ-function describing the resonance of the particle with the field. The imaginary part of it will contain also another δ-function which describes the scattering. These two δ-functions together describe the bremsstrahlung. This last expression completes the theory of the interaction of waves with particle fluctuations and with the fields produced by those fluctuations.

12.7 Spontaneous Scattering Processes

We shall now show that the dielectric permittivity which we have just found describes two effects which are related to one another: (1) absorption of collective waves due to their spontaneous scattering by the plasma particles, including all the effects discussed earlier which so drastically change the scattering cross-sections when there are other particles present in a plasma, and (2) the absorption or amplification of collective waves due to the process which is the inverse of bremsstrahlung with a cross-section which is also changed drastically due to the collective effects produced by the other plasma particles. We leave a detailed discussion of bremsstrahlung to the next chapter and shall here consider the spontaneous scattering processes. Since all the fields we are considering here are random fields – the result is independent of whether the wave field is regular or random as we are dealing with the linear approximation for the wave field – we can use the balance equation from Chap.6 for the scattering of random fields and retain only the terms describing the spontaneous processes (see Eq.(6.14)):

$$\left[\frac{dN_{\mathbf{k}}}{dt}\right]^{sc} = Q_{\mathbf{k}}^{sc} + 2\gamma_{\mathbf{k}}^{sc} N_{\mathbf{k}}, \quad Q_{\mathbf{k}} = \int N_{\mathbf{k}'} w_{\mathbf{p}}^{sc,\alpha}(\mathbf{k},\mathbf{k}') \Phi_{\mathbf{p}}^{\alpha} \frac{d^3\mathbf{p}\, d^3\mathbf{k}}{(2\pi)^6}, \quad (12.32)$$

$$\gamma_{\mathbf{k}}^{sc} = -\frac{1}{2}\int w_{\mathbf{p}}^{sc,\alpha}(\mathbf{k},\mathbf{k}') \Phi_{\mathbf{p}}^{\alpha} \frac{d^3\mathbf{p}\, d^3\mathbf{k}'}{(2\pi)^6}. \quad (12.33)$$

In the case where we are interested only in terms due to spontaneous scattering and due to bremsstrahlung which are linear in the number of waves $N_{\mathbf{k}}$ we can add the two growth rates:

$$\frac{dN_{\mathbf{k}}}{dt} = 2\left(\gamma_{\mathbf{k}}^{Br} + \gamma_{\mathbf{k}}^{sc}\right) N_{\mathbf{k}}, \quad (12.34)$$

where $\gamma_{\mathbf{k}}^{Br}$ is the growth or damping rate due to bremsstrahlung. This is possible only when the two processes can be separated and this can be done only approximately. To separate the two processes we shall focus our interest on such statistical particle fluctuations which have frequencies close to the frequency of the collective plasma oscillations, that is, we shall be interested in that region of particle fluctuations for which $\varepsilon_{\mathbf{k}} = 0$. We shall also assume that the fluctuations are not in resonance with the particles. This means that the term $\varepsilon_{\mathbf{k}}^{NL(5)}$ will not contribute to the scattering processes. However, the quantity $\varepsilon_{\mathbf{k}}^{NL(4)}$ describes, indeed, the scattering processes. We have

$$\gamma_{\mathbf{k}}^{NL(4)} = -\left[\mathrm{Im}\left\{\varepsilon_{\mathbf{k}}^{NL(4)}\right\} \Big/ \frac{\partial \varepsilon_{\mathbf{k}}}{\partial \omega}\right]_{\omega=\omega_{\mathbf{k}}}$$

$$= -2\sum_{\alpha}\frac{e_{\alpha}^4}{m_{\alpha}^2}\int \frac{(\mathbf{k}\cdot\mathbf{k}_1)d^3\mathbf{k}_1}{k^2 k_1^2}\frac{1}{(\omega-(\mathbf{k}\cdot\mathbf{v}))^4}\left[\frac{\partial\varepsilon_{\mathbf{k}}}{\partial\omega}\right]_{\omega=\omega_{\mathbf{k}}}^{-1}$$

$$\times\ \delta\big(\omega_{\mathbf{k}}-\omega_{\mathbf{k}_1}-([\mathbf{k}-\mathbf{k}_1]\cdot\mathbf{v})\big)\left[\frac{\partial\varepsilon_{\mathbf{k}_1}}{\partial\omega_1}\right]_{\omega_1=\omega_{\mathbf{k}_1}}^{-1}\Phi_{\mathbf{p}}^{\alpha}\frac{d^3\mathbf{p}}{(2\pi)^3}. \quad (12.35)$$

By comparing this expression with the quantity $\gamma_{\mathbf{k}}^{sc}$ given by Eq.(12.33) we can find the scattering probability, but comparing it with Eqs.(6.49) and (6.50) we can conclude that the probability found from Eq.(12.35) does not contain all the the polarisation factors, that is, it describes scattering by "bare" particles. In practice the main contribution for Langmuir waves comes from the electrons – due to the factor $1/m_{\alpha}^2$ in Eq.(12.35). For scattering by electrons we have

$$\gamma_{\mathbf{k}}^{NL(4)} \simeq -\frac{e^4}{2m_e^2\omega_{pe}^2}\int \frac{(\mathbf{k}\cdot\mathbf{k}_1)^2 d^3\mathbf{k}_1}{k^2 k_1^2}$$

$$\times\ \delta\big(\omega_{\mathbf{k}}-\omega_{\mathbf{k}_1}-([\mathbf{k}-\mathbf{k}_1]\cdot\mathbf{v})\big)\Phi_{\mathbf{p}}^{\alpha}\frac{d^3\mathbf{p}}{(2\pi)^3}, \quad (12.36)$$

and this expression describes the scattering by "bare" electrons. However, we still must consider the other contributions to the non-linear permittivity. Let us consider expression (12.25) for $\varepsilon_k^{NL(3)}$. In this expression there occur two resonances, $\varepsilon_{k_1} \simeq 0$ and $\varepsilon_{k-k_1} \simeq 0$. We can interchange these two resonances by making the following change of variables: $k_1 \leftrightarrows k - k_1$. We find

$$
\begin{aligned}
\gamma_k^{NL(3)} &= - \left[\operatorname{Im} \left\{ \varepsilon_k^{NL(3)} \right\} \Big/ \frac{\partial \varepsilon_k}{\partial \omega} \right]_{\omega = \omega_k} \\
&= \sum_\alpha \frac{16 e_\alpha^3}{m_\alpha} \int \frac{\pi (\mathbf{k} \cdot \mathbf{k}_1) \, d^4 k_1}{k k_1^2 |\mathbf{k} - \mathbf{k}_1|} \left[\frac{\partial \varepsilon_{k_1}}{\partial \omega_1} \right]_{\omega_1 = \omega_{k_1}}^{-1} \left[\frac{\partial \varepsilon_k}{\partial \omega} \right]_{\omega = \omega_k}^{-1} \\
&\quad \times \delta\big(\omega_k - \omega_{k_1} - ([\mathbf{k} - \mathbf{k}_1] \cdot \mathbf{v})\big) \frac{\varrho_{k,k_1-k}^{(2)}}{\big(\omega - (\mathbf{k} \cdot \mathbf{v})\big)^2}.
\end{aligned}
$$

(12.37)

The contribution from the ions is here negligible. We shall use Eq.(5.51), in which the contribution from the ions was neglected, for the non-linear response $\varrho^{(2)}$. For the sake of simplicity to make it easier to follow the calculations we once more write down Eq.(5.51):

$$
\varrho_{k_1,k_1,k}^{(2)} \simeq \frac{e(\mathbf{k} \cdot \mathbf{k}_1)|\mathbf{k} - \mathbf{k}_1|}{8\pi k m_e \omega_{pe}^2} \left[\varepsilon_{k_1-k}^{(e)} - 1 \right]
$$

(12.38)

Using this equation we find for Langmuir oscillations

$$
\begin{aligned}
\gamma_k^{NL(3)} &\simeq \frac{e^4}{2 m_e^2 \omega_{pe}^2} \int \frac{(\mathbf{k} \cdot \mathbf{k}_1)^2 d^3 k_1}{k^2 k_1^2} \\
&\quad \times \frac{\varepsilon_{k_1-k}^{(e)} - 1}{\varepsilon_{k_1-k}} \delta\big(\omega_k - \omega_{k_1} - ([\mathbf{k} - \mathbf{k}_1] \cdot \mathbf{v})\big) \, \Phi_P^\alpha \frac{d^3 p}{(2\pi)^3}.
\end{aligned}
$$

(12.39)

One sees easily that this term is one of the two terms which describes the interference of the scattering by the central particle and of the scattering by the polarisation "cloud". The second interference term comes from $\varepsilon_k^{NL(2)}$. Making the change of variables $k_1 \leftrightarrows k - k_1$ and carrying out the same kind of calculations we find an expression which differs from Eq.(12.39) only in that k and k_1 change places, that is, $\gamma_k^{NL(2)}$ contains the complex conjugate expression for the polarisation factor; in actual fact, we need to consider the sum $\gamma_k^{NL(3)} + \gamma_k^{NL(2)}$ which is real. Finally, the square of the matrix element for scattering by the polarisation "cloud" comes from $\varepsilon_k^{NL(1)}$. Making the change of variables $k_1 \leftrightarrows k - k_1$ we find that the resonance with the particle fluctuations when their frequency is close to that of the collective oscillations comes solely from the second term of ϱ^{eff}:

$$
\varrho_{k-k_1,k,k_1-k}^{eff} = \varrho_{k-k_1,k,k_1-k}^{(3)} + \frac{8\pi}{i k_1 \varepsilon_{k_1}} \varrho_{k-k_1,k_1}^{(2)} \varrho_{k,k_1-k_1}^{(2)}.
$$

(12.41)

Hence we get

$$\gamma_k^{NL(1)} = \left[\text{Im}\left\{ \varepsilon_k^{NL(1)} \right\} \middle/ \left[\frac{\partial \varepsilon_k}{\partial \omega} \right]^{-1} \right]_{\omega = \omega_k}$$

$$= 64\pi^3 \int \frac{d^3 k_1}{k k_1} \left\{ \varrho_{k-k_1, k_1}^{(2)} \varrho_{k, k_1-k}^{(2)} \left| E^{(0)} \right|^2_{k-k_1} \right.$$

$$\times \left. \left(\frac{\partial \varepsilon_k}{\partial \omega} \right)^{-1} \left(\frac{\partial \varepsilon_{k_1}}{\partial \omega_1} \right)^{-1} \right\}_{\substack{\omega = \omega_k \\ \omega_1 = \omega_{k_1}}}. \tag{12.42}$$

After substituting Eq.(12.16) and (12.35) into this equation and taking into account that

$$\varrho_{k-k_1, k_1}^{(2)} = \frac{e(k \cdot k_1)}{8\pi m_e k_1 \omega_{pe}^2}, \tag{12.43}$$

we find in the case of Langmuir oscillations

$$\gamma_k^{NL(sc)} = -\frac{e^4}{2 m_e^2 \omega_{pe}^2} \int \frac{(k \cdot k_1)^2 d^3 k_1}{k^2 k_1^2} \left| \frac{\varepsilon_{k-k_1}^{(e)} - 1}{\varepsilon_{k-k_1}} \right|^2_{\substack{\omega = \omega_k \\ \omega_1 = \omega_{k_1}}}$$

$$\times \delta\big(\omega_k - \omega_{k_1} - ([k - k_1] \cdot v)\big) \Phi_p^\alpha \frac{d^3 p}{(2\pi)^3}. \tag{12.50}$$

This expression describes the process of scattering by the polarisation "cloud" only.

We have thus obtained a very important result. We introduced earlier the scattering probability *a priori* and used the Einstein relation between the spontaneous and the stimulated processes to describe the interactions. We then found from the non-linear interactions the scattering probability which turned out to be very different from the one when there is no plasma present. Now we calculated independently the scattering probability due to spontaneous processes and the interactions between waves and statistical particle fluctuations and we found the same expression with all the collective effects in the scattering present. This result is not a trivial one since the system may be far from equilibrium and have any arbitrary particle distribution. The change in the scattering cross-section due to collective effects thus shows up in both stimulated and spontaneous scattering processes. The question arises why we can restrict ourselves solely to particle fluctuations close to the resonance for which $\varepsilon \approx 0$ and what will be the contribution from other domains of particle fluctuations. We shall see that this contribution is not small and that it describes bremsstrahlung which is related to spontaneous scattering in a way similar to the relation between spontaneous Cherenkov emission and binary particle collisions – which we have already discussed earlier in detail.

Problems

1. Calculate the non-linear dielectric permittivity for the interaction of electromagnetic waves with particle fluctuation fields and with particle distribution fluctuations. Assume the virtual fields to be longitudinal.
2. Also calculate the non-linear dielectric permittivity for the interaction of electromagnetic waves with particle fluctuation fields and with particle distribution fluctuations, assuming the virtual fields to be transverse.
3. Calculate the spontaneous scattering of Langmuir waves leading to electromagnetic waves.
4. Calculate the spontaneous scattering of electromagnetic waves leading to electromagnetic waves.
5. Show that the spontaneous interactions between resonant and non-resonant waves, which was considered in Chap.11, gives a contribution only to terms proportional to the number of waves.
6. Show that there is no interaction between resonant and non-resonant waves in thermal equilibrium.
7. Find an expression for the non-linear permittivity describing the interaction of waves with fluctuations of highly relativistic particles. Consider the case of a power-law particle distribution, $\bar{\sigma}_p \propto p^{-x}$.

13 Collective Effects in Bremsstrahlung of Plasma Particles

13.1 Introduction. Elementary Concepts of Bremsstrahlung of Colliding Particles

We have shown in earlier chapters that both particle collisions and scattering of waves involve screened particles and that this leads to drastic changes in the collision and scattering cross-sections. There is also a similar change for bremsstrahlung, that is, for the emission occurring when particles collide. This change occurs for wavelengths longer than the screening Debye length. The only way to prove this is by using fluctuation theory. We have both spontaneous and stimulated bremsstrahlung. The stimulated bremsstrahlung is proportional to the number of waves – to their intensity – and thus describes the absorption or amplification of waves due to collisions. For the calculation of bremsstrahlung probabilities we can use the stimulated processes which are sometimes called *inverse bremsstrahlung* since they describe absorption, that is, the process which is the inverse of the emission process. However, we should bear in mind that the final result of absorption or amplification in a non-equilibrium system is the result of the balance between stimulated bremsstrahlung absorption and stimulated bremsstrahlung emission.

We shall here consider systems in an arbitrary non-equilibrium state with an arbitary particle distribution. To calculate the absorption or amplification we shall use the expression for the non-linear permittivity for waves interacting with particle fluctuations which we calculated in Chap.12. We shall consider the fluctuations here in a general way whereas in the previous chapter we considered only those fluctuations which corresponded to resonance for the collective wave fields. The main contribution will come from that part of the particle fluctuations which are outside this resonance. This is similar to the relation between the change in the particle distribution due to collisions and that due to spontaneous Cherenkov emission. Here we are interested in the relation between spontaneous scattering and bremsstrahlung. In order to get a better understanding of this relation we remind ourselves of the classical result for the bremsstrahlung by isolated particles: the probability for bremsstrahlung is the product of the probability for the appearance of a given harmonic of the field of the colliding particle and the probability for the scattering of that harmonic leading to the wave which is emitted in the bremsstrahlung. The fields created in the collisions are just the virtual fields. If the field is close to the wave field it is changed into a propagating wave field rather than a virtual field and the scattering leads to another real wave. This is the spontaneous scattering process. We see thus that the scattering amplitude enters inevitably in the bremsstrahlung amplitude.

As we have seen that the collective effects in a plasma drastically change the scattering we must inevitably expect that the bremsstrahlung is also changed drastically. We can understand this change qualitatively as follows. Let us consider bremsstrahlung of waves with wavelengths larger than the screening length. Not only are the particles in the collisions changing their trajectories which gives rise to the usual bremsstrahlung, but also the polarisation clouds of the colliding particles are displaced creating a variable dipole moment serving as an additional source of bremsstrahlung. We expect thus that even collisions of very heavy particles may lead to bremsstrahlung – this is similar to the possibility for the scattering of very heavy particles in a plasma. These processes are most effective when the wavelength of the emitted wave is larger than the screening length as in that case the dipole moment appearing in the collisions is varying coherently. We should once again bear in mind that a proper consideration of these processes is possible only in the framework of fluctuation theory since each particle is involved both in the collisions and in the screening of the fields of the other particles. Before starting on a detailed theoretical discussion we shall obtain the main results qualitatively from the physical picture of a "dressed" particle. To do this we must first remind ourselves of the usual concept of bremsstrahlung by isolated – "bare" – particles. Another point is that in a bremsstrahlung process all types of waves can be emitted, including electromagnetic waves and longitudinal waves – such as Langmuir and ion-sound waves. As we decided to consider here only longitudinal waves we shall first give some results and necessary modifications of the usual bremsstrahlung processes involving longitudinal waves – that is, the emission of longitudinal waves in collisions of "bare" particles. As everywhere before, we shall assume that the particle velocities are non-relativistic.

We shall start by reminding ourselves of the usual process of bremsstrahlung of electromagnetic waves by collisions of isolated – "bare" – particles in a vacuum. The main emission process is the one where a light particle, say an electron, collides with a heavy particle, say an ion, and the light particle accelerates or decelerates in the collision, that is, the light particle changes its trajectory. One often calls the incident light particle the projectile. The light and heavy particles we shall distinguish, respectively, by subscripts, or superscripts, α and β. We can use the classical formula for dipole emission. The power w_ω emitted by the particle α in the frequency range $d\omega$ is given by the formula

$$w_\omega = \frac{8\pi e_\alpha^2}{3c} \left[\frac{d^2\mathbf{r}}{dt^2} \right]_\omega^2 = \frac{8\pi e_\alpha^4}{3m_\alpha^2 c^3} \left(\mathbf{E}_\omega^\beta \right)^2, \tag{13.1}$$

where $\mathbf{E}_\omega^\beta$ is the Fourier component of the electrostatic field of particle β at the position of particle α. We can clearly see from this expression that the square of the mass of the projectile α occurs in the denominator and the emission is usually negligible when the projectile is heavy. We emphasise this point as this is not the case in a plasma and the polarisation cloud of a heavy particle can lead to a large bremsstrahlung emission.

Returning to the problem of emission by "bare" particles we may mention that the field $\mathbf{E}^\beta$ determines the acceleration of the light particle caused by the heavy particle. Assume that at the minimum distance ϱ between the particles in their collision (see Fig.13.1) the kinetic energy of the projectile is still much larger than the potential energy of the interaction with the heavy particle. The change in the trajectory of the light particle in the collision will then be small. In the field $\mathbf{E}^\beta$ we can then to a first approximation substitute for $\mathbf{r}$ its value, $\mathbf{r} = \rho + \mathbf{v}t$, for an unperturbed projectile trajectory:

$$\mathbf{E}^\beta = e_\beta \frac{\mathbf{r}}{r^3} = -\mathrm{i}e_\beta \int \frac{\mathbf{q}\, d^3\mathbf{q}}{2\pi^2 q^2} e^{\mathrm{i}(\mathbf{q}\cdot\mathbf{r})}$$

$$= -\mathrm{i}e_\beta \int \frac{\mathbf{q}\, d^3\mathbf{q}}{2\pi^2 q^2} e^{\mathrm{i}(\mathbf{q}\cdot\rho)+\mathrm{i}(\mathbf{k}\cdot\mathbf{v})t}. \tag{13.2}$$

The Fourier component of this field,

$$\mathbf{E}_\omega^\beta = -\mathrm{i}e_\beta \int \frac{\mathbf{q}\, d^3\mathbf{q}}{2\pi^2 q^2} e^{\mathrm{i}(\mathbf{q}\cdot\rho)} \delta\big(\omega + (\mathbf{q}\cdot\mathbf{v})\big)$$

$$= -\frac{\mathrm{i}e_\beta}{v} \int \frac{\mathbf{q}\, d^2\mathbf{q}_\perp\, e^{\mathrm{i}(\mathbf{q}_\perp\cdot\rho)}}{2\pi^2\big(q_\perp^2 + \omega^2/v^2\big)}, \tag{13.3}$$

occurs in Eq.(13.1). In Eq.(13.3) we have split $\mathbf{q}$ – the wavevector of the virtual field – into two components, a component $\mathbf{q}_\perp$ perpendicular to the velocity of the projectile α and a component $(\mathbf{q}\cdot\mathbf{v})/v$ parallel to the projectile velocity. The main contribution to Eq.(13.3) comes from $q_\perp < 1/\varrho$ since for $q_\perp \gg 1/\varrho$ the rapidly oscillating exponential factor gives a zero result. The largest Fourier component are therefore those with $\omega \ll v/\varrho$ and evaluating the integral in Eq.(13.3) we find

$$\mathbf{E}_\omega^\beta \simeq \frac{e_\beta\rho}{\pi v\varrho^2}. \tag{13.4}$$

Substituting this result into Eq.(13.1) we get

$$w_\omega = \frac{8e_\alpha^4 e_\beta^2}{3\pi c^3 m_\alpha^2 \varrho^2 v^2}. \tag{13.5}$$

Rather than considering the emission in a single particle collision let us now discuss the emission for a light particle propagating in a plasma through homogeneously distributed heavy particles. We shall here neglect collective effects and use Eq.(13.4). The light particle will undergo a certain number

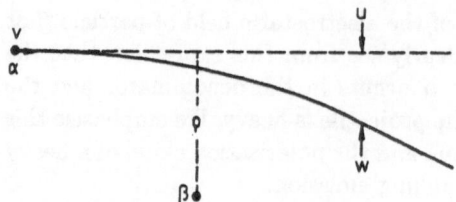

Fig. 13.1. Collision of a projectile α and a heavy particle β; u: unperturbed trajectory, w: weakly perturbed trajectory

of collisions with heavy particles per unit time. We shall assume that the emission in a particular collision is independent of the emission in the previous collision. To find the power emitted by the light particle per unit time we must then multiply w_ω by the number of heavy particles β which the light particle meets in unit time. This number is for ions at distances between ϱ and $\varrho + d\varrho$ from the light particle equal to $2\pi n_\beta v\varrho\, d\varrho$ so that the emitted power Q_ω will be given by the equation

$$Q_\omega = \int w_\omega n_\beta v_\alpha\, 2\pi\varrho\, d\varrho = \frac{16 e_\alpha^4 e_\beta^2 n_\beta}{3 c^3 m_\alpha^2 v} \ln \frac{\varrho_{\max}}{\varrho_{\min}}, \tag{13.6}$$

which is the well known formula for the usual bremsstrahlung. We have given this derivation to show where exactly in its derivation a mistake is made. Let us discuss the possible values of $\varrho_{\max}$ and $\varrho_{\min}$. The value of $\varrho_{\min}$ is determined by the simplification used in the derivation of Eq.(13.6) that we can use the approximation of distant collisions:

$$\frac{e_\alpha e_\beta}{\varrho_{\min}} \simeq \tfrac{1}{2} m_\alpha v^2. \tag{13.7}$$

More important is the value of $\varrho_{\max}$. Equation (13.4) is valid for $\omega \ll v/\varrho$ and one possible value for $\varrho_{\max}$ is $\varrho_{\max} \simeq v/\omega$. It was believed for a long time that another value should be $\varrho_{\max} \simeq d$ since the field is screened at these distances and we should thus have $\varrho_{\max} \simeq \min\{v/\omega, d\}$. However, at this point we just want to mention that the change in the dipole moment of the screening shell has not been taken into account, and this emission mechanism is the most important one for wavelengths larger than the screening length.

13.2 Qualitative Description of Collective Effects in Bremsstrahlung of Particles in a Plasma

Let us now consider the bremsstrahlung process for $\varrho \gg d$. To simplify matters we shall assume that the light particle is a fast particle so that it will be "undressed" – it will not have a screening shell. This particle displaces the shell of the heavy particle and creates a dipole moment. Let the charge of the heavy particle – the ion – be $-Ze$ and this will also be the value of the total charge in its screening shell, but with the opposite sign. Let the charge of the light particles contributing to the screening of the ion be $Z_{\text{eff}}e$ – the other part, $(Z - Z_{\text{eff}})e$, is the contribution from the heavy particles to the screening. For wavelengths much larger than the screening radius this charge is to a first approximation a point charge. To find the emission by the shell we must interchange the particles in the previous discussion. Now the field of the projectile changes the position of the sreening charge. In the first equality in Eq.(13.1) we must change e_α to $Z_{\text{eff}}e$ and for the charge to mass ratio substitute the electron charge to mass ratio e/m_e – assuming that the light screening particles are electrons. Finally, we must take into account that the field acting on the light particles in the screening shell is determined by the charge e_α of the projectile. We therefore get

$$Q_\omega = \frac{16 e_\alpha^2 (Z_{\text{eff}}e)^2 e^2}{3 m_e^2 c^3 v} \ln \frac{\varrho_{\max}}{\varrho_{\min}}. \tag{13.8}$$

It is reasonable now to put $\varrho_{\min} \simeq d$ and $\varrho_{\max} \simeq v/\omega$. We see that Eq.(13.8) depends on the masses and the charges of the colliding particles in quite a different way from Eq.(13.6) – we remind ourselves that in Eq.(13.6) we have $e_\beta = -Ze$. We must mention that now the mass of the projectile does not enter into the final expression. A heavy projectile will produce the same emission as a light projectile if its velocity is the same. This results changes the whole emission process qualitatively. Not only is there now the possibility for bremsstrahlung emission by a heavy particle but there are also important changes in the bremsstrahlung by a light particle. Indeed, if the projectile is an electron its emission and the emission by the electrons in the screening shell of the heavy particle will interfere with each other. It is well known that the dipole emission in collisions between particles with equal charge to mass ratios is zero to a first approximation. This means that the just mentioned interference will have as a result the cancellation of that part of the emission which is due to the screening electrons of the heavy ion. The emission will correspond to that by a partially "bare" ion which has only a screening charge $Z_{\text{eff}}^i e = (Z - Z_{\text{eff}}^e)e$. To obtain the correct expression for the emission at wavelengths larger than the screening length by an electron as projectile colliding with an ion we must therefore substitute into Eq.(8.6) $e_\beta = Z_{\text{eff}}^i e$ rather than $e_\beta = Ze$. This shows the substantial changes in a plasma not only of the bremsstrahlung process in ion-ion collisions but also in electron-ion collisions. Of course, one must prove all what we have just said using fluctuation theory. However, the final result that the picture we have just painted is, indeed, the correct one is one of the most important conclusions of non-linear plasma kinetics. We have given here the simplest possible example. In the general case one must take into account the presence of screening shells for both of the colliding particles. In the collisions the particles may emit any kind of waves, not just the electromagnetic waves which we considered in order to have easy reference to the well known classical result. In what follows we shall deal with bremsstrahlung of longitudial waves. However, we must first describe the emission of longitudinal waves by "bare" particles.

13.3 Bremsstrahlung of Longitudinal Waves by Bare Particles

The longitudinal waves always have wavelengths larger than the screening length and collective effects in bremsstrahlung are therefore always important for them. The velocity of light should not occur in the expression for the bremsstrahlung of longitudinal waves; however, the changes in the bremsstrahlung due to collective effects are qualitatively the same for longitudinal and for electromagnetic waves and can be expressed in terms of the appropriate effective charges and masses of the particles – the mass of the light particles in the shell will appear. In the present section we consider the emission of longitudinal waves by a light particle changing its trajectory under the influence of a heavy particle, that is, the emission corresponding to the classical description of bremsstrahlung. The equation of motion for the light particle has the form

$$m_\alpha \frac{d^2\mathbf{r}}{dt^2} = e_\alpha \mathbf{E}^\beta = -ie_\alpha e_\beta \int \frac{\mathbf{q}\, d^3q}{2\pi^2 q^2}\, e^{i(\mathbf{q}\cdot\mathbf{r})-i(\mathbf{q}\cdot\mathbf{r}_\beta)}, \tag{13.9}$$

where $\mathbf{r}_\beta$ is the position of the ion β. For distant collisions we can on the right-hand side put $\mathbf{r} = \mathbf{v}t + \delta\mathbf{r}$ and assume that $\delta\mathbf{r}$ is small. We then have

$$\delta\mathbf{r} \simeq \frac{ie_\alpha e_\beta}{2\pi^2 m_\alpha} \int \frac{\mathbf{q}\, d^3q}{q^2(\mathbf{q}\cdot\mathbf{v})^2}\, e^{i(\mathbf{q}\cdot\mathbf{v})t-i(\mathbf{q}\cdot\mathbf{r}_\beta)}. \tag{13.10}$$

The charge density $\varrho_{\mathbf{k},\omega}$ produced by an arbitrary moving charge and the change in it, $\delta\varrho_{\mathbf{k},\omega}$, due to the change in the particle trajectory are

$$\left. \begin{aligned} \varrho_{\mathbf{k},\omega} &= \frac{e_\alpha}{(2\pi)^4} \int e^{-i\left(\mathbf{k}\cdot\mathbf{r}(t)\right)+i\omega t}\, dt, \\[2mm] \delta\varrho_{\mathbf{k},\omega} &\simeq \frac{ie_\alpha}{(2\pi)^4} \int (\mathbf{k}\cdot\delta\mathbf{r})\, e^{i\omega t}\, dt, \end{aligned} \right\} \tag{13.11}$$

or, if we take Eq.(13.10) into account

$$\begin{aligned} \delta\varrho_{\mathbf{k},\omega} &\simeq \frac{e_\alpha^2 e_\beta}{\pi m_\alpha} \int \frac{(\mathbf{k}\cdot\mathbf{q})d^3q}{(2\pi)^4 q^2(\mathbf{q}\cdot\mathbf{v})^2}\, e^{-i(\mathbf{q}\cdot\mathbf{r}_\beta)}\, \delta\bigl(\omega + (\mathbf{q}\cdot\mathbf{v})\bigr) \\[2mm] &= \frac{e_\alpha}{(2\pi)^3\omega^2} \left(\mathbf{k}\cdot\frac{e_\alpha \mathbf{E}_\omega^\beta}{m_\alpha} \right). \end{aligned} \tag{13.12}$$

We have here expressed the result in terms of the acceleration $e_\alpha \mathbf{E}_\omega^\beta / m_\alpha$ of the projectile. Using the Poisson equation,

$$\delta E_{\mathbf{k},\omega} = \frac{4\pi\, \delta \varrho_{\mathbf{k},\omega}}{ik\varepsilon_{\mathbf{k},\omega}}, \quad \mathbf{E}_{\mathbf{k},\omega} = \frac{\mathbf{k}}{k} E_{\mathbf{k},\omega}, \tag{13.13}$$

we find the field which occurs here and the work w it does on the charge density $\delta\varrho$:

$$
\begin{aligned}
w &= \int (\delta\mathbf{E} \cdot \delta\mathbf{j})\, d^3\mathbf{r}\, dt = -(2\pi)^4 \int \delta E_{\mathbf{k},\omega} (\mathbf{k} \cdot \delta\mathbf{j}_{-\mathbf{k},-\omega}) \frac{d^3\mathbf{k}\, d\omega}{k} \\
&= -(2\pi)^4 \int \frac{\omega}{k} \delta E_{\mathbf{k},\omega} \delta\varrho_{-\mathbf{k},-\omega}\, d^3\mathbf{k}\, d\omega.
\end{aligned}
\tag{13.14}
$$

In deriving the last result we used the equation of continuity.

The only contribution comes from the imaginary part,

$$
\begin{aligned}
\operatorname{Im}\left\{\frac{1}{\varepsilon_{\mathbf{k},\omega}}\right\} &= -\frac{i\pi\omega}{|\omega|} \delta\left(\operatorname{Re}\{\varepsilon_{\mathbf{k},\omega}\}\right) \\
&= -\frac{i\pi\omega}{|\omega|} \left[\delta(\omega - \omega_\mathbf{k}) + \delta(\omega + \omega_\mathbf{k})\right] \left[\frac{\partial\varepsilon_{\mathbf{k},\omega}}{\partial\omega}\right]_{\omega=\omega_\mathbf{k}}^{-1},
\end{aligned}
\tag{13.15}
$$

which leads to

$$
\begin{aligned}
w &= e_\alpha^2 \int \frac{|\omega|}{\omega^3} \left|\left(\frac{\mathbf{k}}{k} \cdot \frac{e_\alpha}{m_\alpha} \mathbf{E}_\omega^\beta\right)\right|^2 \delta(\varepsilon_{\mathbf{k},\omega})\, d^3\mathbf{k}\, d\omega \\
&= e_\alpha^2 \int \frac{2\, d^3\mathbf{k}}{3\omega_\mathbf{k}^3} \left[\frac{\partial\varepsilon_{\mathbf{k},\omega}}{\partial\omega}\right]_{\omega=\omega_\mathbf{k}}^{-1} \left|\frac{e_\alpha}{m_\alpha} \mathbf{E}_{\omega_\mathbf{k}}^\beta\right|^2.
\end{aligned}
\tag{13.16}
$$

For longitudinal waves it is more convenient to introduce the power emitted in the range $d^3\mathbf{k}$ rather than in the range $d\omega$. The integration over $\mathbf{k}$ in Eq.(13.16) is restricted by the domain in which longitudinal waves exist. Equation (13.16) differs from Eq.(13.1) only in that it contains the component of the acceleration $e_\alpha \mathbf{E}^\beta / m_\alpha$ along the polarisation unit vector of the longitudinal waves rather than the polarisation unit vector of transverse waves. The numerical factor $\frac{1}{3}$ in Eq.(13.16) appears after averaging over angles. For Langmuir waves we have $\partial\varepsilon/\partial\omega \simeq 2/\omega_{\mathrm{pe}}$ and the maximum wavenumber is determined by the Debye radius. The energy emitted in longitudinal waves therefore contains the cube of the thermal velocity rather than the cube of the light velocity:

$$w \approx \tfrac{1}{3} e_\alpha^2 \left(\frac{e_\alpha}{m_\alpha} \mathbf{E}_{\omega_{\mathrm{pe}}}^\beta\right)^2 \frac{1}{v_{Te}^3}. \tag{13.17}$$

We now calculate the power emitted in the range dk, where $k = |\mathbf{k}|$, for a particle moving through a plasma and on its way colliding with many ions. Writing $w = \int w_k\, dk$ to introduce a quantity w_k in Eq.(13.16) and using Eq.(13.12) we find

$$Q_k = \int w_k v n_\beta \, 2\pi \varrho \, d\varrho = \frac{16 e_\alpha^4 e_\beta^2 k^2 n_\beta}{3 m_\alpha \omega_k^3 v} \left[\frac{\partial \varepsilon_k}{\partial \omega}\right]_{\omega = \omega_k}^{-1} \ln \frac{\varrho_{max}}{\varrho_{min}}. \quad (13.18)$$

We can compare this expression with Eq.(13.6).

For the following discussion we must introduce the probability of bremsstrahlung by a particle α with initial momentum $\mathbf{p}$ when it collides in its path with randomly and homogeneously distributed particles β. Under those conditions we are interested in the emitted power – the energy emitted per unit time – but not in the total emission in a single collision, as we were for Eq.(13.14). To find the emitted power we must integrate over t and average over the positions of the ions β:

$$Q = -\int (\mathbf{j} \cdot \mathbf{E}) \, d^3\mathbf{r} \, n_\beta \, d^3\mathbf{r}_\beta$$

$$= -(2\pi)^3 \int \frac{\omega'}{k} \, \delta E_{\mathbf{k},\omega} \delta \varrho_{-\mathbf{k},-\omega'} \, e^{i\omega t - i\omega' t} \, d\omega \, d\omega' d^3\mathbf{k} \, n_\beta \, d^3\mathbf{r}_\beta$$

$$= 4\pi i (2\pi)^3 \int \frac{\omega' d^3\mathbf{k} \, d\omega \, d\omega'}{k^2 \varepsilon_{\mathbf{k},\omega}} \, n_\beta \, e^{i\omega t - i\omega' t} \int \delta \varrho_{\mathbf{k},\omega} \delta \varrho_{-\mathbf{k},-\omega'} \, d^3\mathbf{r}_\beta.$$

One can easily calculate the last integral, using the first equality of Eq.(13.12):

$$\int \delta \varrho_{\mathbf{k},\omega} \delta \varrho_{-\mathbf{k},-\omega'} \, d^3\mathbf{r}_\beta = \frac{e_\alpha^2 e_\beta^2 \delta(\omega - \omega')}{\pi^2 (2\pi)^5 m_\alpha^2}$$

$$\times \int \frac{(\mathbf{k} \cdot \mathbf{q})^2}{q^4 \omega^4} \, \delta(\omega + (\mathbf{q} \cdot \mathbf{v})) \, d^3\mathbf{q}, \quad (13.19)$$

and we therefore get finally

$$Q = \frac{2 e_\alpha^4 e_\beta^2}{\pi^2 m_\alpha^2} \int \frac{(\mathbf{k} \cdot \mathbf{q})^2 n_\beta}{k^2 q^4} \, \delta(\omega + (\mathbf{k} \cdot \mathbf{q})) \left[\frac{\partial \varepsilon_{\mathbf{k},\omega}}{\partial \omega}\right]_{\omega = \omega_k}^{-1} d^3\mathbf{k} \, d^3\mathbf{q}. \quad (13.20)$$

By a simple integration this expression leads again to Eq.(13.18) with $\ln(q_{max}/q_{min})$ instead of $\ln(\varrho_{max}/\varrho_{min})$; the quantity q is obviously related to ϱ by $q \simeq 1/\varrho$ and $q_{max} \simeq 1/\varrho_{min}$, $q_{min} \simeq 1/\varrho_{max}$. The quantity q plays the rôle of the momentum transferred from one particle to another in the emission process. We can introduce a bremsstrahlung probability $w_{\mathbf{p},\mathbf{p}'}^{\alpha,\beta}(\mathbf{k}, \mathbf{q})$ per particle, normalised per unit phase volume $d^3\mathbf{k}/(2\pi)^3$ of the emitted waves as well as normalised per unit phase volume $d^3\mathbf{q}/(2\pi)^3$ of the transferred momentum. To define this probability we use the expression for the emitted power,

$$Q_{\mathbf{p}} = \int \omega_k w_{\mathbf{p},\mathbf{p}'}^{\alpha,\beta}(\mathbf{k}, \mathbf{q}) \Phi_{\mathbf{p}'}^\beta \frac{d^3\mathbf{k} \, d^3\mathbf{p}' \, d^3\mathbf{q}}{(2\pi)^9}. \quad (13.21)$$

From Eq.(13.20) we then find that

$$w_{\mathbf{p},\mathbf{p}'}^{\alpha,\beta}(\mathbf{k},\mathbf{q}) = \frac{8(2\pi)^4 e_\alpha^4 e_\beta^2}{m_\alpha^2 q^2 \omega_{\mathbf{k}}^4} \delta(\omega_{\mathbf{k}} + (\mathbf{q}\cdot\mathbf{v})) \left[\frac{\partial \varepsilon_{\mathbf{k}}}{\partial \omega}\right]_{\omega=\omega_{\mathbf{k}}}^{-1} \frac{(\mathbf{k}\cdot\mathbf{q})^2}{k^2 q^2}. \quad (13.22)$$

This probability describes the bremsstrahlung of "bare" particles and in some sense is, strictly speaking, never correct. However, it can serve as a reference expression to show the kind of changes introduced by collective effects.

13.4 Bremsstrahlung Probabilities and Balance Equations

If in a bremsstrahlung process the momentum of particle β is changing by an amount $\mathbf{q}$ from an initial value $\mathbf{p}'$ to a final value $\mathbf{p}' - \mathbf{q}$, the momentum of particle α must change by an amount $\mathbf{k} - \mathbf{q}$ from an initial value $\mathbf{p}$ to a final value $\mathbf{p} - \mathbf{k} + \mathbf{q}$ since the emitted wave takes away a momentum $\mathbf{k}$. We can therefore write the energy conservation in an elementary bremsstrahlung process in the form

$$\varepsilon_{\mathbf{p}}^\alpha + \varepsilon_{\mathbf{p}'}^\beta = \varepsilon_{\mathbf{p}-\mathbf{k}+\mathbf{q}}^\alpha + \varepsilon_{\mathbf{p}'-\mathbf{q}}^\beta + \omega_{\mathbf{k}}. \quad (13.23)$$

If both the transferred momentum and the momentum of the emitted wave are small as compared to the particle momenta we find from Eq.(13.23) that

$$([\mathbf{q}-\mathbf{k}]\cdot\mathbf{v}) - (\mathbf{q}\cdot\mathbf{v}') + \omega_{\mathbf{k}} = 0. \quad (13.24)$$

The discussion of the previous section was restricted to the case where the charge β was fixed so that $\mathbf{v}' = 0$, as well as to the dipole approximation for the emission, that is, to the case where $k \ll q$. In that case Eq.(13.24) becomes (see Eq.(13.20))

$$(\mathbf{q}\cdot\mathbf{v}) + \omega_{\mathbf{k}} = 0.$$

We must also mention that the expressions in the previous section describe the power emitted by a single particle α. If we are interested in the power emitted by unit volume of the plasma we must integrate these expressions after multiplying them by the distribution function of the particles α:

$$Q = \int Q_{\mathbf{p}} \Phi_{\mathbf{p}}^\alpha \frac{d^3\mathbf{p}}{(2\pi)^3}. \quad (13.25)$$

Let us now write down the balance equation which takes into account both spontaneous and stimulated bremsstrahlung emission. The number of waves increases due to spontaneous and stimulated emission processes:

$$\left[\frac{dN_{\mathbf{k}}}{dt}\right]_+ = \int w_{\mathbf{p},\mathbf{p}'}^{\alpha,\beta}(\mathbf{k},\mathbf{q}) \Phi_{\mathbf{p}}^\alpha \Phi_{\mathbf{p}'}^\beta (N_{\mathbf{k}}+1) \frac{d^3\mathbf{q}\, d^3\mathbf{p}\, d^3\mathbf{p}'}{(2\pi)^9}. \quad (13.26)$$

The number of waves decreases due to the stimulated absorption processes:

$$\left[\frac{dN_{\mathbf{k}}}{dt}\right]_{-} = -\int w_{\mathbf{p},\mathbf{p}'}^{\alpha,\beta}(\mathbf{k},\mathbf{q})\, \Phi_{\mathbf{p}-\mathbf{k}+\mathbf{q}}^{\alpha}\Phi_{\mathbf{p}'-\mathbf{q}}^{\beta}\, N_{\mathbf{k}}\, \frac{d^3q\, d^3p\, d^3p'}{(2\pi)^9}. \qquad (13.27)$$

Expanding in terms of the emitted and transferred momenta we find

$$\frac{dN_{\mathbf{k}}}{dt} = \left[\frac{dN_{\mathbf{k}}}{dt}\right]_{+} + \left[\frac{dN_{\mathbf{k}}}{dt}\right]_{-} = Q_{\mathbf{k}} + 2\gamma_{\mathbf{k}}N_{\mathbf{k}}, \qquad (13.28)$$

where $Q_{\mathbf{k}}$ is the spontaneous emission power of a unit volume of plasma per unit phase volume $d^3k/(2\pi)^3$ divided by the energy $\omega_{\mathbf{k}}$ of a single quantum:

$$Q = \int Q_{\mathbf{k}}\omega_{\mathbf{k}}\, \frac{d^3k}{(2\pi)^3}. \qquad (13.29)$$

The quantity Q is, of course, the same as the one introduced earlier (see Eqs.(13.21) and (13.25)):

$$Q_{\mathbf{k}} = \int w_{\mathbf{p},\mathbf{p}'}^{\alpha,\beta}(\mathbf{k},\mathbf{q})\, \Phi_{\mathbf{p}}^{\alpha}\Phi_{\mathbf{p}'}^{\beta}\, \frac{d^3q\, d^3p\, d^3p'}{(2\pi)^3}. \qquad (13.30)$$

The quantity $\gamma_{\mathbf{k}}$ determines the damping or amplification of the waves due to the inverse bremsstrahlung processes:

$$\gamma_{\mathbf{k}} = \frac{1}{2}\int w_{\mathbf{p},\mathbf{p}'}^{\alpha,\beta}(\mathbf{k},\mathbf{q})$$
$$\times \left[\left((\mathbf{k}-\mathbf{q})\cdot\frac{\partial\Phi_{\mathbf{p}}^{\alpha}}{\partial\mathbf{p}}\right)\Phi_{\mathbf{p}'}^{\beta} + \left(\mathbf{q}\cdot\frac{\partial\phi_{\mathbf{p}'}^{\beta}}{\partial\mathbf{p}'}\right)\Phi_{\mathbf{p}}^{\alpha}\right]\frac{d^3q\, d^3p\, d^3p'}{(2\pi)^9}. \qquad (13.31)$$

In all these expressions the total probabilities occur which should take into account all collective effects. If necessary we shall in what follows give $\gamma_{\mathbf{k}}$ the superscript "Br" to indicate that we are interested in the bremsstrahlung processes, that is, we may write $\gamma_{\mathbf{k}}^{\mathrm{Br}}$. The spontaneous scattering processes are described by the equatins given in Chap.12 and also by Eq.(6.14). We remind ourselves that for spontaneous scattering we have

$$\left[\frac{dN_{\mathbf{k}}}{dt}\right]^{\mathrm{sc}} = Q_{\mathbf{k}}^{\mathrm{sc}} + 2\gamma_{\mathbf{k}}^{\mathrm{sc}}N_{\mathbf{k}}, \quad Q_{\mathbf{k}} = \int N_{\mathbf{k}'}w_{\mathbf{p}}^{\mathrm{sc},\alpha}(\mathbf{k},\mathbf{k}')\, \Phi_{\mathbf{p}}^{\alpha}\, \frac{d^3p\, d^3k'}{(2\pi)^6}, \qquad (13.32)$$

$$\gamma_{\mathbf{k}}^{\mathrm{sc}} = -\frac{1}{2}\int w_{\mathbf{p}}^{\mathrm{sc},\alpha}(\mathbf{k},\mathbf{k}')\, \Phi_{\mathbf{p}}^{\alpha}\, \frac{d^3p\, d^3k'}{(2\pi)^6}. \qquad (13.33)$$

If we take into account only terms linear in $N_{\mathbf{k}}$ and if we can separate the spontaneous scattering and the bremsstrahlung we have

$$\frac{dN_{\mathbf{k}}}{dt} = 2\left(\gamma_{\mathbf{k}}^{\mathrm{Br}} + \gamma_{\mathbf{k}}^{\mathrm{sc}}\right)N_{\mathbf{k}}. \qquad (13.34)$$

Using the bremsstrahlung probabilities which we have introduced here we can also find the equations describing the change in the distribution functions of the particles which take part in the bremsstrahlung process. We leave the derivation of those equations and of the global conservation laws as exercises.

13.5 Bremsstrahlung Probabilities for Particles in a Plasma Including Collective Effects

We shall now formulate the exact relations and probabilities for bremsstrahlung by arbitrarily distributed plasma particles taking into account all collective effects which are important for bremsstrahlung. We shall use the expressions which we have found earlier for the non-linear interactions between the collective plasma waves and the statistical particle fluctuations in a plasma. We shall collect all the terms which according to Eq.(13.24) contain the δ-function describing the conservation of energy in the bremsstrahlung processes:

$$\delta\big(\omega_{\mathbf{k}} - ([\mathbf{k} - \mathbf{q}] \cdot \mathbf{v}) + (\mathbf{q} \cdot \mathbf{v}')\big). \tag{13.35}$$

We shall start with the expression for $\varepsilon_{\mathbf{k}}^{\mathrm{NL}(1)}$ (see Eq.(12.14)) and, in particular, with the contribution to it from the cubic non-linear response. Putting $\mathbf{k}_1 = \{\mathbf{q}, \omega_1\}$ we get from Eq.(5.45):

$$\mathrm{Im}\left\{\frac{1}{i}\, \varrho_{\mathbf{k}_1,\mathbf{k},-\mathbf{k}_1}^{(3)}\right\} = -\sum_{\alpha} \frac{\pi e_{\alpha}^4}{2q^2 k m_{\alpha}^2} \int \frac{(\mathbf{k} \cdot \mathbf{q})^2}{\big(\omega_{\mathbf{k}} - (\mathbf{k} \cdot \mathbf{v})\big)^4}$$
$$\times \left([\mathbf{k} - \mathbf{q}] \cdot \frac{\partial \Phi_{\mathbf{p}}^{\alpha}}{\partial \mathbf{p}}\right) \delta\big(\omega - \omega_1 - ([\mathbf{k} - \mathbf{q}] \cdot \mathbf{v})\big) \frac{d^3 p}{(2\pi)^3}. \tag{13.36}$$

Using this expression in Eq.(12.14) we find

$$\gamma_{\mathbf{k}}^{\mathrm{NL}(1)} = -\left[\mathrm{Im}\left\{\varepsilon_{\mathbf{k}}^{\mathrm{NL}(1)}\right\} \Big/ \frac{\partial \varepsilon_{\mathbf{k}}}{\partial \omega}\right]_{\omega = \omega_{\mathbf{k}}}$$
$$= \sum_{\alpha,\beta} \int \frac{8\pi^4 e_{\alpha}^2 e_{\beta}}{m_{\alpha}^2 \big(\omega_{\mathbf{k}} - (\mathbf{k} \cdot \mathbf{v})\big)^4} \frac{d^3 q}{q^2 \big|\varepsilon_{\mathbf{q},(\mathbf{q} \cdot \mathbf{v}')}\big|^2}$$
$$\times \delta\big(\omega_{\mathbf{k}} - (\mathbf{q} \cdot \mathbf{v}') - ([\mathbf{k} - \mathbf{q}] \cdot \mathbf{v})\big) \left[\frac{\partial \varepsilon_{\mathbf{k}}}{\partial \omega}\right]_{\omega = \omega_{\mathbf{k}}}^{-1}$$
$$\times \frac{(\mathbf{k} \cdot \mathbf{q})^2}{k^2 q^2} \left([\mathbf{k} - \mathbf{q}] \cdot \frac{\partial \Phi_{\mathbf{p}}^{\alpha}}{\partial \mathbf{p}}\right) \Phi_{\mathbf{p}'}^{\beta} \frac{d^3 p\, d^3 p'}{(2\pi)^6}. \tag{13.37}$$

Comparing this equation with Eq.(13.31) we find an expression for the bremsstrahlung probability – it turns out to be that part of it which is not due to the oscillations of the polarisation cloud:

$$w^{\alpha,\beta}_{\mathbf{p},\mathbf{p}'}(\mathbf{k},\mathbf{q}) = \frac{16\pi(2\pi)^3 e_\alpha^4 e_\beta^2}{m_\alpha^2 q^2 \left|\varepsilon_{\mathbf{q},(\mathbf{q}\cdot\mathbf{v}')}\right|^2} \frac{1}{(\omega_{\mathbf{k}} - (\mathbf{k}\cdot\mathbf{v}))^4} \frac{(\mathbf{k}\cdot\mathbf{q})^2}{k^2 q^2}$$

$$\times\; \delta\big(\omega_{\mathbf{k}} - (\mathbf{q}\cdot\mathbf{v}') - ([\mathbf{k}-\mathbf{q}]\cdot\mathbf{v})\big) \left[\frac{\partial\varepsilon_{\mathbf{k}}}{\partial\omega}\right]^{-1}_{\omega=\omega_{\mathbf{k}}} . \tag{13.38}$$

This expression becomes the same as the expression we obtained earlier independently – neglecting the oscillations of the polarisation cloud – if we put $v' = 0$, $(\mathbf{k}\cdot\mathbf{v}) \ll \omega_{\mathbf{k}}$ and neglect the effect of the screening of the field of the heavy particle β (one must then put the dielectric permittivity in the denominator in Eq.(13.22) equal to unity). The last two approximations, in fact, show the domain where Eq.(13.22) is applicable. Finally, this discussion shows that if v' is not equal to zero we are dealing with dynamic screening. However, this screening effect does not take into account the oscillations of the screening charge.

We shall now consider the contribution from $\varepsilon_{\mathbf{k}}^{NL(4)}$ (see Eq.(12.28)) which we have already considered earlier when discussing spontaneous scattering. Here we consider the case when $\varepsilon_{\mathbf{k}}$ is arbitrary and not necessarily close to zero, as we assumed when discussing spontaneous scattering. We shall show that the same expression also describes the bremsstrahlung, but again only that part of it which is not connected with the oscillations of the screening shell. The point is that Eq.(13.31) for $\gamma_{\mathbf{k}}^{Br}$ contains two terms. The first of them contains the derivative with respect to the momentum of the particle α and was already obtained in the form of Eq.(13.37). We shall show that the second term can be found from Eq.(12.28) for $\varepsilon_{\mathbf{k}}^{NL(4)}$. To do this we write the imaginary part of $1/\varepsilon_{\mathbf{k}}$ in the form ($\mathbf{k}_1 = \{\mathbf{q},\omega_1\}$):

$$\mathrm{Im}\left\{\frac{1}{\varepsilon_{\mathbf{k}_1}}\right\} = -\frac{\mathrm{Im}\{\varepsilon_{\mathbf{k}_1}\}}{|\varepsilon_{\mathbf{k}_1}|^2}$$

$$= \sum_\beta \frac{4\pi^2 e_\beta^2}{q^2} \int \frac{\delta(\omega_1 - (\mathbf{q}\cdot\mathbf{v}'))}{\left|\varepsilon_{\mathbf{q},(\mathbf{q}\cdot\mathbf{v}')}\right|^2} \left(\mathbf{q}\cdot\frac{\partial\Phi^\beta_{\mathbf{p}'}}{\partial\mathbf{p}'}\right) \frac{d^3p}{(2\pi)^3}, \tag{13.39}$$

and this leads to

$$\gamma_{\mathbf{k}}^{NL(4)} = -\left[\mathrm{Im}\left\{\varepsilon_{\mathbf{k}}^{NL(4)}\right\}\Big/\frac{\partial\varepsilon_{\mathbf{k}}}{\partial\omega}\right]_{\omega=\omega_{\mathbf{k}}}$$

$$= \sum_{\alpha,\beta} \frac{8\pi e_\alpha^4 e_\beta^2}{m_\alpha^2} \int \frac{(\mathbf{k}\cdot\mathbf{q})^2}{k^2 q^2} \frac{d^3q}{q^2\left|\varepsilon_{\mathbf{q},(\mathbf{q}\cdot\mathbf{v}')}\right|^2} \frac{1}{(\omega_{\mathbf{k}} - (\mathbf{k}\cdot\mathbf{v}))^4}$$

$$\times\; \delta\big(\omega_{\mathbf{k}} - (\mathbf{q}\cdot\mathbf{v}') - ([\mathbf{k}-\mathbf{q}]\cdot\mathbf{v})\big) \left[\frac{\partial\varepsilon_{\mathbf{k}}}{\partial\omega}\right]^{-1}_{\omega=\omega_{\mathbf{k}}}$$

$$\times\; \Phi_{\mathbf{p}}^\alpha \left(\mathbf{q}\cdot\frac{\partial\Phi^\beta_{\mathbf{p}'}}{\partial\mathbf{p}'}\right) \frac{d^3p\, d^3p'}{2\pi^6}. \tag{13.40}$$

Comparing this expression with Eq.(13.38) we can verify that Eq.(13.40) corresponds to the second term of Eq.(13.31) with the same expression (13.38) for the probability.

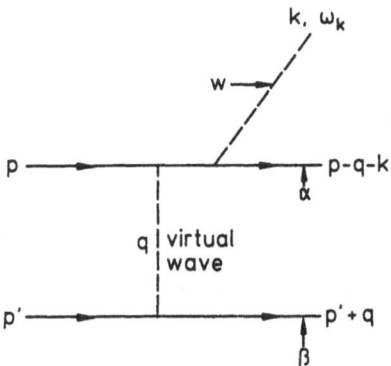

Fig. 13.2. The normal bremsstrahlung process diagram; v: virtual wave, w: bremsstrahlung wave

It is not surprising that this derivation leads to the same expressions for the damping rate of the spontaneous scattering processes and for the damping rate of the bremsstrahlung processes; it just shows the profound relations between these two processes. As regards the relations between spontaneous Cherenkov emission and collisions, spontaneous scattering corresponds to that part of the fluctuations which are close to resonance, corresponding to the relation $\varepsilon_{k_1} \approx 0$. As the integration is carried out over all possible values of k_1, including the resonance one, both processes are of the same order of magnitude and the only difference lies in logarithmic factors. As a rule, the total expression for bremsstrahlung has a much larger logarithm which means that the integration over the whole of the non-resonant domain of k_1 gives a much larger contribution than the integration over the resonant domain. We show in the diagram of Fig.13.2 the bremsstrahlung process described by Eqs.(13.37), (13.38), and (13.4). Resonance corresponds to the case when the virtual wave becomes a real propagating wave.

We shall now consider the other terms in the expression for $\varepsilon_k^{NL(1)}$ (see Eq.(12.14)) and, in particular, the contribution of the imaginary part of $1/\varepsilon_{k-k_1}$ to ϱ^{eff}. We obtain an expression which is similar to Eq.(13.39):

$$\text{Im}\left\{\frac{1}{\varepsilon_{k-k_1}}\right\} = \sum_\alpha \frac{4\pi^2 e_\alpha^2}{|k-q|^2} \int \left|\varepsilon_{k-q,([k-q]\cdot v)}\right|^{-2}$$

$$\times \delta(\omega - \omega_1 - ([k-q]\cdot v)) \left(\left[k-q\right]\cdot \frac{\partial\Phi_p^\alpha}{\partial p}\right) \frac{d^3p}{(2\pi)^3}. \qquad (13.41)$$

If there is no Cherenkov resonance we obtain through integrating by parts:

$$\varrho_{\mathbf{k},-\mathbf{k}_1}^{(2)} = \frac{|\mathbf{k} - \mathbf{k}_1|}{k} \varrho_{\mathbf{k}_1,\mathbf{k}-\mathbf{k}_1}^{(2)}$$

$$= \frac{|\mathbf{k} - \mathbf{k}_1|}{k} \varrho_{\mathbf{k}_1,\mathbf{k}-\mathbf{k}_1}^{(2)*} + 2\mathrm{i}\,\mathrm{Im}\left\{\varrho_{\mathbf{k},-\mathbf{k}_1}^{(2)}\right\}. \tag{13.42}$$

We leave the derivation of the first of these equalities as an exercise. We have split off the term with ϱ^* intending to get in the corresponding expression for the probability the absolute square of ϱ. For the moment we leave alone the last term in Eq.(13.42); we shall calculate its contribution later on. As far as the first term on the right-hand side of Eq.(13.42) is concerned we must mention that we have already obtained an approximate expression for it in Eqs.(5.49) and (5.51). Taking these arguments into account we obtain the following expression for the contribution from the imaginary part of $1/\varepsilon_{\mathbf{k}-\mathbf{k}_1}$ to $\varepsilon_{\mathbf{k}}^{NL(1)}$ – we denote it by $\gamma_{\mathbf{k}}^{NL(1)'}$:

$$\gamma_{\mathbf{k}}^{NL(1)'} = (8\pi)^2 \left[\frac{\partial \varepsilon_{\mathbf{k}}}{\partial \omega}\right]_{\omega=\omega_{\mathbf{k}}}^{-1} \int \frac{d^4 k_1}{k^2} \left|\varrho_{\mathbf{k}_1,\mathbf{k}-\mathbf{k}_1}^{(2)}\right|^2 \left|E^{(0)}\right|_{\mathbf{k}_1}^2 \mathrm{Im}\left\{\frac{1}{\varepsilon_{\mathbf{k}-\mathbf{k}_1}}\right\}$$

$$= (8\pi)^3 \left[\frac{\partial \varepsilon_{\mathbf{k}}}{\partial \omega}\right]_{\omega=\omega_{\mathbf{k}}}^{-1} \sum_{\alpha,\beta} 0_\alpha^2 0_\beta^2 \int \frac{\left|\varrho_{\mathbf{q},(\mathbf{q}\cdot\mathbf{v}');\mathbf{k}-\mathbf{q},([\mathbf{k}-\mathbf{q}]\cdot\mathbf{v})}^{(2)}\right|^2}{\left|\varepsilon_{\mathbf{q},(\mathbf{q}\cdot\mathbf{v}')}\right|^2 \left|\varepsilon_{\mathbf{k}-\mathbf{q},([\mathbf{k}-\mathbf{q}]\cdot\mathbf{v})}\right|^2}$$

$$\times \frac{d^3 q}{k^2 q^2} \delta(\omega - (\mathbf{q}\cdot\mathbf{v}') - ([\mathbf{k}-\mathbf{q}]\cdot\mathbf{v}))$$

$$\times \left(\frac{\mathbf{k}-\mathbf{q}}{|\mathbf{k}-\mathbf{q}|^2} \cdot \frac{\partial \Phi_{\mathbf{p}}^\alpha}{\partial \mathbf{p}}\right) \Phi_{\mathbf{p}'}^\beta \frac{d^3 p\, d^3 p'}{(2\pi)^6}. \tag{13.43}$$

This expression is symmetric under the substitution $\alpha \leftrightarrows \beta$, $\mathbf{q} \leftrightarrows \mathbf{k} - \mathbf{q}$, $\mathbf{p} \leftrightarrows \mathbf{p}'$ so that for concrete values of α and β both terms appear in the expression for the growth or damping rate of the bremsstrahlung (see Eq.(13.31)). If we ignore all other contributions the corresponding expression for the probability will be

$$w_{\mathbf{p},\mathbf{p}'}^{\alpha,\beta}(\mathbf{k},\mathbf{q}) = \frac{2(8\pi)^3 (2\pi)^3 e_\alpha^2 e_\beta^2}{k^2 q^2 |\mathbf{k}-\mathbf{q}|^2} \frac{\left|\varrho_{\mathbf{q},(\mathbf{q}\cdot\mathbf{v}');\mathbf{k}-\mathbf{q},([\mathbf{k}-\mathbf{q}]\cdot\mathbf{v})}^{(2)}\right|^2}{\left|\varepsilon_{\mathbf{q},(\mathbf{q}\cdot\mathbf{v}')}\right|^2 \left|\varepsilon_{\mathbf{k}-\mathbf{q},([\mathbf{k}-\mathbf{q}]\cdot\mathbf{v})}\right|^2}$$

$$\times \delta(\omega - (\mathbf{q}\cdot\mathbf{v}') - ([\mathbf{k}-\mathbf{q}]\cdot\mathbf{v})) \left[\frac{\partial \varepsilon_{\mathbf{k}}}{\partial \omega}\right]_{\omega=\omega_{\mathbf{k}}}^{-1}. \tag{13.44}$$

This expression describes only the bremsstrahlung of the polarisation cloud. This interpretation follows from the fact that in the case of resonance this process corresponds to spontaneous scattering by only the polarisation charge.

We show in Fig.13.3 the diagram of the process described by the probability (13.44). We see clearly from this diagram that the virtual waves can have either a momentum $\mathbf{q}$ or a momentum $\mathbf{k} - \mathbf{q}$. This explains the above mentioned symmetry under a substitution $\mathbf{q} \leftrightarrows \mathbf{k} - \mathbf{q}$. We proved in the case

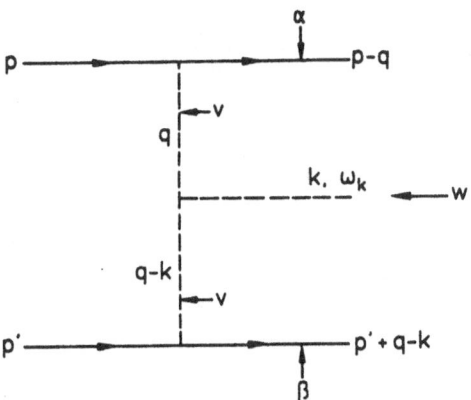

Fig. 13.3. Diagram of bremsstrahlung by the screening charge; v: virtual waves, w: bremsstrahlung wave

of scattering that one should add the amplitudes for the process rather than the probabilities. The same is true for bremsstrahlung. In fact, the summation of the amplitudes takes into account the interference between the two bremsstrahlung processes; it can change the total probability by an order of magnitude. We shall therefore write the probability (13.44) in terms of the amplitude $M^{\alpha,\beta}$ and later on find all the interference terms. We put

$$
\begin{aligned}
w^{\alpha,\beta}_{\mathbf{p},\mathbf{p}'}(\mathbf{k},\mathbf{q}) &= 16\pi(2\pi)^3 e_\alpha^2 e_\beta^2 \left[\frac{\partial \varepsilon_\mathbf{k}}{\partial \omega}\right]^{-1}_{\omega=\omega_\mathbf{k}} \\
&\times \left|M^{\mathrm{Br}}\right|^2 \delta(\omega_\mathbf{k} - (\mathbf{q}\cdot\mathbf{v}') - ([\mathbf{k}-\mathbf{q}]\cdot\mathbf{v})),
\end{aligned}
\tag{13.45}
$$

where in the case of Eq.(13.44) we have $M^{\mathrm{Br}} = M^{\alpha,\beta}$ with

$$
M^{\alpha,\beta} = -\frac{(\mathbf{k}\cdot\mathbf{q})}{kq|\mathbf{k}-\mathbf{q}|} \frac{\varrho^{(2)}_{\mathbf{q},(\mathbf{q}\cdot\mathbf{v}');\mathbf{k}-\mathbf{q},([\mathbf{k}-\mathbf{q}]\cdot\mathbf{v})}}{\varepsilon_{\mathbf{q},(\mathbf{q}\cdot\mathbf{v}')}\varepsilon_{\mathbf{k}-\mathbf{q},([\mathbf{k}-\mathbf{q}]\cdot\mathbf{v})}}.
\tag{13.46}
$$

We have included here the minus sign in order to obtain all interference terms correctly – in Eq.(13.44) the square of $M^{\alpha,\beta}$ occurs, but the interference terms will contain the products of the various matrix elements. Similarly, we can write the probability (13.38) in terms of the matrix element M^α with

$$
M^\alpha = \frac{e_\alpha}{m_\alpha q \varepsilon_{\mathbf{q},(\mathbf{q}\cdot\mathbf{v}')}} \frac{1}{(\omega_\mathbf{k} - (\mathbf{k}\cdot\mathbf{v}))^2} \frac{(\mathbf{k}\cdot\mathbf{q})}{kq}.
\tag{13.47}
$$

The sum (13.37) contains also terms in which the indices α and β occur in the opposite order. It describes the emission by particles β under the action of the field of the particles α. We can neglect this term if the particle β is a heavy one. However, since collective effects change all the bremsstrahlung processes in a plasma in a very significant way we must not use the idea that the main bremsstrahlung process occurs when light and heavy particles collide, but also consider the bremsstrahlung processes occurring in the collisions of two heavy or two light particles. We shall therefore take into account the bremsstrahlung process related to the acceleration of particle β in the collisions. Moreover, up to this moment we had not assumed that particle β is a heavy particle. In changing α to β we must change at the same time $\mathbf{v}$ to $\mathbf{v}'$, as we decided to denote the velocity of particle α by $\mathbf{v}$ and the velocity of particle β by $\mathbf{v}'$. The δ-function will then be changed and in order to get back to the form (13.45) we must change the momentum of the virtual wave from $\mathbf{q}$ to $\mathbf{k} - \mathbf{q}$. In this way we get thus the matrix element M^β due to the acceleration of particle β in the form:

$$M^\beta = \frac{e_\beta}{m_\beta |\mathbf{k} - \mathbf{q}| \varepsilon_{\mathbf{k}-\mathbf{q}, ([\mathbf{k}-\mathbf{q}] \cdot \mathbf{v})}} \frac{1}{(\omega_\mathbf{k} - (\mathbf{k} \cdot \mathbf{v}'))^2} \frac{(\mathbf{k} \cdot [\mathbf{k} - \mathbf{q}])}{k|\mathbf{k} - \mathbf{q}|}. \tag{13.48}$$

At this moment we may assume – and we shall prove this later – that the total probability is described by Eq.(13.45) with the matrix element M^{Br} given by the equation

$$M^{\mathrm{Br}} = M^\alpha + M^{\alpha,\beta} + M^\beta. \tag{13.49}$$

To prove this last equation we must find all interference terms. Let us first turn to the second term in Eq.(13.42), the discussion of which we had decided to postpone. We denote the growth rate described by this term by $\gamma_\mathbf{k}^{\mathrm{NL}(1)''}$. We find

$$\gamma_\mathbf{k}^{\mathrm{NL}(1)''} = \frac{2(8\pi)^2}{k} \left[\frac{\partial \varepsilon_\mathbf{k}}{\partial \omega} \right]_{\omega=\omega_\mathbf{k}}^{-1} \int |E^{(0)}|_{\mathbf{k}_1}^2$$

$$\times \frac{\mathrm{Im}\left\{ \varrho_{\mathbf{k}, -\mathbf{k}_1}^{(2)} \right\}}{|\mathbf{k} - \mathbf{k}_1|} \mathrm{Re} \left\{ \frac{\varrho_{\mathbf{k}_1, \mathbf{k}-\mathbf{k}_1}^{(2)}}{\varepsilon_{\mathbf{k}-\mathbf{k}_1}} \right\} d^4 k_1. \tag{13.50}$$

From the definition of the quadratic non-linear response we have

$$\mathrm{Im}\left\{ \varrho_{\mathbf{k}, -\mathbf{k}_1}^{(2)} \right\} = -\int \sum_\alpha \frac{\pi e_\alpha^3}{2kk_1 m_\alpha} \delta(\omega - \omega_1 - ([\mathbf{k} - \mathbf{k}_1] \cdot \mathbf{v}))$$

$$\times \frac{(\mathbf{k} \cdot \mathbf{k}_1)}{(\omega - (\mathbf{k} \cdot \mathbf{v}))^2} \left([\mathbf{k} - \mathbf{k}_1] \cdot \frac{\partial \Phi_\mathbf{p}^\alpha}{\partial \mathbf{p}} \right) \frac{d^3 p}{2\pi)^3}. \tag{13.51}$$

Substituting Eqs.(13.51) and (12.16) into Eq.(13.50) we find

$$\gamma_{\mathbf{k}}^{NL(1)''} = 8\pi(2\pi)^3 e_\alpha^2 e_\beta^2 \left[\frac{\partial \varepsilon_{\mathbf{k}}}{\partial \omega}\right]_{\omega=\omega_{\mathbf{k}}}^{-1} \sum_\alpha \delta\big(\omega_{\mathbf{k}} - (\mathbf{q}\cdot\mathbf{v}') - ([\mathbf{k}-\mathbf{q}]\cdot\mathbf{v})\big)$$

$$\times\, 2\,\mathrm{Re}\Bigg\{-\frac{8\pi e_\alpha(\mathbf{k}\cdot\mathbf{q})}{m_\alpha k^2 q^3 |\mathbf{k}-\mathbf{q}|\varepsilon_{\mathbf{q},(\mathbf{q}\cdot\mathbf{v}')}}$$

$$\times\, \frac{1}{(\omega_{\mathbf{k}} - (\mathbf{k}\cdot\mathbf{v}'))^2}\,\frac{\varrho_{\mathbf{q},(\mathbf{q}\cdot\mathbf{v}');\mathbf{k}-\mathbf{q},([\mathbf{k}-\mathbf{q}]\cdot\mathbf{v})}^{(2)}}{\varepsilon_{\mathbf{q},(\mathbf{q}\cdot\mathbf{v}')}\varepsilon_{\mathbf{k}-\mathbf{q},([\mathbf{k}-\mathbf{q}]\cdot\mathbf{v})}}\Bigg\}$$

$$\times\, \left([\mathbf{k}-\mathbf{q}]\cdot\frac{\partial \Phi_{\mathbf{p}}^\alpha}{\partial \mathbf{p}}\right)\Phi_{\mathbf{p}'}^\beta\,\frac{d^3\mathbf{p}\,d^3\mathbf{p}'\,d^3\mathbf{q}}{(2\pi)^9}. \tag{13.52}$$

One sees easily that Eq.(13.52) contains the products of the matrix elements $M^{\alpha,\beta}$ and M^α. The interference terms do, indeed, appear, but there is in γ^{NL} no second term involving the derivative $(\mathbf{q}\cdot[\partial\Phi_{\mathbf{p}'}^\beta/\partial\mathbf{p}'])$. On the other hand, Eq.(13.52) contains the product of the matrix elements $M^{\alpha,\beta}$ and M^β. Indeed, in the sum over α and β in Eq.(13.52) there appears a term in which the actual values of α and β are interchanged. In this term one must make the complete interchange $\mathbf{v} \leftrightarrows \mathbf{v}'$, and $\mathbf{q} \leftrightarrows \mathbf{k} - \mathbf{q}$, as well as $\alpha \leftrightarrows \beta$. In that case we now have no term involving the derivative $([\mathbf{k}-\mathbf{q}]\cdot[\partial\Phi_{\mathbf{p}}^\alpha/\partial\mathbf{p}])$. All these absent terms can be found from $\varepsilon_{\mathbf{k}}^{NL(3)}$ (see Eq.(12.25)). Integrating by parts we can write this expression in the form

$$\varepsilon_{\mathbf{k}}^{NL(3)} = \sum_\alpha \frac{16 e_\alpha}{m_\alpha}\int \frac{(\mathbf{k}\cdot\mathbf{k}_1)d^3\mathbf{k}_1}{k_1^2 k|\mathbf{k}-\mathbf{k}_1|}\,\frac{\varrho_{\mathbf{k},\mathbf{k}_1-\mathbf{k}}^{(2)}}{\varepsilon_{\mathbf{k}_1}\varepsilon_{\mathbf{k}_1-\mathbf{k}}}\,\frac{1}{(\omega - (\mathbf{k}\cdot\mathbf{v}))^2}$$

$$\times\, \delta\big(\omega - \omega_1 - ([\mathbf{k}-\mathbf{k}_1]\cdot\mathbf{v})\big)\Phi_{\mathbf{p}}^\alpha\,\frac{d^3\mathbf{p}}{(2\pi)^3}. \tag{13.53}$$

Using the first equality in Eq.(13.42) and the relation $\varrho_{-\mathbf{k},\mathbf{k}}^{(2)} = \varrho_{\mathbf{k},-\mathbf{k}}^{(2)*}$ we find

$$\varrho_{\mathbf{k},\mathbf{k}_1-\mathbf{k}}^{(2)} = \frac{k_1}{k}\,\varrho_{\mathbf{k}_1,\mathbf{k}-\mathbf{k}_1}^{(2)*}. \tag{13.54}$$

Using also Eq.(13.39) we obtain

$$\gamma_k^{NL(3)} = -\text{Im}\left\{\varepsilon_k^{NL(3)}\right\}\bigg/\left[\frac{\partial\varepsilon_k}{\partial\omega}\right]_{\omega=\omega_k}^{-1}$$

$$= \sum_{\alpha,\beta} 8\pi(2\pi)^3 e_\alpha^2 e_\beta^2 \left[\frac{\partial\varepsilon_k}{\partial\omega}\right]_{\omega=\omega_k}^{-1} \int \frac{d^3q}{(2\pi)^3}$$

$$\times \left[-\frac{8\pi}{kq|k-q|}\frac{\varrho_{q,(q\cdot v');k-q,([k-q]\cdot v)}^{(2)*}}{\varepsilon_{q,(q\cdot v')}^*\varepsilon_{k-q,([k-q]\cdot v)}^*}\right]$$

$$\times \left[\frac{8\pi e_\alpha(k\cdot q)}{m_\alpha kq^2\varepsilon_{q,(q\cdot v')}}\frac{1}{(\omega_k - (k\cdot v))^2}\right]\Phi_p^\alpha\left(q\cdot\frac{\partial\Phi_{p'}^\beta}{\partial p'}\right)$$

$$\times \delta\big(\omega_k - (q\cdot v') - ([k-q]\cdot v)\big)\frac{d^3p\,d^3p'}{(2\pi)^6}. \tag{13.55}$$

This expression contains the product of $M^{\alpha,\beta*}$ and M^α and by interchanging α and β we see that it also contains the product of $M^{\alpha*}$ and M^β; we must add this to Eq.(13.52). However, we also need the term involving $2\text{Re}\{M^{\alpha,\beta}\} = M^{\alpha,\beta} + M^{\alpha,\beta*}$. It is found from the last expression $\gamma_k^{NL(2)}$ (see Eq.(12.22)) which we have not yet used. Integrating by parts we find for $\varepsilon_k^{NL(2)}$:

$$\varepsilon_k^{NL(2)} = \sum_\beta \frac{16e_\beta^2}{m_\beta k^2}\int\frac{(k[k-k_1])\,d^4k_1}{k_1|k-k_1|}\frac{1}{(\omega-(k\cdot v))^2}\frac{\varrho_{k_1,k-k_1}^{(2)}}{\varepsilon_{k_1}\varepsilon_{k-k_1}}$$

$$\times \delta(\omega_1 - (k\cdot v'))\,\Phi_{p'}^\beta\frac{d^3p'}{(2\pi)^3}. \tag{13.56}$$

Using the expression for the imaginary part of

$$\frac{1}{\varepsilon_{k-k_1}}$$

(see Eq.(13.41)) we find the term involving the product of $M^{\alpha,\beta}$ and M^β from Eq.(13.55) through the substitution $\alpha \leftrightarrows \beta$, $v \leftrightarrows v'$, and $q \leftrightarrows k-q$. If we make the same subsitution in Eq.(13.56) we get Eq.(13.55) in which $M^{\alpha,\beta}$ occurs instead of $M^{\alpha,\beta*}$. We have thus proved the existence in the general expression of all the interference terms between the matrix elements $M^{\alpha,\beta}$ and M^α, M^β. That leaves us only to prove the existence of the interference terms between the matrix elements M^α and M^β. This term, indeed, exists in Eq.(13.56) if we take into account the last term, so far not yet used, connected with the imaginary part of

$$\varrho_{k_1,k-k_1}^{(2)}$$

in Eq.(13.56) and the imaginary part of

$$\varrho_{k,k_1-k}^{(2)}$$

in Eq.(13.53) for which we can use Eqs.(13.42) and (13.51). We thus get

$$
\mathrm{Im}\left\{\varrho_{\mathbf{k},\mathbf{k}_1-\mathbf{k}}^{(2)}\right\} = -\sum_\beta \frac{\pi e_\beta^3}{2k|\mathbf{k}-\mathbf{k}_1|} \int \frac{(\mathbf{k}\cdot[\mathbf{k}-\mathbf{k}_1])}{(\omega-(\mathbf{k}\cdot\mathbf{v}))^2}
$$
$$
\times \, \delta(\omega_1-(\mathbf{k}_1\cdot\mathbf{v}')) \left(\mathbf{k}\cdot\frac{\partial\Phi_{\mathbf{p}'}^\beta}{\partial\mathbf{p}'}\right) \frac{d^3\mathbf{p}'}{(2\pi)^3}. \tag{13.57}
$$

This gives, indeed, the term with the double real part of the product of M^α and M^β and the interchange $\alpha \leftrightarrows \beta$, $\mathbf{q} \leftrightarrows \mathbf{k}-\mathbf{q}$, and $\mathbf{v} \leftrightarrows \mathbf{v}'$ gives both terms in the expression for γ^{NL}.

We have thus proved rigorously that all bremsstrahlung mechanisms interfere with one another, that the probabiliuty is given by Eq.(13.45) in which the total matrix element is given by Eq.(13.49) and is the sum of the matrix elements (13.47) and (13.48) of the usual bremsstrahlung of colliding particles and the matrix element (13.46) of the emission by the polarisation charges. From the physical point of view it is very important that the bremsstrahlung is changed drastically in a plasma as compared to the bremsstrahlung of single isolated particles.

13.6 Analysis of the Bremsstrahlung Cross-sections for Particles in a Plasma

We shall consider here the radical changes in the bremsstrahlung cross-sections in a plasma using as an example the Langmuir wave bremsstrahlung. Let us first remind ourselves that the usual bremsstrahlung for electron-electron collisions is very small when collective effects are not taken into account and that it is well known that it is equal to zero in the dipole approximation. We can obtain this from the expression we get when we take into account only the matrix elements M^α and M^β considering the limit as $k \ll q$ and $(\mathbf{k}\cdot\mathbf{v}) \ll \omega_{\mathbf{k}}$ and assuming that the particles are not screened $(\varepsilon \simeq 1)$. We find in that case

$$
M^\alpha + M^\beta \simeq \frac{1}{q\omega_{\mathbf{k}}^2}\frac{(\mathbf{k}\cdot\mathbf{q})}{kq}\left(\frac{e_\alpha}{m_\alpha}-\frac{e_\beta}{m_\beta}\right), \tag{13.58}
$$

which equals zero when $e_\alpha/m_\alpha = e_\beta/m_\beta$. We must also mention that for ion-ion collisions the matrix elements of the usual bremsstrahlung are small as they contain the ion mass in the denominator.

An interesting point is that if we take the collective effects into account for bremsstrahlung both the electron-electron and the ion-ion collisions can make significant contributions to the bremsstrahlung and the bremsstrahlung in the case of electron-ion collisions is also changed.

To find simpler expressions for the bremsstrahlung cross-sections in a plasma we shall as a first approximation neglect only the Doppler corrections $(\mathbf{k} \cdot \mathbf{v})$ to the frequency of the emitted wave. We then have

$$M^{\alpha} \simeq \frac{e_{\alpha}}{m_{\alpha}\omega_{\mathbf{k}}^2 q \varepsilon_{\mathbf{q},(\mathbf{q}\cdot\mathbf{v}')}} \frac{(\mathbf{k} \cdot \mathbf{q})}{kq}, \tag{13.59}$$

$$M^{\beta} \simeq \frac{e_{\beta}}{m_{\beta}\omega_{\mathbf{k}}^2 |\mathbf{k}-\mathbf{q}| \varepsilon_{\mathbf{k}-\mathbf{q},([\mathbf{k}-\mathbf{q}]\cdot\mathbf{v})}} \frac{(\mathbf{k} \cdot [\mathbf{k}-\mathbf{q}])}{k|\mathbf{k}-\mathbf{q}|}. \tag{13.60}$$

In order to calculate the matrix element $M^{\alpha,\beta}$ we shall find an approximate expression for the quadratic non-linear response assuming that only the light particles – the electrons – make a contribution to it and again neglecting the Doppler corrections to the frequency of the bremsstrahlung waves:

$$
\begin{aligned}
\varrho^{(2)}_{\mathbf{k}_1,\mathbf{k}-\mathbf{k}_1} \simeq \; & -\frac{e^3}{2k_1|\mathbf{k}-\mathbf{k}_1|} \int \frac{1}{\omega_{\mathbf{k}} - (\mathbf{k}\cdot\mathbf{v}) + \mathrm{i}0} \left[\left(\mathbf{k}_1 \cdot \frac{\partial}{\partial \mathbf{p}}\right) \right. \\
& \times \frac{1}{\omega - \omega_1 - ([\mathbf{k}-\mathbf{k}_1]\cdot\mathbf{v}) + \mathrm{i}0} \left([\mathbf{k}-\mathbf{k}_1] \cdot \frac{\partial}{\partial \mathbf{p}}\right) \\
& + \left([\mathbf{k}-\mathbf{k}_1] \cdot \frac{\partial}{\partial \mathbf{p}}\right) \frac{1}{\omega_1 - (\mathbf{k}_1 \cdot \mathbf{v}) + \mathrm{i}0} \left.\left(\mathbf{k}_1 \cdot \frac{\partial}{\partial \mathbf{p}}\right)\right] \Phi^{\mathrm{e}}_{\mathbf{p}} \frac{d^3\mathbf{p}}{(2\pi)^3} \\
\simeq \; & \frac{e(\mathbf{k}\cdot\mathbf{k}_1)|\mathbf{k}-\mathbf{k}_1|}{8\pi k_1 \omega_{\mathbf{k}}^2 m_{\mathrm{e}}} \left\{ \varepsilon^{(\mathrm{e})}_{\mathbf{k}-\mathbf{k}_1} - 1 \right\} \\
& + \frac{e(\mathbf{k}\cdot[\mathbf{k}-\mathbf{k}_1])k_1}{8\pi|\mathbf{k}-\mathbf{k}_1|\omega_{\mathbf{k}}^2 m_{\mathrm{e}}} \left\{ \varepsilon^{(\mathrm{e})}_{\mathbf{k}_1} - 1 \right\},
\end{aligned}
\tag{13.61}
$$

where e is the electron charge. From Eqs.(13.61) and (13.46) we find

$$
\begin{aligned}
M^{\alpha,\beta} \simeq \; & \frac{e}{m_{\mathrm{e}}\omega_{\mathbf{k}}^2 q \varepsilon_{\mathbf{q},(\mathbf{q}\cdot\mathbf{v}')}} \frac{(\mathbf{k} \cdot \mathbf{q})}{kq} \left[\frac{\varepsilon^{(\mathrm{e})}_{\mathbf{k}-\mathbf{q},([\mathbf{k}-\mathbf{q}]\cdot\mathbf{v})} - 1}{\varepsilon_{\mathbf{k}-\mathbf{q},([\mathbf{k}-\mathbf{q}]\cdot\mathbf{v})}} \right] \\
& + \frac{e}{m_{\mathrm{e}}\omega_{\mathbf{k}}^2 |\mathbf{k}-\mathbf{q}| \varepsilon_{\mathbf{k}-\mathbf{q},([\mathbf{k}-\mathbf{q}]\cdot\mathbf{v})}} \frac{(\mathbf{k} \cdot |\mathbf{k}-\mathbf{q}|)}{k|\mathbf{k}-\mathbf{q}|} \left[\frac{\varepsilon^{(\mathrm{e})}_{\mathbf{q},(\mathbf{q}\cdot\mathbf{v}')} - 1}{\varepsilon_{\mathbf{q},(\mathbf{q}\cdot\mathbf{v}')}} \right].
\end{aligned}
\tag{13.62}
$$

The difference between this matrix element and M^{α} and M^{β} is that now the electron charge and mass enter and that polarisation factors appear in the numerators describing the emission by the polarisation charges.

13.7 Collective Effects in Bremsstrahlung in Electron–Electron, Ion–Ion, and Electron–Ion Collisions

We shall start with a discussion of bremsstrahllung in electron-electron collisions. For this case the total matrix element for bremsstrahlung is

$$M^{Br(e,e)} \simeq \frac{e}{m_e^2\omega_k^2 q\varepsilon_{\mathbf{q},(\mathbf{q}\cdot\mathbf{v}')}} \frac{(\mathbf{k}\cdot\mathbf{q})}{kq} \frac{\varepsilon_{\mathbf{k}-\mathbf{q},([\mathbf{k}-\mathbf{q}]\cdot\mathbf{v})}^{(i)}}{\varepsilon_{\mathbf{k}-\mathbf{q},([\mathbf{k}-\mathbf{q}]\cdot\mathbf{v})}}$$

$$+ \frac{e}{m_e^2\omega_k^2|\mathbf{k}-\mathbf{q}|\varepsilon_{\mathbf{k}-\mathbf{q},([\mathbf{k}-\mathbf{q}]\cdot\mathbf{v})}} \frac{(\mathbf{k}\cdot[\mathbf{k}-\mathbf{q}])}{k|\mathbf{k}-\mathbf{q}|} \frac{\varepsilon_{\mathbf{q},(\mathbf{q}\cdot\mathbf{v}')}^{(i)}}{\varepsilon_{\mathbf{q},(\mathbf{q}\cdot\mathbf{v}')}}. \quad (13.63)$$

If we neglect the contribution from the ions in the linear response the cancellation in the matrix elements is very strong.

We can consider the special case when the electron velocities are musch smaller than the ion thermal veolcity and the case when they are much larger than the ion thermal velocity. In the first case we can approximate the dielectric permittivities by the Debye screening expressions and we have

$$M^{Br(e,e)} \simeq \frac{e}{km_e\omega_k^2} \left[1 + \frac{1}{q^2d^2}\right]^{-1} \left[1 + \frac{1}{|\mathbf{k}-\mathbf{q}|^2d^2}\right]^{-1}$$

$$\times \left[\frac{(\mathbf{k}\cdot\mathbf{q})}{q^2} + \frac{k^2-(\mathbf{k}\cdot\mathbf{q})}{|\mathbf{k}-\mathbf{q}|^2} + \frac{k^2}{q^2|\mathbf{k}-\mathbf{q}|^2d_i^2}\right]. \quad (13.64)$$

We remind ourselves that $1/d^2 = 1/d_i^2 + 1/d_e^2$.

If $k \ll q$ the terms proportional to $(\mathbf{k}\cdot\mathbf{q})$ cancel; this reflects the well known fact that bremsstrahlung is weak in collisions of particles with the same charge to mass ratio. For $q \gg 1/d$ there is an additional small factor k/q in the matrix element. A factor of the order k^2/q^2 occurs in the integral over the transferred momenta and the main contribution will come from the smallest possible k values, that is $k \simeq 1/d$. In the opposite limit, $q \ll 1/d$, the matrix element contains an extra factor qkd^2, that is, in the integral over the transferred momenta an additional factor $q^2k^2d^2$ appears and the main contribution will come from the largest possible q values, that is, again $q \simeq 1/d$. We thus find that the total integral will contain an extra factor k^2d^2 which should be the case for quadrupole emission. However, the coefficient in front of this expression will be different fom the one obtained simply by using only the quadrupole term in the emission of "bare" particles.

If the electron velocities are much larger than the ion velcities we must use for the ion dielectric permittivity the expression ω_{pi}^2/ω^2. The contribution from the last term in Eq.(13.64) becomes negligible and we must in the factor in front of the bracket replace d by d_e. An estimate will lead to the same order of magnitude for the rate of emission as earlier, but with a different numerical factor.

We must emphasise that in the case of electron-electron collisions the collective effects do not change the order of magnitude for the total rate of emission; however, the differential cross-section and the angular dependence of the rate of emission are changed significantly. Also, the numerical factor in front of the expression for the total emission rate is changed by the collective effects.

We shall now turn to bremsstrahlung in ion-ion collisions. In that case the matrix elements M^α and M^β are small and the total matrix element is determined by Eq.(13.62). The most interesting case is the one where the velocity is smaller than the ion thermal velocity. For this case we have

$$M^{\mathrm{Br(i,i)}} = \frac{ek}{m_e \omega_k^2 d_e^2} \left[q^2 + \frac{1}{d^2} \right]^{-1} \left[|\mathbf{k} - \mathbf{q}|^2 + \frac{1}{d^2} \right]^{-1}. \tag{13.65}$$

Again, for $k \ll q$ and $q \gg 1/d$ the matrix element contains an extra factor $k/q^3 d^2$ and the main contribution comes from the smallest values of the transferred momentum $q \simeq 1/d$ while for $q \ll 1/d$ a factor qkd^2 occurs and the main contribution again comes from $q \simeq 1/d$. A factor $k^2 d^2$ occurs in the power emitted. The rate of bremsstrahlung in ion-ion collisions is thus of the same order of magnitude as that in electron-electron collisions. This is a very large effect as in the case when there is no plasma the bremsstrahlung in ion-ion collisions was smaller than the bremsstrahlung in electron-electron collisions by a factor $(m_e/m_i)^2$. For $kd \gg m_e/m_i$ the collective effects increased the rate of bremsstrahlung by several orders of magnitude. It is still not large since the bremsstrahlung in electron-electron collisions is not large. However, for Langmuir waves the factor kd can even be of the order of $\frac{1}{3}$.

We shall finally consider the bremsstrahlung in electron-ion collisions. Let α be an electron and β be an ion. We can then neglect the M^β matrix element and we find

$$M^{\mathrm{Br(e,i)}} = \frac{e}{m_e \omega_k^2 \varepsilon_{\mathbf{q},(\mathbf{q} \cdot \mathbf{v}')}} \frac{(\mathbf{k} \cdot \mathbf{q})}{kq^2} \frac{\varepsilon^{(i)}_{\mathbf{k}-\mathbf{q},([\mathbf{k}-\mathbf{q}] \cdot \mathbf{v})}}{\varepsilon_{\mathbf{k}-\mathbf{q},([\mathbf{k}-\mathbf{q}] \cdot \mathbf{v})}}$$
$$- \frac{e}{m_e \omega_k^2 \varepsilon_{\mathbf{k}-\mathbf{q},([\mathbf{k}-\mathbf{q}] \cdot \mathbf{v})}} \frac{(\mathbf{k} \cdot [\mathbf{k}-\mathbf{q}])}{k|\mathbf{k}-\mathbf{q}|^2} \frac{\varepsilon^{(e)}_{\mathbf{q},(\mathbf{q} \cdot \mathbf{v}')} - 1}{\varepsilon_{\mathbf{q},(\mathbf{q} \cdot \mathbf{v}')}}. \tag{13.66}$$

The first term represents the usual bremsstrahlung by an electron and the bremsstrahlung due to the oscillations of its polarisation charge and the second term represents the bremsstrahlung by the polarisation charge of the ion. We have already defined in Eq.(6.52) the effective polarisation charges for scattering processes. Here we can also introduce the effective polarisation charges for bremsstrahlung. The difference now appears only in that the effective polarisation charges occur in the differences of the wavevectors and the frequencies of the emitted and the virtual waves, determined by the transferred momena $\mathbf{q}$, rather than in those for the incident and the scattered waves. Since the result is determined by integration over all transferred momenta the final conclusion will depend on the values of the transferred momenta which make the largest contributions. We shall therefore analyse a few special cases.

The first special case is the one when both the electron and the ion velocities are much smaller than the ion thermal velocity. The ion will then be able to screen both the colliding particles and we can use the Debye screening approximation for the dielectric permittivities of both the electrons and the ions. In the case when $k \ll q$ we have for the total amplitude

$$M^{\mathrm{Br}(e,i)} = \frac{e}{m_e \omega_k^2} \frac{(\mathbf{k} \cdot \mathbf{q})}{kq^2} \left[1 + \frac{1}{q^2 d^2} \right]^{-1} . \tag{13.67}$$

This result is the same as that for the usual bremsstrahlung when we neglect the polarisation charge of the electron and assume that the field of the ion is screened by the total Debye length. The final expression appears to be the result of cancellations in the total matrix element.

We shall now consider another limit when the ion velocity is still much smaller than the ion thermal velocity but the electron velocity is larger than the ion-sound velocity. In this case we can neglect the ion polarisation in those expressions for the dielectric permittivity in which the electron velocity enters, that is, we can neglect $\varepsilon^{(i)}_{\mathbf{k}-\mathbf{q},([\mathbf{k}-\mathbf{q}]\cdot\mathbf{v})} - 1$ as compared to 1. In the term of the matrix element which describes the emission by the screened electron we can put $\varepsilon^{(i)}_{\mathbf{k}-\mathbf{q},([\mathbf{k}-\mathbf{q}]\cdot\mathbf{v})}$ equal to 1:

$$
\begin{aligned}
M^{\mathrm{Br}(e,i)} &= \frac{e}{m_e \omega_k^2 \varepsilon_{\mathbf{q},(\mathbf{q}\cdot\mathbf{v}')}} \frac{(\mathbf{k} \cdot \mathbf{q})}{kq^2} \frac{\varepsilon^{(e)}_{\mathbf{q},(\mathbf{q}\cdot\mathbf{v}')}}{\varepsilon^{(e)}_{-\mathbf{q},-(\mathbf{q}\cdot\mathbf{v})}} \\
&= \frac{e}{m_e \omega_k^2 \varepsilon^{(e)}_{-\mathbf{q},-(\mathbf{q}\cdot\mathbf{v})}} \frac{(\mathbf{k} \cdot \mathbf{q})}{kq^2} \frac{T_i}{T_i + ZT_e} .
\end{aligned}
\tag{13.68}
$$

In this expression the effective ion charge enters in the form defined by Eq.(6.7). The screening of the electrons is determined by the electron Debye radius rather than the total Debye radius since the electron velocity is much larger than the ion-sound velocity and much smaller than the electron thermal velocity. We can write

$$M^{\mathrm{Br}(e,i)} = \frac{e}{m_e \omega_k^2} \frac{(\mathbf{k} \cdot \mathbf{q})}{kq^2} \left[1 + \frac{1}{q^2 d_e^2} \right]^{-1} \frac{Z^{(i)}_{\mathrm{eff}}}{Z} . \tag{13.69}$$

We must mention that in Eq.(13.68) only the electron velocity and not, as in the classical expression for the usual bremsstrahlung, the ion velocity occurs in the dielectric permittivity. This makes a qualitative difference for the emission rate since for velocities much larger than the electron thermal velocity there will be no screening, and this will not happen for the classical expression. In this case we have

$$\varepsilon^{(e)}_{-\mathbf{q},-(\mathbf{q}\cdot\mathbf{v})} \simeq \varepsilon^{(e)}_{-\mathbf{q},\omega} \simeq 1 - \frac{\omega^2_{pe}}{\omega_k} - \frac{3q^2 v^2_{Te}}{\omega^2_{pe}} \simeq -\frac{3q^2 v^2_{Te}}{\omega^2_{pe}}.$$

The appearance of q^2 in the denominator shows that the main contribution comes from small values of q, that is, very distant collisions. The main part of the emission is produced by the oscillating charge of the ion. The integral does not diverge for small values of the transferred momentum q since the smallest values of q are determined by the conservation of energy in the elementary bremsstrahlung process. This conservation law is $\omega_k = -(\mathbf{q} \cdot \mathbf{v})$ for the approximations we are using so that we have $q_{min} = \omega_k/v$. This means that a small factor v^2_{Te}/v^2 will occur in the denominator of the matrix element which means that the power of the bremsstrahlug will be larger by a factor

$$\frac{v^4}{v^4_{Te}} \gg 1 \tag{13.70}$$

than in the case when collective effects are not taken into account. The reason for this is that the nature and the magnitude of the bremsstrahlung is completely changed in a plasma: the emission is determined by the oscillation of the screening charge of the ion and for $v \gg v_{Te}$ the electrons of this screening shell come in resonance with the Langmuir oscillations.

The examples we have given here are sufficient to give an impression of the rôle played by collective effects in bremsstrahlung. We must emphasise that the general expression (13.66) is valid when we are not using the dipole approximation. In conclusion we mention that similar effects appear for bremsstrahlung of electromagnetic waves with wavelengths longer than the Debye screening length in the case when

$$\omega^t_k < \omega_{pe} \frac{c}{v_{Te}}. \tag{13.71}$$

Finally, we must remark that all effects considered are valid for regular waves and can be applied when we consider damping rates of coherent structures.

Problems

1. Calculate the bremsstrahlung probabilities for electromagnetic waves in electron-ion collisions in a plasma, taking collective effects into account.
2. Calculate the bremsstrahlung probabilities for electromagnetic waves in ion-ion collisions in a plasma, taking collective effects into account.
3. Calculate the bremsstrahlung probabilities for electromagnetic waves in electron-electron collisions in a plasma, taking collective effects into account.

4. Prove the validity of Eq.(13.42).
5. Find the change in the particle distributions in the bremsstrahlung processes.
6. Prove the conservation of the total energy of particles and waves in the bremsstrahlung processes.
7. Prove the conservation of the total momentum of particles and waves in the bremsstrahlung processes.

14 Some Remarks on Plasma Non-linear Kinetics in External Magnetic Fields. Conclusion

14.1 Summary of Problems Considered

In this book we have so far restricted ourselves to particular collective motions – electrostatic modes – and we have mainly considered Langmuir and ion-sound oscillations. However, for those collective modes we have tried to show as fully as possible what is known about their different non-linear interactions and non-linear kinetic processes. We have considered all that is known about all the interactions and relations between the various non-linearities. We want now to emphasise that the modes considered are representatives of many other modes which can occur in actual plasmas. For practical applications one needs therefore only to know the whole spectrum of the modes which may occur in a system and their interactions. Inclusion of a broad spectrum of modes will also broaden the spectrum of interactions, but not that of the kind of interactions – which we have considered in detail. The behaviour of all other interactions can be formulated by complete analogy of those we have considered so far. However, there may be specific points in which these interactions may be different in some way from what we have considered earlier and in the present chapter we shall try to look at those points.

First of all, we must bear in mind the restrictions on the considerations which we have presented so far: (1) we assumed that there were no external magnetic or electric fields present; (2) we considered only longitudinal collective modes and only longitudinal statistical particle fluctuations; (3) we assumed that all particle velocities were non-relativistic; and (4) we restricted our considerations to weak non-linearities – non-linearities which are quadratic or cubic in the fields.

We intentionally considered a very simple case in order to make the general principles of non-linear kinetics as clear as possible. However, we found that even for this simple case there are many aspects of non-linear interactions. Moreover, we were unable to avoid brief excursions into some of the fundamental problems of present-day physics such as stochastisation, self-organisation, and interactions between random and regular motions. This last topic was studied partly phenomenologically since we were unable to enter here into a discussion of how regular motions become random or of the problems of dynamic chaos. We could only discuss a single approach for the case when one type of motion is known to be regular while another type of motion is known to be random and for that case we were able to find possible interactions between them, including the growth of regular motions in the presence of random motions. This may be called self-organisation, although there exist many complicated problems of self-organisation which are not covered by the simple approach which we have given here. We did, however, consider almost all possible processes for the very simple case to which we restricted ourselves and it is interesting to note that a simple system of charged particles interacting through Couloimb forces can show such a rich behaviour and have such a broad spectrum of possible physical processes. We shall now summarise the main features of these physical processes.

1. The combined action by the system of charged particles which forms a plasma is able radically to change the cross-sections of such basic physical processes as particle collisions, scattering of waves, and emission of waves by particles. This point was emphasised throughout all we have said so far. The change in the cross-sections does not need a high density or a large value of the potential energy of the interactions. The effect is produced by fluctuations and by the screening of the charges. We also illustrated that various physical processes in a large system of charged particles will inevitably be related to one another – spontaneous Cherenkov emission is related to binary collisions, and spontaneous scattering is related to bremsstrahlung. Sometimes these relations are such that we cannot separate the processes involved in them.

2. A specific feature of plasmas is that collective modes will often and very easily be excited and that the macroscopic properties of the plasma are determined by the microscopic level of these collective modes and by the non-linear interactions of these modes with one another and with the particles.

3. The nature of all the non-linear interactions is the same, they are related to one another, and they change from one type to another when the intensity of the collective modes changes. We have examined the relations which exist between decay processes and stimulated scattering processes, and between those two and the modulational interactions. Sometimes these relations are so strong that one cannot separate the processes.

4. Plasmas are the simplest systems in which turbulence can be studied. The concept of weak turbulence can be formulated in a closed, non-contradictive manner and it is in good agreement with many existing experiments. The idea that strong turbulence is the random behaviour of self-organised dissipative structures has many advantages when one wants to understand various experimental problems which have cropped up.

5. Plasmas are the simplest systems to study the problems of stochastisation and the randomisation of regular motions.

6. In many cases a plasma is an open system and it can behave globally as a self-organised system even though also in a plasma the self-organising processes occur on a microscopic level. Some very general problems of modern physics can thus be analysed by experiments in plasmas.

7. The main microscopic processes responsible for the interactions between the particles and the collective modes are very simple and include resonant interactions between collective modes and particles, scattering of modes by particles, collisions, and emission processes during those collisions. The term "particle" here, in fact, means collective excitations since they appear surrounded by their screening shells. In the case of slow particle motions the plasma looks rather like a collection of neutral atoms, rather than a collection of charged particles because each particle is screened and can be considered to be an elementary excitation of the plasma which in many aspects behaves like a neutral atom. In the case of high velocities the shells may be stripped off and the particles will behave more like charged particles.

All the main features of what we have discussed are very general and can be taken over to the general case when there are many different modes present in the plasma and not only those we have discussed so far in the various examples considered.

In the present concluding chapter we do not want to give a list of the various experimental or other special problems – we did that already in Chap.1. We here wish to point out some general physical aspects of the problems which one may encounter when one extends the discussion to the general case where there are many different collective modes in the system.

14.2 External Magnetic Fields

In the presence of a magnetic field, even a uniform one, of strength $\mathbf{H}_0$, the number of possible collective modes increases drastically. Apart from the frequencies ω_{pe} and ω_{pi} the characteristic cyclotron frequencies – or gyrofrequencies – ω_{He} and ω_{Hi} appear:

$$\omega_{H\alpha} = \frac{e_\alpha H_0}{m_\alpha c}. \qquad (14.1)$$

A single particle will gyrate around the magnetic field lines with the frequency (14.1). A new type of characteristic velocity – and therefore new spatial scales – also appears: apart from the thermal velocities v_{Te} and v_{Ti} and the ion-sound velocity v_s the Alfvén velocity v_A appears:

$$v_A = c \frac{\omega_{Hi}}{\omega_{pi}} = \frac{H_0}{\sqrt{4\pi n_i m_i}} = \frac{v_s}{\beta}, \qquad \beta = \frac{4\pi n_i T_e}{H_0^2}. \qquad (14.2)$$

We can now distinguish between a low-pressure plasma when $\beta \ll 1$, that is, when the kinetic pressure nT is small as compared to the magnetic pressure $H_0^2/4\pi$, and a high-pressure plasma when $\beta \gg 1$, that is, when the kinetic pressure is large as compared to the magnetic pressure. For a low-pressure plasma we have always $v_A \gg v_s$.

The third complication is that magnetic fields lead to an anisotropic propagation of the collective modes, that is, the frequency of the modes will depend on the angle θ between the wavevector $\mathbf{k}$ and the direction of the external magnetic field. The first feature of the presence of magnetic fields is thus the presence of a great variety of collective modes. We show some of them in Fig.14.1 for the case when $\omega_{He} \gg \omega_{pe}$, $v_s \ll v_A$, and $\cos\theta \gg \sqrt{m_e/m_i}$, where θ is the angle between $\mathbf{k}$ and $\mathbf{H_0}$. However, this feature does not qualitatively change the general picture we have painted of the non-linear kinetics. Only the number of interacting modes has increased quantitatively.

One of the most important physical features of the presence of magnetic fields is that particles rotate in magnetic fields and this creates the possibility that rotational motion will appear not only in the linear behaviour of the modes but also in the non-linear responses and interactions. This is the second point we want to emphasise. The presence of rotational motion in non-linearities creates much greater possibilities for the appearance of stable structures. Indeed, rotation in structures may stabilise them due to the conservation of angular momentum. One must now take centrifugal forces into account in the balance between forces. Self-contraction may also be stabilised. This creates a possibility for the existence of stable vortex structures.

A third important point is the qualitative difference between the structures of magnetised modes and those of unmagnetised modes. We use the term "magnetised" here for the case when the size or the wavelength across the magnetic field is much smaller than the Larmor radius, $v_{T\alpha}/\omega_{H\alpha}$, of the particles. Magnetised particles move almost directly along the magnetic field lines and they also only change the component of their velocity along the magnetic field lines. To a first approximation only the component $E\cos\theta$ of the electric field acts on the particles. This is the reason why the factor $\cos\theta$ appears in the dispersion relations for the magnetised Langmuir mode which has a frequency $\omega_{pe}\cos\theta$ and for the magnetised ion-sound mode which has a frequency $kv_s\cos\theta$. If $\cos\theta \ll \sqrt{m_e/m_i}$ the magnetised Langmuir waves have a frequency close to that of the so-called lower-hybrid frequency $\omega_{\mathbf{k}}^{LH}$:

$$\omega_{\mathbf{k}}^{LH} = \sqrt{\omega_{He}\omega_{Hi}}. \tag{14.3}$$

Often the whole branch of the magnetised Langmuir osillations together with its limit (14.3) is called the lower-hybrid mode.

The main point we want to make about the magnetised modes is that the structures they form are usually stretched out along the magnetic field lines. They look rather like "cigars" and this term is sometimes used for them. If there is no magnetic field the structures are usually dipole structures like

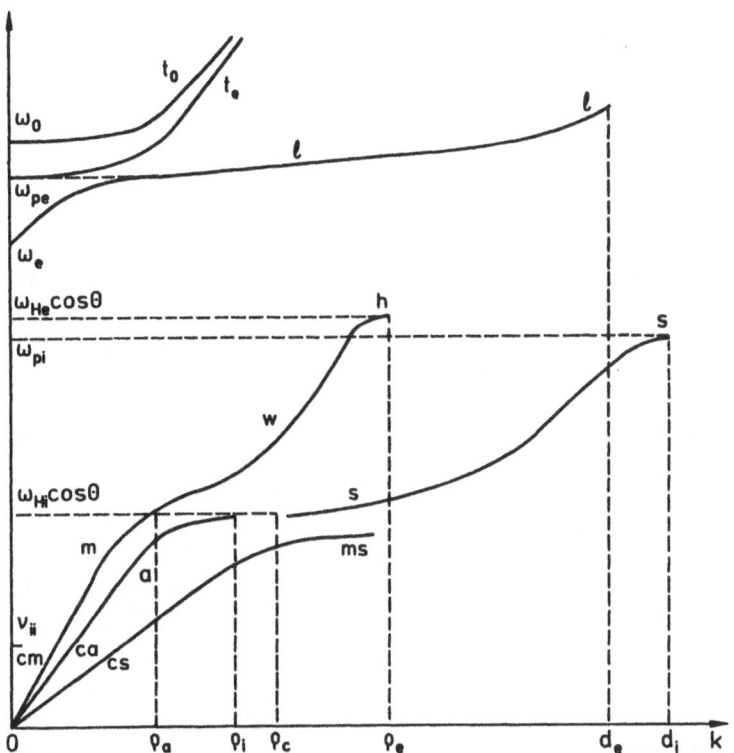

Fig. 14.1. Qualitative picture of the frequency-wavenumber dependence for plasma modes if there is an external magnetic field present; d_e and d_i are, respectively, the electron and ion Debye lengths; ϱ_e and ϱ_i are, respectively, the electron and ion Larmor radii; $\varrho_c = \omega_{He}/c$; $\varrho_a = \omega_{Hi}/c$; s: unmagnetised ion-sound mode; ms: magnetised ion-sound mode; a: Alfvén mode; m: fast magnetosound mode; mc: collisional fast magnetosound mode; w: whistler mode; h: longitudinal cyclotron mode; ℓ: Langmuir mode; t_o: ordinary electromagnetic mode; t_e: extraordinary electromagnetic mode; $\omega_o = \omega_{pe} - \frac{1}{2}\omega_{He}$; $\omega_e = \omega_{pe} + \frac{1}{2}\omega_{He}$

"pancakes" with a dipole moment oriented at right angles to the plane of the "pancake". In a magnetic field the magnetised structures form "cigars" with a dipole moment oriented along its major axis which lies along the magnetic field.

The fourth point we wish to make is that for magnetised modes the resonance conditions are significantly altered – they become a set of cyclotron resonances:

$$\left. \begin{aligned} \omega_k - (\mathbf{k} \cdot \mathbf{h})(\mathbf{v}_\alpha \cdot \mathbf{h}) - \nu\omega_{H\alpha} &= 0, \\ \nu = 0, \pm 1, \pm 2, \pm 3, \ldots, \qquad \mathbf{h} &= \frac{\mathbf{H}_0}{H_0}, \end{aligned} \right\} \qquad (14.4)$$

where $\mathbf{h}$ is the unit vector along the magnetic field. The resonance condition $\nu = 0$ corresponds to Cherenkov resonance for magnetised particles. A strong field thus modifies even the resonance in which the cyclotron frequency does not occur since in the resonance condition the component of the wavevector along the magnetic field appears. Only the first few cyclotron resonances, $\nu = \pm 1$, $\nu = \pm 2$ are important for magnetised particles; the interaction decreases rapidly with increasing number of the resonance. Formally the set of resonances has the same form (14.4) for unmagnetised particles, but then the main contribution comes from the very large values of ν. In that case one must sum the interaction over many cyclotron resonances. The result of this summation is that to a first approximation the resonance has the same form as when there is no magnetic field:

$$\omega_{\mathbf{k}} = (\mathbf{k} \cdot \mathbf{v}_\alpha). \tag{14.5}$$

This is obvious since for the unmagnetised case the particle will interact with the field before its trajectory is curved by the magnetic field. The great variety of resonance conditions creates a new possibility when some of the modes may be simultaneously in resonance with different kinds of particles. The best known example is given by the lower-hybrid waves. In that case the ions are unmagnetised and the electrons are magnetised. This means that the unmagnetised resonance condition can be satisfied by the ions and the Cherenkov resonance $\nu = 0$ by the electrons. This is important for at least two reasons. The ions can excite the lower-hybrid waves resonantly and they can, through another resonance, transfer their energy to the electrons. This may be a very efficient mechanism to convert the energy of hydrodynamic motions into the energy of accelerated electrons. The other point is that there could exist structures which would be amplified by ions and damped by electrons. This creates new possibilities for the formation of self-organising structures.

The non-linear resonance interactions are also changed by the magnetic field. For example, in the case when both modes in a scattering process are magnetised the resonant condition has the form:

$$\omega_{\mathbf{k}} - \omega_{\mathbf{k}'} - ([\mathbf{k} - \mathbf{k}'] \cdot \mathbf{h})(\mathbf{v}_\alpha \cdot \mathbf{h}) - \nu \omega_{H\alpha} = 0. \tag{14.6}$$

This also creates new possibilities for the existence of simultaneous scattering by different particles.

The fifth point is that the presence of many modes can qualitatively change the rates of decay processes if some mode can take part simultaneously in two or more decay processes. Such decays are known as connected decays. The channel of a fast decay can in that case increase the rate of a slow decay. The equations describing connected decays are similar to those describing connected chemical reactions.

The sixth point is that the presence of many modes creates more possibilities for the coexistence of regular and random modes and this means a more rapid appearance of self-organising processes.

The seventh and last point is that the modulational interactions may change qualitatively. In the case of Langmuir waves only the density deple-tions can develop when there is no magnetic field. If there is a magnetic field present the frequencies of the modes depend on the magnetic field strength. Therefore, in a magnetic field modulational interactions can develop in which new magnetic fields, δH, can be excited, usually in an opposite direction to the external magnetic field – a diamagnetic kind of perturbation. This is, for example, the case for Alfvén vortices and cyclotron structures.

14.3 Plasma Inhomogeneities

The presence of inhomogeneities in the density and in the magnetic field produce new kinds of particle motions – they drift in such inhomogeneities. This produces the possibility of the existence of new collective modes: drift modes. They are often unstable. If the density is inhomogeneous their drift velocity v_D is determined by the thermal velocity $v_{T\alpha}$ multiplied by the ratio of the Larmor radius ϱ_α to the scale L of the inhomogeneity:

$$v_D = v_{T\alpha} \frac{\varrho_\alpha}{L} = \frac{v_{T\alpha}^2}{L\omega_{H\alpha}}. \tag{14.7}$$

If the density gradient is at right angles to the magnetic field the frequency of the drift oscillations, $\omega_{\mathbf{k}}^D$, will depend on the component of the wavevector at right angles both to the magnetic field and to the density gradient (we shall takes this as the y-direction):

$$k_y = \frac{(\mathbf{h} \cdot [\mathbf{k} \wedge \nabla n]) L}{n}, \tag{14.8}$$

and we have:

$$\omega_{\mathbf{k}}^D = \frac{k_y v_D}{1 + k_y^2 \varrho_e^2}. \tag{14.9}$$

The same dispersion relation is valid for the so-called Rossby waves in a thin layer of a rotating gas – when the Coriolis force plays the rôle of the magnetic field. This is very important as it enables one to describe in a similar manner, on the one hand, collective vortex structures in planetary atmospheres and, on the other hand, drift-wave structures in installations for magnetic confinement of plasmas. Such structures are two-dimensional vortices. A new kind of non-linearity – the vector non-linearity – is here im-portant. An example of an equation with vector non-linearities is the Chernii-Obuchov equation – sometimes called the Hasegawa-Mima equation. The first name comes because Chernii and Obuchov were the first to use this equation for the description of vortices in planetary atmospheres whereas the second name comes from the fact that Hasegawa and Mima were the first to use it to describe drift waves in plasmas. As this equation is widely used we shall just quote it here without derivation in order to show the new kind of non-linear interactions it contains. We give the equation for the dimensionless potential $\phi_{k,\omega}$ of the drift waves:

$$\left(\omega - \omega_k^D\right)\phi_{k,\omega} = -\frac{i}{2}\int \Lambda_{k_1,k_2}\phi_{k1,\omega_1}\phi_{k_2,\omega_2}\,d^2k_1\,d^2k_2\,d\omega_1\,d\omega_2$$
$$\times\,\delta(\omega - \omega_1 - \omega_2)\,\delta(k - k_1 - k_2), \qquad (14.10)$$

where

$$\Lambda_{k_1,k_2} = \frac{([k_1 \wedge k_2]\cdot h)}{1 + |k_1 + k_2|^2}. \qquad (14.11)$$

One can see that interactions between harmonics which propagate in the same direction are forbidden. This is the reason why the non-linear structures which follow from this equation are always two-dimensional. At the present time a large amount is known not only about the regular structures described by Eq.(14.10) but also about the Kolmogorov spectrum of its weak turbulence as well as about the interactions between small-scale vortex turbulence and large-scale regular structures.

Another important role played by the inhomogeneities is the creation of regions for wave conversion by the inhomogeneities and of regions where wave energy is accumulated at the inhomogeneities.

Another new kind of modes which we must mention in connection with inhomogeneities are surface waves. There are non-linear interactions not only of surface waves with each other, but also between surface waves and volume waves.

The last point – and an important one – is that the statistical particle fluctuations are changed by inhomogeneities. This is very important for wavevectors of the order of the scale of the inhomogeneities and it changes not only the particle collisions but also the power of the bremsstrahlung.

14.4 Electromagnetic Effects

Among the electromagnetic effects there are some which even in the absence of magnetic fields are responsible to produce currents in a plasma and hence magnetic fields produced by those currents. A classical example is the case when initially only electrostatic oscillations are excited in the plasma. We can show that in general such oscillations will excite currents in the plasma. This effect is electromagnetic, as it is determined by the parameter v/c. We shall now illustrate the physical mechanism of this kind of mechanism. Consider the field of the Langmuir waves:

$$\mathbf{E} = \mathrm{Re}\left\{\mathcal{E}\,e^{-i\omega_{pe}t}\right\}, \qquad (14.12)$$

where the amplitude field $\mathcal{E}$ consists of at least several harmonics – the effect does not occur when there is just a single harmonic with a definite value of k. The equation for the electron motion in this field contains oscillations δv_e of the electron velocity in this field:

$$m_e \frac{d\delta v_e}{dt} = e\mathbf{E}, \qquad \delta v_e \simeq \frac{e}{m_e \omega_{pe}} \operatorname{Re}\left\{i\mathcal{E} e^{-i\omega_{pe}t}\right\}. \tag{14.13}$$

We have here taken into account that the fastest process corresponds to changes in time with the plasma frequency. We can use the equation of continuity for the electrons to find the variations δn_e in the electron density:

$$\frac{\partial \delta n_e}{\partial t} = -n_0 \operatorname{div} \delta v_e, \qquad \delta n_e = \frac{en_0}{m_e \omega_{pe}^2} \operatorname{Re}\left\{\operatorname{div}\mathcal{E} e^{-i\omega_{pe}t}\right\}. \tag{14.13}$$

Averaging the product of the variation δn_e in the density and the variation δv_e in the velocity over the time scale of a plasma period gives us the rotational part of the current which is excited:

$$\mathbf{j} = \overline{e\,\delta n_e\,\delta v_e} = \frac{e^3 n_0}{m_e^2 \omega_{pe}^3} \overline{\operatorname{Re}\left\{i\mathcal{E} e^{-i\omega_{pe}t}\right\} \operatorname{Re}\left\{\operatorname{div}\mathcal{E} e^{-i\omega_{pe}t}\right\}}$$

$$= \frac{ie^3 n_0}{4m_e \omega_{pe}^3} \left[\mathcal{E} \operatorname{div}\mathcal{E}^* - \mathcal{E}^* \operatorname{div}\mathcal{E}\right] = \frac{ie}{4\pi m_e \omega_{pe}} \operatorname{curl}[\mathcal{E} \wedge \mathcal{E}^*]. \tag{14.15}$$

Neglecting the displacement current and using Ampère's law we find

$$\operatorname{curl}\mathbf{H} = \frac{4\pi}{c}\mathbf{j}, \qquad \nabla^2\mathbf{H} = \operatorname{curl}\operatorname{curl}\mathbf{H} = i\nabla^2 \frac{e[\mathcal{E} \wedge \mathcal{E}^*]}{4\pi cm_e \omega_{pe}}, \tag{14.16}$$

and by removing the Laplacian operator ∇^2 through integration,

$$\mathbf{H} = i\frac{e}{4\pi cm_e \omega_{pe}}[\mathcal{E} \wedge \mathcal{E}^*]. \tag{14.17}$$

We see thus that a purely electrostatic field can excite a magnetic field. This is of interest for applications to laser-plasma interactions when strong Langmuir turbulence is excited as well as for various astropysical applications. Due to a gyroscopic effect in the linear response the magnetic fields which are excited in this way will produce a change in the equation of motion for the Langmuir field:

$$i\frac{\partial \mathcal{E}}{\partial t} + \frac{3v_{Te}^2}{2\omega_{pe}}\nabla^2\mathcal{E} = \frac{\omega_{pe}}{2n_0}\delta n\,\mathcal{E} + \frac{ie}{2m_e c}[\mathcal{E} \wedge \mathbf{H}]. \tag{14.18}$$

The excitation of magnetic fields changes the modulational instability. The change in the growth rates is not very large when the plasma temperature is not too high but there will appear magnetic fields and they follow the exitation of longitudinal modulational perturbations. A magnetic field is thus excited with the standard growth rate of the modulational interactions and one can estimate the magnitude of the magnetic field which is excited by using Eq.(14.17). If we substitute Eq.(14.16) into Eq.(14.17) and use for δn the estimated value $|\mathcal{E}|^2/4\pi T$ we find that the effect of the excitation of magnetic fields is of relative order v_T^2/c^2. Other electromagnetic effects in a non-relativistic plasma are of the same order of magnitude.

14.5 Relativistic Effects

Rather than considering weakly relativistic effects of the order of v^2/c^2 we shall now discuss effects for highly relativistic particles for which $\gamma \gg 1$, where

$$\gamma = \left\{1 - \frac{v^2}{c^2}\right\}^{-1/2}.$$

The first point we want to make is that there is now a possibility of a charged plasma. The attraction between currents is smaller than the repulsion between charges only by a factor v^2/c^2 so that the cancellation of the repulsion between relativistic electrons which move in the same direction may occur for a small ratio of the ion to the electron density:

$$\frac{n_i}{n_e} = \frac{1}{\gamma^2}. \tag{14.19}$$

This means that various charged structures can be formed in such a plasma. Among those we have the toroidal structures of relativistic electrons which have been obtained experimentally and have been found to be rather stable entities. The magnetic fields produced by such structures may be larger than the external magnetic fields. Experimentally it was found that if these structures were formed in an external magnetic field the magnetic field strength on the centre of the axis of the structure was in the direction opposite to that of the external magetic field. Another possibility are cylindrical structures which are called E-layers because the electrons produce the charge of the layer.

The magnetic fields produced by some relativistic electron beams with the appropriate transverse dimensions may be so high that the Larmor radius of a beam particle may be comparable with or even smaller than the transverse size of the beam. The transverse structure of the beam is in that case determined by the self-consistent motion of the beam particles in the magnetic field created by themselves. Structures may appear when in the rest frame of the beam particles the electromagnetic energy of the self-fields exceeds the total particle rest energy. Such a structure looks like a bunch of electromagnetic energy rather than a bunch of particles. This possibility arises due to the collective nature of the fields produced by the relativistic electrons. The radiation by such structures can be large; the beam system then becomes an open system. Radiation losses may lead to radiative self-contraction.

Another important application is a relativistic plasma where the relativistic particles do not have a directed velocity. The most interesting case is the relativistic electron-positron plasma which can be produced experimentally and which is believed to be the source of the pulsar emission. An important feature of a relativistic electron-positron plasma with equal electron and positron distributions is that the quadratic non-linear responses are equal to zero – since the electrons and the positrons have the same mass, but opposite charge. This means that all their structures can be described by a non-linear Schrödinger equation. An important point here is that there are no ion-sound waves in an electron-positron plasma which means the loss of an extra degree of freedom. The creation of soliton structures – especially in a strong magnetic field which forces the particles to move one-dimensionally along the magnetic field – will be a very common phenomenon in an electron-positron plasma. Most of the various modes of an electron-positron plasma and their non-linear interactions have been the subject of detailed studies.

14.6 Quantum Effects

Quantum effects become important for a degenerate plasma which may occur in solid-state plasmas and in the plasma occurring in white dwarf stars. Another possibility of the appearance of quantum effects is in a very dense plasma when $\hbar/mv$ may be of the order of the Debye screening length v/ω_{pe} or when the temperature of the thermal particles is sufficiently low so that

$$\hbar\omega_{\mathrm{pe}} > T, \tag{14.20}$$

which corresponds to high densities and low temperatures. If we have $T \simeq 1$ keV it corresponds to densities higher than 10^{26} cm^{-3}. For even higher densities such that

$$\hbar\omega_{\mathrm{pe}} > m_e c^2, \tag{14.21}$$

the plasma oscillations may create electron-positron pairs. There is a threshold for this process. There is no threshold for the creation of neutrino-antineutrino pairs and the creation of such pairs by plasma oscillations may be important for star cooling.

The last point we wish to stress is the importance of radiative corrections to the quasi-linear interactions and to the particle accelerations. The collective effects of a change in the cross-sections are here also very important. These processes may transfer a part of the oscillation energy of the order of $e^2/3\pi\hbar c \cong 10^{-4}$ to relativistic particles. This is an important process in the generation of cosmic rays; however, this would need a separate discussion which might well fill another book if we want to consider all these problems.

14.7 Conclusion. Acknowledgements

Even the brief display of possible non-linear processes in a plasma which we were able to give in our presentation shows the complexity and great variety of the possible interactions. This also is clear from the available experimental data. Chemistry and physics parted company a long time ago because it became clear that chemistry was a different field of research and involved specific chemical processes. The great variety of non-linear plasma processes suggests that non-linear plasma kinetics is also a special field of research. However, since all non-linear problems are closely related to fundamental problems in physics non-linear plasma kinetics remains in the frontline of current research. Applying the general results to particular problems needs in any case a strong physical intuition to choose, from the many processes which can occur simultaneously, the ones which are the most important. This is a broad field for applying good physics and it especially needs good physicists. This work needs, I feel, a great deal of "physical art" or, more precisely, the art of a physicist with a good physical intuition.

This book contains many new scientific results published quite recently in collaboration with R. Bingham of the Rutherford-Appleton Laboratory in the United Kingdon, U. de Angelis of the University of Naples in Italy, and K. Spatschek of the University of Dusseldorf in Germany. It has been a source of much pleasure to work with them and to discuss with them these new aspects of non-linear plasma kinetics. I should like to express my gratitude to Professor G. Ecker, the editor of the series in which this will be published, for his many efforts to improve the contents, and for many helpful suggestions and discussions. Many scientific results concerning the modulational interactions and applications of these interactions in astrophysics were obtained in collaboration with Professor D.ter Haar. The author is grateful to him for many discussions of these problems.

Glossary

d Debye radius of the plasma, defined by Eq.(2.25).

d_e electron Debye radius.

d_i ion Debye radius.

$d_{1,2}$ $= \delta(\mathbf{k} - \mathbf{k}_1 - \mathbf{k}_2) \, d^4 k_1 \, d^4 k_2$.

$d_{1,2,3}$ $= \delta(\mathbf{k} - \mathbf{k}_1 - \mathbf{k}_2 - \mathbf{k}_3) \, d^4 k_1 \, d^4 k_2 \, d^4 k_3$.

$d^4 k \equiv d^3 k \, d\omega$ volume element in 4D wavevector space.

e electron charge.

e_α charge of particle of kind α.

$e_{\text{eff}}, \, e_{\text{eff}}^{(\alpha)}$ effective screening charge of particle of kind α; $e + e_{\text{eff}}^{(e)} = Z_{\text{eff}}^{(e)} e$;
      $e_{\text{eff}}^{(i)} = Z_{\text{eff}}^{(i)} e$.

E energy density of particles.

E $= \int \epsilon_p \Phi_p \, d^3 p / (2\pi)^3$.

E_k magnitude of the Fourier component of the electrostatic field.

E_k^{NR} field of non-resonant waves.

E_k^{R} field of resonant waves.

$|E|_k^2$ correlation function of stationary and homogeneous electrostatic
      fields.

$|E|_k^2$ $= |E|_{k,\omega}^2$.

$|E|_k^2(\mathbf{r}, t)$ correlation function of slightly inhomogeneous and slightly
      non-stationary electrostatic fields.

$|E|_k^2(\mathbf{r}, t)$ $= \int \langle \delta E_k^+ \delta E_{-k+\kappa}^- \rangle \, e^{i(\boldsymbol{\kappa} \cdot \mathbf{r}) - i\nu t} \, d^4 \kappa, \; \kappa = \{\boldsymbol{\kappa}, \nu\}, \; d^4 \kappa = d^3 \kappa \, d\nu$.

E_k^+ positive-frequency Fourier component of the electrostatic field.

E_k^- negative-frequency Fourier component of the electrostatic field.

E^{v} magnitude of virtual electric field.

E_α energy density of particles.

E^+ positive-frequency Fourier component of the electrostatic field.

E^- negative-frequency Fourier component of the electrostatic field.

$E_1^\pm$ $\equiv E_{k_1}^\pm$.

$E_2^\pm$ $\equiv E_{k_2}^\pm$.

$E_3^\pm$ $\equiv E_{k_3}^\pm$.

$\mathbf{E}(\mathbf{r}, t)$ electrostatic field strength.

$\mathbf{E}_k$ $= E_k \mathbf{k} / k$.

$\mathbf{E}_{k,\omega}$ Fourier component of the electrostatic field.

$\mathcal{E}(x, t)$ amplitude of one-dimensional Langmuir field.

$\mathcal{E}^{\text{sol}}$ amplitude of soliton field.

$\mathcal{E}_0$ amplitude of pump field or amplitude of soliton.

$\mathcal{E}(\mathbf{r}, t)$ amplitude of Langmuir field.

$\mathcal{E}(\mathbf{r}, t)$ $= \int \mathbf{k}\, E_{\mathbf{k}}^{+}\, e^{-i(\omega - \omega_{\text{pe}})t + i(\mathbf{k}\cdot\mathbf{r})}\, dk/k$.

$f_{\mathbf{p}}$ particle distribution function.

$f_{\mathbf{p}}^{\alpha}$ particle distribution function; $\alpha = \{e, i\}$ for electrons and ions, respectively.

H total – that is, integrated over all space – energy of regular plasma perturbations in modulational interactions.

H^{reg} total – that is, integrated over all space – energy of regular plasma perturbations in modulational interactions.

$\widehat{I}_{\mathbf{p}}^{\text{coll}}$ particle collision integral.

$\widehat{I}_{\mathbf{p}}^{\text{coll}\,\alpha}$ particle collision integral.

$I_{\mathbf{p}}^{\text{QL}}$ quasi-linear collision integral.

$k \equiv \{\mathbf{k}, \omega\}$ four-dimensional (4D) wavevector of the field.

$k = |\mathbf{k}|$ modulus of the wavevector of the field.

k_* one step in wavenumber in the energy transfer process.

$\mathbf{k}$ wavevector.

m_{e} electron mass.

m_{i} ion mass.

m_{α} mass of particle of kind α.

M^{Br} total bremsstrahlung matrix element in collisions between a particle of kind α and a particle of kind β.

$M^{\text{Br}(\alpha,\beta)}$ total matrix element of bremsstrahlung in collisions between a particle of kind α and a particle of kind β.

$M_{\mathbf{k},\mathbf{k}'}$ scattering matrix element.

$M_{\mathbf{k},\mathbf{k}'}^{\text{sc}}$ scattering matrix element.

$M_{\mathbf{k},\mathbf{k}'}^{\text{tr}}$ scattering matrix element.

M^{α} matrix element of the bremsstrahlung due to the change in the trajectories of a particle of kind α.

M^{β} matrix element of the bremsstrahlung due to the change in the trajectories of a particle of kind β.

$M^{\alpha,\beta}$ matrix element of bremsstrahlung due to the mutual change in the screening charges of both particles, one of kind α and one of kind β.

n_{b} density of a one-dimensional beam.

n_{b} $= \int \Phi_v^{\text{b}}\, dv$.

n_{e} electron density.

n_{i} ion density.

n_{α} density of particle of kind α.

n_{α} $= \int f_{\mathbf{p}}^{\alpha}\, d^3\mathbf{p}/(2\pi)^3$.

n_0 averaged electron density.

N total number of waves.

N $= \int N_{\mathbf{k}}\, d^3\mathbf{k}/(2\pi)^3$.

N_d number of particles in Debye sphere.

N_{de} number of electrons in Debye sphere.

N_{di} number of ions in Debye sphere.

N_k spectral density of number of waves.

N_k^σ spectral density of number of waves.

N^σ total number of waves.

$\mathbf{p}$ particle momentum.

$\mathbf{P}$ total – that is, integrated over all space – momentum of regular plasma perturbations in modulational interactions.

$\mathbf{P}$ momentum density of particles.

$\mathbf{P}$ $= \int \mathbf{p} \Phi_\mathbf{p} \, d^3\mathbf{p}/(2\pi)^3$.

$\mathbf{P}^{reg}$ total – that is, integrated over all space – momentum of regular plasma perturbations in modulational interactions.

$\mathbf{P}_\alpha$ momentum density of particles.

Q power emitted by unit plasma volume.

Q $= \int Q_\mathbf{p} \, d^3\mathbf{p}/(2\pi)^3$.

Q_k^{NL} non-linear emission power density in range $d^3\mathbf{k}$.

$Q_\mathbf{p}$ power emitted per particle.

$Q_\mathbf{p}^{sc}$ power scattered by one particle.

T_e electron temperature.

T_i ion temperature.

u soliton velocity; drift velocity of electrons.

$v_{ph}^\sigma = \omega_k^\sigma/k$ phase velocity of collective mode of kind σ.

v_s ion-sound velocity.

v_{Te} electron thermal velocity.

v_{Te} averaged electron velocity for non-equilibrium electron distribution, defined by Eq.(2.79).

v_{Ti} ion thermal velocity.

$\mathbf{v} = \mathbf{p}/\varepsilon_\mathbf{p}$ particle velocity.

$\mathbf{v}_{gr}^\sigma = d\omega_k^\sigma/d\mathbf{k}$ group velocity of collective mode of kind σ.

V_{Te} averaged electron velocity for non-equilibrium electron distribution, defined by Eq.(2.85).

w_k energy of longitudinal waves in the wavenumber range dk during a particle collision.

$w_\mathbf{p}(\mathbf{k})$ probability for Vavilov-Cherenkov emission.

$w_\mathbf{p}^{\alpha,\sigma}(\mathbf{k})$ probability for Vavilov-Cherenkov emission; $\alpha = \{e, i\}$, $\sigma = \{\ell, s, t\}$.

$w_\mathbf{p}(\mathbf{k}, \mathbf{k}')$ probability for the scattering of waves of kind σ by particles of kind α with a conversion to waves of kind σ'.

$w_\mathbf{p}^{\sigma,\sigma'}(\mathbf{k}, \mathbf{k}')$ probability for the scattering of waves of kind σ by particles of kind α with a conversion to waves of kind σ'.

$w_\mathbf{p}^{\sigma,\sigma'\,(\alpha)}(\mathbf{k}, \mathbf{k}')$ probability for the scattering of waves of kind σ by particles of kind α with a conversion to waves of kind σ'.

$w_{\mathbf{p},\mathbf{p}'}^{\alpha,\beta}(\mathbf{k})$ probability for collision of particle of kind α with particle of kind β in a plasma.

$w_{p,p'}^{\alpha,\beta}(\mathbf{k},\mathbf{q})$ probability for bremsstrahlung in collision of a particle of kind α with a particle of kind β with a transfer $\mathbf{q}$ of momentum from the particle of kind α to the particle of kind β and emission of a wave with momentum $\mathbf{k}$.

$w_{\sigma',\mathbf{k}'}^{\sigma'',\sigma}(\mathbf{k})$ probability for the decay of a wave of kind σ' into a wave of kind σ'' and a wave of kind σ.

w_ω energy of electromagnetic waves in the frequency range $d\omega$ during a particle collision.

W energy density of collective oscillations.

W energy density of waves.

W $= \int W_k\, dk$.

W $= \int \omega_\mathbf{k} N_\mathbf{k}\, d^3\mathbf{k}/(2\pi)^3$.

W_{cr} critical value of energy density, which separates the resonant non-linear and the modulational non-linear interactions.

W_k energy density of isotropic turbulence and energy density for waves of kind σ in the wavenumber range from k to $k + dk$.

W_k^σ energy density of isotropic turbulence and energy density for waves of kind σ in the wavenumber range from k to $k + dk$.

W^σ energy density of collective oscillations.

W^σ energy density of waves.

$-Ze$ ion charge.

α $= \{e, i\}$ for electrons and ions, respectively.

$\gamma_\mathbf{k}$ linear damping or growth rate.

$\gamma_\mathbf{k}^m$ non-linear damping or growth rate for non-linear modulational interactions.

$\gamma_\mathbf{k}^{\mathrm{NL(dec)}}$ non-linear damping or growth rate for decay.

$\gamma_\mathbf{k}^{\mathrm{NL(sc)}}$ non-linear damping or growth rate for stimulated scattering.

$\gamma_\mathbf{k}^{\mathrm{NL}}$ non-linear damping or growth rate.

$\gamma_\mathbf{k}^\sigma$ linear damping or growth rate of collective mode of kind σ.

γ_{str} coherent damping or growth rate of non-linear structures.

$\langle \delta E_\mathbf{k}^+ \delta E_{\mathbf{k}'}^- \rangle = -|E^+|_\mathbf{k}^2\, \delta(\mathbf{k}+\mathbf{k}'),\ |E|^2 = |E^+|_\mathbf{k}^2 + |E^-|_{-\mathbf{k}}^2$.

$\delta\mathbf{E}$ strength of fluctuating electric field.

$\delta(\mathbf{k}) \equiv \delta(\mathbf{k})\,\delta(\omega)$ δ-function for 4D wavevector.

δn deviation of density from its average value or fluctuation in density, in which case we have $\langle \delta n \rangle = 0$.

$\langle \delta n \rangle$ variation in density produced by random fields interacting with. regular fields.

δn_{reg} variation in density produced by regular fields.

δn_{sol} variation in density produced by a soliton.

δv_b spread of velocities of a one-dimensional beam.

$\delta\omega^d$ frequency shift due to dispersion.

$\delta\omega_\mathbf{k}^{\mathrm{NL}}$ non-linear frequency shift.

$\delta\omega^{\mathrm{NL}}$ non-linear frequency broadening.

$\varepsilon_\mathbf{k} = \varepsilon_{\mathbf{k},\omega} = \varepsilon$ linear dielectric permittivity for longitudinal fields.

$\varepsilon_{\mathbf{k}}$ $\qquad = \varepsilon_{\mathbf{k}}^{(e)} + \varepsilon_{\mathbf{k}}^{(i)} - 1.$

$\varepsilon_{\mathbf{k}}^{NL}$ $\qquad$ non-linear plasma dielectric permittivity.

$\varepsilon_{\mathbf{k}}^{(\alpha)}$ $\qquad$ linear dielectric permittivity for particles of kind α, $\alpha = \{e, i\}$.

$\varepsilon_{\mathbf{p}}$ $\qquad$ particle energy.

ε_{1+2} $\qquad = \varepsilon_{\mathbf{k}_1 + \mathbf{k}_2}.$

$\lambda = 2\pi/k$ $\quad$ wavelength.

Λ $\qquad$ Coulomb logarithm.

ν_{eff} $\qquad$ effective turbulent collision frequency.

ν_{ei} $\qquad$ electron-ion collision frequency.

ν_{ii} $\qquad$ ion-ion collision frequency.

Π $\qquad$ momentum density of waves.

Π $\qquad = \int \mathbf{k} N_{\mathbf{k}} \, d^3\mathbf{k}/(2\pi)^3.$

Π^{σ} $\qquad$ momentum density of waves.

$\varrho_{1,2,3}^{\text{eff}}$ $\qquad$ effective non-linear response function.

$\varrho_{\mathbf{k}}^{NL}$ $\qquad$ Fourier component of non-linear charge density.

$\varrho_{\mathbf{k}_1,\mathbf{k}_2}^{(2)}$ $\qquad$ non-linear quadratic plasma response function.

$\varrho_{\mathbf{k}_1,\mathbf{k}_2,\mathbf{k}_3}^{(3)}$ non-linear cubic plasma response function.

$\varrho_{1,2}^{(2)}$ $\qquad$ non-linear quadratic plasma response function.

$\varrho_{1,2,3}^{(3)}$ $\qquad$ non-linear cubic plasma response function.

$\varrho_{1,2}^{(2)\,(\alpha)}$ $\qquad$ non-linear quadratic response function of particles of kind α.

$\varrho_{1,2,3}^{(3)\,(\alpha)}$ $\qquad$ non-linear cubic response function of particles of kind α.

σ $\qquad = \{\ell, s, t\}$ for Langmuir, ion-sound, and transverse waves, respectively.

ϕ $\qquad$ electrostatic potential; $\mathbf{E} = -\nabla\phi.$

Φ_v^b $\qquad$ distribution function of a one-dimensional beam.

$\Phi_{\mathbf{p}} = \langle f_{\mathbf{p}} \rangle$ $\quad$ averaged distribution function for random fields.

$\Phi_{\mathbf{p}} = \langle f_{\mathbf{p}} \rangle$ $\quad$ slowly varying part of distribution function for regular fields.

$\Phi_{\mathbf{p}}^{NR}$ $\qquad$ distribution function of non-resonant particles.

$\Phi_{\mathbf{p}}^{R}$ $\qquad$ distribution function of resonant particles.

$\Phi_{\mathbf{p}}^{\alpha} = \langle f_{\mathbf{p}}^{\alpha} \rangle$ $\quad$ averaged distribution function of particles of kind α for random fields.

$\Phi_{\mathbf{p}}^{\alpha} = \langle f_{\mathbf{p}}^{\alpha} \rangle$ $\quad$ slowly varying part of distribution function of particles of kind α for regular fields.

ω $\qquad$ frequency.

$\omega_{\mathbf{k}}^{\sigma}$ $\qquad$ frequency of the collective mode of kind σ; $\sigma = \{\ell, s, t\}$ for Langmuir, ion-sound, and transverse waves, respectively.

ω_{pe} $\qquad$ electron Langmuir frequency.

ω_{pi} $\qquad$ electron ion Langmuir frequency.

Index